# Springer Earth System Sciences

More information about this series at http://www.springer.com/series/10178

Guillermo W. Rougier · Agustín G. Martinelli ·
Analía M. Forasiepi

# Mesozoic Mammals from South America and Their Forerunners

Springer

Guillermo W. Rougier
Department of Anatomical Sciences
and Neurobiology
University of Louisville
Louisville, KY, USA

Agustín G. Martinelli
Museo Argentino de Ciencias Naturales
"Bernardino Rivadavia"-CONICET
Sección Paleontología de Vertebrados
Buenos Aires, Argentina

Analía M. Forasiepi
Instituto Argentino de Nivología,
Glaciología y Ciencias Ambientales
Centro Científico Tecnológico-CONICET
Mendoza, Argentina

ISSN 2197-9596 ISSN 2197-960X (electronic)
Springer Earth System Sciences
ISBN 978-3-030-63860-3 ISBN 978-3-030-63862-7 (eBook)
https://doi.org/10.1007/978-3-030-63862-7



This Springer imprint is published by the registered company Springer Nature Switzerland AG
The registered company address is: Gewerbestrasse 11, 6330 Cham, Switzerland

*Si no me equivoco, las heterogéneas piezas que he enumerado se parecen a Kafka [...], pero si Kafka no hubiera escrito, no la percibiríamos; vale decir, no existiría [...]. El hecho es que cada escritor crea a sus precursores. Su labor modifica nuestra concepción del pasado, como ha de modificar el futuro.*

*[If I am not mistaken, the heterogeneous pieces I have listed resemble Kafka [...], but if Kafka had never written a line, we would not perceive this quality; that is to say, it would not exist [...]. The fact is that every writer creates his own precursors. His work modifies our perception of the past, as it will modify the future.]*

*Jorge L. Borges*
*Kafka y sus Precursores, 1951*

# Preface

As evolutionary biologists, we live in exciting and promising times. The material basis of our science is growing at an unprecedented rate; fossils preserving structures we thought inaccessible like fur, ear ossicles, endocasts at the beginning of our careers are now routinely reported. Finally, technological and methodological advances originating outside our field have been transformed into powerful new techniques allowing us to tackle complex 3D reconstructions, chemical analyses, growth rates, dated phylogenies, etc. In the midst of this ever-changing milieu, there is never a good time to write a book like this; the moment we put down the pen, or more correctly, we stop typing the book, it is, at best, outdated, and at worst not reflecting what we think anymore. It is much like parenthood, if there is never a good time, anytime is equally good. The tragic circumstances and disruptions brought about by the 2020 COVID-19 pandemic forced us to stay homebound and gave us the final impetus to finish this lengthy project. These are the perhaps inauspicious circumstances at the birth of this project. Despite an exponential growth of our knowledge of Mesozoic mammals in general and South American ones in particular, we have never been as unsure about some of the most fundamental questions sculpting the shape of early mammalian research. The phylogenetic framework of mammals and their close relatives is in a perennial state of flux. The minimal age of Mammalia and its place of origin changes dramatically depending on the preferred phylogeny, and discoveries of new forms from typically underrepresented portions of the globe like Madagascar, Africa, or South America demonstrate that major new branches of the mammalian tree are yet to be discovered. In paleontology, the staunching materialistic maxim of the 2000s, "what you have is more important than what you believe", is painfully appropriate. Our current interpretations and answers to cardinal questions on early mammalian evolution cannot be anything but the most unrefined take of a reality barely constrained by a handful of precious fossils. The tangible basis of our studies—a specimen more, or two—can easily make a laboriously constructed cladogram completely obsolete. This is not a discouraging fact; it is a measure of the magnitude of the task awaiting future generations of enterprising field paleontologists. If a new specimen does not have the potential to change our understanding of the evolutionary process that took place in that lineage, why even bother risking limb and fortune in the field? The potentially disruptive effect of discovery, the sudden change, the new implications for time and place of origin of a group that logically follows a new cladogram, these are the dynamic, ever-changing trading commodities of paleontology. We do make extensive use of cladograms in this book; they are necessary, but finicky and seldom consensual guiding tools. Some of these trees are already outdated, others surely will be very soon. In this book, we present what we believe is our best interpretation of the mammalian fossil record of the Mesozoic in South America; the guiding principle, however, is what we have: a surprising and rapidly growing fossil record.

The simplistic, yet comforting, view that a handful of iconic characters of recent mammals, like mammary glands, hair, three middle ear ossicles, and a single dentary condyle, were essential and transformative has lost its methodological appeal and has been uncontroversially falsified by current phylogenies. They all preceded the origin of the crown clade Mammalia. In hindsight, it is only logical to expect that those features that so easily differentiate mammals from any other living vertebrate have a long phylogenetic history developing successively in the many, now pruned, branches of non-mammalian synapsids. The extinction of all those ancient lineages creates a very long branch, artificially accumulating diagnostic features allowing for easy recognition of a mammal among other living vertebrates. Although, much—certainly much, much more—is left to be found in the Mesozoic of South America, we are at a stage where there is a handful of localities of different ages, representing several distinct groups; some can be relatively easily integrated into the known diversity from northern continents ("triconodonts", *Vincelestes*, for example), while others are of Gondwanan or South American distribution (henosferids, meridiolestidans, gondwanatherians). The geographical distribution of those South American Mesozoic localities is very disparate, with most of them hailing from the southern tip of the continent, Patagonia, and a smattering of others outside Argentina. This is admittedly a very incomplete state of affairs; however, we are confident that the recognition of some major groups will survive the test of time, that the ubiquitous presence of some of them during certain times of the geological past (meridiolestidans/mesungulatids in the Late Cretaceous) records a successful and dominant mammalian group. In addition to isolated teeth, the quintessential Mesozoic mammal remains, South America has provided a few fundamental cranial and postcranial specimens that are central to the discussion of important mammalian transformations. The recent discovery of brasilodontid cynodonts, and a trove of other small derived non-mammaliaform cynodonts, in Southern Brazil and Argentina has opened a vibrant new avenue of research into the morphological background for the origin of mammals. The new finds in Brazil highlight how much of the mammaliaform and early mammalian evolution is just the continuation of trends and building upon character systems established very early—Mid- to Late Triassic at least—in these diverse and crucial mammalian forerunners. As Jorge L. Borges quote brilliantly shows, ancestors only have a reality in retrospect, looking back from the comfy armchair of the present. Unjustly, by an almost perverse inversion of evolutionary logic, we call these mammalian forerunners mammal-like cynodonts. Those characters that make them mammal-like are the characters that appeared for the first time in those archaic lineages and we, mammals, inherited as symplesiomorphies; they are not mammalian characters at all! Mammals are the inheritors and beneficiaries of a distinguished evolutionary past documenting radical transformations setting the stage for the dramatic explosion in diversity and ecomorphology the mammalian cynodonts effected during the Mesozoic and Cenozoic.

We hope to provide here comprehensive documentation of the fossil record of Mesozoic mammals and closest relatives in South America, tracing them further across the K/Pg boundary in the few cases where those lineages survived the cataclysmic end of the Mesozoic. This effort has been an opportunity to critically revise earlier views in a somewhat coherent and comprehensive framework, without delving exhaustively into all the pros and cons of alternative ideas. On occasion, this book may be seen as idiosyncratic, given that full-fledged discussion of decisions may not always be present; it does, however, represent a consensual view of the authors, or at least the opinion of the most stubborn of us! We hope this book offers a quick reference for the South American mammalian record, an occasional irksome question, or a stimulating idea.

The authors, Agustín G. Martinelli, Analía M. Forasiepi and Guillermo W. Rougier (left to right) at the Museo Argentino de Ciencias Naturales "Bernardino Rivadavia" (Buenos Aires, Argentina) in February 2018. Photo by Alejandro Kramarz

Louisville, USA
Buenos Aires, Argentina
Mendoza, Argentina

Guillermo W. Rougier
Agustín G. Martinelli
Analía M. Forasiepi

# Acknowledgements

We are very grateful for the assistance of a number of colleagues who reviewed parts of this manuscript and through many years discussed with us the vagrancies of mammalian evolution and preservation: John R. Wible, James Hopson, Fuzz Crompton, Michael Novacek, Farish Jenkins (Jr.), Brian Davis, Richard L. Cifelli, Zhe-Xi Luo, Robin M. Beck, Jin Meng, José F. Bonaparte, Craig B. Wood, Rosendo Pascual, David Krause, Marina Bento Soares, Cesar L. Schultz, Ana Maria Ribeiro, Pamela Gill, Colin Palmer, Emily Rayfield, Ian Corfe, Fernando Abdala, Ross D. E. MacPhee, Francisco J. Goin, Alberto Garrido, Marcelo Sánchez-Villagra, Mariano Bond, Alejandro Kramarz, Marcelo S. de la Fuente, Marcelo Reguero, Francisco J. Prevosti, Tom Rich, John Flynn, Diego Pol, Leandro Gaetano, Julia Desojo, Lucas Fiorelli, Yamila Gurovich, Laura Chornogubsky, Martin Ezcurra, Giuseppe Leonardi, Téo Oliveira, Rachel O'Meara, Simon D. Kay, Alexander Vargas, Sergio Soto-Acuña, Marcelo Leppe, Roy Fernández, Felipe Suazo Lara, Jonatan Kaluza, Tomaz Melo, Pedro Fonseca, Mauricio Schmitt, Voltaire Paes Neto, Morgan Guignard, Heitor Francischini, Marcel Lacerda, Paulo Romo de Vivar, Pablo Rodrigues, Leonardo Kerber, Martín Hechenleitner, Christian Kammerer, Marcos Sales, Sergio Dias-da-Silva, Felipe Pinheiro, Thiago Marinho, Rogerio Rubert, Átila A. da Rosa, William Nava, and Pedro Buck. This book is a personal account of our experiences in the field, as all three of us see our protracted and dedicated effort to collect specimens as central to our understanding of early mammalian evolution and as a fundamental aspect of our "ethos" as paleontologists. This emphasis in fieldwork and collecting, to move the field forward by the sheer weight of the specimens, is perhaps the most enduring positive influence on the three of us by the late José F. Bonaparte.

Many technicians, students, and volunteers have been instrumental in the success of our fieldwork program. Nothing would be the same, or even possible, without the inexhaustible enthusiasm and skill of Leandro Canessa, who has been our right hand through years of fieldwork, a calming influence in moments of crisis, an affable team member who has personally found many great specimens. In the account of his reliable help, we reluctantly forgive his crippling distaste of garlic. Pablo Puerta, Diego Pol, Sebastian Apesteguía, and Ruben Cúneo have been instrumental in securing crucial support and the logistics for these expeditions; they provided scouting of localities, obtained permits from authorities and landowners, in addition to actively taking part in many expeditions. We also thank Raúl Gómez, Lucas Apella-Guiscafre, Guillermo (Willy) Turazzini, Andres Lires, Leandro Gaetano, Juliana Sterli, Paula Muzzopappa, Andres Giallombardo, Magalí Cardenas, Mauricio S. Cardozo, Fernando Garberoglio, Tony Harper, Marcos Becerra, Rocio Belen-Vera, Maximiliano Delocca, Jonatan Kaluza, Fernando Abdala, Marcelo Isasi, Leonardo Pazo, Kerin Cleason, Robert Hill, Domingo Notao, and a large number of more occasional collaborators.

Several institutions and funding agencies have supported our work, including the University of Louisville (UofL), Consejo Nacional de Investigaciones Científicas y Técnicas (CONICET-Argentina), Museo Argentino de Ciencias Naturales "Bernardino Rivadavia" (MACN), National Science Foundation (NSF), National Geographic Society (NGS), Sepkoski Grant (Paleontological Society), Instituto Argentino de Nivología, Glaciología y Ciencias Ambientales (IANIGLA, CCT-CONICET, Mendoza), Museo Paleontológico "Egidio Feruglio" (MPEF), Museo Municipal de Lamarque (MML), Museo de Ciencias Naturales "Carlos Ameghino" (MPCA), Conselho Nacional de Desenvolvimento Científico e Tecnológico (CNPq-Brazil), Universidade Federal do Rio Grande do Sul (UFRGS), and Centro de Pesquisas Paleontologicas L. I. Price (CPPLIP-UFTM).

We are very grateful to the curators and staff who assisted during collection visits: Alejandro Kramarz (MACN), Eduardo Ruigómez (MPEF), Susan K. Bell (AMNH), William Simpson (FMNH), Danieli Sanches Venturini (MVP), Marco Brandalise de Andrade (MCP), Cesar Schultz (UFRGS), Gabriela Cisterna (PULR), Emilio Vaccari (PULR), Carlos Nunes Rodrigues (MMACR), Rodrigo Machado (MCT), Belarmino Stefanello (MMACR), Marcelo Reguero (MLP), Jaime Powell (PVL), Rodrigo González (PVL), Ana Maria Ribeiro (MCN/FZBRS), Jorge Ferigolo (MCN/FZBRS).

Jorge Blanco produced the exquisite illustrations of the extinct mammals in this book; our sincere recognition for his hard work. Silvina Lassa, Federico González (CCT-CONICET, Mendoza), and Fabián Tricárico (MACN) assisted with the MEB pictures, and Flávio Lopes (UFRGS) with some photographs of specimens in Chap. 3.

For photographs used in the chapters, we especially thank: Erraín Casamiquela and Julia Heredia for Fig. 1.1; Mariano Bond for Fig. 1.4a; Pablo Puerta for Fig. 1.4b–c and 2.15c; Giuseppe Leonardi for Fig. 2.13; Alberto C. Garrido for Figs. 2.19 and 2.20a–b; Felipe Suazo Lara for Fig. 2.25b; Jorge L. Blanco for Fig. 2.27d–e; Ross D. E. MacPhee for Fig. 2.30a–b; Marcelo Reguero for Fig. 2.30c; Sergio Soto-Acuña for Fig. 8.7; Heitor Francischini for Fig. 9.4; and Pedro V. Buck for Figs. 9.5 and 9.8.

Special thanks from GWR. I want to thank James A. Hopson and John R. Wible for the enormous positive influence they had in my career, particularly early on; both served as models of what a scientist should be. My life-long research agenda was set by those late afternoon discussions about *Vincelestes* in the blight of Chicago's 1989–1990 winter. Wolfgang Maier opened the doors of his fascinating embryological collection so I could finally stop being just a paleontologist and become an anatomist; the few months I spent in Tübingen permanently shaped my career and interests. Michael J. Novacek gave me the opportunity to work on specimens I could hardly have dreamt about in the thrilling intellectual environment of the American Museum of Natural History and the intoxicating experience that is NYC. All of you, please accept my thanks and lasting gratitude. Special thanks from AGM and AMF. Our work in this book is dedicated to the memory of José F. Bonaparte, one of the pillars of the knowledge of South American Mesozoic mammals as well as non-mammalian cynodonts. We are fortunate to have worked with him in the last decades before his recent decease.

We are deeply grateful for the assistance and time of Cynthia Corbitt and Brian Davis who reviewed and edited the text for the ever-present Spanglish and clarity. We express our enormous gratitude for their help and time.

A special thank goes to Rajan Muthu, Project Coordinator and Book Production at Springer who, with kindness and patience, gave us multiple extensions to finalize this book.

# Contents


**1 Introduction** . . . . . . . . . . 1
1.1 Mesozoic Mammals . . . . . . . . . . 1
1.2 First Findings: Humble Beginnings . . . . . . . . . . 3
1.3 Abbreviations . . . . . . . . . . 13
1.3.1 Institutional Abbreviations . . . . . . . . . . 13
1.3.2 Anatomical Abbreviations . . . . . . . . . . 14
1.3.3 Other Abbreviations . . . . . . . . . . 14
1.4 Nomenclature . . . . . . . . . . 15
References . . . . . . . . . . 19

**2 The Fossil Record of South American Mesozoic Mammals and Their Close Relatives** . . . . . . . . . . 25
2.1 Introduction . . . . . . . . . . 25
2.1.1 South American Gondwanan Mammals and the Paleobiogeographical Context . . . . . . . . . . 31
2.2 Fossiliferous Localities from South America . . . . . . . . . . 33
2.2.1 Triassic . . . . . . . . . . 33
2.2.2 Jurassic . . . . . . . . . . 45
2.2.3 Cretaceous . . . . . . . . . . 49
2.2.4 Cenozoic . . . . . . . . . . 79
2.3 Concluding Remarks . . . . . . . . . . 89
References . . . . . . . . . . 90

**3 The Origin and the Radiation of Early Mammals: A Southern Perspective** . . . . . . . . . . 127
3.1 Introduction . . . . . . . . . . 127
3.2 Non-mammaliaform Cynodonts . . . . . . . . . . 128
3.2.1 Cynognathia . . . . . . . . . . 131
3.2.2 Probainognathia . . . . . . . . . . 140

3.3 Concluding Remarks . . . . . 149
References . . . . . 153

**4 Australosphenidans** . . . . . 163
4.1 Introduction . . . . . 164
4.2 Systematics . . . . . 169
4.3 Concluding Remarks . . . . . 179
References . . . . . 181

**5 "Triconodonts"** . . . . . 187
5.1 Introduction . . . . . 187
5.2 Systematics . . . . . 190
5.3 Concluding Remarks . . . . . 195
References . . . . . 196

**6 Dryolestoids** . . . . . 201
6.1 Introduction . . . . . 202
6.2 South American Dryolestoids . . . . . 204
6.3 Systematics . . . . . 210
6.4 Unnamed SA Dryolestoids and Corollaries on Distribution and Age . . . . . 247
6.4.1 Dryolestoids from Bolivia (Maastrichtian) . . . . . 247
6.4.2 Dryolestoids from Antarctica (Eocene) . . . . . 247
6.4.3 Putative Oldest Mesungulatid from Lago Los Barreales (Coniacian), Patagonia . . . . . 249
6.4.4 Mesungulatid indet. from Cerro Yeso (Maastrichtian), Patagonia . . . . . 249
6.4.5 Meridiolestida indet. from Paso Córdoba (Campanian), Patagonia . . . . . 249
6.5 Concluding Remarks . . . . . 251
References . . . . . 254

**7 Stem Therians** . . . . . 261
7.1 Introduction . . . . . 261
7.2 Systematics . . . . . 263
7.3 Concluding Remarks . . . . . 280
7.4 Vignette by GWR: The Pact of La Amarga, or How Careers Are Determined in Science! . . . . . 282
References . . . . . 284

**8 Allotheria: Gondwanatherians and Multituberculates** . . . . . 289
8.1 Introduction . . . . . 289
8.2 Systematics . . . . . 294
8.2.1 Gondwanatheria *Indet.* . . . . . 308
8.2.2 Mammalia *Incertae Sedis* . . . . . 310

8.3 Concluding Remarks ... 313
References ... 318

**9 Ichnology: Mammalian/Mammal-Like Footprints** ... 325
9.1 Introduction ... 325
9.2 Systematics ... 327
9.3 Concluding Remarks ... 339
References ... 340

**10 The South American Mesozoic Record and Early Evolution of Mammals** ... 345
10.1 Paleogeography, Distribution, and Paleoecology ... 346
10.2 Paleobiological Corollaries from the SA Mammals ... 362
10.3 Perspectives and Future Directions ... 369
References ... 373

# Chapter 1
# Introduction

*...porque te hago saber, Sancho, que la boca sin muelas es como Molino sin piedra, y en mucho más se ha de estimar un diente que un diamante.*
[...you must know Sancho, that a mouth without teeth is like a mill without a stone, and that a diamond is not so precious as a tooth.]
Miguel de Cervantes Saavedra
*El Ingenioso Hidalgo Don Quijote de la Mancha*, 1605

**Abstract** Mammals from ancient times, living in past ecosystems surrounded by dinosaurs, toothed birds, fast running crocodyliforms, among other unfamiliar creatures, are a fascinating subject. These archaic mammals include the old, distant, relatives of the lineages that gave rise to the modern radiation of mammals and eventually ourselves. Mesozoic mammals are known from the Northern Hemisphere since the eighteenth century, while the first record of these creatures in the Southern Hemisphere came about half a century ago. As usual for Mesozoic mammals, the bulk of the SA species are solely known by jaw fragments and isolated teeth; but the collections also hold footprints, and a few beautifully preserved skulls and partial skeletons. In this chapter, we present a historical account of the first steps in the knowledge of SA Mesozoic mammals, the main personalities that helped to get it established as a distinct discipline, and provide definitions for some of the major mammalian groups.

**Keywords** Mammalia · Mesozoic · Discoveries · South America

## 1.1 Mesozoic Mammals

Mesozoic mammals were part of ecosystems that included many Recent groups familiar to anyone casually interested in nature, such as turtles, lizards, crocodilians, and birds, as well as the emblematic extinct non-avian dinosaurs and other archaic vertebrates (Chap. 2). Mammals evolved in a systematically diverse world, amidst dynamic geography that had a deep and lasting impact on the biological milieu that



G. W. Rougier et al., *Mesozoic Mammals from South America and Their Forerunners*, Springer Earth System Sciences, https://doi.org/10.1007/978-3-030-63862-7_1

gave rise not only to a varied array of extinct groups, but also to the roots of the nearly 6,500 living species (Burgin et al. 2018). Mammals have the distinction of being the only major group represented by almost four fossil genera per each living one (O'Leary et al. 2013). The bulk of this prolific fossil record comes from the younger Cenozoic where mammals are relatively abundant to the point that it is commonly referred to as the "Age of Mammals". On the other hand, fossils from the Mesozoic are rare, often represented by a few, precious, isolated teeth, with an uneven temporal, geographical, and environmental representation. They are, however, treasured, allowing us to understand the rudiments of how mammals evolved during more than 200 million years. Our knowledge of mammalian diversity, phylogeny, and ecology during the Mesozoic still is disarmingly incomplete, especially when compared to the Cenozoic fossil record. Findings of Mesozoic mammals mostly trace the evolutionary journey of the group in the northern continents, which historically benefitted from longer and more sustained collecting efforts and studies. However, the findings in the southern continents, in general, and South America (SA) in particular, bear a great phylogenetic import to understand the intriguing puzzle of mammal origin and evolution. Non-mammaliaform cynodonts, closely related to mammals, are found in the Middle to Late Triassic of Argentina and Brazil, the latter providing one of the most outstanding discoveries of the last decades (Chap. 3).

Mesozoic mammals are known to occur from about a dozen localities of Argentina, Brazil, Bolivia, Chile, and perhaps Peru spanning from the latest Early Jurassic to the Late Cretaceous, and some of those archaic lineages unexpectedly survive the cataclysmic end of the Cretaceous Period, remaining as minority elements in the Paleocene–early Miocene of SA. There are about 30 recognized species distributed in several distinctive lineages, including australosphenidans, eutriconodonts, "amphilestids", dryolestids, meridiolestidans, gondwanatherians, multituberculates, and potentially other poorly known taxa (Chaps. 4–9). Among these finds, several species are based on fragmentary and isolated elements that result in an uncertain taxonomic position. On the other hand, and offering a stark contrast, other species are represented by exquisitely preserved material that plays a pivotal role in understanding mammalian character transformations during the Mesozoic and stretching the known diversity beyond the traditional canon established by Holarctic material. Each new material from the fossil record is a valuable piece from a complex puzzle that has to be integrated into the dynamic evolutionary history of our early ancestors framed by co-occurring physical factors, such as tectonics, geography, and climate (Chap. 10).

The purpose of this book is to summarize the published paleontological information about the Mesozoic mammal record in South America, including the taxa belonging to archaic Mesozoic lineages surviving the K/Pg extinction, deep into the Cenozoic, and a background of their immediate non-mammalian relatives.

## 1.2 First Findings: Humble Beginnings

Almost 35 years ago, an isolated mammalian molar from the Upper Cretaceous (Maastrichtian) Los Alamitos Formation, Patagonia, Argentina, was reported by Bonaparte and Soria (1985). The material MACN-RN 01 served as the holotype of *Mesungulatum houssayi* from the collections of the Museo Argentino de Ciencias Naturales "Bernardino Rivadavia" and was found by a volunteer, Marcelo Rougier (a brother of GWR), during one of Bonaparte's yearly expeditions to Patagonia in search of Mesozoic vertebrates. The specimen was the single mammalian specimen collected in the 1983 field season, and it was originally understood as a condylarth-eutherian mammal. However, by the time the paper was in press, a sizable collection of additional specimens recovered during the 1984 field season became available. The additional material made the original interpretation unlikely, and it was promptly re-interpreted as a non-tribosphenic mammal possibly related to dryolestoids, but this realization was too late to either withdraw or modify the original paper (Bonaparte and Soria 1985, postscript). This find overshadowed other mammalian remains of uncertain stratigraphic and/or age that had been considered with reservations as Mesozoic (Bonaparte 1978; Lillegraven et al. 1979). The molar of *Mesungulatum* was closely associated with ornithischian and titanosaur sauropod articulated specimens in a well-controlled geological setting (e.g., Bonaparte et al. 1984; Andreis 1987). This was the first unquestionable osseous remain of a Mesozoic mammal from SA.

The findings from the Los Alamitos locality had a great impact on the scientific community; however, the presence of Mesozoic mammals in the continent was not unexpected. By then, Mesozoic mammals were considered to be likely present on the basis of therians collected from Peru and Bolivia (later shown to be Paleogene in age; Gayet et al. 1991; Muizon 1991; Sigé et al. 2004), and on footprints from Patagonia, Argentina (Casamiquela 1961, 1964). Metatherians and eutherians collected from the Laguna Umayo, Peru, Vilquechico Formation (Grambast et al. 1967; Sigé 1968) were first interpreted as coming from the Upper Cretaceous and consequently considered osseous remains of Mesozoic mammals (e.g., Marshall et al. 1985; Marshall and Muizon 1988). However, further stratigraphic and magnetostratigraphic studies re-interpreted the fossil-bearing levels and assigned those mammals to the Upper Paleocene–Eocene interval (Sigé et al. 2004). Similarly, metatherians collected from Tiupampa, Bolivia, were first mentioned in the literature as having being collected in the Upper Cretaceous El Molino Formation (e.g., Marshall et al. 1983), but later reassigned to the Santa Lucía Formation, of Paleocene age (Tiupampa SALMA) (e.g., Gayet et al. 1991; Muizon 1991).

Interestingly, the first insights about SA Mesozoic mammals were not bones but footprints reported in the sixties by Casamiquela (1961, 1964) from Estancia Laguna Manantiales locality, Santa Cruz, Patagonia, Argentina. These tracks from Jurassic beds, named *Ameghinichnus patagonicus*, were found in slabs also bearing a variety of dinosaur and other reptilian footprints; this was "*...un bello símbolo, la prueba de la coexistencia [...] de dinosaurios y mamíferos en la Patagonia...*"—a beautiful

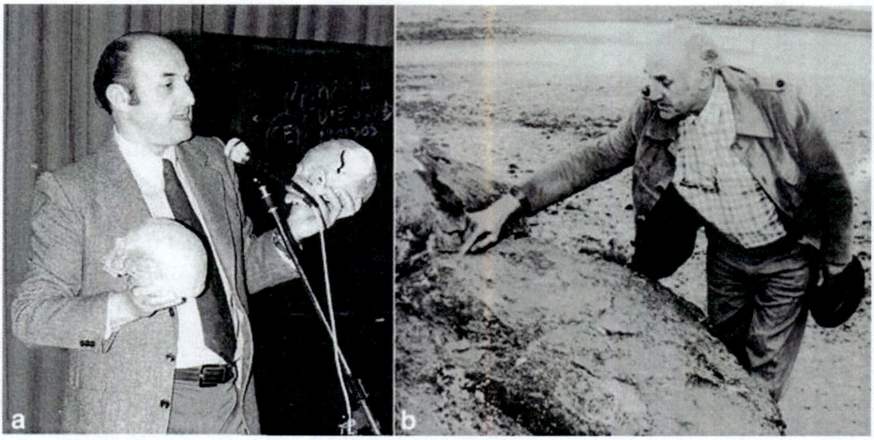

**Fig. 1.1** Rodolfo M. Casamiquela discovered and described the first Mesozoic mammal remains from South America, the footprints *Ameghinichnus patagonicus*. R. M. Casamiquela in a conference in the 1980s (picture from Erraín Casamiquela) (**a**). R. M. Casamiquela at Bajo de Yamnago, Meseta de Somuncurá, Patagonia, Argentina, indicating a big sacred rock, known as "La Vieja", where native Tehuelches acknowledged good luck after large guanaco hunting, 2006 (picture from Museo Naturalístico, Antropológico e Histórico "Jorge Gerhold") (**b**)

symbol, the proof of coexistence [...] of dinosaurs and mammals in Patagonia—(Casamiquela 1961: 225). More than 100 numbered specimens of *Ameghinichnus* have been collected, ranging from single footprints to lengthy tracks with dozens of individual tracks, and it is the most abundant ichnotaxon from the site (de Valais 2009). Rodolfo M. Casamiquela (Ingeniero Jacobacci, December 11, 1932–Cipolleti, December 4, 2008) was a scientist of many talents, expert in paleontology, anthropology, history, and linguistics, including the Tehuelche—Aonikenk—and Mapuche first nation's SA languages (Fig. 1.1). He published about 400 papers and 24 books, and funded the Museo Naturalístico, Antropológico e Histórico "Jorge Gerhold" in Ingeniero Jacobacci, Río Negro, Patagonia, Argentina. In the field of paleontology, many of his findings and publications, including those on Jurassic mammals, still stand as important references, even under the changing light of new paradigms (Haller 2009).

Footprints similar to *Ameghinichnus* were later discovered from paleodunes of the Lower Cretaceous Botucatu Formation (São Bento Group, Paraná Basin), in the famous region of Araraquara, São Paulo State, Brazil (Leonardi 1981). These tracks were discovered by the paleontologist Giuseppe Leonardi and received the name *Brasilichnium elusivum* (Leonardi 1981; Chap. 9).

It is the discovery by José F. Bonaparte, 20 years later, of the vertebrate association at Los Alamitos Formation, including *Mesungulatum houssayi*, which became foundational for our understanding of the Mesozoic mammalian fauna from SA. At the time it represented most of the material basis for the entire southern continents of the former Gondwana. José F. Bonaparte (Rosario, June 14, 1928–Mercedes,

February 18, 2020) was an untiring and tenacious field worker, resulting in a legacy of more than 50 new dinosaurs, birds, pterosaurs, crocodyliforms and other archosauromorphs, therapsids, and mammals from the Mesozoic and younger ages (Fig. 1.2). This long list of discoveries earned him the moniker of "Master of the Mesozoic" by the renowned paleontologist Robert Bakker, something he enjoyed and was proud of. While young, Bonaparte founded the Museo Popular de Ciencias Naturales "Carlos Ameghino" in 1947 in Mercedes, Buenos Aires, Argentina. Afterwards, he moved to the Instituto Miguel Lillo of the Universidad Nacional de Tucumán in 1959 by invitation of Osvaldo Reig. In that institution, he obtained the title Doctor *Honoris Causa* in 1974 because of his contributions to the Mesozoic of SA, especially to the tetrapod faunas from the Triassic. In 1978, Bonaparte moved to Buenos Aires, where he was the Director of the Sección Paleontología de Vertebrados of the Museo Argentino de Ciencias Naturales "Bernardino Rivadavia", until 2004, when he officially retired, but never retired from his passionate work. In 2007, he returned to the city of Mercedes where he started again at the museum he founded and in his last years continued actively working in his house. His extensive production includes more than 150 papers, 7 books, and the discoveries of several now traditional Mesozoic fossil localities in Argentina and Brazil. His passing, some months ago, found him writing a new book about SA Triassic faunas and his idea about the origin of mammals and the relationships of gondwanatherians with glyptodonts, which he had already outlined a few years ago (Bonaparte 2017).

At Los Alamitos, Bonaparte undertook intense multiyear systematic fieldwork, including large groups of students and volunteers. The efforts, supported by the National Geographic Society, ultimately led to the recognition of 16 genera and 18 mammalian species (Bonaparte 1986a, b, 1990, 2002; Krause and Bonaparte 1993; see also Bonaparte and Migale 2010, 2015), although some of them are likely not valid (Rougier et al. 2011a; Averianov et al. 2013; Gaetano et al. 2013). Bonaparte used for the first time a combination of grueling flat-on-your-stomach surface picking (see Chap. 2, Fig. 2.20a) and the then innovative technique of screen-washing, which at the time was being developed in the USA (McKenna 1962; Cifelli 1996) and which with minor modifications is still currently used. He and his crew eventually collected more than 300 mammalian specimens, mostly isolated teeth and thousands of Cretaceous microvertebrates (most of them isolated fish elements) from this locality. Soon after the findings at the Los Alamitos Formation in 1985, new mammalian remains were discovered in the Lower Cretaceous La Amarga Formation, Neuquén, Patagonia, Argentina (Bonaparte 1986a), during an expedition focused on the excavation of the peculiar sauropod dinosaur *Amargasaurus cazaui* with extremely long neural spines (Figs. 1.2a and 1.3) found the year before. The first mammalian bones at La Amarga were discovered by Mr. Martín Vince, Bonaparte's long-term expert technician. The original discovery, however, was not recognized in the field, beyond "small bones", but back in Buenos Aires, when preparation revealed a well-preserved jaw and partial postcranium of a relatively large mammal. In 1986, a large excavation was initiated from the spot where the first bones were found, and when the crew was almost ready to give up, flecks of bone were found embedded in the rock. The quarry was enlarged and after several days, large and mid-sized blocks with bones were collected. Preparation at the MACN (mostly done by GRW, by then a University

**Fig. 1.2** José F. Bonaparte, the "Master of the Mesozoic", discovered and named more than 50 vertebrates from the Mesozoic and early Cenozoic, including the first Mesozoic mammal fossil bones from South America. In the pictures, J. F. Bonaparte at La Amarga, during the excavation of *Amargasaurus cazaui* (MACN-N 15) and at the side of the long neural spines of cervical vertebrae, 1984 (**a**). J. F. Bonaparte at Museo Argentino de Ciencias Naturales "Bernardino Rivadavia" (MACN), 2005 (**b**). Bonaparte's team at MACN, after assembling the skeleton of *Piatnitzkysaurus floresi* for the dinosaur hall; from left to right: José A. Pumares, Orlando Gutierrez, Oscar Donadío, Guillermo W. Rougier, Miguel Soria, Jaime Powell, José L. Gomez, and José F. Bonaparte, 1984. Note that in this first presentation, the tail of *Piatnitzkysaurus* is hanging down; soon after, it was lifted to a more dynamic position (**c**)

**Fig. 1.3** Artistic reconstruction from La Amarga, Neuquén, Argentina, more than 120 Ma ago, with *Vincelestes neuquenianus*, *Amargazaurus cazaui,* and Pterosauria indet. by Jorge L. Blanco

student and preparator for Bonaparte) revealed several skeletons of the cladotherian *Vincelestes neuquenianus*, furnishing the first relatively complete mammal from the SA Mesozoic (Chap. 7). With two data points, Los Alamitos and La Amarga, the first rudiments of a biostratigraphic context were established. As it was clear from the onset that nothing similar to *Vincelestes* had been found in Los Alamitos, some major change had occurred between the Early and Late Cretaceous.

Although Bonaparte found the first Mesozoic mammals relatively late in his career, he had an extensive background that facilitated his tackling of this new rising field. In the sixties and seventies, Bonaparte collected and studied a wide diversity of non-mammaliaform cynodonts from Triassic rocks of La Rioja, San Juan, and Mendoza in western Argentina (reviewed in Bonaparte and Migale 2010, 2015). Initially, Bonaparte worked as a preparator and as a field hand of the expeditions organized by the Argentinean biologist and paleontologist Osvaldo Reig from Tucumán University. Under the tutelage of Reig and occasionally sponsored by Prof. Alfred S. Romer from Harvard University, Bonaparte progressively became an expert on non-mammalian cynodonts. These complex and advanced mammalian forerunners are among the most abundant and best-preserved worldwide. Some like *Probainognathus* and *Therioherpeton* are emblematic and integral to most discussions on the origin of early mammals (Romer 1970; Bonaparte and Barberena 2001; Martinelli et al. 2017). Bonaparte's prolific production on non-mammalian cynodonts and Mesozoic mammals established him as an expert in the field. In the late 90s, Bonaparte accepted a temporary position in the Fundação Zoobotânica of Rio Grande do Sul, and subsequently in the early 2000s, he moved to the Universidade Federal do Rio Grande do Sul, Porto Alegre, Brazil. His renewed interest in the early evolution of mammals and their origin resulted in a new bout of fieldwork in the Triassic outcrops of Southern Brazil. Over a number of years, the expeditions led by Bonaparte provided more than 70 new specimens of, among others, non-mammaliaform cynodonts, rhynchocephalians, procolophonians, and basal dinosaurs, from Rio Grande do Sul in Southern Brazil. Five new cynodont species were described, several of them with a highly derived and mammal-like anatomy, changing implications on the diversity, morphology, and biostratigraphy of mammaliaform precursors (Bonaparte and Migale 2010, 2015; Martinelli and Soares 2016; Chap. 3). The details of these massive collections are only now just beginning to be fully explored and will likely be influential for years to come.

Rosendo Pascual (Mendoza, July 10, 1923–La Plata, December 23, 2012) was an Argentine vertebrate paleontologist, the head of the department of Vertebrate Paleontology at the Museo de La Plata (Fig. 1.4). He was the advisor and director of several generations of mostly Cenozoic paleontologists that to date constitute the bulk of the paleontological community in Argentina. Pascual studied the gamut of fossil SA endemic mammals, including metatherians, xenarthrans, native ungulates, rodents, sirenians, carnivores, monotremes, and in the last years of his career, he was involved in the study of Mesozoic mammalian lineages that survived the Cretaceous extinction: gondwanatherians and dryolestoids. He focused on systematics, evolution, biostratigraphy, biochronology, and biogeography, with a vast output captured in more than 170 papers. He recognized the South American Land Mammals Ages

**Fig. 1.4** Rosendo Pascual, the "Cenozoic Mammal Fellow", dedicated his work to the study of South American mammals, mostly from the Cenozoic but also older ages, including finding and describing the first non-Australasian monotreme. In the pictures, R. Pascual at Gran Hondonada, Chubut, Argentina, in an expedition from the Museo de La Plata, 1982 (picture from Mariano Bond) (**a**). R. Pascual and Pablo Puerta enjoying a mate in the windy Patagonian steppe (**b**) and in an airplane, on their way to Punta Peligro (**c**), field season from 1996 (pictures from Pablo Puerta)

(SALMAs), establishing a geologic timescale for the Cenozoic in SA based on the succession of mammal associations. In the words of the renowned paleontologist George Gaylord Simpson (1984: 206), his senior colleague, Rosendo Pascual was an "eminent paleomammalogist". In the nineties, the discovery in Patagonia of the first non-Australian monotreme (Pascual et al. 1992) provided shocking new evidence of the biogeographic commonality of the evolution of the SA mammalian fauna and those of other Gondwanan masses. The discovery was the result of fieldwork led by Rosendo Pascual at Punta Peligro, Salamanca Formation, Chubut, Patagonia, Argentina, which, in addition to monotreme material, also recovered the first post-Cretaceous remains of the gondwanatherians (Scillato-Yané and Pascual 1985) and dryolestoid meridiolestidans (Gelfo and Pascual 2001; Rougier et al. 2011a). The survival of these archaic lineages and their relative abundance within faunas with majority composition of groups with NA, or at least Holarctic, origin (i.e., Eutheria and Metatheria) showed for the first time the complexity of the early Cenozoic SA biota and the milder effect of the Cretaceous transition in southern latitudes (Rougier et al. 2011a).

Three scientists, Rodolfo M. Casamiquela, José F. Bonaparte, and Rosendo Pascual, each with his own focus, strengths, and weaknesses, are the pillars—a tripartite foundation—for the early stages of study of the SA Mesozoic mammals. More recently, additional localities, filling either temporal or systematic gaps, have been discovered, helping to piece together a somewhat more complete framework in which mammalian evolution can be studied in the general context of large-scale phylogeny. However, finds of SA Mesozoic mammals from places other than Argentina are still extremely limited and confined to scant remains in Bolivia, Brazil, Chile, and eventually Peru (Leonardi 1981; Mourier et al. 1986; Bertini et al. 1993; Gayet et al. 2001; Castro et al. 2018; Goin et al. 2020). The SA fossil record of Mesozoic mammals, close stem relatives, and their surviving lineages form isolated pieces of the larger Gondwanan puzzle, which is being gradually improved by findings in continental Africa, Madagascar, India, Australia, and Antarctica (e.g., Archer et al. 1985; Sigogneau-Russell 1991, 1995; Flannery et al. 1995; Rich et al. 1997, 1999, 2001; Heinrich 1998; Flynn et al. 1999; Rich and Vickers-Rich 2004; Anantharaman et al. 2006; Goin et al. 2006; Prasad et al. 2007; Wilson et al. 2007; Krause et al. 2014, 2020; Martinelli et al. 2014).

Despite a long and venerable paleontological tradition in Argentina, which started as a well-established and prime scientific discipline in the 1800s by outstanding local and European figures like Florentino Ameghino (Luján, September 18, 1854–La Plata, August 6, 1911), Francisco P. Moreno (Buenos Aires, May 31, 1852–Buenos Aires, November 22, 1919), and Hermann Burmeister (Stralsund, January 5, 1807–Buenos Aires, May 2, 1892), the discovery of Mesozoic mammals is relatively recent. This gap is not due to the lack of suitable rocks, paleoenvironments, or exposures; there are vast expanses of Mesozoic rocks ranging in age from the Triassic to the Cretaceous that form a continuous garland to the pampas from the northwest near the border with Bolivia, along the Precordillera, and covering a large percentage of the Patagonian desert. Perhaps, the delay in finding Mesozoic mammals in SA was due to their relative fragility and small size (although several SA Mesozoic mammals are actually quite large, opossum-like), or perhaps they were obscured by the novel, diverse, and exquisite dinosaur fossil record that captures the attention of most specialists. To find Mesozoic mammals requires a focus and a set of techniques at odds with those prevalent in the exploration for dinosaurs. With the intensification of prospecting in Mesozoic rocks, the findings of mammals in SA and other Gondwanan places were only a matter of time. The application of search techniques developed with the aim to recover small, isolated elements, such as the collection of sediments, screen-washing (Fig. 1.5), and picking under a binocular microscope, greatly increased the chances of finding the small but durable mammalian teeth that constitute the lion's share of the Mesozoic mammal record. Other labor-intensive techniques, such as opening up of quarries and systematic careful reduction of boulders with a focus on microvertebrates (Fig. 1.6), have radically changed the quantity and quality of the specimens recovered over the last decades. In recent years, new taxa have been collected from the Jurassic (Rauhut et al. 2002; Rougier et al. 2007a, b; Gaetano and Rougier 2011, 2012) and Cretaceous (e.g., Pascual et al. 2000; Rougier et al. 2009a, b, 2011b; Harper et al. 2019), and new insights have been gained from

**Fig. 1.5** Different techniques for finding microvertebrates. Screen-washing helps to process large quantities of sediments, cleaning the sediment with water, and sieving the rocks and fossils with different size meshes. The final search is done in the lab under a binocular microscope. In (**a**), Tony Harper transporting large quantities of sediments from La Colonia Formation, Mirasol Chico, Chubut, Argentina, 2018. In (**b**), Daniel Hernández and Agustín G. Martinelli screen-washing sediments from the Lohan Cura Formation, Picún Leufú, Neuquén, Argentina, 1999. In (**c**), Magalí Cardenas searching for larger mammalian bones from the La Colonia Formation, while sediments are drying over a plastic tarp after the screen-washing, 2001. In (**d**) and (**e**), Leandro Canessa screen-washing sediments from La Colonia Formation in a barrel of water with gentle movements and drying the processed sediments in the same mesh to reduce the breakage of fossil material, 2020

**Fig. 1.6** Different techniques for finding microvertebrates. Breaking stones has the potential to find small specimens in articulation. In (**a**), quarry at Cerro Cóndor locality, Cañadón Asfalto Formation, Chubut, Argentina, during the extraction of several large blocks, 2019. In (**b**), Juan M. Leardi and Guillermo (Willy) Turazzini taking large blocks from the quarry at Cerro Cóndor locality, 2010. In (**c**), Juliana Sterli reducing rocks to small pieces while searching for small bones, 2003. A good eye in the field is crucial for finding new material and localities. In (**d**), Agustín G. Martinelli and Leandro Canessa after finding the first mammal specimen in the Canela locality, Cañadón Asfalto Formation, Chubut, Argentina, 2015. Little precious mammal bones are cataloged and carefully wrapped at the end of the day in the field, for transportation to the lab, preparation, and study. In (**e**), Raúl O. Gómez and Leandro Gaetano cataloging fossils from the Cañadón Asfalto Formation, at the local school from Cerro Cóndor, Chubut, Argentina, 2009

Gondwanan lineages that persisted into the Cenozoic (e.g., Gelfo and Pascual 2001; Pascual et al. 2002; Goin et al. 2004, 2006; Rougier et al. 2012; Wible and Rougier 2017). In a few decades, the field of Mesozoic mammals in SA has been transformed from the finding of a single tooth to a burgeoning discipline, with an increasingly large material basis and by an active community involved in the global discourse of early mammalian evolution.

## 1.3 Abbreviations

### *1.3.1 Institutional Abbreviations*

| | |
|---|---|
| CONICET | Consejo Nacional de Investigaciones Científicas y Técnicas, Argentina |
| CPAP | Paleontological Collection of Antarctica and Patagonia, Instituto Antártico Chileno, Punta Arenas, Chile |
| CRILAR | Centro Regional de Investigaciones Científicas y Transferencia Tecnológica de La Rioja, Anillaco, La Rioja, Argentina |
| LIEB-PV | Laboratorio de Investigaciones en Evolución y Biodiversidad, Vertebrate Paleontology Collection, Esquel, Chubut, Argentina |
| LPP | Laboratório de Paleoecologia e Paleoicnologia, Departamento de Ecología e Biología Evolutiva, Universidade Federal de São Carlos, São Carlos, Brazil |
| MACN | Museo Argentino de Ciencias Naturales "Bernardino Rivadavia" (A, Ameghino Collection; CH, Chubut Collection; N, Neuquén Collection; PV, Vertebrate Paleontology Collection; RN, Río Negro Collection), Buenos Aires, Argentina |
| MCT | Museu de Ciências da Terra, Rio de Janeiro, Brazil |
| MCF-PVPH | Museo "Carmen Funes", Vertebrate Paleontology Collection, Plaza Huincul, Neuquén, Argentina |
| MCP-PV | Museu de Ciências e Tecnologia, Pontifícia Universidade Católica de Rio Grande do Sul, Porto Alegre, Brazil |
| MHNC | Museo de Historia Natural "Alcide d'Orbigny", Cochabamba, Bolivia |
| MHNSR-PV | Museo de Historia Natural de San Rafael, Vertebrate Paleontology Collection, San Rafael, Mendoza, Argentina |
| MLP | Museo de La Plata, Buenos Aires, La Plata, Argentina |
| MMACR-PV-T | Museu Municipal "Aristides Carlos Rodrigues", Paleovertebrates, Triassic Collection, Candélaria, Rio Grande do Sul, Brazil |
| MML-PV | Museo Municipal de Lamarque, Vertebrate Paleontology Collection, Río Negro, Lamarque, Argentina |
| MPCA | Museo Provincial "Carlos Ameghino", Cipolletti, Río Negro, Argentina |

| | |
|---|---|
| MPEF-PV | Museo Paleontológico "Egidio Feruglio", Vertebrate Paleontology Collection, Trelew, Chubut, Argentina |
| MVP | Museu "Vicente Pallotti", Santa Maria, Rio Grande do Sul, Brazil |
| PULR | University of La Rioja, Paleontology Collection, La Rioja, Argentina |
| PVL | Instituto Miguel Lillo, San Miguel de Tucumán, Argentina |
| UFRJ-DG | Departamento de Geologia, Universidade Federal do Rio de Janeiro, Rio de Janeiro, Brazil |
| UFRGS-PV-T | Laboratório de Paleontologia, Vertebrate Paleontology Collection, Triassic, Universidade Federal Rio Grande do Sul, Porto Alegre, Brazil |
| UNPSJB-PV | Universidad Nacional de la Patagonia "San Juan Bosco", Vertebrate Paleontology Collection, Comodoro Rivadavia, Chubut, Argentina |
| URC (M/R) | Museu de Paleontologia e Estratigrafia "Prof. Dr. Paulo Milton Barbosa Landim", Instituto de Geociências e Ciências Exatas, Universidade Estadual Paulista Julio de Mesquita Filho, Rio Claro Campus, São Paulo, Brazil |

### *1.3.2 Anatomical Abbreviations*

| | |
|---|---|
| C/c | upper and lower canine |
| I/i | upper and lower incisor (numbers indicate the loci, representing the position from front to back) |
| M/m | upper and lower molar (numbers indicate the loci, representing the position from front to back) |
| Mf/mf | upper and lower molariform (numbers indicate the loci, representing the position from front to back) |
| P/p | upper and lower premolar (numbers indicate the loci, representing the position from front to back) |
| Pmf/pmf | upper and lower premolariform (numbers indicate the loci, representing the position from front to back) |

### *1.3.3 Other Abbreviations*

| | |
|---|---|
| AZ | Assemblage Zone |
| Ma | million years |
| NA | North America |
| SA | South America |
| SALMA | South American Land Mammal Age |
| SANU | South American Native Ungulate |

## 1.4 Nomenclature

The taxonomic nomenclature in the text aims to reflect monophyletic groups (Fig. 1.7). There are controversies in the literature about the monophyly of some of the groups treated in the text from which we indicate alternative positions. Additionally, some traditional "groups" have been consistently recovered as paraphyletic in cladistic analysis; in those cases, the definition is adjusted to reflect monophyly or

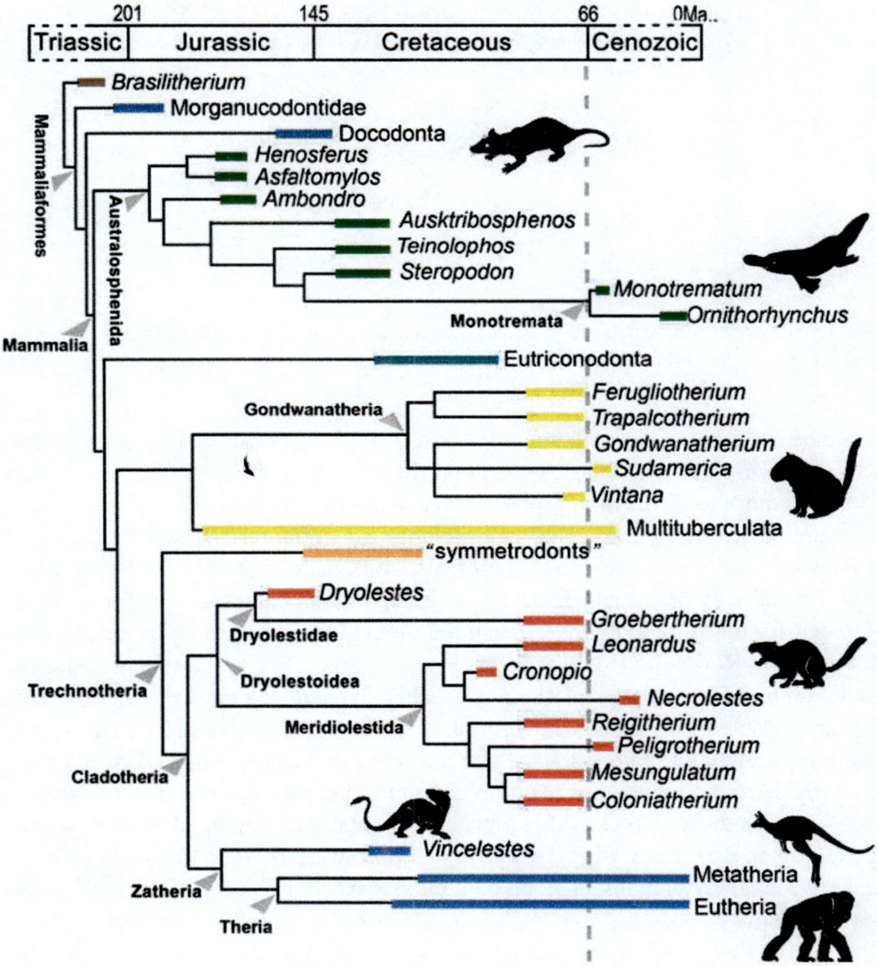


**Fig. 1.7** Phylogenetic hypothesis calibrated with geological timescale showing the main groups of South American Mesozoic mammals (based on Krause et al. 2014 with some modifications following Rougier et al. 2011b)

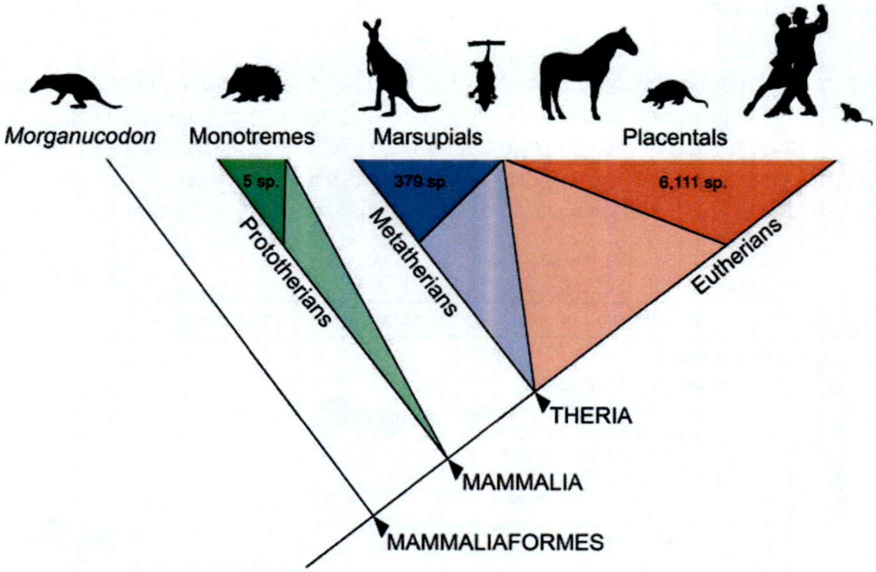


**Fig. 1.8** Phylogenetic relationships of the major groups of living mammals (monotremes, marsupials, and placentals)

the name is indicated between quotation marks if it is retained because of its usefulness to provide a reference to a general portion of the mammalian tree or a distinct quick reference to a general morphology (such as "triconodonts").

Mammaliaformes is the clade that includes the common ancestor of the fossil *Morganucodon*, the crown group mammals, and all its descendants (Rowe 1988; Figs. 1.7 and 1.8). Mammalia is the crown group that includes the common ancestor of extant species of monotremes, marsupials, and placentals, and all its descendents (Rowe 1988; Rowe and Gauthier 1992; Figs. 1.7 and 1.8). Prototheria includes the crown group Monotremata (platypus and echidnas) and taxa closer to them than to therians (adapted from Kielan-Jaworowska et al. 2004). Eutheria includes the crown group Placentalia (armadillos, cats, humans, etc.) and closely related extinct taxa; similarly, Metatheria includes the crown group Marsupialia (opossums, kangaroos, koalas, etc.) and any extinct species more closely related to marsupials than to placentals (Rougier et al. 1998; Figs. 1.7 and 1.8). This cluster of definitions takes care of the living diversity of mammals, and their closest extinct relatives. These are the taxa we commonly refer to as "mammals", one way or the other, and form the core of what informs our view of what a mammal is. Nature does not come pre-defined for us; there is no intrinsic, Platonic archetype of a mammal. These names, and those that follow, are simple handles we use to communicate and hopefully simplify and reduce the diversity to a manageable size. This artificial definition of what a mammal is should make it very clear that we see no intrinsic difference between the earliest mammals and their immediate forerunners, or between mammaliaforms and slightly more basal non-mammaliaform cynodonts. These definitions are arbitrary snippets of the bushy synapsid tree. None of these definitions make any reference to morphological characters, either; in a given study, a character may be associated with a clade, but the diagnostic value of such a character depends on the taxa included and the scorings of other characters in that particular study. A restudy of the problem including more or different taxa, amended or increased data, will likely result in a different optimization of the characters even if the basic tree topology is maintained. There is no ultimate "Mammalian Character", just a set of diagnostic features for the included taxa, using a given database and using a set of specific analytical methods.

Several groups closely related to Mammalia (stem mammals) and many extinct mammals are unique to the fossil record. Some of those groups, which are mentioned in the text, are listed below.

Docodonts are stem mammals (Fig. 1.7). Docodonta is the group that includes taxa more closely related to *Docodon* and *Simpsonodon* than morganucodonts, *Shuotherium,* and the Late Triassic "symmetrodont" mammaliaforms (such as *Kuehneotherium, Woutersia,* and *Tikitherium*) (Luo and Martin 2007).

Australosphenida is the clade that includes the common ancestor of *Ambondro, Ausktribosphenos*, living monotremes, and all its descendants (Fig. 1.7). It includes all extinct taxa more closely related to living monotremes than to *Shuotherium* or living therians (Luo et al. 2001, 2002).

Gondwanatheria is the clade composed of *Ferugliotherium, Sudamerica*, their common ancestor, and all of its descendants (Chimento et al. 2014; Krause et al. 2014, 2020). However, the gondwanatherian affinities of *Ferugliotherium* and *Trapalcotherium* are still uncertain (Chap. 8) and consequently this definition should be considered with caution.

Trechnotheria is a clade originally erected by McKenna (1975) to include "symmetrodonts" and Cladotheria (Fig. 1.7). It is defined as the common ancestor of *Zhangheotherium* (and by extrapolation, the monophyletic group of Spalacotheriidae) and the crown Theria, plus all of its descendants (Luo et al. 2002).

Cladotheria is the group that includes the common ancestor of dryolestoids and living therians, plus all of its descendants (McKenna 1975; Luo et al. 2002) (Fig. 1.7).

In the traditional sense, Dryolestoidea is the clade that includes *Amphitherium*, Paurodontidae, and Dryolestidae (McKenna 1975). Depending on the phylogenetic topologies obtained, Meridiolestida (see below) is grouped in a monophyletic clade with Dryolestidae and Paurodontidae (Rougier et al. 2011b), or these are stem groups of Theria (Rougier et al. 2012). Dryolestida is the group that includes any dryolestoid more closely related to *Dryolestes* than to *Peligrotherium* (Rougier et al. 2011b). Meridiolestida is the group that includes any dryolestoid more closely related to *Peligrotherium* than to *Dryolestes* or *Paurodon* (Rougier et al. 2011b). Within Meridiolestida, Mesungulatoidea includes the last common ancestor of *Reigitherium, Mesungulatum*, and *Peligrotherium* plus all its descendants (Rougier et al. 2011b) (Fig. 1.7).

Zatheria is the clade that includes the last common ancestor of *Peramus* and living marsupials and placentals, plus all of its descendants (McKenna 1975; Luo et al. 2002).

Theria groups the common ancestor of Metatheria (including marsupials) and Eutheria (including placentals), plus all of its descendants (e.g., Rowe 1988, 1993; Rougier et al. 1998; Rowe and Gauthier 1992; Luo et al. 2002; Fig. 1.7). Depending on the year of the production of the paper, definitions based on characters are frequent in the literature. In this sense, Tribosphenida was originally defined by McKenna (1975) to "reflect the view that their [i.e., eutherians, metatherians, and some stem taxa] acquisition of a protocone is synapomorphous and is meant to be the cladistic taxonomic equivalent of Simpson's (1936: 8) descriptive term tribosphenic" (McKenna 1975: 27). Tribosphenida is equivalent to Boreosphenida (Luo et al. 2001), a name erected to emphasize the "dual origin of tribosphenic mammals" (see also Kielan-Jaworowska et al. 2004).

Despite the intensive work in the last decades on mammalian phylogeny, including both fossils and molecules (e.g., Meredith et al. 2011; O'Leary et al. 2013), disputes about the phylogenetic position of some extinct groups persist. These controversial issues have consequences in the calibration of nodes and cladogenetic events required to understand mammal evolution (e.g., Cifelli and Davis 2013). One of the disputes that has been recently addressed is the relationships of Multituberculata and Haramiyida and their implication for the time and place of the origin of Mammalia (e.g., Zheng et al. 2013; Zhou et al. 2013; Meng et al. 2017; Huttenlocker et al. 2018).

Multituberculates are nested within Mammalia (e.g., Rowe 1988; Luo 2007; Huttenlocker et al. 2018; King and Beck 2020). They are known by hundreds of species, some of them represented by exquisite fossil material from Asia, as well as from Europe, NA, and partial dentition from SA, dating from the Middle Jurassic to the late Eocene (Kielan-Jaworowska et al. 2004; Kielan-Jaworowska 2013). Haramiyidans are known by material from the Late Triassic–Late Jurassic of Europe and Asia (e.g., Luo et al. 2002, 2017; Zheng et al. 2013; Zhou et al. 2013; Meng et al. 2017). Multituberculates and haramiyidans have complex occlusal molar surfaces with multiple cusp rows. According to some authors, multituberculates and haramiyidans are sister taxa (e.g., Hahn 1973; Hahn et al. 1989; Zheng et al. 2013) and have been grouped together in controversial "Allotheria" (e.g., Zheng et al. 2013; Bi et al. 2014; Meng et al. 2018) (Fig. 1.9a), sometimes including also gondwanatherians (Krause et al. 2014; see also Chap. 8). This implies that the origin of Mammalia took place at least during the Late Triassic, that is, 40–50 Ma before the mammalian radiation, and consequently implying a long fuse model for mammal diversification (Archibald and Deutschman 2001).

Alternatively, some authors exclude haramiyidans from Mammalia and regard their similarities with multituberculates as convergent (e.g., Butler 2000; Zhou et al. 2013; Luo et al. 2015; Huttenlocker et al. 2018; see also King and Beck 2020) (Fig. 1.9b). In this scenario, the documented origin of Mammalia is not earlier than the latest Early–Middle Jurassic, where accepted mammals have been found (e.g., Rauhut et al. 2002; Rougier et al. 2007a; Luo et al. 2011) and mammalian diversification follows an explosive model (Archibald and Deutschman 2001). Clearly, the fossils can provide only minimal ages for phylogenetic calibrations; more realistic ages for most clades can be estimated via probabilistic approaches, in particular, using Bayesian methods (e.g., Barba-Montoya et al. 2017; King and Beck 2020).

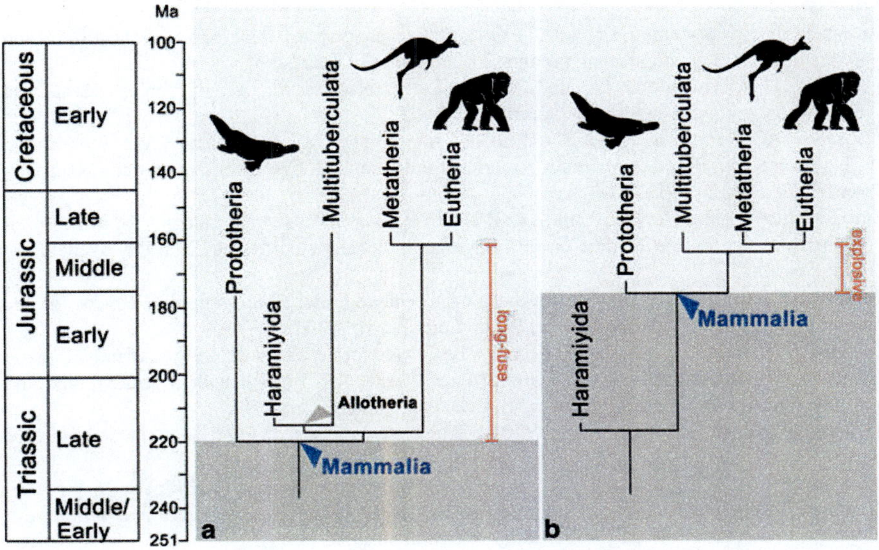


**Fig. 1.9** Alternative interpretations on the relationships and timing of the origin of Mammalia (modified from Cifelli and Davis 2013): in (**a**), the long fuse model for mammal diversification, in (**b**), the explosive model

Given how labile node age estimates are to the introduction of new, particularly older, material, it is likely that our current estimates will be quickly adjusted.

Similar competing hypotheses are present regarding the timing of the origin of placental mammals by the inclusion, or exclusion, in the crown group of plesiomorphic stem taxa. Some authors root the origin of placental mammals deep in the Mesozoic (e.g., Nessov et al. 1998; Archibald et al. 2001; Kielan-Jaworowska et al. 2004; Lee and Beck 2015). Alternatively, others place the placental origin in the vicinity of the K/Pg boundary (e.g., Wible et al. 2007; Beck and Lee 2014), close to the beginning of the Cenozoic and the start of the massive diversification pattern as recorded by the fossil evidence (see also Archibald and Deutschman 2001; Springer et al. 2003; Lyson et al. 2019).

## References


Anantharaman S, Wilson GP, Das Sarma DC, Clemens WA (2006) A possible Late Cretaceous “haramiyidan” from India. J Vertebr Paleontol 26:488–490

Andreis RR (1987) Part I -Stratigraphy and paleoenvironment. In: Bonaparte JF (ed) The Late Cretaceous fauna of Los Alamitos, Patagonia, Argentina. Rev Mus Argent Cienc Nat “B Rivadavia”, Paleontología 3:103–110

Archer M, Flannery TF, Ritchie A, Molnar RE (1985) First Mesozoic mammal from Australia, an Early Cretaceous monotreme. Nature 318:363–366

Archibald JD, Deutschman DH (2001) Quantitative analysis of the timing of the origin and diversification of extant placental orders. J Mammal Evol 8:107–124

Archibald JD, Averianov AO, Ekdale EG (2001) Late Cretaceous relatives of rabbits, rodents, and other extant eutherian mammals. Nature 414:62–65

Averianov AO, Martin T, Lopatin AV (2013) A new phylogeny for basal Trechnotheria and Cladotheria and affinities of South American endemic Late Cretaceous mammals. Naturwissenschaften 100:311–332

Barba-Montoya J, dos Reis M, Yang Z (2017) Comparison of different strategies for using fossil calibrations to generate the time prior in Bayesian molecular clock dating. Mol Phylogenet Evol 114:386–400

Beck RMD, Lee MSY (2014) Ancient dates or accelerated rates? Morphological clocks and the antiquity of placental mammals. Proc R Soc Lond B 281:20141278

Bertini RJ, Marshall LG, Gayet M, Brito P (1993) Vertebrate faunas from the Adamantina and Marília formations (Upper Baurú Group, Late Cretaceous, Brazil) in their stratigraphic and paleobiogeographic context. Neues Jahrb Geol Paläontol 188:71–101

Bi S, Wang Y, Guan J, Sheng X, Meng J (2014) Three new Jurassic euharamiyidan species reinforce early divergence of mammals. Nature 514:579–584

Bonaparte JF (1978) El Mesozoico de América del Sur y sus tetrápodos. Op Lilloana 26:1–596

Bonaparte JF (1986a) History of the terrestrial Cretaceous vertebrates of Gondwana. 4°Congreso Argentino de Paleontología y Bioestratigrafía, Mendoza, Actas 2:63–95

Bonaparte JF (1986b) Sobre *Mesungulatum houssayi* y nuevos mamíferos cretácicos de Patagonia. 4° Congreso Argentino de Paleontología y Bioestratigrafía, Mendoza, Actas 2:48–61

Bonaparte JF (1990) New Late Cretaceous mammals from the Los Alamitos Formation, Northern Patagonia. Natl Geogr Res 6:63–93

Bonaparte JF (2002) New Dryolestida (Theria) from the Late Cretaceous of Los Alamitos, Argentina, and paleogeographical comments. Neues Jahrb Geol Paläontol 224:339–371

Bonaparte JF (2017) Notas sobre el origen de los gliptodontes. PE-APA 18(2):R13

Bonaparte JF, Barberena MC (2001) On two advanced carnivorous cynodonts from the Late Triassic of Southern Brazil. Bull Mus Comp Zool 156:59–80

Bonaparte JF, Migale LA (2010) Protomamíferos y mamíferos Mesozoicos de América del Sur. Museo de Ciencias Naturales Carlos Ameghino de Mercedes, Buenos Aires. 1° Edición

Bonaparte JF, Migale LA (2015) Protomamíferos y mamíferos Mesozoicos de América del Sur. Fundación de Historia Natural Felix de Azara, Buenos Aires. 2° Edición

Bonaparte JF, Soria MF (1985) Nota sobre el primer mamífero del Cretácico Argentino, Campaniano-Maastrichtiano (Condylarthra). Ameghiniana 21:177–183

Bonaparte JF, Franchi MR, Powell JE, Sepúlveda EC (1984) La Formación Los Alamitos (Campaniano–Maastrichtiano) del sudoeste de Río Negro, con descripción de *Kritosaurus australis* nov. sp. (Hadrosauridae). Significación paleobiogeográfica de los vertebrados. Rev Asoc Geol Argent 39:284–299

Burgin CJ, Colella JP, Kahn PL, Upham NS (2018) How many species of mammals are there? J Mammal 99:1–14

Butler PM (2000) Review of the early allotherian mammals. Acta Palaeontol Pol 45:317–342

Casamiquela RM (1961) Sobre la presencia de un mamífero en el primer elenco (icnológico) de vertebrados del Jurásico de la Patagonia. Physis 22:225–233

Casamiquela RM (1964) Estudios icnológicos. Problemas y métodos de la icnología con aplicación al estudio de pisadas mesozoicas (Reptilia, Mammalia) de la Patagonia. Colegio Industrial Pío IX, Buenos Aires

Castro MC, Goin FJ, Ortiz-Jaureguizar E, Vieytes EC, Tsukui K, Ramezani J, Batezelli A, Marsola JCA, Langer MC (2018) A Late Cretaceous mammal from Brazil and the first radioisotopic age for the Bauru Group. Royal Soc Open Sci 5:180482

Chimento NR, Agnolín FL, Novas FE (2014) The bizarre "metatherians" *Groeberia* and *Patagonia*, late surviving members of gondwanatherian mammals. Hist Biol 27:603–623

Cifelli RL (1996) Techniques for recovery and preparation of microvertebrate fossils. Oklahoma Geol Surv Spec Publ 96–4:1–36

Cifelli RL, Davis BM (2013) Jurassic fossils and mammalian antiquity. Nature 500:160–161

de Valais S (2009) Ichnotaxonomic review of *Ameghinichnus*, a mammal ichnogenus from the La Matilde Formation (Middle Jurassic), Santa Cruz Province, Argentina. Zootaxa 2203:1–21

Flannery TF, Archer M, Rich TH, Jones R (1995) A new family of monotremes from the Cretaceous of Australia. Nature 377:418–420

Flynn JJ, Parrish JM, Rakotosamimanana B, Simpson WF, Wyss AR (1999) A Middle Jurassic mammal from Madagascar. Nature 401:57–60

Gaetano LC, Rougier GW (2011) New materials of *Argentoconodon fariasorum* (Mammaliaformes, Triconodontidae) from the Jurassic of Argentina and its bearing on triconodont phylogeny. J Vertebr Paleontol 31:829–843

Gaetano LC, Rougier GW (2012) First amphilestid from South America: A molariform from the Jurassic Cañadón Asfalto Formation, Patagonia, Argentina. J Mammal Evol 19:235–248

Gaetano LC, Marsicano CA, Rougier GW (2013) A revision of the putative Late Cretaceous triconodonts from South America. Cretac Res 46:90–100

Gayet M, Marshall LG, Sempere T (1991) The Mesozoic and Paleocene vertebrates of Bolivia and their stratigraphic context: A review. In: Suarez-Soruco R (ed) Fósiles y facies de Bolivia. Vertebrados. Rev Tec YPFB 12:393–433

Gayet M, Marshall LG, Sempere T, Meunier FJ, Cappetta H, Rage J-C (2001) Middle Maastrichtian vertebrates (fishes, amphibians, dinosaurs and other reptiles, mammals) from Pajcha Pata (Bolivia). Biostratigraphic, palaeoecologic and palaeobiogeographic implications. Palaeogeogr Palaeoclimatol Palaeoecol 169:39–68

Gelfo JN, Pascual R (2001) *Peligrotherium tropicalis* (Mammalia, Dryolestida) from the early Paleocene of Patagonia, a survival from a Mesozoic Gondwanan radiation. Geodiversitas 23:369–379

Goin FJ, Vieytes EC, Vucetich MG, Carlini AA, Bond M (2004) Enigmatic mammal from the Paleogene of Peru. In: Campbell KE (ed) The Paleogene mammalian fauna of Santa Rosa, Amazonian Peru. Nat Hist Mus Los Angeles County, Sci Ser 40:145–153

Goin FJ, Reguero MA, Pascual R, Koenigswald W von, Woodburne MO, Case JA, Vieytes C, Marenssi SA, Vizcaíno SF (2006) First gondwanatherian mammal from Antarctica. In: Francis JE, Pirrie D, Crame JA (eds) Cretaceous–Tertiary high-latitude palaeoenvironments, James Ross Basin, Antarctica. Geological Society of London, Spec Publ 258:135–144

Goin FJ, Martinelli AG, Soto-Acuña S, Vieytes EC, Manríquez LME, Fernández RA, Pino JP, Trevisa C, Kaluza J, Reguero MA, Leppe M, Ortiz H, Rubilar-Rogers D, Vargas AO (2020) First Mesozoic mammal from Chile: the southernmost record of a Late Cretaceous gondwanatherian. Bol Mus Nac Hist Nat Chile 69:5–31

Grambast L, Martínez M, Mattauer M, Thaler L (1967) *Perutherium altiplanense* nov. gen., nov. sp., premier mammifère Mésozoïque d'Amérique du Sud. C R Acad Sci 264:707–710

Hahn G (1973) Neue Zahne von Haramiyiden aus der deutschen Ober-Trias und ihre Beziehungen zu den Multituberculaten. Palaeontographica A 142:1–15

Hahn G, Sigogneau-Russell D, Wouters G (1989) New data on Theroteinidae: their relations with Paulchoffatiidae and Haramiyidae. Geol Palaeontol 23:205–215

Haller MJ (2009) Necrológica. Dr. Rodolfo Magin Casamiquela 11/12/32–5/12/2008. Rev Asoc Geol Argent 64:555–558

Harper T, Parras A, Rougier GW (2019) *Reigitherium* (Meridiolestida, Mesungulatoidea) an enigmatic Late Cretaceous mammal from Patagonia, Argentina: morphology, affinities, and dental evolution. J Mammal Evol 26:447–478

Heinrich W-D (1998) Late Jurassic mammals from Tendaguru, Tanzania, East Africa. J Mammal Evol 5:269–290

Huttenlocker AK, Grossnickle DM, Kirkland JI, Schultz JA, Luo Z-X (2018) Late-surviving stem mammal links the lowermost Cretaceous of North America and Gondwana. Nature 558:108–112

Kielan-Jaworowska Z (2013) In persuit of early mammals. Indiana University Press, Bloomington

Kielan-Jaworowska Z, Cifelli RL, Luo Z-X (2004) Mammals from the age of dinosaurs. Origins, evolution, and structure. Columbia University Press, New York

King B, Beck RMD (2020) Tip dating supports novel resolutions of controversial relationships among early mammals. Proc R Soc Lond B 287:20200943

Krause DW, Bonaparte JF (1993) Superfamily Gondwanatherioidea: a previously unrecognized radiation of multituberculate mammals in South America. Proc Natl Acad Sci USA 90:9379–9383

Krause DW, Hoffmann S, Wible JR, Kirk EC, Schultz JA, von Koenigswald W, Groenke JR, Rossie JB, O'Connor PM, Seiffert ER, Dumont ER, Holloway WL, Rogers RR, Rahantarisoa RJ, Kemp AD, Andriamialison H (2014) First cranial remains of a gondwanatherian mammal reveal remarkable mosaicism. Nature 515:512–517

Krause DW, Hoffmann S, Hu Y, Wible JR, Rougier GW, Kirk EC, Groenke JR, Rogers RR, Rossie JB, Schultz JA, Evans AR, von Koenigswald W, Rahantarisoa LJ (2020) Skeleton of a Cretaceous mammal from Madagascar reflects long-term insularity. Nature 581:421–427

Lee MSY, Beck RMD (2015) Mammalian evolution: a Jurassic spark. Curr Biol 25:R753–R773

Leonardi G (1981) Novo icnogênero de tetrápode Mesozóico da Formação Botucatu, Araraquara, SP. An Acad Bras Cienc 53:795–805

Lillegraven JA, Kielan-Jaworowska Z, Clemens WA (1979) Mesozoic mammals: the first two thirds of mammalian history. University of California Press, Berkeley

Luo Z-X (2007) Transformation and diversification in early mammal evolution. Nature 450:1011–1019

Luo Z-X, Martin T (2007) Analysis of molar structure and phylogeny of docodont genera. Bull Carnegie Mus Nat Hist 39:27–47

Luo Z-X, Cifelli RL, Kielan-Jaworowska Z (2001) Dual origin of tribosphenic mammals. Nature 409:53–57

Luo Z-X, Kielan-Jaworowska Z, Cifelli RL (2002) In quest for a phylogeny of Mesozoic Mammals. Acta Palaeontol Pol 47:1–78

Luo Z-X, Yuan C-X, Meng Q-J, Ji Q (2011) A Jurassic eutherian mammal and divergence of marsupials and placentals. Nature 476:442–445

Luo Z-X, Gatesy SM, Jenkins FA Jr, Amaral WW, Shubin NH (2015) Mandibular and dental characteristics of Late Triassic mammaliaform *Haramiyavia* and their ramifications for basal mammal evolution. Proc Natl Acad Sci USA 112:E7101–E7109

Luo Z-X, Meng Q-J, Grossnickle DM, Liu D, Neander AI, Zhang Y-G, Ji Q (2017) New evidence for mammaliaform ear evolution and feeding adaptation in a Jurassic ecosystem. Nature 548:326–329

Lyson TR, Miller IM, Bercovici AD, Weissenburger K, Fuentes AJ, Clyde WC, Hagadorn JW, Butrim MJ, Johnson KR, Fleming RF, Barclay RS, Maccracken SA, Lloyd B, Wilson GP, Krause DW, Chester SGB (2019) Exceptional continental record of biotic recovery after the Cretaceous-Paleogene mass extinction. Science 366:977–983

Marshall LG, de Muizon C (1988) The dawn of the age of Mammals in South America. Natl Geogr Res 4:23–55

Marshall LG, de Muizon C, Sigé B (1983) Late Cretaceous mammals (Marsupialia) from Bolivia. Geobios 16:739–745

Marshall LG, de Muizon C, Sigé B (1985) The "Rosetta Stone" for mammalian evolution in South America. Nat Geogr Res 1:274–288

Martinelli AG, Soares MB (2016) Evolution of South American cynodonts. Mus Argent Cienc Nat B Rivadavia Contribuciones 6:183–197

Martinelli AG, Chornogubsky L, Abello A, Goin FJ, Reguero M (2014) The first non-therian dryolestoid from Antarctica. 2014 SCAR Open Science Conference, Auckland, Abstracts:432

Martinelli AG, Eltink E, Da-Rosa ÁAS, Langer MC (2017) A new cynodont (Therapsida) from the *Hyperodapedon* Assemblage Zone (upper Carnian–Norian) of southern Brazil improves the Late Triassic probainognathian diversity. Pap Palaeontol 3:401–423

McKenna MC (1962) Collecting small fossils by washing and screening. Curator 5:221–235

McKenna MC (1975) Toward a phylogenetic classification of the Mammalia. In: Luckett WP, Szalay FS (eds) Phylogeny of the primates. Plenum Publishing Corporation, New York, pp 21–46

Meng Q-J, Grossnickle MD, Liu D, Zhang Y-G, Neander AI, Ji Q, Luo Z-X (2017) New gliding mammaliaforms from the Jurassic. Nature 548:291–296

Meng J, Bi S, Zheng X, Wang X (2018) Ear ossicle morphology of the Jurassic euharamiyidan *Arboroharamiya* and evolution of mammalian middle ear. J Morphol 279:441–457

Meredith RW, Janecka JE, Gatesy J, Ryder OA, Fisher CA, Teeling EC et al (2011) Impacts of the Cretaceous terrestrial revolution and K-Pg extinction on mammal diversification. Science 334:521–524

Mourier T, Jaillard E, Laubacher G, Noblet C, Pardo F, Sigé B, Taquet P (1986) Découverte de de restes dinosauriens et mammalien d'âge Crétacé Supérieur à la base des couches rouges du synclinal de Bagua (Andes nord-Péruviennes): Aspects stratigraphiques, sédimentologiques et paléogéographiques concernant la régression fini-Crétacée. Bull Soc Géol France 2:171–175

Muizon C de 1991. La fauna de mamíferos de Tiupampa (Paleoceno inferior, Formación Santa Lucia), Bolivia. In: Suarez-Soruco R (ed) Fósiles y facies de Bolivia. Vertebrados. Rev Tec YPFB 12:575–624

Nessov LA, Archibald JD, Kielan-Jaworowska Z (1998) Ungulate-like mammals from the Late Cretaceous of Uzbekistan and a phylogenetic analysis of Ungulatomorpha. Bull Carnegie Mus Nat Hist 34:40–88

O'Leary MA, Bloch JI, Flynn JJ, Gaudin TJ, Giallombardo A, Giannini NP et al (2013) The placental mammal ancestor and the post-K-Pg radiation of placentals. Science 339:662–667

Pascual R, Archer M, Ortiz-Jaureguizar E, Prado JL, Godthelp H, Hand SH (1992) First discovery of monotremes in South America. Nature 356:704–705

Pascual R, González P, Ardolino A, Puerta PF (2000) A highly derived docodont from the Patagonian Late Cretaceous: evolutionary implications for Gondwanan mammals. Geodiversitas 22:395–414

Pascual R, Goin FJ, Balarino L, Udrizar Sauthier DE (2002) New data on the Paleocene monotreme *Monotrematum sudamericanum*, and the convergent evolution of triangulate molars. Acta Paleontol Pol 47:487–492

Prasad GVR, Verma O, Sahni A, Krause DW, Khosla A, Parmar V (2007) A new Late Cretaceous gondwanatherian mammal from central India. Proc Indian Natl Sci Acad 73:17–24

Rauhut OWM, Martin T, Ortiz-Jaureguizar E, Puerta P (2002) A Jurassic mammal from South America. Nature 416:165–168

Rich TH, Vickers-Rich P (2004) Diversity of Early Cretaceous mammals from Victoria, Australia. Bull Am Mus Nat Hist 285:36–53

Rich TH, Vickers-Rich P, Constanine A, Flannery TF, Kool L, van Klaveren NA (1997) A tribosphenic mammal from the Mesozoic of Australia. Science 278:1438–1442

Rich TH, Vickers-Rich P, Constanine A, Flannery TF, Kool L, van Klaveren NA (1999) Early Cretaceous mammals from Flat Rocks, Victoria, Australia. Rec Queen Vic Mus 106:1–35

Rich TH, Flannery TF, Trusler P, Kool L, van Klaveren NA, Vickers-Rich P (2001) A second tribosphenic mammal from the Mesozoic of Australia. Rec Queen Vic Mus 110:1–10

Romer AS (1970) The Chañares (Argentina) Triassic reptile fauna. VI. A chiniquodontid cynodont with incipient squamosal-dentary jaw articulation. Breviora 344:1–18

Rougier GW, Wible JR, Novacek MJ (1998) Implications of *Deltatheridium* specimens for early marsupial history. Nature 396:459–463

Rougier GW, Martinelli AG, Forasiepi AM, Novacek MJ (2007a) New Jurassic mammals from Patagonia, Argentina: a reappraisal of australosphenidan morphology and interrelationships. Am Mus Novit 3566:1–54

Rougier GW, Garrido A, Gaetano L, Puerta P, Corbitt C, Novacek MJ (2007b) First Jurassic triconodont from South America. Am Mus Novit 3580:1–17

Rougier GW, Chornogubsky L, Casadio S, Paéz Arango N, Giallombardo A (2009a) Mammals from the Allen Formation, Late Cretaceous, Argentina. Cretac Res 30:223–238

Rougier GW, Forasiepi AM, Hill RV, Novacek MJ (2009b) New mammalian remains from the Late Cretaceous La Colonia Formation, Patagonia, Argentina. Acta Palaeontol Pol 54:195–212

Rougier GW, Gaetano L, Drury BR, Colella R, Gómez RO, Páez Arango N (2011a) A review of the Mesozoic mammalian record of South America. In: Calvo J, Porfiri J, González Riga B, Dos Santos D (eds) Dinosaurios y paleontología desde América Latina. Editorial de la Universidad Nacional de Cuyo, Mendoza, pp 195–214

Rougier GW, Apesteguía S, Gaetano LC (2011b) Highly specialized mammalian skulls from the Late Cretaceous of South America. Nature 479:98–102

Rougier GW, Wible JR, Beck RMD, Apesteguía S (2012) The Miocene mammal *Necrolestes* demonstrates the survival of a Mesozoic nontherian lineage into the late Cenozoic of South America. Proc Natl Acad Sci USA 109:20053–20058

Rowe TB (1988) Definition, diagnosis, and origin of Mammalia. J Vertebr Paleontol 8:241–264

Rowe TB (1993) Phylogenetic systematics and the early history of mammals. In: Szalay FS, Novacek MJ, McKenna MC (eds) Mammal phylogeny: Mesozoic differentiation, multituberculates, monotremes, early therians, and marsupials. Springer-Verlag, New York, pp 129–145

Rowe TB, Gauthier JA (1992) Ancestry, paleontology, and definition of the name Mammalia. Syst Biol 41:372–378

Scillato-Yané GJ, Pascual R (1985) Un peculiar Xenarthra del Paleoceno Medio de Patagonia (Argentina). Su importancia en la sistemática de los Paratheria. Ameghiniana 21:316–318

Sigé B (1968) Dents de micromammifères et fragments de coquilles d'oeufs de dinosauriens dans la faune de vertébrés du Crétacé Supérieur de Laguna Umayo (Andes Péruviennes). C R Acad Sci 267:1495–1498

Sigé B, Sempere T, Butler RF, Marshall LG, Crochet J-Y (2004) Age and stratigraphic reassessment of the fossil-bearing Laguna Umayo red mudstone unit, SE Peru, from regional stratigraphy, fossil record, and paleomagnetism. Geobios 37:771–794

Sigogneau-Russell D (1991) First evidence of Multituberculata (Mammalia) in the Mesozoic of Africa. Neues Jahrb Geol Paläontol 1991:19–125

Sigogneau-Russell D (1995) Two possibly aquatic triconodont mammals from the Early Cretaceous of Morocco. Acta Palaeontol Pol 40:149–162

Simpson GG (1936) Studies of the earliest mammalian dentitions. The Dental Cosmos 78:791–800

Simpson GG (1984) Discoverers of the lost word. An Account of some of those who brought back to life South American mammals long buried in the abyss of time. Yale University Press, New Haven and London

Springer MS, Murphy WJ, Eizirik E, O'Brien SJ (2003) Placental mammal diversification and the Cretaceous-Tertiary boundary. Proc Natl Acad Sci USA 100:31056–31061

Wible JR, Rougier GW (2017) Craniomandibular anatomy of the subterranean meridolestidan *Necrolestes patagonensis* Ameghino, 1891 (Mammalia, Cladotheria) from the Early Miocene of Patagonia. Ann Carnegie Mus 84:183–251

Wible JR, Rougier GW, Novacek M, Asher RJ (2007) Cretaceous eutherians and Laurasian origin for placental mammals near the K/T boundary. Nature 447:1003–1006

Wilson GP, Das Sarma DC, Anantharaman S (2007) Late Cretaceous sudamericid gondwanatherians from India with paleobiogeographic considerations of Gondwanan mammals. J Vertebr Paleontol 27:521–531

Zheng X, Bi S, Wang X, Meng J (2013) A new arboreal haramiyid shows the diversity of crown mammals in the Jurassic period. Nature 500:199–202

Zhou C-F, Wu S, Martin T, Luo Z-X (2013) A Jurassic mammaliaform and the earliest mammalian evolutionary adaptations. Nature 500:163–167

# Chapter 2
# The Fossil Record of South American Mesozoic Mammals and Their Close Relatives

*South America is a place I love, and I think, if you take it right through from Darien to Fuego, it's the grandest, richest, most wonderful bit of earth upon this planet... Why shouldn't somethin' new and wonderful lie in such a country? And why shouldn´t we be the men to find it out?*
Sir Arthur Conan Doyle
*The Lost World*, 1912

**Abstract** The South American fossil record of Mesozoic mammals and close relatives is one of the best for Gondwana. Early mammals and relatives are found in about a dozen localities in Argentina, Brazil, Bolivia, Chile, and presumably Peru, including a broad sample of non-mammaliaform cynodonts of the Triassic age. Mesozoic mammals span from the latest Early Jurassic to the latest Cretaceous, furthermore some of those archaic lineages unexpectedly survived the end of the Cretaceous period, remaining as minority elements in the Paleocene–Miocene faunal associations. The fossiliferous localities bearing these fossils are presented in this chapter, highlighting the geological setting, age, and their faunal associations.

**Keywords** Mammalia · Vertebrates · Triassic · Jurassic · Cretaceous · Cenozoic

## 2.1 Introduction

Current geographical terms do not always refer to logical divisions when applied over long periods of time. As such, South America as currently conceived is mostly a product of Late Mesozoic events; while its current physical connection to North America is a recent incident (e.g., Coates and Stallard 2013, 2015), contacts with other landmasses and splits have been integral to the biogeographical history of the continent. During the late Paleozoic–early Mesozoic there was a unique landmass, Pangea, around Equatorial latitudes (Fig. 2.1). The connection of SA to the rest of Pangea persisted until the Jurassic, when this large continental mass was rapidly broken by the formation of the South Atlantic Ocean. By the end of the Early Cretaceous, SA was almost isolated, but it retained a limited connection with Antarctica and Australia (Smith et al. 1994; Scotese 2004; Wilf et al. 2013; Defler 2019). Their



G. W. Rougier et al., *Mesozoic Mammals from South America and Their Forerunners*, Springer Earth System Sciences, https://doi.org/10.1007/978-3-030-63862-7_2

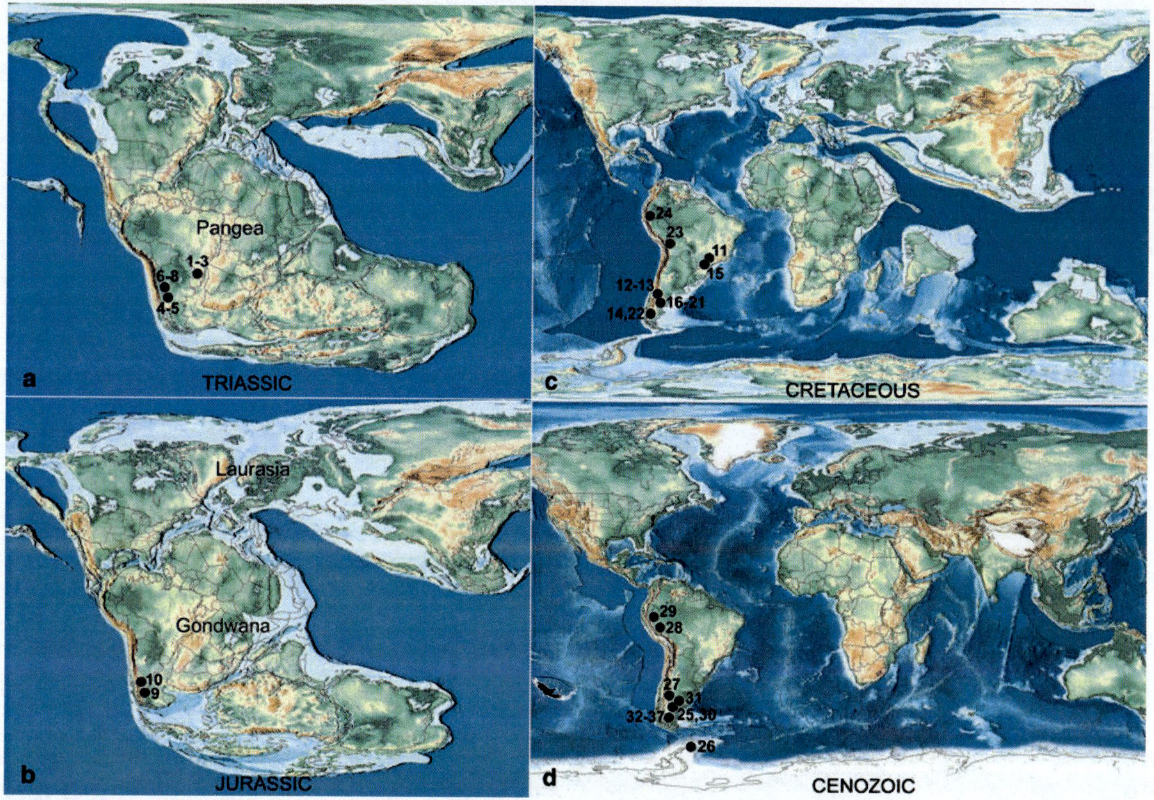


**Fig. 2.1** Paleogeographic maps that illustrate the changing distribution of the continents and the different locations of South America in particular through time. Numbers indicate the fossiliferous localities mentioned in the text (abbreviations as in Fig. 2.2). South America (SA) as part of the supercontinent Pangea at the beginning of the Mesozoic (Late Triassic map taken from Scotese 2013a) (**a**); SA as part of the supercontinent Gondwana (Middle Jurassic map taken from Scotese 2013b) (**b**) and as part of Occidental Gondwana at the end of the Mesozoic (Late Cretaceous map taken from Scotese 2013c) (**c**); SA isolated from other landmasses at the mid-Cenozoic (middle/late Miocene map taken from Scotese 2013d) (**d**)

complete separation was not achieved until much later, the late Oligocene (Pfuhl and McCave 2005), defined by deep water circulation between the Pacific and Atlantic Oceans, a consequence of the formation of the Drake Passage.

In SA, principal tectonic events during the Mesozoic included convergence of the Pacific margin and participation in the breakup via rifting of initially Pangea and later Gondwana, in relation to the opening of the South Atlantic Ocean (Wilf et al. 2013). This activity resulted in intense magmatism, represented by extensive lavas, ignimbrites, and tuffaceous sediments, including occasionally fossiliferous beds (Fig. 2.2). In southern SA, these volcanic rocks were the source for most of the post-Jurassic sediments deposited in continental basins, primarily Marifil, Lonco Trapial, and equivalent formations in Patagonia (e.g., Aragón et al. 1996; Iglesias et al. 2011; Cúneo et al. 2013) and in the Serra Geral Group, Paraná Basin, Brazil (e.g., Rossetti et al. 2018).

Argentina and Brazil are major sources in SA of Mesozoic terrestrial vertebrates, including non-mammalian cynodonts and mammals, with several worldwide significant sites (Figs. 2.3, 2.4, 2.5, 2.6, 2.7, 2.8, 2.9, 2.10, 2.11, 2.12, 2.13, 2.14, 2.15, 2.16, 2.17, 2.18, 2.19, 2.20, 2.21, 2.22, 2.23, 2.24, 2.25, 2.26 and 2.27). Triassic fossiliferous outcrops have been identified in the west of Argentina and south of Brazil (Figs. 2.3, 2.4, 2.5, 2.6, 2.7 and 2.8), while Patagonia in Southern Argentina has provided our best record of Jurassic and Cretaceous taxa (Figs. 2.9, 2.10, 2.11, 2.12, 2.13, 2.14, 2.15, 2.16, 2.17, 2.18, 2.19, 2.20, 2.21, 2.22, 2.23, 2.24, 2.25, 2.26 and 2.27). The recent discovery of Cretaceous mammals in Chile has expanded their record to the southernmost part of the continent (Goin et al. 2020).

Major sedimentary basins in Argentina with a significant record of Triassic vertebrates include the epiclastic Ischigualasto-Villa Unión Basin (Permian–Triassic; Groeber and Stipanicic 1953; Stipanicic and Bonaparte 1979), Cuyo Basin (Permian–Neogene), and San Rafael Block (Upper Permian–Upper Triassic; also considered a division of the Cuyo Basin; see Monti and Franzese 2016). In Brazil, sedimentary sequences include well-preserved and highly diverse vertebrate associations (e.g., Langer et al. 2007; Martinelli et al. 2017a; Schultz et al. 2020), enclosed within the red-beds of the Santa Maria Supersequence, Paraná Basin (Ordovician–Cretaceous) (Milani et al. 1994; see also Menegazzo et al. 2016), exposed at the Rio Grande do Sul State (Zerfass et al. 2003; Horn et al. 2014). In addition, a few Triassic vertebrates have been found in the Antofagasta region, Chile, in the Estratos El Bordón, but so far it is mostly represented by archosaurs (e.g., *Chilenosuchus forttae*; Casamiquela 1980; Rubilar et al. 2002; Desojo 2003). The Triassic fossil vertebrates from Chile are still poorly known and no synapsids have yet been found. In contrast, in Argentina and Brazil, Triassic non-mammaliaform cynodonts are taxonomically diverse, represented by conspicuous material that illustrates about 20 million years of evolution, preceding the origin of Mammaliaformes (Chap. 3).

Three large sedimentary basins represent the Patagonian Jurassic: the seasonally arid continental Somuncura-Cañadón Asfalto Basin of Central Patagonia (Cúneo et al. 2013; Figari et al. 2015); the nearshore San Jorge Basin, which has produced a rich fossiliferous sequence extending to the late Cenozoic; and the predominantly marine Magallanes-Austral Basin that persisted into the Cenozoic (Malkowski et al.

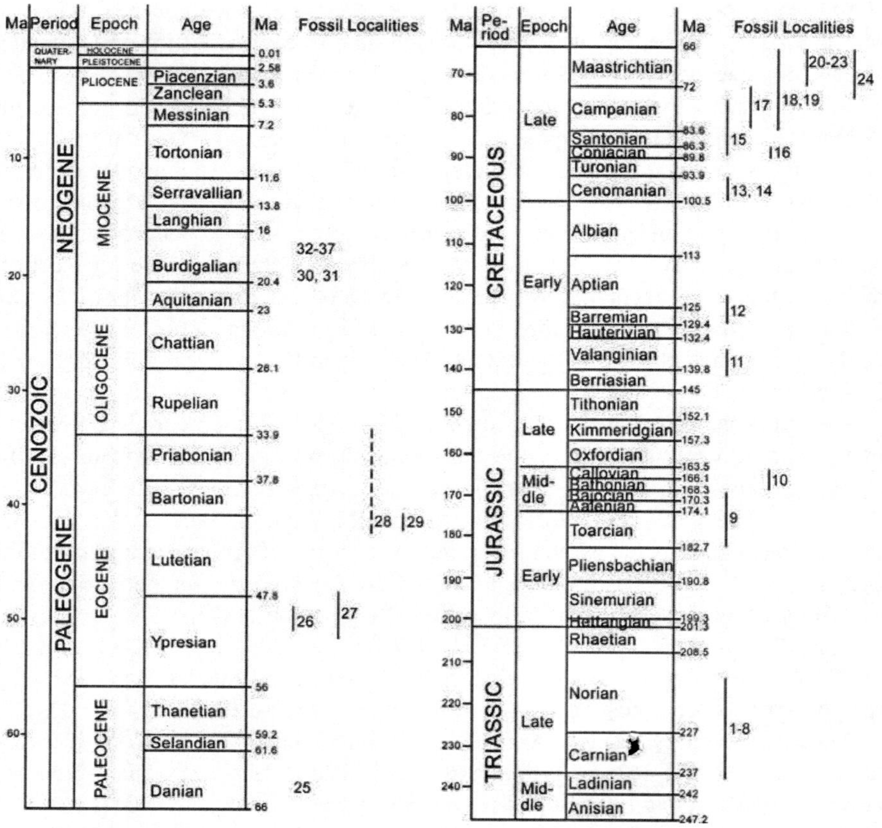


**Fig. 2.2** Stratigraphic distribution of the South American fossiliferous localities with non-mammaliaform cynodonts and Mesozoic mammal lineages. TRIASSIC: 1, Candelária region, Rio Grande do Sul, Brazil; 2, Faxinal do Soturno region, Rio Grande do Sul, Brazil; 3, Santa Maria region, Rio Grande do Sul, Brazil; 4, Puesto Viejo, Mendoza, Argentina; 5, Uspallata, Mendoza, Argentina; 6, Ischigualasto-Talampaya Parks, San Juan and La Rioja, Argentina; 7, Los Colorados, La Rioja, Argentina; 8, El Carrizal, San Juan, Argentina. JURASSIC: 9, Queso Rallado, Chubut, Argentina; 10, Laguna Manantiales, Santa Cruz, Argentina. CRETACEOUS: 11, São Bento, Araraquara, São Paulo, Brazil; 12, La Amarga, Neuquén, Argentina; 13, La Buitrera, Río Negro, Argentina; 14, Tres Lagos, Santa Cruz, Argentina; 15, Santo Anastácio, São Paulo, Brazil; 16, Los Barreales Lake, Neuquén, Argentina; 17, Paso Córdoba, Río Negro, Argentina; 18, Ea. Los Alamitos, west slope of Cerro Cuadrado, Río Negro, Argentina; 19, Cerro Tortuga, Río Negro, Argentina; 20, Mirasol Chico Canyon, Chubut, Argentina; 21, Ingeniero Jacobacci, Río Negro, Argentina; 22, Río de Las Chinas Valley, Última Esperanza Province, Chile/Alta Vista and La Anita farms, Santa Cruz, Argentina; 23, Pajcha Pata, Cochabamba, Bolivia; 24, Synclinal de Bagua, Peru. CENOZOIC: 25, Punta Peligro, Chubut Argentina; 26, Seymour Locality IAA 90/1, Seymour Island, Antarctic Peninsula; 27, La Barda, Chubut, Argentina; 28, Santa Rosa, Ucayali, Peru; 29, Contamana, Loreto, Peru; 30, Gran Barranca and 31, Gaiman Chubut, Argentina; 32, Monte Observación; 33, La Cueva; 34, Monte León; 35, Estancia La Costa; 36, Killik Aike Norte, and 37, 8 km south of Coy Inlet, Santa Cruz, Argentina. Chronology adjusted on the International Chronostratigraphic Chart 2016

2016; Fosdick et al. 2016). Patagonian Jurassic faunas are perhaps less cosmopolitan than the Triassic faunas, being partially distinguishable at low taxonomic levels from those of other parts of Pangea (Bonaparte 1979; see also Wilf et al. 2013). The distinctiveness of some Jurassic faunas, considering the absence of oceanic barriers, implies that the distribution of some taxa was constrained by intra-continental climatic zonation or particular geographic features (Wilf et al. 2013). For example, among mammals a Gondwanan assemblage is recognized with australosphenidans found at the Cañadón Asfalto Formation, sharing a close common ancestor with Malagasy and Australian taxa, including monotremes (Chap. 4).

The Neuquén Basin at northwest Patagonia was predominantly marine during the Jurassic. However, its Pacific connection was terminated by the Early Cretaceous (Spalletti and Franzese 2007), with an increase in continental sedimentation. Many fossiliferous stratigraphic units from the Lower to Upper Cretaceous have resulted in a rich source of marine vertebrates and continental tetrapods (e.g., Bonaparte 1996a; Leanza et al. 2004; Wilf et al. 2013) (Fig. 2.14).

Marine transgressions during the Late Cretaceous/Paleocene (Campanian–Danian) resulted in the break up of the main continental mass of Patagonia into a large series of smaller islands that favored the development of numerous mixed freshwater to shallow marine coastal deposits (Malumian and Náñez 2011), containing abundant fossil material (e.g., Allen, La Colonia, Los Alamitos, Lefipán, Salamanca formations) (e.g., Bonaparte et al. 1984, 1993; Martinelli and Forasiepi 2004; Goin et al. 2006a; Rougier et al. 2009a, b, 2011a; Gasparini et al. 2015) (Figs. 2.12, 2.14). These deposits were initially jointly identified as the "Senoniano Lacustre" (Wichmann 1924), or the "Kawas Sea" (Casamiquela 1978; Apesteguía 2002; O'Gorman 2016). Episodic immigration from NA to SA and vice versa occurred apparently during the Campanian–Maastrichtian (Bonaparte 1986a, 1996a; the oddly named "First Great Turnover" sensu Pascual et al. 2001; Pascual and Ortiz-Jaureguizar 2007; Woodburne et al. 2014a, 2014b; see the alternative view in Ezcurra and Agnolín 2012). Bonaparte (1986a) firstly claimed that teiid lizards, protoceratopsid, hadrosaur, and pachycephalosaur dinosaurs, and therian mammals (metatherians and eutherians) are groups of Laurasian origin that entered into SA from NA. Particularly, the description of *Kritosaurus australis* (Bonaparte et al. 1984), a genus that was known from NA, but now considered a junior synonym of the SA hadrosaur *Secernosaurus koerneri* (Prieto-Marquez and Salinas 2010; Becerra et al. 2018), as well as a large number of non-therian mammals (e.g., allotherians and dryolestoids; Chaps. 6 and 8) from the Los Alamitos Formation were fundamental pillars to draw the paleobiogeographical model for both Americas. The (putative) lack of records of Late Cretaceous therian mammals in SA supported the view that they entered from the north at the beginning of the Cenozoic or very late in the Cretaceous (but see comments below). The still poorly known fossil record of lizards in the Mesozoic of SA and the conflicting phylogenetic position of teiioids and other lizards (e.g., iguanians) do not help to illustrate a clear biogeographic pattern of SA during the Cretaceous (e.g., Simões et al. 2017). In turn, the growing knowledge of Late Cretaceous faunas from Brazil and Argentina, and other parts of Gondwana (e.g., Africa, Australia) highlight a more complex paleobiogeographic scenario than previously thought (Novas 2009; Ezcurra and Agnolín

2012). The absence of several NA Late Cretaceous groups (e.g., tyrannosaurid, therizinosaurid, oviraptosaurid theropods, and marginocephalian ornithischians) in SA (Ezcurra and Agnolín 2012), as well as the absence of gondwanatherian and meridiolestidan mammals (with possible exceptions, Fox 1985; see Bonaparte 2002) and abelisaurid dinosaurs in NA, undercuts the idea of a land bridge between NA and SA by the end of the Cretaceous.

Although non-therian mammals are dominant in the Mesozoic record from SA, there are few, extremely fragmentary, and some poorly studied specimens that were alleged to belong to therians: a fragmentary jaw with one premolar from the Adamantina Formation, Brazil (Bertini et al. 1993), an isolated large-sized tooth, also from the Adamantina Formation, named *Brasilestes stardusti* (Castro et al. 2018), isolated teeth from the El Molino Formation, Bolivia (Gayet et al. 2001), and a fragment of a tiny tooth from the Upper Cretaceous of the Bagua syncline, Peru (Mourier et al. 1986). The fragmentary nature of the specimens in some cases precludes a well-supported taxonomical determination, while in other cases, they can be re-interpreted as meridiolestidan dryolestoids (see below and Chap. 6). None of these putative Cretaceous therians from SA have been included in a cladistics analysis or any other kind of testable phylogenetic study. As a consequence, the oldest definite SA therian to date is a marsupial relative (Metatheria) from Danian strata, ~65 Ma, recovered at Grenier Farm (Lefipán Formation) near Paso del Sapo, central-western Patagonia, Chubut, Argentina (Goin et al. 2006a). The specimen is an isolated lower molar of *Cocatherium lefipanum* and it is likely related to polydolopimorphians, a group potentially known from the Late Cretaceous of NA (Case et al. 2005), but more certainly recorded from the Paleocene and Eocene deposits of SA (see Goin et al. 2006a, 2016), and early–middle Eocene of Antarctica (Case et al. 1988; Goin et al. 1999).

During most of the Cenozoic (beginning at the late Oligocene), SA was a large island continent, separated from other landmasses (e.g., Simpson 1980; Pascual and Ortiz-Jaureguizar 2007; Defler 2019). This geographic isolation likely triggered the evolution of an endemic SA terrestrial biota. Mammals were dominated by eutherians and metatherians, but relictual members of the Gondwanan biota persisted. The K/Pg major extinction event seems not to have affected as severely the mammalian communities in SA as it did in other landmasses or other major taxonomic groups (Pascual et al. 2001; Pascual and Ortiz- Jaureguizar 2007; Rougier et al. 2009a, 2012), considering that monotremes, gondwanatherians, and dryolestoids are registered in the Paleogene and dryolestoids appear to survive until the early Neogene (Rougier et al. 2012; Wible and Rougier 2017). Major events in SA during the Cenozoic are related to the progressive fragmentation of the terrain forming western Gondwana by the opening of the central Atlantic Dorsal (Malumián and Náñez 2011), with the concomitant westward drifting of SA, subduction of the eastern tip of the Pacific plate and disruption of the Austro-Antarctic connection by differential angular movement and drift rate with SA (Fig. 2.1). These major events had a long-lasting effect on the continent including increased tectonic activity and the rise of the Andean mountain range (e.g., Ramos 2009), formation of the Antarctic circumpolar current (Pfuhl and McCave 2005), and consequent global climatic modifications (e.g., glaciations, climate differentiation, and temperate fluctuations; Zachos et al. 2001; Mudelsee et al. 2014; de Vleeschouwer et al. 2017). By the Neogene, SA physically reconnected with North America via the establishment of the Panama Isthmus (Coates and Stallard 2013, 2015, and references there), with a consequent major biotic movement termed the Great American Biotic Interchange (GABI) that defined nowadays vertebrate composition in both Americas (e.g., Woodburne 2010; Cione et al. 2015).

### *2.1.1 South American Gondwanan Mammals and the Paleobiogeographical Context*

Excluding the uncertain therians from the Late Cretaceous of Brazil (Bertini et al. 1993; Castro et al. 2018), Bolivia (Gayet et al. 1993, 2001), and Peru (Mourier et al. 1986, 1988), the origin of all major mammalian lineages from the Mesozoic of SA can be traced to prior to the opening of the South Atlantic in a Pangean or Gondwanan setting (Bonaparte 1986a; Pascual 2006; Rougier et al. 2011a). Therefore, it is likely that the roots of the SA Mesozoic mammalian fauna can be found in taxa that had at least a pan-Gondwanan distribution, or even Pangeic.

The Jurassic *Asfaltomylos patagonicus* and *Henosferus molus* (Rauhut et al. 2002; Martin and Rauhut 2005; Rougier et al. 2007a; see also Chap. 4) are basal members of the endemic Gondwanan group Australosphenida, which also includes living monotremes (Luo et al. 2002). Their position on the tree indicates an origin for australosphenidans before the opening of the South Atlantic, but they are potentially compatible with an origin after the separation of Laurasia and Gondwana (Rougier et al. 2011a). However, their very early age (Early- to early Middle Jurassic; Cúneo et al. 2013) could also suggest a missing record of basal members of this lineage in the landmasses that were to become part of Laurasia. The "triconodont" *Argentoconodon fariasorum* (Rougier et al. 2007b; Gaetano and Rougier 2011; see also Chap. 5) shows close affinities with the Jurassic *Volaticotherium antiquus* (Meng et al. 2006) from China and the Early Cretaceous *Ichthyoconodon jaworowskorum* from Morocco (Sigogneau-Russell 1995). The widely disparate locations of these taxa and the minimal age for the last common ancestor suggest a Pangeic distribution with a pre-Early Jurassic origin. A review of the Cañadón Asfalto Formation vertebrate and paleobotanical evidence (Escapa et al. 2008a; Cúneo et al. 2013) suggests that the root of the assemblage is very old, most likely pre-establishment of the South Atlantic Igneous Province (Foulger 2018).

Dryolestoids are major components of the SA fauna; appropriately, *Mesungulatum houssayi*, the first Mesozoic mammal named from SA, is a member of this dryolestoid radiation in SA (Bonaparte and Soria 1985; Bonaparte 1986b; Bonaparte and Migale 2010, 2015; see also Chap. 6). In SA, dryolestoids are by far the most diverse and abundant Mesozoic mammalian group (e.g., Bonaparte 1986a, b, c, 1990, 1994, 2002; Rougier et al. 2009a, b). Some taxa (e.g., *Groebertherium* spp.) appear to be more closely related to Jurassic Laurasian dryolestids than other forms from SA (Rougier et al. 2011b, 2012). All other dryolestoids cluster together in a large grouping of endemic forms: Meridiolestida (Rougier et al. 2011b), which can be nested within the Northern hemisphere dryolestoids, or might be closer to the therian stem (Rougier et al. 2011b, 2012). Either one of these alternative phylogenetic positions is labile to minor data changes and at present considered as equally likely. Consequently, the stock from which the SA dryolestoids derived must share a common ancestor with the Laurasian forms prior to the opening of the South Atlantic during the Jurassic/Early Cretaceous. Some Early Cretaceous forms from Africa appear to show similarities with those from the Cretaceous of SA (Sigogneau-Russell 1989, 1991; Sigogneau-Russell and Ensom 1998; Sigogneau-Russell et al. 1990; Bonaparte 2002; Chimento et al. 2016). However, there are no formal phylogenetic analyses including both SA and African forms in a single study; as such, despite the convincing similarities the idea is yet to be tested and corroborated. Regardless of the ultimate rooting of the SA radiation of dryolestoids, a Jurassic origin is likely, accepting the current phylogenetic hypothesis for the clade that includes *Groebertherium* (Rougier et al. 2011b, 2012). Controversial alternative views for the SA taxa have been advanced (Averianov et al. 2013; see Chap. 6), that although unlikely, would root the SA radiation even deeper in the Jurassic and would guarantee them an origin in a Pangeic framework. Although the limited fossil record makes relying on our current understanding of the biogeography of the Mesozoic mammals of the Southern continents a highly tentative affair, SA is the best represented of all the landmasses of the former Gondwanan supercontinent. The SA record suggests that the mammalian lineages produced endemic taxa that flourished during the Cretaceous, which in some instances survived the K/Pg extinction event. Examples of survivors are meridiolestidan dryolestoids (*Peligrotherium tropicalis* in the Paleocene, *Necrolestes mirabilis* and *N. patagonensis* in the early Miocene of Patagonia, and an indeterminate taxon in the Eocene of Antarctica; Bonaparte et al. 1993; Gelfo and Pascual 2001; Páez Arango 2008; Martinelli et al. 2014; Wible and Rougier 2017) and gondwanatherians (*Sudamerica ameghinoi* in the Paleocene and *Greniodon sylvaticus* in the Eocene of Patagonia, and perhaps indeterminate taxa in the Eocene of Antarctica and the Neotropics; Campbell et al. 2004; Goin et al. 2004a, 2006b; Antoine et al. 2012, 2016). Other possible records of gondwanatherians in the Eocene and Miocene of Argentina have been suggested (Chimento et al. 2016) and are discussed in Chap. 8. Monotremes themselves are not found in the Mesozoic of SA, however, australosphenidans (Rauhut et al. 2002; Rougier et al. 2007a) are recovered as basal members of the monotreme clade, therefore implying that they, or members of their stem, were present in the early SA faunas and can be seen as yet another group that survived the Cretaceous extinction. In support of this view is the Cretaceous record of monotremes in Australia (e.g., Archer et al. 1985; Rich et al. 2001b, 2016; Musser 2003) and the presence of the ornithorhynchid *Monotrematum sudamericanum* (Pascual et al. 1992, 2002; Forasiepi and Martinelli 2003) in the Paleocene of Patagonia. It is possible, but less likely, that Monotremata originated in Australia, Antarctica, or other southern landmasses, and *Monotrematum* represents a Paleocene migration into South America of already differentiated monotremes.

Bonaparte's (1986a) hypothesis of a Late Cretaceous ingression of therians from North America is still the most likely explanation for the origin of the bulk of the

Cenozoic mammalian therian fauna; however, Mesozoic mammal faunas from central and northern SA are almost non-existent and understanding the dynamics of these extinct creatures in this broad geographical region without a fossil record is largely tentative. The survival of archaic lineages in the Paleocene of Patagonia and Antarctica (Rougier et al. 2000; Gelfo and Pascual 2001; Gelfo et al. 2007; Goin et al. 2004a, 2006b; Martinelli et al. 2014), but perhaps not in the rest of this continent, suggests a provincial division of a northern and a southern SA (Pascual 2006). The extensive Maastrichtian–Paleocene marine ingression stretching from Venezuela to Argentina (Zambrano 1987; Gayet et al. 1993; Wilson and Arens 2001; Malumián and Náñez 2011; Gianni et al. 2018) along the western margin of the continent determined a Late Cretaceous–early Paleogene Patagonian archipelago (Casadio 1998), which likely explains the Paleocene provincialism (Rougier et al. 2011a). Later Cenozoic assemblages from the Neotropics (Goin et al. 2004a; Antoine et al. 2012) could perhaps suggest removal of the previous biogeographic barrier. However, given the presence of at least some meridiolestidan remains in the Cretaceous of Bolivia (Gayet et al. 1993, 2001; Rougier et al. 2011a) it is likely that the putative provincialism of Patagonia is an artifact of the more scarce extra-Patagonian Cretaceous/Paleocene record. Despite the intrinsic difficulties of fossil collecting in tropical latitudes (poor exposures, lower preservation potential, etc.), fieldwork in the late Mesozoic/early Cenozoic of northern and tropical SA should be a focal point for future research and exploration, which could dramatically change our views of SA vertebrate evolution during the Mesozoic.

We present in this chapter a summary of the fossiliferous SA localities that have yielded early mammalian relatives, Mesozoic mammals, and the Cenozoic surviving members of those archaic lineages. We complement our comments on mammals with a summary of the stratigraphic background, ages, and associated vertebrate fauna, hoping to provide the rudiments of pictures of the past SA ecosystems.

## 2.2 Fossiliferous Localities from South America

### *2.2.1 Triassic*

#### 2.2.1.1 State of Rio Grande Do Sul, Brazil; Santa Maria Supersequence, Lower—Upper Triassic

Brazil's non-mammaliaform cynodont record is restricted to the Triassic of the state of Rio Grande do Sul. These are Triassic red-beds exposed along an east–west-oriented belt of ~500 km in the center of the state (Schultz et al. 2000; Langer et al. 2007), and includes the well-known localities of Candelária, Faxinal do Soturno, and Santa Maria (Fig. 2.3). Traditionally, the Triassic red-beds were divided into the Sanga do Cabral, Santa Maria, and Caturrita formations, all of them part of the Paraná Basin (Andreis et al. 1980). However, several problems in the lithological

**Fig. 2.3** Fossiliferous localities from the Triassic of South America with non-mammaliaform cynodonts. Abbreviations (as in Fig. 2.2): 1, Candelária region, Rio Grande do Sul, Brazil; 2, Faxinal do Soturno region, Rio Grande do Sul, Brazil; 3, Santa Maria region, Rio Grande do Sul, Brazil; 4, Puesto Viejo, Mendoza, Argentina; 5, Uspallata, Mendoza, Argentina; 6, Ischigualasto-Talampaya Parks, San Juan and La Rioja, Argentina; 7, Los Colorados, La Rioja, Argentina; 8, El Carrizal, San Juan, Argentina

characterization of the units resulted in the abandonment of this nomenclature. In contrast, other authors recognize two main divisions: the Sanga do Cabral (lower) and Santa Maria (upper) Supersequences (Zerfass et al. 2003). More recently, the Santa Maria Supersequence was divided into three depositional sequences with vertebrates (from base to top): Pinheiros-Chiniquá, Santa Cruz, and Candelária sequences (Horn et al. 2014). The Santa Maria Supersequence begins with massive conglomerates and sandstone deposited by rivers of low sinuosity and high energy that were succeeded by deposits from shallow lakes and ponds. It continues in the sequence with sandstones and conglomerates deposited by low sinuosity and multilateral river channels, followed by massive or laminated pelitic layers, deposited in flood plains and with isolated channels. The top of the unit is composed of meandering fluvial sandstones (Zerfass et al. 2003, 2004; Horn et al. 2014).

The very rich fossil content of the Santa Maria Supersequence (see also Chap. 3) is divided into individual fossil assemblages (e.g., Schultz et al. 2000, 2020; Langer et al. 2007; Horn et al. 2014). The Pinheiro-Chiniquá Sequence includes the *Dinodontosaurus* Assemblage Zone (AZ), which is widely exposed in the regions of Vale Verde, Venâncio Aires, Candelária (including Pinheiro), Agudo, Dona Francisca, and Chiniquá, among others. The *Dinodontosaurus* AZ is characterized by the presence of procolophonians (e.g., *Candelaria barbouri*), hundreds of dicynodonts (e.g., *Dinodontosaurus* sp. and *Stahleckeria potens*), several archosauromorphs (e.g., the rhynchosaur *Brasinorhynchus mariantensis*, indeterminate proterochampsids, the pseudosuchians *Prestosuchus chiniquensis, Decuriasuchus quartacolonia*) (e.g., von Huene 1935, 1942; Langer et al. 2007; Lacerda et al. 2016), in addition to at least ten cynodont species (Martinelli and Soares 2016; Martinelli et al. 2017a; see Chap. 3). Based on its fossil content (Martinelli et al. 2017a), radiometric dating of the overlying Santa Cruz Sequence (Philipp et al. 2018), and correlations with the Chañares Formation of Argentina based on the content of tetrapods (e.g., Ezcurra et al. 2017), the *Dinodontosaurus* AZ is considered Ladinian–early Carnian in age.

The Santa Cruz Sequence includes the *Santacruzodon* AZ. Outcrops of this unit are restricted to the region of Santa Cruz do Sul, Vera Cruz, and Venâncio Aires. This assemblage is represented mostly by cynodonts, including the traversodontids *Santacruzodon hopsoni, Menadon besairiei,* and *Massetognathus* sp., and the probainognathians *Chiniquodon* sp. and *Santacruzgnathus abdalai* (e.g., Abdala and Ribeiro 2003; Melo et al. 2015; Martinelli et al. 2016). Other records include a proterochampsid archosauriform, a pseudosuchian, and a dicynodont (e.g., Lacerda et al. 2015; Martinelli et al. 2016). The *Dinodontosaurus* AZ is considered early Carnian (e.g., Horn et al. 2014; Martinelli et al. 2016; Philipp et al. 2018).

The Candelária Sequence includes the *Hyperodapedon* (older) and the *Riograndia* (younger) AZs (Fig. 2.4). The *Hyperodapedon* AZ is principally recognized in Candelária, Santa Maria, São Pedro do Sul, São João de Polêsine, and Venancio Aires regions, while the *Riograndia* AZ is represented in Candelária (i.e., Sesmaria do Pinhal in Botucaraí area; Fig. 2.4a), Faxinal do Soturno (Fig. 2.4c), and Água Negra, among others. Two zircon U/Pb dates, one from the *Hyperodapedon* AZ (Alemoa in Santa Maria) and another from the *Riograndia* AZ (Linha São Luiz in Faxinal do Saturno), provided a maximum deposition age of $233.23 \pm 0.73$ Ma (late Carnian) and

**Fig. 2.4** Bonaparte's expedition at Sesmaria do Pinhal outcrops, Candelária, Rio Grande do Sul, Brazil; Upper Triassic of the Santa María Supersequence, 2001 (**a**). José F. Bonaparte and Agustín G. Martinelli searching for small cynodonts at Candelária, 2005 (**b**). Bonaparte's expedition at Linha São Luiz outcrops, Faxinal do Soturno, Rio Grande do Sul, Brazil; Upper Triassic of the Santa María Supersequence, 2001 (**c**). *Riograndia guaibensis* (UFRGS-PV-1319-T), right dentary (**d**) and left maxilla (UFRGS-PV-788-T) (**e**), from Linha São Luiz site, Faxinal do Soturno, Rio Grande do Sul, Brazil

225.42 ± 0.37 Ma (Norian), respectively (Langer et al. 2018). The *Hyperodapedon* AZ is correlated with the Argentinean Ischigualasto Formation (Langer et al. 2007), which bears some of the best-known and oldest dinosaurs worldwide. The cynodonts are well-documented, including at least eight species of traversodontids and probainognathians (Abdala and Ribeiro 2010; Martinelli and Soares 2016; Martinelli et al. 2017b; see Chap. 3). Other faunal components include temnospondyls and stereospondyls, sphenodonts (*Clevosaurus hadroprodon*), hundreds of rhynchosaur specimens (e.g., *Hyperodapedon* spp. and *Teyumbaita sulcognathus*), aetosaurs (e.g., *Aetobarbakinoides brasiliensis*), rauisuchians (e.g., *Rauisuchus tiradentes*), dinosauromorphs (e.g., *Ixalerpeton polesinensis*), and basal dinosaurs (e.g., *Staurikosaurus pricei, Gnathovorax cabreirai, Saturnalia tupiniquim, Pampadromeus barberenai, Buriolestes schultzi*) (e.g., Barberena 1982; Langer et al. 1999, 2007; Langer and Schultz 2000; Montefeltro et al. 2010; Cabreira et al. 2011, 2016; Dias-da-Silva et al. 2011; Desojo et al. 2012; Hsiou et al. 2019; Pacheco et al. 2019).

The *Riograndia* AZ particularly has provided a great diversity of small-sized vertebrates, including a basal lepidosauromorph (e.g., *Cargninia enigmatica*; Bonaparte et al. 2010), a procolophonian (*Soturnia caliodon*; Cisneros and Schultz 2003), several specimens of sphenodontians (e.g., *Lanceirosphenodon ferigoloi* and the clevosaurid *Clevosaurus brasiliensis*; Bonaparte and Sues 2006; Romo de Vivar et al. 2020), and several cynodonts, such as *Brasilodon quadrangularis, Riograndia guaibensis* (the most abundant cynodont in the AZ; Fig. 2.4d–e), and *Irajatherium hernandezi* (Bonaparte et al. 2003, 2005; Martinelli et al. 2005; see Chap. 3). Among larger sized animals, there are records of temnospondyls, phytosaurs, dinosaurs (e.g. *Guaibasaurus candelariensis, Macrocollum itaquii*), and dicynodonts (e.g., *Jachaleria candelariensis*) (e.g., Araújo and Gonzaga 1980; Bonaparte et al. 1999, 2010; Müller et al. 2018; Martinelli et al. 2020).

#### 2.2.1.2 Puesto Viejo, Mendoza Province, Argentina; Puesto Viejo Group, Middle–Lower Upper Triassic

The Puesto Viejo Group consists of up to 1000 m of alluvial and fluvial sediments and volcanic material in the form of rhyolitic pyroclastic flows, andesitic lavas, and shallow intrusives (González Díaz 1964; Monti and Franzese 2016), restricted to the San Rafael Block, Mendoza, western Argentina (Figs. 2.3 and 2.5). The Puesto Viejo Group was divided into two formations: Quebrada de los Fósiles (older) and Río Seco de la Quebrada (younger) (Stipanicic et al. 2007). The basal unit that comprises the traditional "Agua de los Burros Local Fauna" (Bonaparte 1982) includes pleuromeian and sphenopsid plant remains, fish scales, dicynodonts, and an archosauriform (i.e., *Koilamasuchus gonzalezdiazi*) (e.g., Bonaparte 1981; Zavattieri and Papú 1993; Ezcurra et al. 2010a). The Río Seco de la Quebrada Formation comprises the traditional "Puesto Viejo Local Fauna" (Bonaparte 1982) and includes a significant tetrapod assambleage useful for biostratigraphic correlations (e.g., Martinelli et al. 2017a). The Río Seco de la Quebrada Formation has provided the cynodonts *Cynognathus crateronotus* and *Diademodon tetragonus* of Pangeic distribution (e.g.,

**Fig. 2.5** Puesto Viejo area, San Rafael Department, Mendoza, Argentina; general view of the outcrops of the Puesto Viejo Group (**a**). Paleontologist Marcelo S. de la Fuente, organizing a field expedition to the outcrops, 2010 (**b**)

Abdala 1996; Martinelli et al. 2009), the basal traversodontid *Pascualgnathus polanskii* (Bonaparte 1966), and dicynodonts (e.g., *Vinceria vieja* and "*Kannemeyeria*"; Bonaparte 1966; Domnanovich and Marsicano 2012).

The "Agua de los Burros Local Fauna" was correlated with the *Lystrosaurus* AZ, while the "Puesto Viejo Local Fauna" with the *Cynognathus* AZ of South Africa, corresponding to the Early and Middle Triassic ages (e.g., Bonaparte 1966; Stipanicic et al. 2007; Martinelli et al. 2017a). However, SHRIMP 238U/206Pb age from a rhyolitic ignimbrite between the Quebrada de los Fósiles and Río Seco de la Quebrada formations indicated $235.8 \pm 2.0$ Ma, early Carnian age (Ottone et al. 2014). Its fossil content (basal traversodontids, *Cynognathus crateronotus*, *Diademodon tetragonus*), the lack of shared faunal elements with other well-sampled units from SA (e.g., Chañares Formation in Argentina and Chiniquá-Pinheiros in Brazil), and the correlations with the Middle Triassic units of South Africa strongly suggest caution while inferring the age of the Puesto Viejo Group (see discussion in Martinelli et al. 2017a). Further radiometric datings and fossils will be necessary to address this controversy.

#### 2.2.1.3 Uspallata, Mendoza Province, Argentina; Uspallata Group, Cuyo Basin, Middle–Upper Triassic

The Cuyo Basin is relatively small and it is located in the north of Mendoza, Argentina (Fig. 2.3). This is a rift basin developed at the beginning of the Triassic

Fig. 2.6 Uspallata, Mendoza, Argentina; general view of the outcrops of the Cerro de las Cabras Formation, 2000 (**a**). Bonaparte's team excavating the quarry where several specimens of *Andescynodon mendozensis* and *Rusconiodon mignonei* were found, 2000; from left to right: Agustín G. Martinelli, Walter Alarcón, and José F. Bonaparte (**b**)

(Kokogian et al. 2001) prior to the western Gondwana breakup. The Triassic sedimentary sequence is represented by the Uspallata Group and includes five formations (from base to top): Río Mendoza, Cerro de las Cabras, Potrerillos, Cacheuta, and Río Blanco formations (Stipanicic 1979). Non-mammalian cynodonts were found in the area of Cerro Bayo de Potrerillos (Fig. 2.6). This vertebrate association, traditionally called "Fauna Local Río Mendoza", was originally considered to come from the Río Mendoza Formation (e.g., Bonaparte 1969), but other authors (e.g., Zavattieri and Arcucci 2007, and references therein) considered that the fossils actually came from the Cerro de Las Cabras Formation (*contra* Bonaparte 2000). The "Fauna Local Río Mendoza" includes the dicynodont *Vinceria andina*, the basal traversodontids *Andescynodon mendozensis* and *Rusconiodon mignonei,* and the probainognathian *Cromptodon mamiferoides* (Bonaparte 1969, 1972a).

The dicynodont genus *Vinceria* was also reported in the Puesto Viejo Group (see above). This taxon was used to correlate the synapsid faunas from northern and southern Mendoza, which also share basal traversodontids.

#### 2.2.1.4 Ischigualasto-Talampaya Park, San Juan and La Rioja Provinces, Argentina; Ischigualasto-Villa Unión Basin, Middle to Upper Triassic

The Ischigualasto-Villa Unión Basin is an Upper Permian–Upper Triassic, continental rift basin developed along western Argentina, with extensive outcrops in

San Juan and La Rioja, northwestern Argentina (Bonaparte 1997; Kokogian et al. 2001; Mancuso and Caselli 2012; Colombi et al. 2017; Desojo et al. 2020). The Ischigualasto-Villa Unión Basin includes the Upper Permian–Lower Triassic red-beds of the Talampaya and Tarjados formations and the Agua de la Peña Group, with the Chañares/Ischichuca, Los Rastros, Ischigualasto, and Los Colorados formations (e.g., Kokogian et al. 2001; Colombi et al. 2017). The Chañares Formation is formed by grayish to brownish tuffs and mudstones interlayered with sandstones, which represent volcaniclastic mudflats and shallow lacustrine systems (e.g., Colombi et al. 2017; Fiorelli et al. 2018) (Fig. 2.7a, b). The Los Rastros Formation includes greenish to blackish laminated mudstones, which represent lacustrine and deltaic systems (e.g., Mancuso and Caselli 2012). The Ischigualasto Formation includes grayish to yellowish sandstones and mudstones, representing fluvial channels, deltaic and lacustrine deposits (e.g., Martínez et al. 2013a; Colombi et al. 2017) (Fig. 2.7c). Finally, the Los Colorados Formation is formed by a succession of reddish sandstones, mudstones, and conglomerates that represent alternated cycles of anastomosed and meandering fluvial systems (e.g., Colombi et al. 2017) (Fig. 2.8).

The succession of Triassic continental tetrapod assemblages from the Agua de la Peña Group is recognized worldwide for the amazing fossil record of basal turtles, diverse non-ornithodiran archosauromorphs (e.g., rhynchosaurs, erpetosuchids, proterochampsids, ornithosuchids, aetosaurids, rauisuchians), and ornithodirans, including several dinosauriforms and the oldest ornithischian and saurischian dinosaurs, several dicynodonts and non-mammaliaform cynodonts (e.g., Romer 1966; Bonaparte 1972b, 1997; Rogers et al. 1993, 2001; Martínez et al. 2011, 2013a; Mancuso et al. 2014; Ezcurra et al. 2017). The bulk of the tetrapod fossil record is restricted to the Chañares, Ischigualasto, and Los Colorados formations. The Chañares Formation has two main faunal associations: the *Tarjadia* and *Massetoganthus-Chanaresuchus* AZs (Ezcurra et al. 2017). The older AZ includes a few cynodonts (cf. *Aleodon* sp. and aff. *Scalenodon* sp.), indeterminate dicynodonts, indeterminate rhynchosaurids, the erpetosuchid *Tarjadia ruthae*, the rauisuchian *Luperosuchus fractus,* and other suchians under study. The *Massetognathus-Chanaresuchus* AZ includes mawsoniids remains, proterochampsids (e.g., *Chanaresuchus bonapartei*), the gracilisuchid *Gracilisuchus stipanicicorum*, several dinosauriforms (e.g., *Lewisuchus admixtus*, *Lagerpeton chanarensis*), dicynodonts (e.g., *Dinodontosaurus* sp.), and cynodonts (e.g., *Probainognathus jenseni* and *Massetognathus pascuali*; Fig. 2.7b; Chap. 3) (e.g., Romer 1966, 1973; Mancuso et al. 2014; Ezcurra et al. 2017 and references therein).

The Ischigualasto Formation includes rhynchosaurus (e.g., *Hyperodapedon sanjuanensis*), proterochampsids (e.g., *Proterochampsa barrionuevoi*), rauisuchians (e.g., *Saurosuchus galilei*), aetosaur (e.g., *Aetosauroides scagliai*), ornithosuchids (e.g., *Venaticosuchus rusconii*), several worldwide famous basal dinosaurs (*Herrerasaurus ischigualastensis*, *Eoraptor lunensis*, *Eodromaeus murphi*, *Pisanosaurus mertii*), and dicynodonts (*Ischigualastia jenseni*) (e.g., Cox 1962; Reig 1963; Casamiquela 1967; Sereno and Novas 1992; Rogers et al. 1993; Sereno et al. 1993; Bonaparte 1997; Martínez et al. 2011, 2013a and references herein). Also, there is an important record of traversodontids and probainognathian cynodonts

**Fig. 2.7** The lower Upper Triassic Chañares Formation (Ischigualasto-Villa Unión Basin) at Talampaya National Park, La Rioja, Argentina, 2014 (**a**). Isolated maxilla with postcanine teeth of the traversodontid cynodont *Massetognathus pascuali* (specimen from the collection of CRILAR), one of the most abundant species of this unit (**b**). Ischigualasto Provincial Park, San Juan, Argentina, with geoforms from the Ischigualasto Formation (Ischigualasto-Villa Unión Basin), and at the back, the red sandstones from the Los Colorados Formation (**c**)

(e.g., *Exaeretodon frenguelli*, *Ischignathus sudamericanus*, and *Ecteninion lunensis*; Chap. 3).

The Los Colorados Formation includes the basal turtle *Palaeochersis talampayensis* (Fig. 2.8c), the rauisuchian *Fasolasuchus tenax*, the aetosaur *Neoaetosauroides engaeus*, the ornithosuchid *Riojasuchus tenuisceps*, crocodylomorphs

**Fig. 2.8** Los Colorados, La Rioja, Argentina. General view of the outcrops of the Los Colorados Formation; Rougier's expedition, 1996 (**a**). Excavation at the Tortuga Site; from left to right: Guillermo W. Rougier, Diego (Caco) Pol, Fernando (Viuti) E. Abdala, Marcelo S. de la Fuente, and Santiago (Vultur) Reuil (**b**). *Palaeochersis talampayensis* (PULR 68), skull, carapace, and postcranial skeleton from the oldest turtles from South America (**c**)

(e.g., *Hemiprotosuchus leali*), several dinosaurs (e.g., *Riojasaurus incertus*, *Zupaysaurus rougieri*, and *Powellvenator podocitus*), and the dicynodont *Jachaleria colorata* (e.g., Bonaparte 1972b, 1981; Rougier et al. 1995; Arcucci and Coria 2003; Sterli et al. 2007; von Baczko and Desojo 2016; Ezcurra 2017), in addition to the tritheledontid cynodont, *Chaliminia musteloides* (Bonaparte 1980; Martinelli and Rougier 2007).

The Ischigualasto-Villa Unión Basin is particularly rich in fossil material (Bonaparte 1997; Martínez et al. 2013a). Although plant, reptile-footprint and vertebrate fossils from this basin were first studied in the first half of the twentieth century (e.g., Bodenbender 1911; Kurtz 1921; von Huene 1931; Cabrera 1943; Frenguelli 1948), only in the late 1950s did this area bring real interest for scientists. The renowned American paleontologist Alfred Romer (1894–1973) from Harvard University started an intensive collection of fossil vertebrates in 1958 (in the Ischigualasto Formation) and 1964–1965 (discovery of the Chañares Formation fauna) (Romer 1973; Jensen 2001; Sereno 2013). The first expedition of Romer in Ischigualasto, in collaboration with the Argentinean colleagues at MACN, resulted in the discovery of important specimens. This success automatically activated the competition by the Argentine paleontologist Osvaldo A. Reig (1929–1992) and his team of the Universidad Nacional de Tucumán, who the same year visited the locality and collected several specimens. Since then, the intense collection led by Reig, Bonaparte, and Casamiquela and the second expedition of Romer resulted in hundreds of specimens in great preservational state and several studies that played a pivotal role in understanding the evolution of vertebrates during the Triassic in southern landmasses (e.g., Reig 1963; Romer 1973; Bonaparte 1997; Jensen 2001; Sereno 2013; Martínez et al. 2013a). According to the report presented by Alfred Romer to the National Commission of Heritage of Argentina, he commented that "*Each paleontologist dreams to find a place, one day, untouched, and full of skulls and skeletons. Almost never this dream is fulfilled. However to our amazement and happiness, the dream came true in Ischigualasto*" (La Nacion 2004; original in Spanish).

Currently, the paleontological sites from the Ischigualasto-Villa Unión Basin are included in the Ischigualasto Provincial Park and Talampaya National Park, San Juan and La Rioja Provinces, respectively. The objective of these national areas is to protect the rich paleontological sites and to enhance the tourism in those remote, fantastic landscapes from central-west Argentina.

#### 2.2.1.5 El Carrizal, San Juan Province, Argentina; Quebrada Del Barro Formation, Marayes-El Carrizal Basin, Upper Triassic

The Marayes Group represents the Triassic sequence of the Marayes-El Carrizal Basin, exposed in southeastern San Juan and northwestern San Luis, Argentina (Bossi 1976; Spalletti et al. 2011) (Fig. 2.3). The Marayes Group is divided into four stratigraphic units (from base to top): the Esquina Colorada, Carrizal, Quebrada del Barro, and Balde de Leyes formations (Bossi 1976; Colombi et al. 2015). Among these units, the Quebrada del Barro Formation bears a tantalizing fossil record, including derived non-mammaliaform probainognathians yet to be fully described (Martínez et al. 2015). This formation is composed of coarse sandstones and conglomerates interbedded with sandy-claystones, interpreted as a distributive fluvial system (Colombi et al. 2015). The described fossil assemblage includes basal turtles, the opisthodontian sphenodont *Sphenotitan leyesi*, basal crocodylomorphs, the lagerpetid dinosauromorph *Dromomeron gigas*, the coelophysid theropod *Lucianovenator bonoi*, and sauropodomorphs (*Ingentia prima*) (Martínez et al. 2013b, 2015, 2016; Martínez and Apaldetti 2017; Apaldetti et al. 2018). The age of this formation is considered Norian.

**Fig. 2.9** Fossiliferous localities from the Jurassic of South America with mammals. Abbreviations (as in Fig. 2.2): 9, Queso Rallado, Chubut, Argentina; 10, Laguna Manantiales, Santa Cruz, Argentina

## 2.2.2 *Jurassic*

### 2.2.2.1 Queso Rallado, Chubut Province, Argentina; Cañadón Asfalto Formation, Toarcian–Bajocian, Lower—Middle Jurassic

The non-marine Mesozoic deposits from central Patagonia, Argentina, were recognized early in the twentieth century (Piatnitzky 1936; Feruglio 1949; Frenguelli 1949) and studied with detail ever since. They comprise the sedimentary sequence of the Cañadón Asfalto Basin, on the Somuncura Massif (Cabaleri et al. 2010a; Cúneo et al. 2013; Figari et al. 2015; Olivera et al. 2015; Volkheimer et al. 2015). The depositional sequence includes fluvial, volcanic, volcaniclastic, and lacustrine deposits in a seasonally dry subtropical environment. The Cañadón Asfalto Basin is a typical rift basin developing in a paleogeographic setting that counts as central actors the split of Pangea, the subduction of the Pacific margins of SA, and older structural features, like the Huincul Dorsal (Mpodozis and Ramos 2008). Ultimately, the opening of the Weddell Sea between the Antarctic Peninsula and Patagonian Cordillera materialized about 160 Ma ago (Ghidella et al. 2007; Ramos and Ghiglione 2008; Ramos 2009). The Cañadón Asfalto Formation and the overlaying Cañadón Calcáreo Formation represent the most extensive exposures of Jurassic continental rocks in SA (Cúneo et al. 2013).

The Cañadón Asfalto Formation is composed of calcareous, siliciclastic, and volcanic deposits. Its depositional environment has been interpreted as a perilitoral flood plain with meandering rivers in association with small, shallow lacustrine bodies (e.g., Musacchio 1995; Silva Nieto et al. 2002; Rougier et al. 2007a; Volkheimer et al. 2008, 2009; Escapa et al. 2008a; Cabaleri et al. 2010a, b, 2013; Olivera et al. 2015).

The age of the Cañadón Asfalto Formation has been traditionally considered Jurassic but a more precise determination was controversial (Cabaleri et al. 2010b), only recently set on firmer ground by U/Pb radiometric dating, with dates between $178.766 \pm 0.092$ Ma and $176.15 \pm 0.12$ Ma (Cúneo et al. 2013) for samples coming relatively low in the sequence. Stratigraphic interpretations of the fluvial-lacustrine system of the Cañadón Asfalto Formation suggested the initiation of the sequence at ca. 179 Ma (early–mid-Toarcian) that most probably continued into the early Middle Jurassic (Aalenian or Bajocian) (Cúneo et al. 2013), as sunken graben basin, or pull-apart basin (Silva Nieto et al. 2002; Mpodozis and Ramos 2008) progressively filled up, transitioning to more terrestrial deposits.

The Cañadón Asfalto Formation preserves an extraordinary fossil record of key events in the evolution of major SA plants, dinosaurs, and mammals (Escapa et al. 2008a). These events include the diversification of the araucarian and cupressaceous conifers and the osmundaceous ferns, major faunal turnover of the sauropodomorph dinosaurs, the first successful radiation of the ornithischian dinosaurs, and the earliest radiation of SA mammals (Escapa et al. 2008a; Cúneo et al. 2013). Fossils include rich assemblages of palynomorphs (Volkheimer et al. 2008, 2009; Olivera et al. 2015), taphofloras (Stipanicic et al. 1968; Stipanicic and Bonetti 1970; Cortés and

Baldoni 1984; Baldoni 1990; Escapa et al. 2008b), calcareous microfossils and invertebrates, including "conchostracans", ostracods, bivalves, gastropods, and insects (Frenguelli 1949; Tasch and Volkheimer 1970; Vallatti 1986; Musacchio et al. 1990; Gallego1994; Musacchio1995; Genise et al. 2002; Volkheimer et al. 2009; Gallego et al. 2010, 2011). The vertebrate record of the Cañadón Asfalto Formation includes exquisite material of anurans (*Notobatrachus reigi*; Báez and Nicoli 2008), turtles (*Condorchelys antiqua*; Sterli 2008; Sterli and de la Fuente 2010; Sterli et al. 2019), sphenodontians (*Sphenocondor gracilis*; Apesteguía et al. 2012), pterosaurs (*Allkaruen koi*; Codorniú et al. 2016), dinosaurs (the sauropods *Patagosaurus fariasi* and *Volkheimeria chubutensis*, the theropods *Piatnitzkysaurus floresi*, the very similar *Condorraptor currumili*, *Eoabelisaurus mefi*, and *Asfaltovenator vialidadi*, the basal ornithischian *Manidens condoriensis*; Bonaparte 1979, 1986d; Rauhut 2005, 2007; Pol et al. 2009, 2011; Pol and Rauhut 2012; Rauhut and Pol 2019), and mammals.

Most of the micro/mid-sized fossils, including the totality of the mammals described to date, were collected at the Queso Rallado quarry (Rauhut et al. 2002) and the neighboring locality Canela (Rougier et al. 2007a, b, 2011a; Gaetano and Rougier 2011, 2012). Queso Rallado is located ~5.5 km west of Cerro Cóndor Village, Chubut, Argentina (Figs. 2.9 and 2.10) and it is interpreted as being low in the stratigraphic section, not too far above the volcanic sediments marking the transition from the underlying Lonco Trapial Formation into the Cañadón Asfalto (Rougier et al. 2007b), in levels that would be near equivalent to those used by Cúneo et al. (2013) to date the formation. Queso Rallado yields a rich vertebrate association with amphibians, turtles, lepidosaurs, pterosaurs, ornithisquian, theropod, and sauropod dinosaurs, in addition to mammals (Rauhut et al. 2002; Rougier et al. 2007a, b; Gaetano and Rougier 2011, 2012). The bone-bearing level is extensive but discontinuous because of fracturing and deformation of the sediments in the area of the quarry, which makes it difficult to follow over distance. The fossils were collected from indurated sediments of variable thickness, reaching up to 60 cm of thin interlaminated tuffaceous and calcareous deposits, very rich in bleached silica from the under- and overlying sediments rich in pyroclastic material. The vertebrate remains are mostly isolated, showing evidence of moderate transport (Rougier et al. 2007a) or disassociation by disturbance of the floor of what is considered a very shallow ephemeral body of water. This environment allows the occasional recovery of partially articulated bones or associated specimens, as in the case of the heterodontosaurid *Manidens condoriensis* (Pol et al. 2011), the "triconodont" *Argentoconodon fariasorum* (Gaetano and Rougier 2011), and turtles (Sterli et al. 2019). Four different mammalian species have been described from the unit. The australosphenidans *Asfaltomylus patagonicus* (Rauhut et al. 2002; Martin and Rauhut 2005) and *Henosferus molus* (Rougier et al. 2007a), along with *Ambondro mahabo* from a slightly younger site in Madagascar (Flynn et al. 1999), are putatively related to Early Cretaceous taxa from Australia (Rich et al. 1997, 1999, 2001a, b) and perhaps ultimately to monotremes (Luo et al. 2002; Kielan-Jaworowska et al. 2004; Rougier et al. 2007a; but see Rich et al. 2002; Woodburne 2003; Woodburne et al. 2003; Rowe et al. 2008; Rich et al. 2016; Chap. 4). The other two are "triconodonts", the eutriconodont *Argentoconodon fariasorum* and the "amphilestid" *Condorodon spanios* (Rougier

**Fig. 2.10** Queso Rallado, Chubut, Argentina. General view of the outcrops of the Cañadón Asfalto Formation (**a**). Left dentary of *Henosferus molus* (MPEF-PV 2357) (**b**) and *Argentoconodon fariasorum* (MPEF-PV 1877), at the fossil site (**c**). Quarry on a bone-bed from the Cañadón Asfalto Formation; from left to right: Juliana Sterli, Guillermo (Willy) F. Turazzini, Diego (Caco) Pol, Marcos Becerra, Ignacio Maniel, Fernando F. Garberoglio, Imanol Yañéz, Juan Leardi, Lorena Austin, Leandro Gaetano, Raúl O. Gómez, Guillermo W. Rougier, 2008 (**d**)

et al. 2007b; Gaetano and Rougier 2011, 2012; Chap. 5), in addition to an as yet undescribed allotherian (Gaetano and Rougier 2012). The middle–late Toarcian age of the associated strata renders these mammals the oldest representatives of their clades, predating their closest relatives by 10–30 Ma and the oldest mammals overall, providing minimal dates for Mammalia as a whole.

#### 2.2.2.2 Laguna Manantiales, Santa Cruz Province, Argentina; La Matilde Formation, Bathonian–Callovian, Middle Jurassic

The Laguna Manantiales is located ~25 km northwest of the Petrified Forest Natural Monument, Santa Cruz, Patagonia, Argentina (de Valais 2009, 2011) (Figs. 2.9 and 2.11). The area includes outcrops of the Bahia Laura Group, consisting of the interdigitating Chon Aike and La Matilde formations (Stipanicic and Reig 1955, 1957; Lesta and Ferello 1972).

The La Matilde Formation includes primary and reworked volcanoclastic sediments, mainly represented by tuffs, tuffites, tuffaceous siltstones, fine-grained sandstones, and thin layers of ignimbrites (Panza and Genini 1998; de Barrio et al. 1999; Melchor et al. 2004). The environment during deposition represents a low energy fluvial plain with swamps and water bodies associated with volcanic activity (Mazzoni et al. 1981; Melchor et al. 2004).

The age of the La Matilde Formation has been accredited on the basis of its fossil content (Stipanicic and Reig 1955, 1957; Stipanicic and Bonetti 1970) and

**Fig. 2.11** Laguna Manantiales, Santa Cruz, Argentina. General view of the outcrops of La Matilde Formation. Joint expedition of Museo Paleontológico "Egidio Feruglio", Trelew and University of Louisville, KY, 2012

radiometric dating from the Chon Aike Formation (summarized in de Barrio et al. 1999). The combined information constrains the age of the La Matilde Formation to the Bathonian interval up to Callovian (Spalletti et al. 1982; de Barrio 1993; Echeveste et al. 2001; Kloster and Gnaedinger 2018).

The fossils from the La Matilde Formation consist of plants (including the famous in situ stumps and seed cones from the Petrified Forest Natural Monument "Madre e Hija" or "Cerro Cuadrado"; e.g., Cúneo and Panza 2008; Falaschi et al. 2011; Kloster and Gnaedinger 2018), invertebrates, vertebrates (i.e., the anuran *Notobatrachus degiustoi*; originally described by Reig in Stipanicic and Reig 1955, 1957; Báez and Nicoli 2004), and ichnites from invertebrates and vertebrates. Among the latter, four ichnites are regarded as of dinosaurian origin (*Delatorrichnus goyenechei, Sarmientichnus scagliai, Wildeichnus navesi,* and *Grallator* isp.) and two are mammalian, or from close mammalian relatives (*Ameghinichnus patagonicus* and *A. manantialensis*) (Casamiquela 1961, 1964a; de Valais 2009, 2011). *Ameghinichnus* was the first evidence of a Mesozoic mammal in SA (Casamiquela 1961, 1964a); it took, however, 20 years to find the first osseous remain of an actual Mesozoic mammal and about 40 to recover the first osseous remains of Jurassic mammals.

## 2.2.3 *Cretaceous*

### 2.2.3.1 Araraquara, São Paulo State, Brazil; Botucatu Formation, Berriasian–Valanginian, Lower Cretaceous

The Botucatu Formation (São Bento Group, Paraná Basin) consists of a large succession of aeolian sandstones that represents one of the largest paleodeserts (commonly named the Botucatu Paleodesert) recovered so far, with approximately 1.300.000 km$^2$ (e.g., Almeida 1954; Milani et al. 1998; Almeida et al. 2012). The Botucatu Formation crops out along the states of Goiás, Mato Grosso, Mato Grosso do Sul, São Paulo, Paraná, Santa Catarina, and Rio Grande do Sul, in Brazil (e.g., Almeida 1954; Milani et al. 1998) and lateral equivalent paleodesert units are found in Uruguay (Perea et al. 2009), Paraguay (Leonardi 1992), Argentina (Sanford and Lange 1960; Salamuni and Bigarella 1967), and even in Namibia (Stollhofen 1999). The age of the Botucatu Formation is inferred as Berriasian–Valanginian (Lower Cretaceous) based on the radiometric dating of the overlaying Serra Geral Formation basalts (U/PB 134.5 $\pm$ 2.1 and 119.3 $\pm$ 0.95 Ma; e.g., Turner et al. 1994; Brückmann et al. 2014).

The fossil record from the Botucatu Formation is mainly based on a huge diversity of ichnofossils and relative sparse records of conchostraceans (Almeida 1950) and silicified coniferous wood (Suguio and Coimbra 1972; Pires et al. 2011). Ichnofossils include invertebrate traces (mainly insects and arachnids as producers), such as *Hexapodichnus, Octopodichnus, Paleohelcura,* and *Taenidium* (Fernandes et al. 1988, 1990; Leonardi et al. 2007), urolites attributed to dinosaurs (Fernandes et al. 2004), and a diverse assemblage of footprints referred to theropod (Carnosauria

**Fig. 2.12** Fossiliferous localities from the Cretaceous of South America with mammals. Abbreviations (as in Fig. 2.2): 11, São Bento, Araraquara, São Paulo, Brazil; 12, La Amarga, Neuquén, Argentina; 13, La Buitrera, Río Negro, Argentina; 14, Tres Lagos, Santa Cruz, Argentina; 15, Santo Anastácio, São Paulo, Brazil; 16, Los Barreales Lake, Neuquén, Argentina; 17, Paso Córdoba, Río Negro, Argentina; 18, Estancia Los Alamitos, west slope of Cerro Cuadrado, Río Negro, Argentina; 19, Cerro Tortuga, Río Negro, Argentina; 20, Mirasol Chico Canyon, Chubut, Argentina; 21, Ingeniero Jacobacci, Río Negro, Argentina; 22, Río de Las Chinas Valley, Última Esperanza Province, Chile/Alta Vista and La Anita farms, Santa Cruz, Argentina; 23, Pajcha Pata, Cochabamba, Bolivia; 24, Synclinal de Bagua, Peru

and Coelurosauria) and ornithopod (Ornithopoda) dinosaurs, and mammals (e.g., Leonardi 1981, 1994; Fernandes and Carvalho 2007, 2008; Leonardi et al. 2007; Francischini et al. 2015; Buck et al. 2017a, b; D'Orazi Porchetti et al. 2018). Regarding mammalian footprints, four ichnospecies have been recognized: *Brasilichnium elusivum* (Leonardi 1981), *B. saltatorium* (Buck et al. 2017a), *B. anaiti* (D'Orazi Porchetti et al. 2018), and *Aracoaraichnium leonardii* (Buck et al. 2017b) (Figs. 2.12 and 2.13). These ichnospecies represent the only mammalian record for the Early Cretaceous of Brazil, from unknown creatures that left their footprints in a hot sandy desert with large aeolian dunes and interdune valleys (Chap. 9).

Most of the ichnofossils known from the Botucatu Formation were collected in sandstone quarries located in the municipality of the Araraquara (São Paulo State). These quarries provided ornamental rocks for houses and streets and consequently, many places of the Araraquara town show traces and footprints of the Early Cretaceous fauna (Leonardi 1980; Francischini et al. 2018). The most significant quarries are São Bento (Corpedras), Califórnia, Cerrito Velho, Cerrito Novo, Santa Águeda, and Chibarro, and most of them are inactive nowadays (e.g., Leonardi and Carvalho 2002; Francischini et al. 2018). They constitute the important "Jazigo Icnofossilífero do Ouro" of the Botucatu Formation. In particular, the São Bento quarry has provided the holotype specimens of the four mammalian ichnospecies.

#### 2.2.3.2 La Amarga, Neuquén Province, Argentina; La Amarga Formation, Barremian–Aptian, Lower Cretaceous

The La Amarga Formation is exposed in northwest Patagonia, Neuquén, Argentina (Fig. 2.12), and constitutes part of the Lower Cretaceous sedimentary filling of the Neuquén Basin (Fig. 2.14). The La Amarga Formation is divided into three units (from base to top): Puesto Antigual, Bañados de Caichigüe, and Piedra Parada Members (Leanza and Hugo 1997). The Barremian–lower Aptian age for the La Amarga Formation is based on stratigraphical relationships, tecto-sedimentary aspects (Leanza and Hugo 1997), and the palynological content (Prámparo and Volkheimer 2002).

The Puesto Antigual Member is mainly represented by coarse-grained sandstones with conglomerate levels, limestones, and mudstones, deposited in a fluvial environment with shallow water lagoons and well-developed paleosol tops (Leanza and Hugo 1997). In the 1980s, a team led by José F. Bonaparte discovered a rich vertebrate association from the La Amarga Formation, specifically from what later would be identified as the Puesto Antigual Member. This includes the crocodyliform *Amargasuchus minor* (Chiappe 1988), a pterodactyloid pterosaur (Montanelli 1987), the spiny dicraeosaurid sauropod *Amargasaurus cazaui* (Salgado and Bonaparte 1991), the titanosaur sauropod *Amargatitanis macni* (Apesteguía 2007), the abelisauroid theropod *Ligabueino andesi* (Bonaparte 1996a), several theropod teeth of uncertain affinities, fragmentary remains of the only known SA stegosaur (*Amargastegos brevicollus* according to Ulansky 2014, but referable to Stegosauria indet. according to Pereda-Suberbiola et al. 2012 and Galton and Carpenter 2016), and the cladotherian

**Fig. 2.13** Outcrops of the Botucatu Formation, São Paulo, Brazil. Giuseppe Leonardi exploring at the quarry São Bento ("Corpedras"), Araraquara Municipality, where the holotypes of *Brasilichnus* ichnospecies were discovered, 2013 (**a**). Giuseppe Leonardi studying mammalian footprints from the Botucatu Formation, in slabs from the streets of Araraquara city, 1986 (**b**). Giuseppe Leonardi and Marcelo A. Fernandes studying dinosaur footprints in a street of the Araraquara city, 2013 (**c**) (pictures from Giuseppe Leonardi)

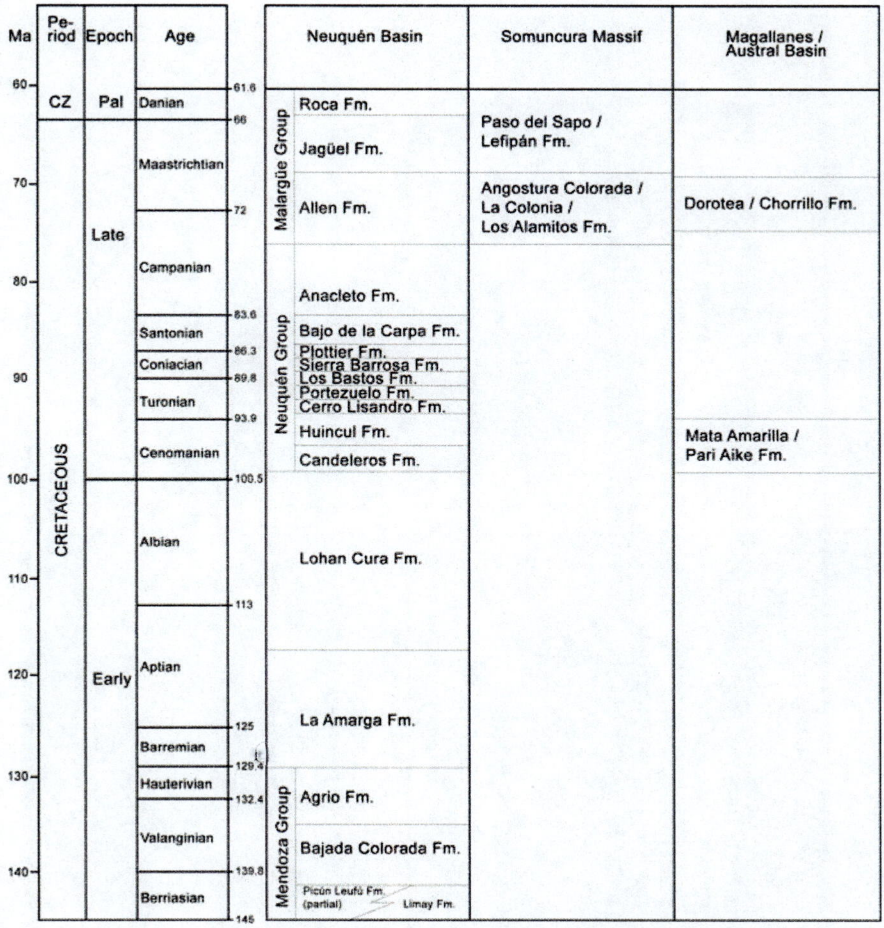


**Fig. 2.14** Stratigraphy of major Cretaceous fossiliferous basins from Patagonia with fossil mammals (see text)

mammal *Vincelestes neuquenianus* (Bonaparte 1986b). The Bañados de Caichigüe Member is composed of white to yellowish lacustrine limestones alternating with black shales and greenish siltstones (Leanza and Hugo 1997; Leanza et al. 2004). These levels have provided charophytes, non-marine ostracods, conchostracans, pollen, and megaspores (e.g., Musacchio 1970, 1971a, b; Volkheimer 1978; Prámparo and Volkheimer 2002; Gallego and Shen 2004; Ottone 2009). The Piedra Parada Member is formed by a thick succession of light brown and reddish, fine- to medium-grained sandstones alternating with pink, reddish, and brown to greenish siltstones, developed in an alluvial system with fluvial channels, swamp lenses, and palaeosol tops (Leanza and Hugo 1997; Leanza et al. 2004). To date, only the rebbachisaurid sauropod *Zapalasaurus bonapartei* has been recovered from this Member (Salgado et al. 2006).

**Fig. 2.15** La Amarga, Neuquén, Argentina. Bonaparte's expedition at the La Amarga Formation; from left to right: Martín Vince, who found the first skeleton of *Vincelestes neuquenianus* and after whom the mammal was named, Guillermo W. Rougier, Luis Chiappe, Silvana Montanelli, José F. Bonaparte, and Ubaldo Bonaparte, 1986 (**a**). Museo Argentino de Ciencias Naturales "Bernardino Rivadavia", hall of dinosaurs, skeleton of *Amargasaurus cazaui* (**b**). Guillermo W. Rougier and Zofía Kielan-Jaworowska, in La Amarga, 1987 (picture by Pablo Puerta) (**c**)

*Vincelestes neuquenianus* was found from the La Amarga locality, 70 km south of Zapala (Figs. 2.12 and 2.15); all the specimens come from a mono-specific assemblage suggesting a family group. Neither the lithology nor the anatomy of *Vincelestes* suggests burrows or burrowing as a dedicated paleobiological specialization. The preservation is exquisite, representing at least eleven individuals, including skulls, jaws, and partially articulated postcrania (Bonaparte 1986b; Bonaparte and Rougier 1987a; Rougier 1993). *Vincelestes* is the only Early Cretaceous mammal from SA and its putative phylogenetic position and preservation make it a mandatory reference for anatomical studies and phylogenetic reconstructions of early mammals (e.g., Rougier et al. 1992; Hopson and Rougier 1993; Luo et al. 2002; Rougier and Wible 2006; Macrini et al. 2007; Crompton et al. 2018).

#### 2.2.3.3 La Buitrera, Río Negro Province, Argentina; Candeleros Formation, Cenomanian, lower Upper Cretaceous

The Neuquén Group (lower Cenomanian –middle Campanian) is a thick fossiliferous succession of continental red-beds of fluvial, aeolian, and shallow lacustrine origin developed in the Neuquén Basin (Fig. 2.14), northwest Patagonia, Argentina (Howell et al. 2005; Aguirre-Urreta and Cristallini 2009; Garrido 2010, 2011; Rojas Vera et al. 2016). The Candeleros Formation is the basal unit of the Neuquén Group and mainly formed by reddish to brownish coarse- and medium-grained sandstones and conglomerates deposited in a fluvial environment, with aeolian influence and deposition of sand dunes dominated locally by desert conditions (Candia Halupczok et al. 2018). These deposits also reflect the general underfilled condition of the basin (Asurmendi et al. 2017). Some brownish siltstones and claystones represent swamp conditions. Paleosols are also frequent (Leanza and Hugo 1997; Hugo and Leanza 2001a; Leanza et al. 2004). The age of the Candeleros Formation has been regarded Cenomanian, likely lower Cenomanian based on stratigraphic correlations (Leanza et al. 2004; Garrido 2010; Candia Halupczok et al. 2018).

The Candeleros Formation provides a very rich vertebrate fossil assemblage, originally identified at the end of the nineteenth century (Calvo et al. 2011). Among fossils sites, La Buitrera, Río Negro, Argentina (Figs. 2.12 and 2.16) and nearby sites, is one of the most promising microvertebrate localities of the Mesozoic from SA.

The Candeleros Formation provided an extended list of well-known dinosaurs: among sauropods, the titanosaur *Andesaurus delgadoi* (Calvo and Bonaparte 1991) and some rare titanosaurs (Calvo 1999; Simón and Calvo 2002), the rebbachisauroid *Nopcsaspondylus alarconensis* (Apesteguía 2007), *Limaysaurus tessonei* (= *Rebbachisaurus tessonei*; Calvo and Salgado 1995; Salgado et al. 2004), and the closely related *Rayososaurus agrioensis* (Bonaparte 1996a). Among theropods, one of the largest worldwide predators, the giant carcharodontosaurid *Giganotosaurus carolinii* (Coria and Salgado 1995), the abelisaurid *Ekrixinatosaurus novasi* (Calvo et al. 2004), the alvarezsaurid *Alnashetri cerropoliciensis* (Makovicky et al. 2012), the unenlagiid *Buitreraptor gonzalezorum* (Makovicky et al. 2005), and the coelurosaurian *Bicentenaria argentina* (Novas et al. 2012a) have been found there. Ornithopod remains have been described by Coria et al. (2007). The reptile association is completed by abundant record of the crocodyliforms *Araripesuchus patagonicus* (Ortega et al. 2000) and *A. buitreraensis* (Pol and Apesteguía 2005), the sphenodontian lepidosaur *Kaikaifilusaurus calvoi* (Simón and Kellner 2003), and *Priosphenodon avelasi* (Apesteguía and Novas 2003); an

Iguanidae indet. (Apesteguía et al. 2005), the snake *Najash rionegrina* (Apesteguía and Zaher 2006) with hindlimbs and sacrum; pterosaurs (Haluza and Canale 2013), and at least two species of chelid turtles assigned to *Prochelidella* (Lapparent de Broin and de la Fuente 2001). Non-reptilian taxa include the pipoid frog *Avitabatrachus uliana* (Báez et al. 2000), the dipnoi fishes *Ceratodus argentinus* and *Atlantoceratodus iheringi* (Apesteguía et al. 2007), and mammals. Poorly preserved dinosaur footprints have been reported from aeolian interdune deposits (Candia Halupczok et al. 2018). More traditional, numerous and better preserved archosaur tracks have been reported in the exposed lakeshore of the Lake Ezequiel Ramos Mexía, including sauropod (*Sauropodichnus giganteus*), theropod (*Abelichnus astigarrae, Bressanichnus patagonicus, Deferrariischnium mapuchensis, Picunichnus benedettoi*), ornithopod (*Limayichnus major*, *Sousaichnium monettae*), and pterosaur (*Pteraichnus* sp. indet.) tracks (Calvo 1991, 1999; Calvo et al. 2011). These beds are also very rich in invertebrate traces (Garrido 2010) and silicified trunks (Ottone 2009 and references therein).

The Candeleros Formation mammalian record is represented by non-tribosphenic mammals recovered at La Buitrera locality, the most abundant of which is *Cronopio dentiacutus,* a dryolestoid related to *Leonardus cuspidatus*, a taxon from the Upper Cretaceous Los Alamitos Formation, and the surviving Miocene *Necrolestes* spp. (Rougier et al. 2011b; Wible and Rougier 2017; see also Chap. 6). In addition, at least one other group with very enigmatic affinities is present (Rougier et al. 2011a), with simplified ever-growing teeth, currently under study.

#### 2.2.3.4 Tres Lagos, Santa Cruz Province, Argentina; Mata Amarilla Formation, Cenomanian, lower Upper Cretaceous

The Magallanes-Austral Basin developed during the Mesozoic in southwestern Patagonia, including portions in current Argentina and Chile (Varela et al. 2012). During the Cretaceous, the basin was affected by continental expansion and was laterally invaded by marine coastal environments (e.g., the Mata Amarilla Formation; including the Pari Aike Formation, which is regarded by some authors as the middle section of the Mata Amarilla Formation; Varela et al. 2012). The Mata Amarilla Formation corresponds to Ameghino's "Sehuenense" (Ameghino 1906). It consists of gray and blackish siltstones and claystones, intercalated with thin whitish and yellowish-gray fine- to medium-grained sandstones and tuff layers (Varela and Poiré 2008; Varela et al. 2012).

Recent U/Pb radiometric dating from a tuff layer of the middle section of the unit provided an age of $96.23 \pm 0.71$ Ma, middle Cenomanian (Varela et al. 2012).

The Mata Amarilla Formation has abundant fossil remains, representing a subtropical temperate climate with marked seasonality (Varela et al. 2018). Fossil flora includes algae, angiosperms, and a petrified forest in life position, the "Bosque Petrificado María Elena", composed of gymnosperms mainly Podocarpaceae, and minority Araucariaceae and Cycdales (e.g., Zamuner et al. 2006; Iglesias et al. 2007a; Martínez et al. 2017; Santamarina et al. 2018). Invertebrates are represented by mollusks

**Fig. 2.16** La Buitrera, Río Negro, Argentina. General view of the outcrops of the Candeleros Formation (**a**). Guillermo W. Rougier with a complete skull of *Cronopio dentiacutus*, 2008 (**b**). Skull of the sphenodont *Priosphenodon avelasi*, the most abundant fossil at the site (**c**). Apesteguía's expedition to La Buitrera; standing from left to right: Lucas Appella-Guiscafre, Dennis Monge, Rocío Belén Vera, Esteban Potenza, Guillermo W. Rougier, Leonardo (Harry) Pazo, Michael W. Caldwell, Sebastián Apesteguía, Eliana Cimorelli, Andrés (Andy) Lires, Marceo Rivero, Laila Toledo, Fernando F. Garberoglio, and kneeling from left to right: Laura Trivigno, and Florencia Fillippini, 2016

(bivalves and gastropods) and perhaps also ammonoids (Griffin and Varela 2012 and literature cited). Vertebrates are abundant and include aquatic taxa such as hybodontiform and neoselachian chondrichthyans, holosteans, teleosteans, dipnoans (*Atlantoceratodus iheringi*; Cione et al. 2007), pleurodiran turtles, crocodyliforms, and plesiosaurs (O'Gorman and Varela 2010; Griffin and Varela 2012; Varela et al. 2012). Ameghino (1893) described teeth assigned to the mosasaur *"Liodon" argentinus* but the specimens are now lost from the museum collections, thus making re-evaluation of its taxonomy difficult (Fernández and Gasparini 2012). Terrestrial vertebrates include probable pterosaur teeth, the theropod abelisauroid *Austrocheirus isasii* (Ezcurra et al. 2010b) and megaraptoran *Orkoraptor burkei* (Novas et al. 2008), the ornithischian *Loncosaurus argentinus* (Coria and Salgado 1996a; however this is a possible *nomen vanum*, see Coria 2016) and *Talenkauen santacrusensis* (Novas et al. 2004), the sauropod *Puertasaurus reuili* (Novas et al. 2005), teeth of the sauropod *"Clasmodosaurus spatula"* (von Huene 1929), and mammals (Martin et al. 2013).

In the locality of Tres Lagos (Figs. 2.12 and 2.17), the fossils are found in estuarine bone-beds. The mammalian remains are to date only mentioned in an abstract and represented by a right lower molar of a "paurodont" dryolestoid, a fragment of a right upper molar of a stem dryolestoid, an incomplete left lower molar of a docodont, and fragmentary teeth attributed to docodonts (Martin et al. 2013). If correct, the "paurodont" molar suggests close affinities to North American and European taxa. The docodont teeth would be the sole record of the group in SA, after the restudy of *Reigitherium* (Rougier et al. 2011b; Harper et al. 2019) vindicated its original attribution to dryolestoids (Bonaparte 1990; Chap. 6). Additionally, if confirmed the

**Fig. 2.17** Tres Lagos, Santa Cruz, Argentina. General view of the outcrops of the Mata Amarilla Formation (**a**). Francisco (Pancho) J. Goin collecting bags of sediments for screen-washing and picking during joint expedition of Museo de La Plata and Universität Bonn, 2007 (**b**)

unpublished specimen from Patagonia would represent the worldwide geologically youngest occurrence of docodonts (Martin et al. 2013).

#### 2.2.3.5 Santo Anastácio, São Paulo State, Brazil; Adamantina Formation, Coniacian–Campanian, Upper Cretaceous

In Brazil, Upper Cretaceous strata crop out extensively. Among them, the well-known Bauru Group is exposed in the southeastern part of the country (e.g., Soares et al. 1980; Menegazzo et al. 2016) and encompasses the most comprehensive fossil record of tetrapods from the Neotropics (Candeiro et al. 2006; Bittencourt and Langer 2011; Martinelli and Teixeira 2015). The Bauru Group has been considered part of the Paraná Basin (e.g., Soares et al. 1980; Milani et al. 1994, 2007) or part of an intracratonic basin, namely the Bauru Basin (e.g., Fernandes and Coimbra 1992, 1996, 2000; Menegazzo et al. 2016). The formal division of the Bauru Group is also problematic, and at the moment, the most widely accepted units are: the Araçatuba (Turonian–Coniacian), Adamantina (=Vale do Rio do Peixe + São José do Rio Preto + Presidente Prudente formations) (Coniacian–Campanian), Uberaba (Coniacian–Campanian), Serra da Galga (Maastrichtian), and Marília (Maastrichtian) formations (e.g., Soares et al. 1980, 2020; Fernandes and Coimbra 2000; Batezelli and Ladeira 2016). In contrast, according to Menegazzo et al. (2016), the Group comprises the Araçatuba, Santo Anastácio, Birigui, São José do Rio Preto, Uberaba, Adamantina, Marília, and Itaqueri formations.

Fossil vertebrates are abundant in the Adamantina Formation (Figs. 2.12 and 2.18) and this unit bears the only unambiguous Cretaceous mammal remains from Brazil. Among the Bauru Group, the Adamantina Formation has the most extensive exposure, covering much of western São Paulo State and Triângulo Mineiro (Minas Gerais State). The Adamantina Formation consists of a succession of fine sandstones intercalated by mudstones and siltstones. It was deposited under distal alluvial fans and alluvial plains subjected to sudden floods in a dry and hot climate, with distinctive rainy and dry seasons (Goldberg and García 2000; Fernandes and Basilici 2009; Carvalho et al. 2010).

Age calibration of the Adamantina Formation is controversial. Fossil charophytes and ostracods suggested Turonian to Santonian age (Dias-Brito et al. 2001), whereas ostracods suggested Campanian–Maastrichtian (Gobbo-Rodrigues et al. 1999). High-precision U/Pb dating for an outcrop assigned to the Adamantina Formation (where the putative mammal *Brasilestes stardusti* was found) in western São Paulo State suggested an upper Coniacian–upper Maastrichtian temporal constraint for the fossil bearings (Castro et al. 2018). However, the age and stratigraphic correlations between the numerous sites of the unit cannot be precisely determined and the synchronicity of the vertebrate association is not necessarily warranted.

The vertebrate fossil content of the Adamantina Formation includes several groups of freshwater fishes (e.g., amiiforms, lepisosteiforms, osteoglossiforms, siluriforms, ceratodontiformes, among other indeterminate actinopterygian and teleostean taxa;

**Fig. 2.18** Outcrops of the Adamantina Formation (Bauru Group); Água Formosa creek, near Marília city, São Paulo, Brazil, where several specimens of the notosuchian *Mariliasuchus amarali* were found (**a**). Anterior part of right dentary with a premolar (URC-M001) of Mammalia *incertae sedis*, in lingual and occlusal views (**b**). Skull of *Adamantinasuchus navae* (holotype, UFRJ-DG 107-R), a notosuchian crocodyliform from the Adamantina Formation (**c**). Skull of *Gondwanasuchus scabrosus* (holotype, UFRJ-DG 408-R), a baurusuchid crocodyliform from the Adamantina Formation, from the same outcrop as *Brasilestes stardusti* (**d**)

e.g., Gayet and Brito 1989; Bertini et al. 1993; Brito et al. 2006, 2017), indeterminate anurans (Carvalho et al. 2003), *Brasiliguana prudensis* and an indeterminate lizard (Candeiro et al. 2009; Nava and Martinelli 2011), the turtles *Bauruemys elegans* (Suárez 1969), *Roxochelys wanderlyi* (Price 1953), *"Podocnemis" brasiliensis* (von Staesche 1937; a possible *nomen dubium* see de la Fuente et al. 2014), *"Podocnemis" harrisi* (Pacheco 1913; also a possible *nomen dubium* see de la Fuente et al. 2014), and *Yuraramirim montealtensis* (Ferreira et al. 2018); and indeterminate anilioid snakes (Zaher et al. 2003; Fachini and Hsiou 2011). Crocodyliforms are highly diverse and extremely abundant, including *Mariliasuchus amarali* (Carvalho and Bertini 1999), *Adamantinasuchus navae* (Nobre and Carvalho 2006) (Fig. 2.18c), *Sphagesaurus huenei* (Price 1950; Pol 2003), *Armadillosuchus arrudai* (Marinho and Carvalho 2009), *Caipirasuchus montealtensis* (Iori et al. 2013), *C. paulistanus* (Iori and Carvalho 2011), *C. stenognathus* (Pol et al. 2014), *C. mineirus* (Martinelli et al. 2018), *Morrinhosuchus luziae* (Iori and Carvalho 2009), *Baurusuchus pachecoi* (Price 1945; Riff and Kellner 2001), *B. salgadoensis* (Carvalho et al. 2005), *B. albertoi* (Nascimento and Zaher 2010), *Stratiotosuchus maxhechti* (Campos et al. 2001), *Gondwanasuchus scabrosus* (Marinho et al. 2013) (Fig. 2.18d), *Campinasuchus dinizi* (Carvalho et al. 2011), *Pissarrachampsa sera* (Montefeltro et al. 2011), *Caryonosuchus pricei* (Kellner et al. 2011), *Montealtosuchus arrudacamposi* (Carvalho et al. 2007), *Pepesuchus deiseae* (Campos et al. 2011), *Barreirosuchus franciscoi* (Iori and García 2012), *Roxochampsa paulistanus* (Pinheiro et al. 2018), among other records. The titanosaur dinosarus are also very abundant, but mostly represented by isolated material. In addition to the species *"Antarctosaurus" brasiliensis* (Arid and Vizzoto 1971; but considered *nomen dubium* see Candeiro et al. 2006), other titanosaurs include *Adamantisaurus mezzalirai* (Santucci and Bertini 2006), *Gondwanatitan faustoi* (Kellner and Azevedo 1999), *"Aeolosaurus" maximus* (Santucci and Arruda-Campos 2011), *Brasilotitan nemophagus* (Machado et al. 2013), and *Maxakalisaurus topai* (Kellner et al. 2006). The theropods are based on isolated material; the record indicates the presence of abelisauroids (e.g., *Thanos simonattoi*), megaraptorans, and few dubious remains of carcharodontosaurids, unenlagiines, and enantiornithine birds (e.g., Bertini 1996; Alvarenga and Nava 2005; Candeiro et al. 2006, 2012; Méndez et al. 2012; Chiappe et al. 2018; Delcourt and Iori 2018).

In addition to the abundant vertebrate record, the Adamantina Formation has provided charophytes, ostracods, bivalves, mollusks, and several ichnofossils, including eggs of turtles, crocodyliforms, and birds (e.g., Azevedo et al. 2000; Dias-Brito et al. 2001; Marsola et al. 2014).

The first mammal remains from the Adamantina Formation was collected after screen-washing in the Santo Anastácio locality, São Paulo State (Bertini et al. 1993). The first osseous Mesozoic mammal from Brazil consists of a fragmentary jaw with a single complete premolar (Fig. 2.18b). The specimen was attributed to Eutheria (Bertini et al. 1993), a viable option, but premolars are rarely diagnostic and non-therian affinities cannot be ruled out (Candeiro et al. 2006; Rougier et al. 2011a). If the original attribution is correct, the mammal from Santo Anastácio may represent the oldest record of a therian mammal in SA.

A possible second mammalian record from General Salgado, São Paulo, also from the Adamantina Formation, includes an isolated large-sized premolar, recently described as *Brasilestes stardusti* (Castro et al. 2018). The specimen was interpreted as a new Tribosphenida (Castro et al. 2018). *Brasilestes* differs from nearly all Mesozoic and Cenozoic compared species, except for superficial resemblances with *Deccanolestes hislopi* from the Maastrichtian of India (Castro et al. 2018). However, the enamel is thin and non-prismatic, the crown and root are almost undifferentiated, without a neck, and the roots are not completely divided. The specimen was found in a locality with a large number of notosuchians, including the hyper-carnivorous baurusuchids and the small-sized herbivorous sphagesaurids (e.g., Carvalho et al. 2005; Godoy et al. 2014; Pol et al. 2014). In our view, the presented evidence is inconclusive to support mammalian affinities of this isolated, bizarre tooth and in fact, we cannot rule out that it may belong to a juvenile of one of these small-sized sphagesaurid notosuchians, with complex tooth morphologies.

#### 2.2.3.6 Los Barreales Lake, Neuquén Province, Argentina; Los Bastos Formation, Coniacian, Upper Cretaceous

The Neuquén Group, Neuquén Basin, northwest Patagonia, Argentina has been intensively studied for more than a century (see Sect. 2.2.3.3); however, the Los Bastos Formation has been only recently recognized as an independent unit (Garrido 2010; Fig. 2.14), previously regarded as part of the Plottier Formation (Danderfer and Vera 1992). The Los Bastos Formation consists of thick beds of red mudstones, with frequent intercalations of yellowish to greenish limestones and fine-grained, well-sorted sandstones, developed in a fluvial environment.

The age of the Los Bastos Formation was estimated by its stratigraphic relationships and correlations and was considered to be lower to middle Coniacian, Upper Cretaceous (Garrido 2010, 2011).

Fossils are scarce in the Los Bastos Formation compared to other units of the Neuquén Group and consist of freshwater mollusks (*Diplodon* sp.), plants, turtle shell fragments, fragmentary dinosaur bones, and a mammal (Garrido 2010; Forasiepi et al. 2012). The mammalian remains consist of an edentulous jaw of perhaps the oldest mesungulatid known (Chap. 6) and it was found at Los Barreales Lake (Figs. 2.12 and

**Fig. 2.19** Edentulous jaw of a Mesungulatidae (MCF-PVPH 412) from Los Barreales Lake, Neuquén, Argentina, Los Bastos Formation; fifth expedition of the Argentina-Canada Dinosaur Project, 2001 (**a**). Bonaparte's expedition to Los Barreales; from left to right: Alberto C. Garrido, Analía M. Forasiepi, José F. Bonaparte, Agustín G. Martinelli, 2002 (**b**)

2.19) in a thin bed of medium- to large-grained sandstones representing an ephemeral fluvial channel.

#### 2.2.3.7 Paso Córdoba, Río Negro Province, Argentina; Anacleto Formation, Campanian, Upper Cretaceous

The Anacleto Formation is the uppermost unit of the Neuquén Group, Neuquén Basin, northwest Patagonia, Argentina (Fig. 2.14). This unit is composed of purple and dark red claystones with intercalation of sandstones and sporadic, small siliceous geodes. The unit represents a low energy fluvial system associated with large flood plains (Hugo and Leanza 2001a, b). The age is considered lower to middle Campanian (Garrido 2010).

Fossil finds are frequent in the Anacleto Formation. They include charophytes, silicified trunks, rhizoliths, and invertebrate traces (Musacchio 1973; Heredia and Salgado 1999; Garrido 2010). Tetrapod ichnites include theropod (Krapovickas and Garrido 2006), perhaps sauropod (Loope et al. 2000), and bird tracks (cf. *Ignotornis* sp. and *Barrosopus slobodai;* Coria et al. 2002b; Calvo 2007).

Among fossil vertebrates, dinosaur remains are abundant: they include the sauropods *Antarctosaurus wichmannianus* (von Huene 1929), *Neuquensaurus australis* (Lydekker 1893; Powell 1986; Salgado et al. 2005), *Laplatasaurus araukanikus* (=*Titanosaurus araukanikus;* von Huene 1929; Gallina and Otero 2015), *Pellegrinisaurus powelli* (Salgado 1996), *Barrosasaurus casamiquelai* (Salgado and Coria 2009), *Pitekunsaurus macayai* (Filippi and Garrido 2008), *Rinconsaurus caudamirus* (Calvo and González Riga 2003), *Narambuenatitan palomoi* (Filippi et al. 2011), *Overosaurus paradasorum* (Coria et al. 2013); the theropods *Abelisaurus comahuensis* (Bonaparte and Novas 1985; Heredia and Salgado 1999), *Aucasaurus garridoi* (Coria et al. 2002a), *Aerosteon riocoloradensis* (Sereno et al. 2008), and the ornithopod *Gasparinisaura cincosaltensis* (Coria and Salgado 1996b; Salgado et al. 1997a). From the locality Auca Mahuevo, north of Neuquén (Fig. 2.20a), several dinosaur nests with *Megaloolithus* eggs have been found, exceptionally preserving embryos and fossilized skins of titanosaurs (Chiappe et al. 1998, 2001; Chiappe and Dingus 2001; Chiappe and Coria 2004; Fig. 2.20b). Auca Mahuevo is one of the most important fossil sites with dinosaur eggs worldwide with exceptional preservation (Lagerstätte). The vertebrate fossil association from the Anacleto Formation also includes the crocodyliform *Pehuenchesuchus enderi* (Turner and Calvo 2005), a possible teiid lizard (Albino 2002), snakes referred to *Dinilysia sp.* (Scanferla and Canale 2007), the long-necked chelid turtle *Yaminuechelys* aff. *Y. maior* (de la Fuente et al. 2015), and an edentulous fragmentary dentary of a dryolestoid (Martinelli and Forasiepi 2004; Rougier et al. 2011a) discovered at Paso Córdoba (Fig. 2.12), originally described as a likely marsupial (Goin et al. 1986).

**Fig. 2.20** Auca Mahuevo, Neuquén, Argentina. Extensive outcrops of the Anacleto Formation (**a**). Sauropod eggs, *Megaloolithus* (**b**). The Lagerstätte from Auca Mahuevo was discovered during 1997–1999 on a joint expedition organized by Luis Chiappe and Rodolfo Coria (pictures from Alberto C. Garrido). Museo Argentino de Ciencias Naturales "Bernardino Rivadavia", hall of dinosaurs, reconstruction of the skull of the theropod *Abelisaurus comahuensis* (**c**)

#### 2.2.3.8 Estancia Los Alamitos, West Slope of Cerro Cuadrado, Río Negro Province, Argentina; Los Alamitos Formation, Campanian–Maastrichtian, Upper Cretaceous

The Los Alamitos Formation is equivalent to the "Senoniano Lacustre" (Wichmann 1924) and it outcrops at northeast Patagonia, Argentina (Figs. 2.12 and 2.14). The sequence is composed of yellow to brownish siltstones, claystones, mudstones, and sandstones, with a minor proportion of conglomerates, breccias, coquines, stromatolites, and some levels with vitric tuffs (Bonaparte et al. 1984; Andreis 1987; Franchi et al. 2001) (Fig. 2.21). It was deposited in a dominant lacustrine environment (Andreis 1987) possibly close to the seashore and under the marine influence (Franchi et al. 2001). The age of the Los Alamitos Formation has been interpreted as Campanian–lower Maastrichtian by its fossil content (Bonaparte et al. 1984).

The lower levels of the Los Alamitos Formation have abundant fossil material. It is possible to correlate with part of the Coli Toro and Allen formations while the upper levels may correlate with the La Colonia, Lefipán, and Jagüel formations (e.g., Page et al. 1999). The fossil record of the Los Alamitos Formation includes a large diversity of palynomorphs: chlorophytes, spores, pollen from conifers and angiosperms (Papú and Sepúlveda 1995), and indeterminate macroscopic plant parts (Bonaparte et al. 1984; Franchi et al. 2001). Invertebrates are frequent and well-preserved, including pelecypods, gastropods (e.g. *Melania*, *Viviparus*, and *Potamolithus*), and ostracods (Bonaparte et al. 1984; Andreis 1987; Forasiepi and Lopez Armengol 1999).

Among vertebrates, fishes are the most abundant, represented by thousands of isolated elements resulting from picking and surface collection. They include batoid chondrichthyans, lepisosteid, siluriform, perciform, and dipnoan Osteichthyes (Cione 1987). Tetrapods are also common, the paleoherpetofauna includes pipid (cf. *Xenopus*) and "leptodactylan" anurans (Báez 1987), cryptodiran and pleurodiran (*Chelus* sp., *Prochelidella* sp., and *Palaeophrynops patagonicus*) turtles (de Broin 1987; Lapparent de Broin and de la Fuente 2001; de la Fuente et al. 2014), madtsoiid snakes (*Alamitophis argentinus, Patagoniophis parvus, Rionegrophis madtsoioides*) (Albino 1987), sphenodontians (*Kawasphenodon expectatus*; Apesteguía 2005), titanosaur sauropods (*Aeolosaurus rionegrinus*) and fragments of eggshells (Powell 1987; Salgado et al. 1997b), indeterminate theropods, and the ornithopod *Secernosaurus koerneri* (=*Kritosaurus australis*; Bonaparte et al. 1984; Bonaparte and Rougier 1987b; Prieto-Marquez and Salinas 2010). Bird remains are poorly preserved and isolated and include *Alamitornis minutes*, cf. *Hesperornithes*, and other indeterminate material (Agnolín and Martinelli 2009).

The first mammal from the Los Alamitos Formation, and probably of South America, was discovered on February 14, 1983 by Marcelo Rougier, a member of J. F. Bonaparte's expedition to the Patagonian Mesozoic (Bonaparte and Soria 1985; Bonaparte and Migale 2010, 2015). The original specimen, a single upper molar (Bonaparte and Soria 1985), was the starting point of what through years of continuous collecting became a very rich association of hundreds of isolated teeth and a few jaw fragments. Bonaparte (1986a, b, c, d, 1990, 1992, 1994, 2002; Bonaparte and Kielan-Jaworowska 1987) recognized the high endemic nature of the association from the site and, by extrapolation, from SA. Bonaparte predicted this fauna to be Gondwanan in nature, a hypothesis yet to be fully confirmed (e.g., Pascual and Ortiz-Jaureguizar 2007) but certainly not presently refuted. Bonaparte and co-authors described a total of 18 mammalian species, including "triconodonts", "symmetrodonts", dryolestoids, ferugliotheriids, and sudamericids (Bonaparte 1986a, b,

**Fig. 2.21** Los Alamitos, Río Negro, Argentina. Bonaparte's expedition at Los Alamitos Formation, in search of small vertebrates and mammal teeth; from left to right: Guillermo W. Rougier, Ubaldo Bonaparte, Diego Rougier, and Igor Gavriloff, 1986 (**a**). Guillermo W. Rougier sieving sediments from the Los Alamitos Formation, 1986 (**b**). Jaime E. Powell transporting materials of *Secernosaurus koerneri* (=*Kritosaurus australis*) protected with plaster; the "Mammal Hill" that provided the first SA Mesozoic mammals is in the background, 1983 (**c**)

c, 1990, 1992, 2002). It is likely, however, that the generic and specific diversity is overestimated (Chaps. 6 and 8) due to the description of different dental positions as separate taxa. We consider that "triconodonts" and "symmetrodonts" are absent from Los Alamitos (Rougier et al. 2011a; Gaetano et al. 2013). Two gondwanatherians are known from Los Alamitos: *Ferugliotherium windhauseni* and

*Gondwanatherium patagonicus*. *Ferugliotherium* was first reported as a possible multituberculate (Bonaparte et al. 1989; Krause and Bonaparte 1993), related to gondwanatherians (Krause and Bonaparte 1993; Pascual et al. 1999; Krause et al. 2014), but allocated to Mammalia *incertae sedis* by some other authors (Pascual et al. 1999; Kielan-Jaworowska et al. 2004). More recently, close relationships with multituberculates have been reasserted (Gurovich and Beck 2009) for *Ferugliotherium, Gondwanatherium,* and allies, an opinion we share (Chap. 8). Moreover, the distant but likely related remains of *Vintana* and *Adalatherium* described from the Cretaceous of Madagascar (Krause et al. 2014, 2020) support allotherian affinities for gondwanatherians, placing them as a sister group of Multituberculata. The uncertainty related to the interrelationships of haramiyidans, gondwanatherians, and multituberculates among themselves and with regard to other basal mammaliaform/mammalian taxa renders the affinities of *Ferugliotherium* and *Gondwanatherium* challenging and inconclusive (Chap. 8).

The most abundant forms in Los Alamitos are meridiolestidans, an endemic group of dryolestoids (Chap. 6). There are four well-defined taxa (*Groebertherium stipanicici, Leonardus cuspidatus, Mesungulatum houssayi,* and *Reigitherium bunodontum*), and no more than five other taxa of questionable taxonomic status (*Casamiquelia rionegrina, Quirogatherium major, Paraungulatum rectangularis, Bondesius ferox,* and *Austrotriconodon mckennai*). These taxa cover a wide array of morphologies and sizes, including minute taxa similar to typical Laurasian dryolestids like *Groebertherium,* and other distinctively endemic forms, such as the large-sized *Mesungulatum* and the small bunodont *Reigitherium* (Rougier et al. 2011a; Harper et al. 2019).

#### 2.2.3.9 Cerro Tortuga, Río Negro Province, Argentina; Allen Formation, Campanian–Maastrichtian, Upper Cretaceous

The Allen Formation is equivalent to the "Senoniano Lacustre" (Wichmann 1924) exposed to north-central Patagonia. The Allen Formation is the basal unit of the Malargüe Group, part of the infill of the Neuquén Basin (Andreis et al. 1974; Fig. 2.14). In the southeast of the basin, the Malargüe Group comprises the Allen, Jagüel, Roca, and El Carrizo formations, while in the northwest the Group is formed by the Loncoche, Jagüel, Roca, and Pircala formations (e.g., Barrio 1990), from base to top, respectively. The Allen Formation corresponds to the first Atlantic transgression ("Kawas Sea"; Casamiquela 1978) into the Neuquén Basin (Wichmann 1927; Uliana and Dellapé 1981). The lower part of the Allen Formation is composed of alternated sandstones and mudstones with small levels of conglomerates, while the upper part is characterized by gray shales covered by limestone and gypsum (Page et al. 1999; Hugo and Leanza 2001a). The environment was first dominated by a fluvial system, becoming subtidal with continental influence at the mid part of the sequence and by a marine regression with the evaporation of shallow lagoons close to the sea, towards the top (Andreis et al. 1974; Uliana and Dellapé 1981; Barrio 1990;

Page et al. 1999). The age of the Allen Formation is upper Campanian—possibly lower Maastrichtian (Hugo and Leanza 2001a).

A very rich fossil association has been recovered from the Allen Formation. This includes megaspores, silicified palms, cycad and conifer trunks, and carbonized plant fragments (Andreis et al. 1991; Ancibor 1995; Del Fueyo 1998; Artabe and Stevenson 1999; Artabe et al. 2004; Ottone 2009). Invertebrates include freshwater mollusks (e.g., *Diplodon* sp., *Biomphalaria* sp., *Paleoanculosa* sp.) (Parras et al. 2004). Vertebrates are represented by chondrichthyans, and siluriform, lepisosteiform, amiid, *cf.* percichthyid, and dipnoan osteichthyes, pipid, and leptodactylid anurans (Martinelli and Forasiepi 2004; Bogan et al. 2009; Bogan and Agnolín 2014). Among non-dinosaur reptiles, there are abundant turtles, both chelids (e.g., *Yaminuechelys gasparinii* and *Prochelidella* sp.) and meiolaniids (de Broin and de la Fuente 1993; Lapparent de Broin and de la Fuente 2001; de la Fuente et al. 2001, 2014), sphenodontians (*Lamarquesaurus cabazai*; Apesteguía and Rougier 2007), plesiosaurs (*Kawanectes lafquenianum*; Gasparini and Goñi 1985), the anilioid *Australophis anilioides* and the madtsoiid snakes *Patagoniophis parvus* and *Alamitophis argentinus* (Martinelli and Forasiepi 2004; Gómez et al. 2008), diverse basal mesoeucrocodylians and neosuchian crocodyliforms (Leanza et al. 2004), and the pterosaur *Aerotitan sudamericanus* (Novas et al. 2012b). The dinosaur record is very abundant and includes the titanosaur *Rocasaurus muniozi* (Salgado and Azpilicueta 2000), *Bonatitan reigi* (Martinelli and Forasiepi 2004), *Panamericansaurus schroederi* (Calvo and Porfiri 2010), *Aeolosaurus rionegrinus* (Salgado and Coria 1993), and *Pellegrinisaurus powelli* (Salgado 1996). Sauropod eggs and fragments of eggshells are particularly abundant at the Bajo de Santa Rosa and Salitral Moreno (Powell 1987; Salgado et al. 2007). Ornithischians are represented by hadrosaurs (e.g., *Willinakaqe salitralensis*; Juárez Valieri et al. 2010; however, possible a *nomen vanum* see Coria 2016; *Lapampasaurus cholinoi*; Coria et al. 2012) and ankylosaur remains (Powell 1987; Salgado and Coria 1996). Theropod dinosaurs are represented by the abelisaurids *Quilmesaurus curriei* (Coria 2001; considered *nomen dubium* by Juárez Valieri et al. 2007) and *Niebla antiqua* (Aranciaga Rolando et al. 2020), the alvarezsaurid *Bonapartenykus ultimus* (Agnolín et al. 2012), the unenlagiid *Austroraptor cabazai* (Novas et al. 2009), teeth of theropods (originally considered as *cf.* carcharodontosaurids by Martinelli and Forasiepi 2004, but possibly belonging to abelisaurs); while birds are known by the carinate *Limenavis patagonica* (Clarke and Chiappe 2001) and the charadriiform *Lamarqueavis australis* (Agnolín 2010).

The mammalian remains, and several of the small vertebrates listed above, come from the Cerro Tortuga, a rich microvertebrate locality (Rougier et al. 2009a; Figs. 2.12 and 2.22). The first known mammals from this locality include the dryolestoids *Groebertherium allensis, Groebertherium stipanicici*, and *Mesungulatum lamarquensis,* and the ferugliotheriid *Trapalcotherium matuastensis* (Rougier et al. 2009a; see Chaps. 6 and 8). High-level mammalian systematics indicates close affinities with taxa from Los Alamitos, but in most cases the species are distinct. The age of the localities of Cerro Tortuga and Los Alamitos have been considered to be close to each other (Martinelli and Forasiepi 2004; Rougier et al. 2009a).

**Fig. 2.22** Cerro Tortuga, Río Negro, Argentina. Rougier's expeditions at outcrops of the Allen Formation, 2008 (**a**). Progressing work at the bone-bed quarry (**b**). Local helper collecting sediments for screen-washing and picking (**c**)

#### 2.2.3.10 Mirasol Chico Canyon and Nearby, Chubut Province, Argentina; La Colonia Formation, Maastrichtian, Upper Cretaceous

The La Colonia Formation is exposed over extensive areas on the Somouncuran Massif, northern central Chubut (Ardolino and Franchi 1996; Pascual et al. 2000; Gandolfo and Cúneo 2005). The unit is mainly composed of a homogeneous sequence of upward-fining sediments integrating sandstones, siltstones, and claystones (Pascual et al. 2000). The sediments of the La Colonia Formation represent different paleoenvironments, including continental to marginal marine, and albupheric/nearshore (Ardolino and Franchi 1996; Pascual et al. 2000; Gandolfo and Cúneo 2005; Cúneo et al. 2014; Gasparini et al. 2015).

The deposition of the La Colonia Formation was traditionally given a Campanian–Maastrichtian age; however, based on the age of the underlying sediments it is more likely its deposition begins in the Maastrichtian extending in some areas into the Cenozoic (Pascual et al. 2000; Guler et al. 2014; see discussion in Gasparini et al. 2015).

Fossils are found on both lips of the Mirasol Chico Canyon, including several localities in the vicinity, such as Bajada Moreno, Buitre Chico Hill, El Uruguayo, Estancia Baibián, Anfiteatro 1, among others (Figs. 2.12 and 2.23). The fossil record includes macro- and micro-paleobotanical fossil material (Gandolfo and Cúneo 2005; Cúneo et al. 2014), indeterminate gastropods, and vertebrates. Well-preserved skeletons are known for the meiolaniform turtle *Patagoniaemys gasparinae* (Sterli and de la Fuente 2011), and particularly the horned theropod dinosaur *Carnotaurus sastrei* (Bonaparte 1985), an exceptional specimen of an almost complete skeleton including the impression of the skin. Mammal-bearing bone-beds have provided thousands of isolated remains dominated by bony fishes, but also including rays, frogs, snakes, turtles, plesiosaurs, theropods, ornithischians, sauropods, birds, and mammals (Rougier et al. 2009b). Pascual et al. (2000) described an isolated jaw of *Reigitherium bunodontum* and new and more complete specimens were recently published from new nearby localities (Harper et al. 2019). Kielan-Jaworowska et al. (2007) named *Argentodites coloniensis* as a possible cimolodontan multituberculate, from the same levels; however the taxonomic validity of the species is questioned (Chap. 8). Later, Rougier et al. (2009b) described a new mesungulatid, *Coloniatherium cilinskii,* represented by jaws, isolated teeth, and a few petrosals. Mammals are relatively abundant but the diversity is low; large-sized mesungulatids form over 90% of the sample in the older localities, likely due to sample bias (e.g., mesh-sizes used during screen-washing). Small dryolestoids, ferugliotheriids, and reigitheriids, however, have been found to be the most abundant mammals in a recently discovered locality (Harper et al. 2019).

Mammals are similar to those from Los Alamitos, but most are distinct at the specific level. Most of the mesungulatids in particular are larger and more robust than those from either Los Alamitos (Los Alamitos Formation) or Cerro Tortuga (Allen Formation). Comparisons of the mammalian fauna from Cerro Tortuga, Los Alamitos, and La Colonia suggest that the latter may be slightly younger than the former two, which would be close in age to each other (Pascual et al. 2000; Martinelli and Forasiepi 2004; Rougier et al. 2009b; Harper et al. 2019).

**Fig. 2.23** Mirasol Chico Canyon, Chubut, Argentina. General view of the outcrops of the La Colonia Formation (**a**). Channel, no more than 50 cm depth, forming a distinct bone-bed, with thousands of bones of microvertebrates and fragments of larger taxa, 2001 (**b–d**). José F. Bonaparte preparing the skull of the horned theropod, *Carnotaurus sastrei* (MACN-CH 894), 1984 (**e**). Pablo Puerta collecting microvertebrates from a bone-bed, 2001 (**f**)

#### 2.2.3.11 Ingeniero Jacobacci, Río Negro Province, Argentina; Coli Toro Formation, Maastrichtian, Upper Cretaceous

Upper Cretaceous fossiliferous outcrops of the Coli Toro Formation are exposed in the vicinities of Ingeniero Jacobacci, north Patagonia, Argentina (Bertels 1969; Volkheimer 1973; Casamiquela 1964b, 1978; Gasparini and Spalletti 1990; Page et al. 1999) (Fig. 2.12). The Coli Toro Formation regarded as Maastrichtian in age (Novas 2009) is represented by sandstones and mudstones deposited in marine, mix, and continental environments, including the presence of small freshwater bodies (Page et al. 1999) (Fig. 2.24).

From marine deposits, the fossils are represented by bivalves, gastropods, ammonites including *Eubaculites*, and plesiosaurs (e.g., *Sulcusuchus erraini*) (Bertels 1969; Riccardi 1974, 1980; Gasparini and Spalletti 1990; Rodríguez et al. 1995; Gasparini and de la Fuente 2000). A mix of environments has contributed algae, angiosperms including silicified trunks, mollusks, the lungfish *Ceratodus* sp. and other fishes, crocodyliforms, turtles, hadrosaur dinosaurs, and a potential mammal (Casamiquela 1964b, 1969a, b, 1978; Volkheimer 1973; Pascual and Bondesio 1976; Coira 1979; Pöthe de Baldis 1984; Novas 2009).

The mammalian remains consist of a fragmentary edentulous lower jaw collected from the Cerro Yeso (in the Estancia Puesto Marileo; Casamiquela 1969a, 1978) and have been more recently referred to as an indeterminate Mesungulatidae, possible a new taxon (Chornogubsky and Gelfo 2011). However, it is worth mentioning that in the area of Cerro Yeso, the Coli Toro Formation is unconformably covered by layers bearing Eocene mammals. Originally, Casamiquela (1978) questioned the

**Fig. 2.24** Cerro Yeso, in the vicinities of Ingeniero Jacobacci, Río Negro, Argentina. General view of the outcrops of the Coli Toro Formation

stratigraphic origin of the specimen, not excluding the possibility that the jaw was transported from the younger levels and perhaps so belongs to some of the endemic SA mammals (Pascual in Casamiquela 1978). A visit to the locality in the company of a helper of Casamiquela, with knowledge of the precise place of collection, confirms that contamination is possible but there are *bona fide* Mesozoic remains in place from the Coli Toro Formation.

#### 2.2.3.12 Río de Las Chinas Valley, Última Esperanza Province, Chile; Dorotea Formation/Alta Vista and La Anita Farms, Santa Cruz Province, Argentina; Chorrillo Formation. Campanian–Maastrichtian, Upper Cretaceous

The Dorotea and Chorrillo formations are units exposed in south Patagonia divided by the geopolitical border between Chile and Argentina. The Dorotea Formation is exposed in Chile and it is equivalent to the La Irene, Chorrillo, Calafate, and Cerro Cazador formations on the Argentinian side (Nullo et al. 2006; Manríquez et al. 2019). Both the Dorotea and Chorrillo formations are part of the Magallanes/Austral Basin, which includes a succession of siliciclastic rocks ranging from the Upper Jurassic to Cenozoic (e.g., Biddle et al. 1986; Arbe 2002; Nullo et al. 2006; Cuitiño et al. 2019) (Fig. 2.12).

On the Chilean side, the Dorotea Formation (Macellari et al. 1989; Vogt et al. 2014; Manríquez et al. 2019) is exposed in the Río de Las Chinas valley (Fig. 2.25), Estancia Cerro Guido, about ~100 km north from Puerto Natales, Última Esperanza Province. In this area, the Dorotea Formation lays over the Tres Pasos Formation and is covered by the Man Aike Formation (Manríquez et al. 2019). The Dorotea Formation comprises 900–1200 m thick deposits of pale sandstones, with layers of conglomerate, siltstones, sandy calcareous concretions, and mudstones (Katz 1963, Manríquez et al. 2019). The unit was deposited in a transitional shallow marine to fluvio-deltaic environment, and also includes intercalation of fluvial and flood-plain systems (Manríquez et al. 2019; Rivera et al. 2020). Two U/Pb datings at two different levels of the Dorotea Formation provided 71.7 ± 1.2 Ma and 74.9 ± 2.1 Ma, suggesting an upper Campanian to lower Maastrichtian time span for the unit (Gutiérrez et al. 2017). Several fossiliferous levels have been found along the stratigraphic profile in the context of a long-term project (starting in 2013) in the Chilean Patagonia by the Chilean Antarctic Institute (Instituto Antártico Chileno) and the University of Chile led by Marcelo Leppe and Alexander O. Vargas. The fossils include records of plants (pteridophytes, gymnosperms, angiosperms), invertebrates (bivalves, gastropods, ammonites), and several groups of vertebrates, such as chondrichthyans, anurans, plesiosaurs, mosasaurs, turtles, dinosaurs, and mammals (e.g., Manríquez et al. 2019; Alarcón-Muñoz et al. 2020; Trevisan et al. 2020; Goin et al. 2020). The mammalian remains from this unit consist of isolated teeth which were the basis to describe *Magallanodon baikashkenke*, the southernmost gondwanatherian for the Mesozoic (Goin et al. 2020; see Chap. 8).

**Fig. 2.25** Río de Las Chinas Valley, Última Esperanza Province, Chile. General view of the outcrops of the Dorotea Formation (right side of the picture), 2020 (**a**). The Chilean team, Roy Fernández (left) and Pedro Vargas (right), looking for small Cretaceous mammals (**b**), 2017 (picture from Felipe Suazo Lara)

On the Argentinean side, the Chorrillo Formation crops out along a narrow belt that starts south to the margin of the Argentino Lake, southwest Santa Cruz (Patagonia, Argentina). The Chorrillo Formation lays over the Alta Vista and La Anita formations and it is covered by the Calafate Formation (Nullo et al. 2006). The whole lithostratigraphic succession represents a regressive episode, with deltaic deposition at the La Anita Formation, braided and meandering fluvial system at the La Irene and Chorrillo formations, and a new marine ingression at the Calafate Formation (Arbe and Hechem 1984; Macellari et al. 1989; Nullo et al. 2006; Tettamanti et al. 2018).

First mentions of fossils from the Chorrillo Formation were reported by the Italian geologist Egidio Feruglio, who described logs and dinosaurs bones (Feruglio 1938, 1944). Much later, in 1980 and during geological explorations in the area of Alta Vista and La Anita farms, the Argentine geologist Francisco Nullo discovered several bones of large dinosaurs and notified José F. Bonaparte. Bonaparte and his crew visited the place in February of 1981 and superficially excavated part of a titanosaur sauropod (Fig. 2.26). A large titanosaur cervical vertebra and a few isolated remains of other dinosaurs (e.g., theropod tooth) were deposited at the MACN-Pv collection, representing the initial work in a remarkable locality (Bonaparte 1996b; Powell 2003). In 2019, a field trip exploration project organized by Fernando Novas (MACN) resulted in the discovery of taxonomically varied faunal and floral associations (Novas et al. 2019). Fossil plants include wood of possible podocarpaceans, palynomorphs, including spores of ferns (e.g., Gleicheniaceae, Osmundaceae, Polypodiaceae, Dicksoniaceae) and pollen grains of gymnosperms (e.g., Podocarpaceae) and angiosperms. Invertebrates include freshwater (ampullariids, pleurocerids, tateids, and physids) and terrestrial (holospirids, bulimulids, and achatinids) gastropods and a few marine bivalves. Among vertebrates, there are amiid and teleost fishes, calyptocephalellid anurans, chelid turtles, snakes (cf. *Rionegrophis madtsoioides* and indeterminate specimens), mosasaurids, the ornithopod *Isasicursor santacruzensis*, the titanosaur *Nullotitan glaciaris* (in part based on the bones originally found by F Nullo and collected by JF Bonaparte in 1980–1981; Fig. 2.26), abelisauroids, megaraptorans, and unenlagiid theropods, the bird *Kookne yeutensis*, dinosaur eggshells, and scarce mammalian remains (Novas et al. 2019).

The mammal specimens include one complete and one partial caudal vertebra (Novas et al. 2019: Fig. 2.31). The taxonomy of the specimens remains indeterminate; because of size, they probably belonged to a small-sized mammal with a long tail (Novas et al. 2019). An isolated tooth of *Magallanodon baikashkenke* has been recently recognized also for the Chorrillo Formation (Chimento et al. 2020).

#### 2.2.3.13 Pajcha Pata, Cochabamba Department, Bolivia; El Molino Formation, Maastrichtian, Upper Cretaceous

The El Molino Formation is exposed in west Central Bolivia (Fig. 2.12). The locality of Pajcha Pata, first discovered in 1989 by Marshall and Sempere, has provided a very rich Late Cretaceous fossil association from the Neotropics (e.g., Gayet et al. 1991, 2001). The El Molino Formation is represented by mudstones, marls, limestones,

**Fig. 2.26** Alta Vista and La Anita farms, Santa Cruz, Argentina. General view of the outcrops of the Chorillo Formation, 1981 (**a**). José F. Bonaparte at the La Anita farm preparing for packing one titanosaur vertebra (MACN-PV 18644), 1981 (**b**). A complete humerus of a titanosaur left in the field (**c**), which was years later collected by Fernando E. Novas. Both titanosaur remains belong to the holotype of *Nullotitan glaciaris*

fine to medium-grained sandstones, and occasionally, evaporites and dolomites with algal laminations (Gayet et al. 1991). The depositional environment was probably estuarine-lagoonal. The age of the El Molino Formation was calibrated by geochronologic data and estimated between 73–60 Ma (Maastrichtian–earliest Paleocene); however, terrestrial fossils at Pajcha Pata have been suggested as Maastrichtian in age (about 68 Ma) (Gayet et al. 2001).

The fossil association from the El Molino Formation is very rich and includes plants (Charophyta), invertebrates (Gastropoda, Bryozoa), and several groups of marine, freshwater, and terrestrial vertebrates. The fish record is abundant and includes Chondrichthyes (Rajiformes and Myliobatiformes) and Osteichthyes (Polypteriformes, Lepisosteiformes, Semionotiformes, Cupleiformes, Pycnodontiformes, Osteoglossiformes, Characiformes, Siluriformes, Aulopiformes, Perciformes, Tetraodontiformes, Salmoniformes, Cypriniformes, Cyprinodontiformes, and Dipnoi) (Gayet et al. 1991, 2001). Tetrapods are represented by anurans (Gymnophiona indet. and Noterpetontidae: *Noterpeton bolivianum*), turtles (Podocnemididae), snakes (Madtsoiidae), crocodiles (Eusuchia: *Dolichochampsa minima*), dinosaurs (indeterminate Theropoda and Sauropoda), and mammals (Gayet et al. 1991, 2001). The El Molino Formation has also provided one of the most extensive dinosaur footprint assemblages in the world (Fig. 2.27d, e). The Cal Orcko fossil site, at 4.5 km from Sucre (Chuquisaca Department, Bolivia), consists of a succession of fossiliferous oolitic limestones, associated with freshwater stromatolites and at least nine levels of dinosaur tracks. It represented an open lacustrine environment, which records more than 5000 individual dinosaur footprints of at least eight species (including sauropods, theropods, ornithopods, and ankylosaurians) (e.g., Meyer et al. 2001, 2018).

The mammalian remains from the El Molino Formation are isolated teeth and are yet to be fully described (Gayet et al. 2001). The authors mentioned for one tooth that "the overall molar morphology is consistent with a tribosphenic mammal, while dryolestoid affinities are suggested" (Gayet et al. 2001: 58). The other tooth was referred to as a possibly P4 of a eutherian (Gayet et al. 2001: Fig. 2.16a–b); however the latter is very unlikely, and possibly all the materials are dryolestoids (Rougier et al. 2011a; further comments in Chap. 6).

#### 2.2.3.14 Synclinal de Bagua, Peru; Fundo El Triunfo Formation, Bagua Group, upper Campanian–Maastrichtian, Upper Cretaceous

The Fundo El Triunfo Formation is the lowest unit of the Bagua Group, followed by the Rentema and Sambimera formations, following Naeser et al. (1991; however, other authors prefer the name of the Chota Formation for the same deposits; Chacaltana et al. 2015). The deposits are exposed in the Bagua synclinal, in the northern Andes of Peru (Fig. 2.12). The Fundo El Triunfo Formation/Chota Formation consists of a transitional sequence from marine to continental conditions, including marly and sandy, and occasionally microconglomeratic red-beds with fossils (Mourier et al. 1986, 1988; Naeser et al. 1991). The fossil content includes isolated remains of chondrichthyans, turtles, titanosaur and theropod dinosaurs, eggshells, and other indeterminate remains (Mourier et al. 1986, 1988), in addition to ostracods, charophytes, and wood fragments (Mourier et al. 1988). Screen-washing activities resulted in the discovery of a partial tooth in the locality "El Pintor", 15 m above the contact between the Celendin (older) and the Fundo El Triunfo/Chota formations. The tooth was described as a molar of a therian mammal (Mourier et al. 1986). Later, Crochet and Sigé (in Mourier et al. 1988), based on the large size of the specimen, suggested a "therian nature of the specimen (placental or more likely marsupial)". However, the fragmentary condition of the material precludes a clear taxonomic determination. In addition, the complex stratigraphy of the region includes units bearing Cenozoic fossil mammals (Naeser et al. 1991), and does not exclude a potential contamination with fossil material from different levels and ages.

**Fig. 2.27** Pajcha Pata, Cochabamba, Bolivia. General view of the outcrops of the El Molino Formation (**a**). Rougier's expedition at Pajcha Pata, looking for microvertebrates in the outcrop (**b**) and processing sediments for screen-washing; from left to right: Andres Giallombardo, local student, and Federico Anaya, 2011 (**c**). Cal Orko ("lime hill", Quechua language), vicinities of Sucre, Chuquisaca. Limestone quarry from the El Molino Formation with diverse dinosaur footprints, 2007 (**d**–**e**) (pictures from Jorge L. Blanco)

In addition to the material published by Mourier et al. (1986, 1988), Vildoso-Morales (1991) presented a preliminary report of Cretaceous mammals from similar deposits (originally as the El Triunfo Member of the Bagua Formation—Vildoso-Morales 1991—now considered Group and Formation, respectively, following Naeser et al. 1991 or the Chota Formation, following Chacaltana et al. 2015). However, the presentation was not substantiated by pictures or specimens. To date, there is no a formal publication in support of the communication, and according to Pascual (1992), the veracity of the account is suspect.

### 2.2.4 *Cenozoic*

The Cenozoic SA mammalian fossil record is extremely rich and varied. We focus here only on those localities documenting archaic lineages recorded in the Mesozoic and surviving as minority members of the therian-dominated Cenozoic fauna.

#### 2.2.4.1 Punta Peligro, Chubut Province, Argentina; Salamanca Formation, Peligran SALMA, Danian, lower Paleocene

The Salamanca Formation is exposed at eastern central Patagonia (Figs. 2.28 and 2.29) and represents the Atlantic marine transgression ("Salamanca Sea") that partially covered Patagonia during the Late Cretaceous–early Paleocene (Feruglio 1949; Andreis et al. 1975; Uliana and Biddle 1988). The Salamanca Formation is divided into the lower Bustamante Member, largely calcareous and deposited in a neritic environment, and the upper Hansen Member deposited in the coastal zone. One of the upper levels of the Hansen Member, the "Banco Negro Inferior" (BNI) or Lower Black Bank (sensu Feruglio 1949), represents the regression of the sea and provided a rich fossil association, one of the earliest from the Cenozoic of Patagonia (Bonaparte et al. 1993; Bond et al. 1995; Pascual et al. 1996; Ortiz-Jaureguizar 1996). The BNI is overlaid by the "Banco Verde" or Green Bank (sensu Feruglio 1949), of glauconitic sands. The Salamanca Formation is mainly composed of fine clastic sediments such as mudstones, claystones, and sandstones. Pelecypod, brachiopod, and echinoid remains occur throughout the sequence. Calcareous concretions, gypsum veins, as well as plant remains, bioturbation, and trails are also common (Andreis et al. 1975). Invertebrates and microplankton suggested that the Salamanca Formation was accumulated in a stable environment of a shallow marine basin under the estuarine influence (Lesta and Ferello 1972; Comer et al. 2015).

The BNI ranges from about 63.2–63.8 Ma (youngest to the east; Woodburne et al. 2014b). The Peligran SALMA, defined by the mammalian association from the BNI, is contained within this interval (Woodburne et al. 2014b; see also Clyde et al. 2014).

Fossils from the Salamanca Formation come mostly from the BNI and include silicified tree trunks, leaves, pollen (Romero 1968; Petriella 1972; Archangelsky 1973, 1976a, b; Archangelsky and Romero 1974; Petriella and Archangelsky 1975; Archangelsky and Zamaloa, 1986; Brea et al. 2005, 2008; Iglesias et al. 2007b;

**Fig. 2.28** Fossiliferous localities from the Cenozoic of South America. Abbreviations (as in Fig. 2.2): 25, Punta Peligro, Chubut Argentina; 26, Seymour Locality IAA 90/1, Seymour Island, Antarctic Peninsula; 27, La Barda, Chubut, Argentina; 28, Santa Rosa, Ucayali, Peru; 29, Contamana, Loreto, Peru; 30, Gran Barranca and 31, Gaiman, Chubut, Argentina; 32, Monte Observación; 33, La Cueva; 34, Monte León; 35, Estancia La Costa; 36, Killik Aike Norte, and 37, 8 km south of Coy Inlet, Santa Cruz, Argentina

**Fig. 2.29** Punta Peligro, Chubut, Argentina. General view of the outcrops of the Salamanca Formation (**a**). Rougier's expedition in search of small bones, 2002 (**b**). Outcrops exposed at the side of the Patagonian Atlantic coast with the Salamanca Hill at the back, 2009 (**c**)

Comer et al. 2015) and the earliest SA evidence of flowering plants (Jud et al. 2018). Most of the abundant vertebrates have been collected by a combination of surface prospecting and screen-washing, and include calyptocephalellid frogs (*Gigantobatrachus casamiquelai* and *Calyptocephalella sabrosa*; Bonaparte et al. 1993; Agnolín 2012; Muzzopappa and Báez 2013; Muzzopappa et al. 2020), the pleurodiran turtles *Yaminuechelys maior* and *Salamanchelys palaeocenica,* and the meiolaniid turtle *Peligrochelys walshae* (Bona 2006; Bona and de la Fuente 2005; Sterli and de la Fuente 2013), the sphenodontian *Kawasphenodon peligrensis* (Apesteguía et al. 2014), the crocodylians *Necrosuchus ionensis, Eocaiman palaeocenicus*, and *Protocaiman peligrensis* (Simpson 1937; Bona 2007; Bona et al. 2018), an avian pellet (Muzzopappa et al. 2020), and a peculiar mammalian association represented by a mixture of therians derived from Laurasian immigrants (Metatheria and Eutheria) together with Gondwanan relicts (Monotremata, Gondwanatheria, and Meridiolestida) (e.g., Bonaparte et al. 1993; Gelfo et al. 2007; Woodburne et al. 2014a, b). Metatherians are known by at least six taxa: two species of Polydolopimorphia (Bonapartheriidae indet. and a generalized Polydolopimorphia), an isolated astragalus probably belonging to a medium-sized sparassodont, at least two "Didelphimorphia" (*Derorhynchus* aff. *D. minutus* and a new species of *Didelphopsis*), and an edentulous fragment of a dentary of an opossum-like taxon, slightly larger than *Derorhynchus* aff. *D. minutus* (Bond et al. 1995; Goin et al. 2002, 2004b; Forasiepi and Rougier 2009). Eutherians are represented by four South American Native Ungulates (SANUs): three Didolodontidae (*Escribania talonicuspis, E. chubutensis,* and *Raulvaccia peligrensis*) and one putative litoptern (the Notonychopidae *Requisia vidmari*) (Bonaparte et al. 1993; Bonaparte and Morales 1997; Gelfo 2007; Gelfo et al. 2007, 2009). The Gondwanan relicts are the only extra-Australian monotreme *Monotrematum sudamericanum* (Pascual et al. 1992; Forasiepi and Martinelli 2003; see Chap. 4), the gondwanatherian *Sudamerica ameghinoi* (Scillato-Yané and Pascual 1985; Gurovich and Beck 2009; see Chap. 8), and the dryolestoid *Peligrotherium tropicalis* (Bonaparte et al. 1993; Gelfo and Pascual 2001; Paéz Arango 2008; see Chap. 6).

#### 2.2.4.2 Seymour Locality IAA 90/1, Seymour Island, Antarctica Peninsula, La Meseta Formation, *Cucullaea* I Fauna, Ypresian, lower Eocene

Paleogene beds of the James Ross Basin are exposed on the Marambio (Seymour) Island of the Antarctic Peninsula (Figs. 2.28 and 2.30). The deposits are mainly marine; however, at the northwest part of the island, terrestrial fossil vertebrates were recovered from the La Meseta (lower Eocene) and the overlaying Submeseta (upper Eocene) Formations (Marenssi et al. 1998; Reguero et al. 2013; Reguero 2016).

The La Meseta Formation includes fine-grained siliciclastic material deposited in deltaic, estuarine, and shallow marine environments (Marenssi et al. 1998). A radiometric dating and calibration of the deposits have suggested between 51 and 49 Ma (Ypresian, Eocene) for the *Cucullaea* I Allomember of the La Meseta Formation (Ivany et al. 2008; Reguero et al. 2013), which contains the mammal-bearing horizon.

Fossil plants from the La Meseta Formation include leaves and tree trunks of Nothofagacea, Dilleniaceae, Myricaceae, Myrtaceae, Lauraceae, and a Grossulariaceae flower, suggesting a mixed mesophytic forest developed under a cool and seasonal climate (Gandolfo et al. 1998a, b). The unit has also provided dung beetle

**Fig. 2.30** Seymour Island (Marambio), Antarctica Peninsula (**a**). Ross D. E. MacPhee searching for fossil mammals in the white continent, American Museum of Natural History expedition, 2015 (**b**) (pictures from Ross D. E. MacPhee); Marcelo Reguero, in the marine deposits from Submeseta Formation, stratigraphically just above the level where the only Antarctic dryolestoid has been found; Instituto Antártico Argentino expedition, 2012 (picture from Marcelo Reguero) (**c**)

brood balls (Laza and Reguero 1993) and a diverse marine invertebrate and vertebrate association (Stilwell and Zinsmeister 1992; Torres et al. 1994; Reguero et al. 2013; Buono et al. 2016). Recently, the first neobatrachian remains were reported for this unit, referred to *Calyptocephalella* sp. (Mörs et al. 2020). Mammals in the *Cucullaea* I Allomember of the La Meseta Formation are represented by metatherians ("Didelphimorphia": *Pauladelphys juanjoi, Derorhynchus minutus,* and Derorhynchidae gen et sp. indet.; Microbiotheria: *Marambiotherium glacialis, Woodburnodon casei*; Polydolopimorphia: *Perrodelphys coquinense, Antarctodolops dailyi*, and *A. mesetaense*, and the indeterminate marsupial *Xenostylos peninsularis*; Woodburne and Zinsmeister 1984; Goin et al. 1999, 2007a; Chornogubsky et al. 2009), a claw of a putative xenarthran, perhaps a sloth (Pilosa) or anteater (Vermilingua) (Reguero et al. 2013; Reguero 2016), the sparnotheriodontid litoptern *Notiolofos arquinotiensis* (Bond et al. 2006), the astrapothere *Antarctodon sobrali* (Bond et al. 2011), and two non-therian taxa: gondwanatherians and a dryolestoid. Gondwanatherians are represented by a rodent-like fragment of dentary with an incisor similar to *Sudamerica ameghinoi* (Goin et al. 2006b). Dryolestoids are known by a single isolated tooth. The material was originally described as a bat or "insectivore" (Goin and Reguero 1993; MacPhee et al. 2008; Reguero et al. 2013), and more recently it was identified as a Dryolestoidea (Martinelli et al. 2014). If this taxonomic interpretation is correct, Antarctica would show the late survival of archaic lineages associated with therians, similar to the better documented Cenozoic of Patagonia (Martinelli et al. 2014).

#### 2.2.4.3 La Barda, Chubut Province, Argentina; Middle Chubut River Volcanic–Pyroclastic Complex, Paso Del Sapo Fauna, Ypresian–Lutetian, lower–middle Eocene

The localities of La Barda and Laguna Fría, in the vicinities of Paso del Sapo, northwestern Patagonia, Argentina (Fig. 2.28) expose a variety of volcaniclastic, intrusive, pyroclastic, and extrusive rocks identified as the Middle Chubut River Volcanic-Pyroclastic complex (Aragón and Mazzoni 1997; Tejedor et al. 2009). A few years ago a diverse assemblage of extinct mammals of early–middle Eocene age (~49.5 Ma; older than 47 and younger than 52 Ma, about the Ypresian/Lutetian boundary) has been recovered from these localities. The association suggested a distinct biochronological unit, the "Sapoan", between the Riochican SALMA and the Vacan subage of the Casamayoran SALMA (Tejedor et al. 2009).

The mammalian association from these localities comprises hundreds of specimens collected by surface picking and screen-washing. This includes about 50 species of mammals, including metatherians, eutherians (dasypodids, bats, "condylarths", and SANUs, such as notoungulates, astrapothers, and litopterns; Tejedor et al. 2009; Lorente et al. 2016), as well as possible gondwanatherians (Goin et al. 2012). The latter group is represented by *Greniodon sylvaticus*, which is based on two isolated molariforms, putatively similar in occlusal dental morphology and histology to ferugliotheriids.

Derorhynchidae *Pauladelphys*, Sparnotheriodontidae-like litoptern, and gondwanatherians are shared (as a major taxonomic group) by the Paso del Sapo association and *Cucullaea* I Allomember from the La Meseta Formation. This correspondence was used to correlate the associations between Patagonia and Antarctica (Tejedor et al. 2009; Goin et al. 2012), introducing the idea that the Paso del

Sapo mammalian faunas represent a continental extension of the Weddellian Biogeographic Province (Tejedor et al. 2009), which by the Late Cretaceous–Paleogene included Antarctica (mainly Western Antarctica), New Zealand, and southeastern Australia (Quattrocchio and Volkheimer 2000).

#### 2.2.4.4 Santa Rosa, Ucayali Department, Peru; Yahuarango Formation; Santa Rosa Local Fauna, middle–upper Eocene, or lower Oligocene

The discovery of the Paleogene association at the Santa Rosa, Peru (Campbell 2004; Fig. 2.28), provided the first insights into a rich vertebrate assemblage for the tropical lowlands of the Amazon Basin, later accompanied by the findings at Contamana (Antoine et al. 2012, 2016). The fossils at Santa Rosa occur in a conglomeratic matrix within massive red clay and sandstones in fluvial beds interpreted as the Yahuarango Formation (Campbell 2004; Bond et al. 2015). The precise age of the unit and fossil-bearing levels remains unclear. According to Bond et al. (2015), it may be upper middle Eocene or upper Eocene; but other authors suggested lower Oligocene or younger (Shockey et al. 2004; Antoine et al. 2012).

The fossil mammals have been described in detail. The fauna includes marsupials and relatives (didelphimorphs, sparassodonts, polydolopimorphians, paucituberculatans, and microbiotherians; Goin and Candela 2004), notoungulates (Shockey et al. 2004), cingulate xenarthrans (Ciancio et al. 2013), caviomorph rodents (Frailey and Campbell 2004), *Perupithecus ucayaliensis,* the earliest and basalmost SA primate (Bond et al. 2015), and bats (Czaplewski and Campbell 2004). In addition, there is a single multicuspidate molariform, chemically degraded, claimed to be a ferugliotheriid (Goin et al. 2004a). If the remain indeed represents a ferugliotheriid, the intriguing specimen would record the persistence of Gondwanatheria in the Neotropics during the Paleogene (Goin et al. 2004a).

#### 2.2.4.5 Contamana, Loreto Department, Peru; Pozo Formation, Coincident with Barrancan Subage of Casamayoran SALMA, Lutetian–Bartonian, middle–upper Eocene

Fossil finding in the Neotropics has multiple difficulties, with dense vegetation, intense surface weathering, and seasonal river flooding (Antoine et al. 2016). The Contamana region, in the Peruvian Amazonia (Fig. 2.28), exposes a fossiliferous Paleogene to Neogene sequence. Among them, the fossiliferous Lower Member of the Pozo Formation is represented by thin levels of siltstone and sandstone of fluvial origin, alternating with thick intervals of paleosols and deposits of a distal floodplain. There is also evidence of levels with marine and/or tidal influence (Antoine et al. 2016). The exposed sequence is upper middle Eocene, around the Lutetian/Bartonian boundary (~41 Ma), with its top likely extending into the upper Eocene (Antoine et al. 2016) and equivalent of the Barrancan subage of the Casamayoran SALMA (Woodburne et al. 2014a, b; Antoine et al. 2016).

The Lower Member of the Pozo Formation has provided a rich fossil association including charophytes, silicified wood, seeds, charcoal, and palynomorphs. The palynoflora indicates the presence of a rainforest, including Fabaceae, Sapotaceae, Malpighiaceae, among others (Antoine et al. 2016). Invertebrates include gastropods and bivalve mollusks, ostracods, and crabs (Yeo et al. 2014). Vertebrates are represented by freshwater stingrays (Potamotrygonidae; Adnet et al. 2014), characiforms, pipid and "leptodactyloid" anurans, anilioid and boid snakes, a lizard (possibly a platynotan), chelonians, crocodyliforms (cf. *Sebecus* sp., gavialoids, and caimanines), a large collection of metatherians (didelphimorphs, sparassodonts, polydolopimorphians, and palaeothentoid paucituberculatans), dasypodids, SANUs (notoungulates, an astrapothere, a litoptern, and a pyrothere), the earliest SA hystricognath rodents (stem Caviomorpha), bats, and one possible gondwanatherian (Antoine et al. 2012). Gondwanatherians are putatively represented by isolated and undescribed teeth (Antoine et al. 2012; Electronic supplementary material). Among them, one element previously reported as a likely gondwanatherian has been re-interpreted as an upper premolar of a polydolopimorphian marsupial; however, two other fragmentary teeth awaiting publication may still represent gondwanatherians, according to Antoine et al. (2016).

#### 2.2.4.6 Gran Barranca and Gaiman, Chubut Province, Argentina; Sarmiento Formation, Colhuehuapian SALMA, Burdigalian, lower Miocene

The Sarmiento Formation is exposed extensively in central Patagonia, Argentina (Fig. 2.28). This unit comprises a thick pyroclastic sequence derived from the rising of the Andes, accumulated on loessic and fluvial plains, and subordinate shallow lakes (Bellosi 2010). The Sarmiento Formation ranges from the middle Eocene (Lutetian, ~ 45 Ma; Vacan subage of the Casamayoran SALMA) to the lower Miocene (Burdigalian, 16.5 Ma; "Pinturan" SALMA). In particular, the locality of Gran Barranca has been intensively studied (e.g., Kay et al. 1999; Madden et al. 2010) providing rich vertebrate associations and a calibrated sequence of the successional faunas from the early Cenozoic in Patagonia.

The Colhue Huapi Member of the Sarmiento Formation (Spalletti and Mazzoni 1979) at Bran Barranca is composed of pyroclastic mudstones, paleosols, and some levels of intraformational conglomerates (Bellosi 2010). These deposits have provided a rich mammal association that is the base for the Colhuehuapian SALMA, with fossil-bearing beds dated as 20.2 to 20 Ma (Ré et al. 2010; Woorburne et al. 2014a).

The Trelew Member of the Sarmiento Formation is exposed along the lower Río Chubut Valley near the village of Gaiman; it is composed mainly of tuffaceous sandstones and mudstones, and dated to about 20 Ma (Kay et al. 2008). Because of similarities in the taxonomic composition with those of the Colhue Huapi Member, both units have been correlated and assigned to the Cohuehuapian SALMA (e.g., Kay et al. 2008; Vucetich et al. 2010; Arnal et al. 2014).

Colhuehuapian floras at Patagonia were characterized by expansion of arid vegetation, with Leguminosae, Calyceraceae, Anacardiaceae, Gramineae, Ephedraceae, and Chenopodiaceae, while forested habitats were restricted to the banks of rivers and streams (Barreda and Palazzesi 2010). Colhuehuapian vertebrates include a long list of typical mammals from the SA Cenozoic associations, including marsupials and close relatives (didelphimorphs, sparassodonts, microbiotherians, and paucituberculatans; Goin et al. 2007b; Goin and Abello 2013), xenarthrans (dasypodids, glyptodontids, and pilosans; Scillato-Yané 1986), SANUs (notoungulates, astrapotheres, and litopterns) (Bond 1986; Croft et al. 2020), caviomorph rodents (Vucetich et al. 2010), primates (MacFadden 1990; Kay 2010), bats (Czaplewski 2010), and one dryolestoid meridiolestidan taxon. This latter taxon is *Necrolestes mirabilis*, known by a fragmentary dentary with a deciduous and replacing permanent tooth (Goin et al. 2007b), providing important data in tooth replacement. Cautiously considered Mammalia *incertae sedis* in the original description (Goin et al. 2007b), *N. mirabilis* was later assigned to Dryolestoidea after a description of the closely related Cretaceous meridiolestidan *Cronopio dentiacutus* (Rougier et al. 2011b) and detailed restudy of the much more complete sister species *Necrolestes patagonensis* (Rougier et al. 2012; Wible and Rougier 2017; see Chap. 6). In addition, from outcrops close to the village of Gaiman, at the south portion of the Chubut River valley, the peculiar mammal *Patagonia peregrina* has been collected. First identified as a rodent-like metatherian (Pascual and Carlini 1987), *P. peregrina* was later considered to be a gondwanatherian (Chimento et al. 2015). The fragmentary nature of the referred specimens and the apomorphic morphology of the species render taxonomic affinities ambiguous (Chap. 8).

#### 2.2.4.7 Monte Observación, La Cueva, Monte León, Estancia La Costa, Killik Aike Norte, 8 km South of Coy Inlet, Santa Cruz Province, Argentina; Santa Cruz Formation, Santacrucian SALMA, Burdigalian, lower Miocene

The Santa Cruz Formation in southern Patagonia, Argentina, is the richest fossiliferous unit of the Cenozoic of SA (e.g., Ameghino 1889; Hatcher 1903; Scott 1913; Simpson 1980; Vizcaíno et al. 2012a). The quantity and quality of fossils include many complete skeletons and the mammalian association from the unit is the base for the Santacrucian SALMA. The Santa Cruz Formation is widely distributed in southern Patagonia (e.g., Marshall 1976; Vizcaíno et al. 2012a; Cuitiño et al. 2016; Figs. 2.28 and 2.31) and is exposed from the Atlantic coast to the Andean foothills and northwest to the headwaters of the Río Pinturas (Perkins et al. 2012). The unit is a continental pyroclastic sequence formed by alluvial deposits of sandstones and mudstones (Perkins et al. 2012). Estuarine levels developed at the base of the unit were inhabited by marginal marine oysters (Griffin and Parras 2012). Outcrops of the

**Fig. 2.31** Monte León, Santa Cruz, Argentina. General view of the outcrops of the Santa Cruz Formation at the Atlantic coast (**a**). Fleagle's expedition, 1994; Analía M. Forasiepi searching for small mammal bones (**b**)

Santa Cruz Formation with Santacucian faunas were dated ~17.22 to ~15.60 Ma at the Río Santa Cruz and ~18.00 to ~16.20 Ma at eastern coastal localities (Burdigalian, upper lower Miocene), while western deposits, with the Notohippidian fauna, were dated between 19.0 and18.0 Ma (Blisniuk et al. 2005; Perkins et al. 2012; Cuitiño et al. 2016).

The vegetation of the Santa Cruz Formation is known by phytoliths, wood and leaf compressions, and silicified trunks, that represent a mixture of grass and arboreal components, defining a mosaic of open temperate humid and semi-arid forests formed by Araucariaceae, Lauraceae, Nothofagaceae, Cunoniaceae, Myrtaceae, Fabaceae, and possibly Proteaceae, and occasional marshlands with a mixture of grasses and forbs (Barreda and Palazzesi 2007; Brea et al. 2012).

Santacrucian vertebrates developed in the context of the SA isolation, before Great American Biotic Interchange (GABI) immigrants entered the southern hemisphere. It is represented by frogs (Neobatrachia; Fernícola and Albino 2012), reptiles (Squamata and Ophidia; Fernícola and Albino 2012), birds (Rheiformes, Tinamiformes, Gruiformes, Anseriformes, Pelecaniformes, Ciconiiformes, Falconiformes, Cariamiformes including Cariamidae and the "terror bird" Phorusrhacidae, the largest known terrestrial carnivorous flightless birds; e.g., Ameghino 1889; Moreno and Mercerat 1891; Degrange et al. 2012), and mammals. The mammalian association is represented by metatherians (Sparassodonta, Microbiotheria, Paucituberculata; Sinclair 1906; Abello et al. 2012; Prevosti et al. 2012), SANUs (Scott 1910, 1912, 1928a, 1937; Cassini et al. 2012; Croft et al. 2020), xenarthrans (Cingulata, Vermilingua, and Pilosa; Scott 1903–1904; Bargo et al. 2012; Vizcaíno et al. 2012b), rodents

(Caviomorpha; Scott 1905; Candela et al. 2012), primates (Platyrrhini; Scott 1928b; Kay et al. 2012), and the meridiolestidan *Necrolestes patagonensis* (Scott 1905; Rougier et al. 2012; Wible and Rougier 2017).

Ameghino recognized *Necrolestes patagonensis* on the basis of a peculiar dentary with dentition (Ameghino 1891) and some postcrania (Ameghino 1894). Later, Scott (1905) described much better preserved material, including three partial specimens, two of them with fairly complete skulls, and substantial postcranial elements. Additional fragmentary material (Tauber et al. 2005) has been more recently assigned to *Necrolestes*, but the publication supplies no additional morphological information. The affinities of *Necrolestes* have been challenging, with opinions diverging widely among a host of researchers (Ameghino 1891; Scott 1905; Leche 1907; Gregory 1910; Romer 1945; Simpson 1945; Saban 1954; Winge 1941; Patterson 1958; Archer 1984; Aplin and Archer 1987; Van Valen 1988; Szalay 1994; Asher et al. 2007; Goin et al. 2007b; Ladevèze et al. 2008; Archer et al. 2011; Rougier et al. 2011b, 2012; Chimento et al. 2012; O'Meara and Thompson 2014; Wible and Rougier 2017). The enigmatic *Necrolestes* was considered closely related to the African golden mole, *Chrysochloris* (Ameghino 1891; Scott 1905; Gregory 1910), palaeanodonts (Saban 1954), a marsupial relative (Patterson 1958; Archer et al. 2011; Ladevèze et al. 2008), and recently it has been grouped together with SA dryolestoids. Despite Van Valen's (1988) first suggestion of affinities between *Necrolestes* and Mesozoic lineages, it was the further preparation of unobserved basicranial structures of the specimens described by Scott (1905) and the finding of skulls of the early Late Cretaceous dryolestoid *Cronopio dentiacutus* that allowed Rougier and colleagues to provide new data in support of non-therian affinities (Rougier et al. 2011b, 2012; Wible and Rougier 2017; see also Chimento et al. 2012; O'Meara and Thompson 2014). *Necrolestes* is the latest surviving member of the Mesozoic Gondwanan dryolestoid radiation, the youngest meridiolestidan on record. A long ghost lineage that includes more than 40 million years exists between *Necrolestes* and the early Paleocene dryolestoid *Peligrotherium tropicalis* (Rougier et al. 2012; O'Meara and Thompson 2014); this substantial lack of record would be shortened if the dryolestoid affinities of the Eocene dryolestoid from Antarctica are confirmed (Martinelli et al. 2014).

## 2.3 Concluding Remarks

The fossil record of Mesozoic mammals in SA is pivotal for the entirety of Gondwana, particularly findings in the Triassic of Brazil and Jurassic and Cretaceous rocks from Patagonia. However, when compared with Laurasia, the brief history of mammal findings in SA, starting in the 1980s, shows we are taking our first steps in a still intricate and obscure evolutionary landscape of enigmatic and bizarre mammals. Compared with the discoveries and studies of Mesozoic mammals in northern continents, which started in the nineteenth century (e.g., Owen 1871; Cope 1882; Osborn

1887; Marsh 1889), we expect that the new discoveries in southern landmasses will provide material that can rewrite the history of Mesozoic mammals.

The distribution of Mesozoic mammals in SA illustrates that the absence of fossil material across different sedimentary basins is the product of biases in the fossil record, produced not only by taphonomic constraints but also by the lack of consistent field effort in search of small-size material. We expect that localities (such as the Patagonian Cañadón Asfalto, Candeleros, Los Alamitos, and La Colonia formations), worked systematically and with adjusted techniques that favor the finding of small-size vertebrates (e.g., screen-washing, quarrying, and rock splitting), have the potential to provide a considerable amount of new material. The Neotropics is still the Pandora's Box for paleontology, since most of what we know is based on the richer and more studied outcrops from Patagonia and surrounding areas of the south. Persistence but also luck would help to discover these Mesozoic, mostly tiny, animals. Continental Mesozoic rocks of SA are drawing a new history for Gondwanan mammals and much of the future work to be done.

## References


Abdala NF (1996) Redescripción del cráneo y reconsideración de la validez de *Cynognathus minor* (Eucynodontia-Cynodonthidae) del Triásico Inferior de Mendoza. Ameghiniana 33:115–126

Abdala NF, Ribeiro AM (2003) A new traversodontid cynodont from the Santa Maria Formation (Ladinian–Carnian), of southern Brazil, with a phylogenetic analysis of Gondwanan traversodontids. Zool J Linn Soc 138:529–545

Abdala NF, Ribeiro AM (2010) Distribution and diversity patterns of Triassic cynodonts (Therapsida, Cynodontia) in Gondwana. Palaeogeogr Palaeoclimatol Palaeoecol 286:202–217

Abello MA, Ortiz-Jaureguizar E, Candela AM (2012) Paleoecology of the Paucituberculata and Microbiotheria (Mammalia, Marsupialia) from the late Early Miocene of Patagonia. In: Vizcaíno SF, Kay RF, Bargo MS (eds) Early Miocene paleobiology in Patagonia. Cambridge University Press, Cambridge, pp 156–172

Adnet S, Salas-Gismondi R, Antoine PO (2014) River stingrays (Chondrichthyes: Potamotrygonidae) from the Middle Eocene of Peruvian Amazonia and an overview of potamotrygonid dental morphology. Naturwissenschaften 101:33–45

Agnolín FL (2010) An avian coracoid from the Upper Cretaceous of Patagonia, Argentina. Stvd Geol Salmant 46:99–119

Agnolín FL (2012) A new Calyptocephalellidae (Anura, Neobatrachia) from the Upper Cretaceous of Patagonia, Argentina, with comments on its systematic position. Stvd Geol Salmant 48:129–178

Agnolín FL, Martinelli AG (2009) Fossil birds from the Late Cretaceous Los Alamitos Formation, Río Negro Province, Argentina. J S Am Earth Sci 27:42–49

Agnolín FL, Powell JE, Novas FE, Kundrát M (2012) New alvarezsaurid (Dinosauria, Theropoda) from uppermost Cretaceous of north-western Patagonia with associated eggs. Cretac Res 35:33–56

Aguirre-Urreta B, Cristallini EO (2009) La cuenca Neuquina, una perspectiva integrada. Rev Asoc Geol Argent 65:237–412

Alarcón-Muñoz J, Soto-Acuña S, Manríquez L, Fernández R, Bajor D, Guevara JP, Suazo Lara F, Leppe M, Vargas AO (2020) Freshwater turtles (Testudines: Pleurodira) in the Upper Cretaceous of Chilean Patagonia. J S Am Earth Sci 102:102652

Albino AM (1987) Part V-The ophidians. In: Bonaparte JF (ed) The Late Cretaceous fauna of Los Alamitos, Patagonia, Argentina. Rev Mus Argent Cienc Nat "B Rivadavia", Paleontología, vol 3, pp 141–146

Albino AM (2002) El lagarto más antiguo de la Argentina. 1° Congreso 'Osvaldo A. Reig' de Vertebradología Básica y Evolutiva e Historia y Filosofía de la Ciencia, Buenos Aires, Abstracts:21

Almeida FFM (1950) Uma faunula de crustáceos bivalvos do Arenito Botucatu no Estado de São Paulo. Divisão de Geología e Mineralogía, Bol 134:1–36

Almeida FFM (1954) Botucatu, um deserto Triássico da América do Sul. Notas Prelim Estudos 80:1–16

Almeida FFM, Assine ML, Carneiro CR (2012) A megadesertificação Mesozoica. In: Hasui Y, Carneiro CR, Almeida FFM, Bartorelli A (eds) Geologia do Brasil. Beca Editora, São Paulo, pp 419–428

Alvarenga H, Nava WR (2005) Aves Enantiornithes do Cretáceo Superior da Formação Adamantina do Estado de São Paulo, Brasil. 2° Congresso Latino-Americano de Paleontología de Vertebrados, Rio de Janeiro, Abstract:20

Ameghino F (1889) Contribución al conocimiento de los mamíferos fósiles de la República Argentina. Acad Nac Cienc Córdoba, Actas 6:1–1027

Ameghino F (1891) Nuevos restos de mamíferos fósiles descubiertos por Carlos Ameghino en el Eoceno Inferior de la Patagonia austral. Especies nuevas adiciones y correcciones. Rev Argent Hist Nat 1:289–328

Ameghino F (1893) Sobre la presencia de vertebrados de aspecto Mesozoico en la formación Santacruceña de la Patagonia Austral. Rev el Jardín Zool Buenos Aires 1:76–84

Ameghino F (1894) Enumération synoptique de espèces de mammifères fossiles des formations éocènes de Patagonie. Acad Nac Cienc Córdoba, Bol 13:259–455

Ameghino F (1906) Les formations sédimentaires du Crétacé Superieur et du Tertiaire de Patagonia, avec an paralèlle entre leurs faunes mammalogiques et celles de l' ancien continent. Anales Mus. Nac. Buenos Aires 15 (ser 3, 8):1–508

Ancibor E (1995) Palmeras fósiles del Cretácico Tardío de la Patagonia Argentina (Bajo de Santa Rosa, Río Negro). Ameghiniana 32:287–299

Andreis RR (1987) Part I -Stratigraphy and paleoenvironment. In: Bonaparte JF (ed) The Late Cretaceous fauna of Los Alamitos, Patagonia, Argentina. Rev Mus Argent Cienc Nat "B Rivadavia", Paleontología, vol 3, pp 103–110

Andreis RR, Iñíguez Rodríguez AM, Lluch JJ, Sabio DA (1974) Estudio sedimentológico de las formaciones del Cretácico Superior del área del Lago Pellegrini (Provincia de Río Negro, República Argentina). Rev Asoc Geol Argent 29:85–104

Andreis RR, Mazzoni M, Spalletti LA (1975) Estudio estratigráfico y paleoambiental de las sedimentitas terciarias entre Pico Salamanca y Bahía Bustamante, Provincia del Chubut, República Argentina. Rev Asoc Geol Argent 30:85–103

Andreis RR, Bossi GE, Montardo DK (1980) O Grupo Rosário do Sul (Triássico) no Rio Grande do Sul, Brasil. 31° Congresso Brasileiro de Geologia, Balneário Camboriú, Anais, vol 2, pp 659–673

Andreis RR, Ancibor E, Archangelsky S, Artabe A, Bonaparte J, Genise J (1991) Asociación de vegetales y animales en estratos del Cretácico Tardío del norte de la Patagonia. Ameghiniana 28:201–202

Antoine PO, Marivaux L, Croft DA, Billet G, Ganerød M, Jaramillo C, Martin T, Orliac MJ, Tejada-Lara J, Altamirano AJ, Duranthon F, Fanjat G, Rousse S, Salas-Gismondi RS (2012) Middle Eocene rodents from Peruvian Amazonia reveal the pattern and timing of caviomorph origins and biogeography. Proc R Soc B 279:1319–1326

Antoine PO, Abello MA, Adnet S, Sierra AJA, Baby P, Billet G, Boivin M, Calderon Y, Candela AM, Chabain J, Corfu F, Croft DA, Ganerod M, Jaramillo C, Klaus S, Marivaux L, Navarrete RE, Orliac MJ, Parra F, Perez ME, Pujos F, Rage JC, Ravel A, Robinet C, Roddaz M, Tejada-Lara JV, Velez-Juarbe J, Wesselingh FP, Salas-Gismondi RS (2016) A 60-million-year Cenozoic history of western Amazonian ecosystems in Contamana, eastern Peru. Gondwana Res 31:30–59

Apaldetti C, Martínez RN, Cerda IA, Pol D, Alcober O (2018) An early trend towards gigantism in Triassic sauropodomorph dinosaurs. Nat Ecol Evol 2:1227–1232

Apesteguía S (2002) Gran Gondwana y la fauna de vertebrados del Mar de Kawas (Campaniano–Maastrichtiano). In: de Castro JC, Dias-Britto D, Musacchio EA (eds), 2° Simpósio sobre el Cretácico de América del Sur, São Pedro, Actas, vol 2, pp 143–147

Apesteguía S (2005) A Late Campanian sphenodontid (Reptilia, Diapsida) from northern Patagonia. CR Palevol 4:663–669

Apesteguía S (2007) The sauropod diversity of the La Amarga Formation (Barremian), Neuquén (Argentina). Gondwana Res 12:533–546

Apesteguía S, Novas FE (2003) Large Cretaceous sphenodontian from Patagonia provides insight into lepidosaur evolution in Gondwana. Nature 425:609–612

Apesteguía S, Rougier GW (2007) A late Campanian sphenodontid maxilla from northern Patagonia. Am Mus Novit 3581:1–11

Apesteguía S, Zaher H (2006) A Cretaceous terrestrial snake with robust hindlimbs and a sacrum. Nature 440:1037–1040

Apesteguía S, Agnolín FL, Lio GL (2005) An early Late Cretaceous lizard from Patagonia, Argentina. CR Palevol 4:311–315

Apesteguía S, Agnolín FL, Claeson K (2007) Review of Cretaceous dipnoans from Argentina (Sarcopterygii: Dipnoi) with descriptions of new species. Rev Mus Argent Cienc Nat "B Rivadavia", ns 9:27–40

Apesteguía S, Gómez RO, Rougier GW (2012) A basal sphenodontian (Lepidosauria) from the Jurassic of Patagonia: new insights on the phylogeny and biogeography of Gondwanan rhynchocephalians. Zool J Linn Soc 166:342–360

Apesteguía S, Gómez RO, Rougier GW (2014) The youngest South American rhynchocephalian, a survivor of the K/Pg extinction. Proc R Soc B 281:20140811

Aplin KP, Archer M (1987) Recent advances in marsupial systematic with a new syncretic classification. In: Archer M (ed) Posums and opossums: studies in evolution, vol 1. Surrey Beatty and Sons and the Royal Zoological Society of New South Wales, Sydney, pp XV–LXXII

Aragón E, Mazzoni MM (1997) Geología y estratigrafía del complejo volcánico piroclástico del Río Chubut medio (Eoceno), Chubut, Argentina. Rev Asoc Geol Argent 52:243–256

Aragón E, Rodriguez AMI, Benialgo A (1996) A calderas field at the Marifil Formation, new volcanogenic interpretation, Norpatagonian Massif, Argentina. J S Am Earth Sci 9:321–328

Aranciaga Rolando M, Cerroni MA, Garcia Marsà JA, Agnolín FI, Motta MJ, Rozadilla S, Brisson Eglí F, Novas FE (2020) A new medium-sized abelisaurid (Theropoda, Dinosauria) from the Late Cretaceous (Maastrichtian) Allen Formation of Northern Patagonia, Argentina. J S A Earth Sci. https://doi.org/10.1016/j.jsames.2020.102915

Araújo DC, Gonzaga TD (1980) Uma nova espécie de *Jachaleria* (Therapsida, Dicynodontia) do Triássico do Brasil. 2° Congreso Argentino de Paleontología y Bioestratigrafía y 1° Congreso Latinoamericano de Paleontología, Buenos Aires, Actas, pp 159–174

Arbe HA (2002) Análisis estratigráfico del Cretácico de la Cuenca Austral. In: Haller MJ(ed) Geología y recursos naturales de Santa Cruz. 15° Congreso Geológico Argentino, El Calafate, Relatorio, vol 1, no 8, pp 103–128

Arbe H, Hechem JJ (1984) Estratigrafía y facies de depósitos marinos profundos del Cretácico superior, Lago Argentino. 9° Congreso Geológico Argentino. Bariloche, Actas 5:7–14

Archangelsky S (1973) Palinología del Paleoceno de Chubut. I. Descripciones sistemáticas. Ameghiniana 10:339–399

Archangelsky S (1976a) Palinología del Paleoceno de Chubut. II. Diagramas polínicos. Ameghiniana 13:43–55

Archangelsky S (1976b) Palinología del Paleoceno de Chubut. III Análisis numérico I Introducción y matrices de similitud. Ameghiniana 13:169–189

Archangelsky S, Romero EJ (1974) Polen de gimnospermas (coníferas) del Cretácico superior y Paleoceno de Patagonia. Ameghiniana 11:217–236

Archangelsky S, Zamaloa MC (1986) Nuevas descripciones palinológicas de las formaciones Salamanca y Bororó, Paleoceno de Chubut (República Argentina). Ameghiniana 23:35–46

Archer M (1984) Origins and early radiations of marsupials. In: Archer M, Clayton G (eds) Vertebrate zoogeography and evolution in Australasia. Hesperian Press, Perth, pp 585–631

Archer M, Flannery TF, Ritchie A, Molnar RE (1985) First Mesozoic mammal from Australia, an Early Cretaceous monotreme. Nature 318:363–366

Archer M, Beck R, Gott M, Hand S, Godthelp H, Black K (2011) Australia's first fossil marsupial mole (Notoryctemorphia) resolves controversies about their evolution and palaeoenvironmental origins. Proc R Soc B 278:1498–1506

Arcucci AB, Coria RA (2003) A new Triassic carnivorous dinosaur from Argentina. Ameghiniana 40:217–228

Ardolino A, Franchi MR (1996) Hoja Geológica 4366-I, Telsen, Provincia del Chubut. Dirección Nacional del Servicio Geológico, Buenos Aires, Bol, vol 215, pp 1–110

Arid FM, Vizzoto LD (1971) *Antarctosaurus brasiliensis*, um novo saurópode do Cretáceo Superior do sul do Brasil. 25° Congresso Brasileiro de Geologia, São Paulo, Anais, vol 1, pp 297–305

Arnal M, Kramarz AG, Vucetich MG, Vieytes EC (2014) A new Early Miocene octodontoid rodent (Hystricognathi, Caviomorpha) from Patagonia (Argentina) and a reassessment of the early evolution of Octodontoidea. J Vertebr Paleontol 34:397–406

Artabe AE, Stevenson DW (1999) Fossil Cycadales of Argentina. Bot Rev 65:219–238

Artabe AE, Zamuner AB, Stevenson DW (2004) The new petrified cycad stems, *Brunoa* gen. nov. and *Worsdellia* gen. nov., from the Cretaceous of Patagonia (Bajo de Santa Rosa, Río Negro Province), Argentina. Bot Rev 70:121–133

Asher RJ, Horovitz I, Martin T, Sánchez-Villagra MR (2007) Neither a rodent nor a platypus: a reexamination of *Necrolestes patagonensis* Ameghino. Am Mus Nov 3546:1–40

Asurmendi E, Sánchez M, Fennell L (2017) Neuquén Group (Upper Cretaceous): a case of underfilled-overfilled cycles in an Andean foreland basin, Neuquén Basin, Argentina. J S Am Earth Sci 80:444–459

Averianov AO, Martin T, Lopatin AV (2013) A new phylogeny for basal Trechnotheria and Cladotheria and affinities of South American endemic Late Cretaceous mammals. Naturwissenschaften 100:311–326

Azevedo SA, Gallo V, Ferigolo J (2000) A possible chelonian egg from the Brazilian Late Cretaceous. An Acad Bras Ciênc 72:187–193

Báez AM (1987) Part III–Anurans. In: Bonaparte JF (ed) The Late Cretaceous fauna of Los Alamitos, Patagonia, Argentina. Rev Mus Argent Cienc Nat "B Rivadavia", Paleontología, vol 3, pp 121–130

Báez AM, Nicoli L (2004) A new look at an old frog: the Jurassic *Notobatrachus* Reig from Patagonia. Ameghiniana 41:257–270

Báez AM, Nicoli L (2008) A new species of *Notobatrachus* (Amphibia, Salientia) from the Middle Jurassic of northwestern Patagonia. J Paleontol 82:372–376

Báez AM, Trueb L, Calvo JO (2000) The earliest known pipoid frog from South America: a new genus from the Middle Cretaceous of Argentina. J Vertebr Paleontol 20:490–500

Baldoni AM (1990) Tafofloras del Jurásico medio de la Patagonia Extraandina. In: Volkheimer W (ed) Bioestratigrafía de los sistemas regionales del Jurásico y Cretácico de América del Sur. Comité Sudamericano del Jurásico–Cretácico, vol 2, pp 313–353

Barberena MC (1982) Uma nova espécie de *Proterochampsa*, *P. nodosa* sp. nov. do Triássico do Brasil. An Acad Bras Ciênc 54:127–141

Bargo MS, Toledo N, Vizcaíno SF (2012) Paleobiology of the Santacrucian sloths and anteaters (Xenarthra, Cingulata). In: Vizcaíno SF, Kay RF, Bargo MS (eds) Early Miocene paleobiology in Patagonia. Cambridge University Press, Cambridge, pp 216–242

Barreda V, Palazzesi L (2007) Patagonian vegetation turnovers during the Paleogene-Early Neogene: origin of arid-adapted floras. Bot Rev 73:31–50

Barreda V, Palazzesi L (2010) Vegetation during the Eocene-Miocene interval in central Patagonia: a context of mammal evolution. In: Madden RH, Carlini AA, Vucetich MG, Kay RF (eds) The paleontology of Gran Barranca: evolution and environmental change through the Middle Cenozoic of Patagonia. Cambridge University Press, Cambridge, pp 375–382

Barrio CA (1990) Paleogeographic control of Upper Cretaceous tidal deposits, Neuquén Basin, Argentina. J S Am Earth Sci 3:31–49

Batezelli A, Ladeira FSB (2016) Stratigraphic framework and evolution of the Cretaceous continental sequences of the Bauru, Sanfranciscana, and Parecis basins, Brazil. J S Am Earth Sci 65:1–24

Becerra MG, Paulina-Carabajal A, Cruzado-Caballero P, Taborda JRA (2018) First endocranial description of a South American hadrosaurid: the neuroanatomy of *Secernosaurus koerneri* from the Late Cretaceous of Argentina. Acta Paleontol Pol 63:693–702

Bellosi ES (2010) Physical stratigraphy of the Sarmiento Formation (middle Eocene–lower Miocene) at Gran Barranca, central Patagonia. In: Madden RH, Carlini AA, Vucetich MG, Kay RF (eds) The paleontology of Gran Barranca: evolution and environmental change through the Middle Cenozoic of Patagonia. Cambridge University Press, Cambridge, pp 19–31

Bertels A (1969) Estratigrafía del límite Cretácico-Terciario en Patagonia Septentrional. Rev Asoc Geol Argent 24:41–54

Bertini RJ (1996) Evidências de Abelisauridae (Carnosauria: Saurischia) do Neo-Cretáceo da Bacia do Paraná. 4° Simposio do Cretáceo do Brasil, Águas de São Pedro, Bol:267–271

Bertini RJ, Marshall LG, Gayet M, Brito P (1993) Vertebrate faunas from the Adamantina and Marília formations (Upper Bauru Group, Late Cretaceous, Brazil) in their stratigraphic and paleobiogeographic context. Neues Jahrb Geol Paläontol 188:71–101

Biddle KT, Uliana MA, Mitchum JRR, Fitzgerald MG, Wrighr RC (1986) The stratigraphic and structural evolution of the central and eastern Magallanes Basin, southern South America. In: Allen PA, Homewood P (eds) Foreland basins, vol 8. International Association of Sedimentologists Special Publications, Ghent, pp 41–61

Bittencourt JS, Langer MC (2011) Mesozoic dinosaurs from Brazil and their biogeographic implications. An Acad Bras Ciênc 73:23–60

Blisniuk PM, Stern LA, Chamberlain CP, Idleman B, Zeitler KP (2005) Climatic and ecologic changes during Miocene surface uplift in the Southern Patagonian Andes. Earth Planet Sci Lett 230:125–142

Bodenbender G (1911) Constitución geológica de la parte meridional de la Provincia de La Rioja y regiones limítrofes. Bol Acad Nac Cienc Córdoba 19:5–21

Bogan S, Agnolín FL (2014) First record of the genus *Angolabatis* (Rajiformes, Hypsobatidae) from the Late Cretaceous of Patagonia, Argentina. Hist Nat 4:5–12

Bogan S, Taverne L, Agnolín FL (2009) First fossil record of an amiid fish (Halecomorphi, Amiidae) from the Latest Cretaceous of Patagonia, Argentina, and comments on the status of *Pappichthys patagonica* Ameghino, 1906 (Teleostei, Osteoglossidae). Bull Inst Rsci Nat Belg, Sci Terre 80:163–170

Bona P (2006) Paleocene (Danian) chelid turtles from Patagonia, Argentina: Taxonomic and biogeographic implications. Neues Jahrb Geol Paläontol 241:303–323

Bona P (2007) Una nueva especie de *Eocaiman* Simpson (Crocodylia, Alligatoridae) del Paleoceno Inferior de Patagonia. Ameghiniana 44:435–445

Bona P, de la Fuente MS (2005) Phylogenetic and paleobiogeographic implications of *Yaminuechelys maior* (Staesche, 1929) new comb. A large long-necked chelid turtle from the Early Paleocene of Patagonia. J Vertebr Paleontol 25:569–582

Bona P, Ezcurra MD, Barrios F, Fernandez Blanco MV (2018) A new Palaeocene crocodylian from southern Argentina sheds light on the early history of caimanines. Proc R Soc B 285:20180843

Bonaparte JF (1966) Una nueva "fauna" Triásica de Argentina. (Therapsida: Cynodontia - Dicynodontia). Consideraciones filogenéticas y paleobiogeográficas. Ameghiniana 4:243–296

Bonaparte JF (1969) Dos nuevas "faunas" de reptiles triásicos de Argentina. Gondwana Stratigraphy, IUGS, Coloquio Mar del Plata, 1967:283–302

Bonaparte JF (1972a) *Cromptodon mamiferoides*, Galesauridae de la Formación Río Mendoza, Mendoza, Argentina (Therapsida-Cynodontia). Ameghiniana 9:343–353

Bonaparte JF (1972b) Los tetrápodos del sector superior de la Formación Los Colorados, La Rioja, Argentina (Triásico Superior). Op Lilloana 22:1–183

Bonaparte JF (1979) Dinosaurs: a Jurassic assemblage from Patagonia. Science 205:1377–1378

Bonaparte JF (1980) El primer ictidosaurio (Reptilia-Therapsida) de América del Sur, *Chaliminia musteloides*, del Triásico Superior de La Rioja, Argentina. 2° Congreso Argentino de Paleontología y Bioestratigrafía y 1° Congreso Latinoamericano de Paleontología, Buenos Aires, Actas 1:123–133

Bonaparte JF (1981) Descripción de *Fasolasuchus tenax* y su significado en la sistemática y evolución de los Thecodontia. Rev Mus Argent Cienc Nat "B Rivadavia", Paleontología 3:56–100

Bonaparte JF (1982) Faunal replacement in the Triassic of South America. J Vertebr Paleontol 2:362–371

Bonaparte JF (1985) A horned Cretaceous carnosaur from Patagonia. Natl Geogr Res 1:149–151

Bonaparte JF (1986a) History of the terrestrial Cretaceous vertebrates of Gondwana. 4° Congreso Argentino de Paleontología y Bioestratigrafía, Mendoza, Actas 2:63–95

Bonaparte JF (1986b) Sobre *Mesungulatum houssayi* y nuevos mamíferos cretácicos de Patagonia. 4° Congreso Argentino de Paleontología y Bioestratigrafía, Mendoza, Actas, vol 2, pp 48–61

Bonaparte JF (1986c) A new and unusual Late Cretaceous mammal from Patagonia. J Vertebr Paleontol 6:264–270

Bonaparte JF (1986d) Les Dinosaures (Carnosaures, Allosauridés, Sauropodes, Cétiosauridés) du Jurassique Moyen de Cerro Cóndor (Chubut, Argentine). Annls Paléont (Vert) 72(247–289):326–386

Bonaparte JF (1990) New Late Cretaceous mammals from the Los Alamitos Formation, northern Patagonia. Natl Geogr Res 6:63–93

Bonaparte JF (1992) Una nueva especie de Triconodonta (Mammalia), de la Formación Los Alamitos, Provincia de Río Negro y comentarios sobre su fauna de mamíferos. Ameghiniana 29:99–110

Bonaparte JF (1994) Approach to the significance of the Late Cretaceous mammals of South America. Berlin Geowiss Abt E 13:31–44

Bonaparte JF (1996a) Cretaceous tetrapods of Argentina. In: Arratia G (ed) Contributions of southern South America to vertebrate paleontology. Münch Geowiss Abh A Paläontol, vol 30, pp 73–130

Bonaparte JF (1996b) Dinosaurios de América del Sur. Museo Argentino de Ciencias Naturales "Bernardino Rivadavia", Buenos Aires, Argentina (2nd edn)

Bonaparte JF (1997) El Triásico de San Juan-La Rioja, Argentina y sus dinosaurios. Talleres Gráficos de Digital Tech SRL

Bonaparte JF (2000) Comentarios críticos sobre el Triásico Inferior de Puesto Viejo y Potrerillos. Acad Nac Cienc Córdoba, Bol 64:147–152

Bonaparte JF (2002) New Dryolestida (Theria) from the Late Cretaceous of Los Alamitos, Argentina, and paleogeographical comments. Neues Jahrb Geol Paläontol 224:339–371

Bonaparte JF, Kielan-Jaworowska Z (1987) Late Cretaceous dinosaur and mammal faunas of Laurasia and Gondwana. In: Currie PM, Koster EH (eds) 4° Symposium on Mesozoic Terrestrial Ecosystems, Short Papers, Occasional Papers of the Tyrrell Museum of Palaeontology, vol 3, pp 24–29

Bonaparte JF, Migale LA (2010) Protomamíferos y mamíferos Mesozoicos de América del Sur. Museo de Ciencias Naturales Carlos Ameghino de Mercedes, Buenos Aires. 1° Edition

Bonaparte JF, Migale LA (2015) Protomamíferos y mamíferos Mesozoicos de América del Sur. Fundación de Historia Natural Felix de Azara, Buenos Aires. 2° Edition

Bonaparte JF, Morales J (1997) Un primitivo Notonychopidae (Litopterna) del Paleoceno Inferior de Punta Peligro, Chubut, Argentina. Est Geol 53:183–286

Bonaparte JF, Novas FE (1985) *Abelisaurus comahuensis*, Carnosauria del Cretácico Tardío de Patagonia. Ameghiniana 21:259–265

Bonaparte JF, Rougier GW (1987a) Mamíferos del Cretácico Inferior de Patagonia. 4° Congreso Latinoamericano de Paleontologia. Santa Cruz de la Sierra 1:343–359

Bonaparte JF, Rougier GW (1987b). Part VII -The Hadrosaurs. In: Bonaparte JF (ed) The Late Cretaceous fauna of Los Alamitos, Patagonia, Argentina. Rev Mus Argent Cienc Nat "B Rivadavia", Paleontología, vol 3, pp 63–169

Bonaparte JF, Soria MF (1985) Nota sobre el primer mamífero del Cretácico Argentino, Campaniano–Maastrichtiano (Condylarthra). Ameghiniana 21:177–183

Bonaparte JF, Sues H-D (2006) A new species of *Clevosaurus* (Lepidosauria: Rhynchocephalia) from the Upper Triassic of Rio Grande do Sul, Brazil. Palaeontology 49:917–923

Bonaparte JF, Franchi MR, Sepúlveda EG (1984) La Formación Los Alamitos (Campaniano–Maastrichtiano) del sudeste de Río Negro, con descripción de *Kritosaurus australis* n. sp. (Hadrosauridae). Significado paleogeográfico de los vertebrados. Rev Asoc Geol Argent 39:284–299

Bonaparte JF, Krause DW, Kielan-Jaworowska Z (1989) *Ferugliotherium windhauseni* Bonaparte, the first known multituberculta from Gondwanaland. J Vertebr Paleontol 1 9:14A

Bonaparte JF, Van Valen LM, Kramarz A (1993) La fauna local de Punta Peligro, Paleoceno Inferior, de la Provincia de Chubut, Patagonia, Argentina. Evol Monogr 14:1–61

Bonaparte JF, Ferigolo J, Ribeiro AM (1999) A new Early Triassic saurischian dinosaur from Rio Grande Do Sul State, Brazil. In: Tomida Y, Rich TH, Vickers-Rich P (eds) Proceedings of the second gondwanan dinosaur symposium, National Science Museum, Tokyo, pp 89–109

Bonaparte JF, Martinelli AG, Schultz CL, Rubert R (2003) The sister group of mammals: Small cynodonts from the Late Triassic of southern Brazil. Rev Bras Paleontol 5:5–27

Bonaparte JF, Martinelli AG, Schultz CL (2005) New information on *Brasilodon* and *Brasilitherium* (Cynodontia, Probainognathia) from the Late Triassic of southern Brazil. Rev Bras Paleontol 8:25–46

Bonaparte JF, Schultz CL, Soares MB, Martinelli AG (2010) La fauna local de Faxinal do Soturno, Triásico Tardío de Rio Grande do Sul, Brasil. Rev Bras Paleontol 13:233–246

Bond M (1986) Los ungulados fósiles de Argentina: Evolución y paleoambientes. 4° Congreso Argentino de Paleontología y Bioestratigrafía, Mendoza, Actas, vol 2, pp 173–185

Bond M, Carlini AA, Goin FJ, Legarreta L, Ortiz-Jaureguizar E, Pascual R, Uliana MA (1995) Episodes in South American land mammal evolution and sedimentation: Testing their apparent concurrence in a Paleocene succession from Central Patagonia. 6° Congreso Argentino de Paleontología y Bioestratigrafía, Trelew, Actas, pp 47–58

Bond M, Reguero MA, Vizcaíno SF, Marenssi SA (2006) A new "South American ungulate" (Mammalia: Litopterna) from the Eocene of the Antarctic Peninsula. In: Francis JE, Pirrie D, Crame JA (eds) Cretaceous–Tertiary high-latitude palaeoenvironments, James Ross Basin, Antarctica. Geological Society of London, Special Publications, vol 258, pp 163–176

Bond M, Kramarz A, Macphee RD, Reguero MA (2011) A new astrapothere (Mammalia, Meridiungulata) from La Meseta Formation, Seymour (Marambio) Island, and a reassessment of previous records of Antarctic astrapotheres. Am Mus Novit 3718:1–16

Bond M, Tejedor MF, Campbell KE, Chornogubsky L, Novo N, Goin FJ (2015) Eocene primates of South America and the African origins of NewWorld monkeys. Nature 520:538–541

Bossi GE (1976) Geología de la Cuenca de Marayes-El Carrizal (Provincia de San Juan, República Argentina). 6° Congreso Geológico Argentino, Bahía Blanca, Actas, vol 1, pp 23–38

Brea M, Matheos S, Zamuner A, Ganuza D (2005) Análisis de los anillos de crecimiento del bosque fósil de Víctor Szlápelis, Terciario Inferior del Chubut, Argentina. Ameghiniana 42:407–418

Brea M, Zamuner AB, Matheos SD, Iglesias A, Zucol AF (2008) Fossil wood of the Mimosoideae from the Early Paleocene of Patagonia, Argentina. Alcheringa 32:427–441

Brea M, Zucol AF, Iglesias A (2012) Fossil plant studies from late Early Miocene of the Santa Cruz Formation: Paleoecology and paleoclimatology at the passive margin of Patagonia, Argentina. In: Vizcaíno SF, Kay RF, Bargo MS (eds) Early Miocene paleobiology in Patagonia. Cambridge University Press, Cambridge, pp 104–128

Brito RJ, Amaral CRL, Machado LP (2006) A ictiofauna do Grupo Bauru, Cretáceo Superior da Bacia Bauru, sudeste do Brasil. In: Gallo V, Brito PM, Silva HM, Figueroa FJ (eds) Paleontologia de vertebrados: Grandes temas e contribuições cientificas. Interciência, Rio de Janeiro, pp 133–143

Brito PM, Nava WR, Martinelli AG (2017) A new fossil amiid (Holostei: Halecomorphi) from the Late Cretaceous Adamantina Formation, southeastern Brazil, with comments on western Gondwana amiid. Cretac Res 77:39–43

Buck PV, Ghilardi AM, Peixoto BCPM, dos Fernandes LBR, Fernandes MA (2017a) A new tetrapod ichnotaxon from Botucatu Formation, Lower Cretaceous (Neocomian), Brazil, with comments on fossil track preservation on inclined planes and local paleoecology. Palaeogeogr Palaeoclimatol Palaeoecol 466:21–37

Buck PV, Ghilardi AM, dos Fernandes LBR, Fernandes MA (2017b) A new ichnotaxon classification of large mammaliaform trackways from the Lower Cretaceous Botucatu Formation, Paraná Basin, Brazil. Palaeogeogr Palaeoclimatol Palaeoecol 485:377–388

Brückmann M, Hartmann LA, Tassinari CCG, Sato K, Baggio SB (2014) The duration of magmatism in the Serra Geral Group, Paraná volcanic province. In: Hartmann LA, Sergio B, Baggio B (eds) Metallogeny and crustal evolution of the Serra Geral Group. Gráfica da UFRGS, Porto Alegre, pp 507–518

Buono MR, Fernández MS, Reguero MA, Marenssi SA, Santillana SN, Mörs T (2016) Eocene basilosaurid whales from the La Meseta Formation, Marambio (Seymour) Island, Antarctica. Ameghiniana 53:296–315

Cabaleri NG, Volkheimer W, Armella C, Gallego O, Silva Nieto D, Páez M, Cagnoni M, Ramos A, Panarello H, Koukharsky M (2010a) Estratigrafía, análisis de facies y paleoambientes de la Formación Cañadón Asfalto en el depocentro jurásico Cerro Cóndor, provincia del Chubut. Rev Asoc Geol Argent 66:349–367

Cabaleri NG, Volkheimer W, Silva Nieto DG, Armella C, Cagnoni M, Hauser N, Matteini M, Pimentel MM (2010b) U-Pb ages in zircons from Las Chacritas and Puesto Almada members of the Jurassic Cañadón Asfalto Formation, Chubut Province, Argentina. 2°South American Symposium of Isotope Geology, Brasilia, Abstracts:190–193

Cabaleri NG, Benavente CA, Monferran MD, Narváez PL, Volkheimer W, Gallego OF, Do Campo MD (2013) Palaeoenvironmental and climatic significance of the Puesto Almada Member (Upper Jurassic, Cañadón Asfalto Formation), at the Fossati Sub-basin, Patagonia Argentina. Sediment Geol 296:103–121

Cabreira SF, Schultz CL, Bittencourt JS, Soares MB, Fortier DC, Silva LR, Langer MC (2011) New stem-sauropodomorph (Dinosauria, Saurischia) from the Triassic of Brazil. Naturwissenschaften 98:1035–1040

Cabreira SF, Kellner AWA, Dias-da-Silva S, Roberto-da-Silva L, Bronzati M, Marsola JCA, Müller RT, Bittencourt JS, Batista BJA, Raugust T, Carrilho R, Brodt A, Langer MC (2016) A unique Late Triassic dinosauromorph assemblage reveals dinosaur ancestral anatomy and diet. Curr Biol 26:1–6

Cabrera A (1943) El primer hallazgo de terápsidos en la Argentina. Notas Mus La Plata, Paleont 8:317–331

Calvo JO (1991) Huellas fósiles de dinosaurios en la Formación Río Limay (Albiano–Cenomaniano), Picún Leufú, Provincia del Neuquén, Argentina. (Ornithischia-Saurischia: Sauropoda-Teropoda). Ameghiniana 28:241–258

Calvo JO (1999) Dinosaurs and other vertebrates of the Lake Ezequiel Ramos Mexía Area, Neuquén-Patagonia, Argentina. In: Tomida Y, Rich TH, Vickers-Rich P (eds) Proceedings of the Second Gondwanan Dinosaur Symposium, National Science Museum, Tokyo, pp 13–45

Calvo JO (2007) Mesozoic reptilian tracks. In: Gasparini Z, Coria RA, Salgado L (eds) Patagonian Mesozoic reptiles. Indiana University Press, Bloomington, pp 314–334

Calvo JO, Bonaparte JF (1991) *Andesaurus delgadoi* n.g. n.sp. (Saurischia, Sauropoda) dinosaurio Titanosauridae de la Formación Río Limay (Albiano–Cenomaniano), Neuquén, Argentina. Ameghiniana 28:303–310

Calvo JO, González Riga BJ (2003) *Rinconsaurus caudamirus* gen. et sp nov., a new titanosaurid (Dinosauria, Sauropoda) from the Late Cretaceous of Patagonia, Argentina. Rev Geol Chile 30:333–353

Calvo JO, Porfiri JD (2010) *Panamericansaurus schroederi* gen. nov. sp. nov. un nuevo Sauropoda (Titanosauridae-Aeolosaurini) de la Provincia del Neuquén, Cretácico Superior de Patagonia, Argentina. Brazilian Geogr J Geosci Human Resmed 1:100–115

Calvo JO, Salgado L (1995) *Rebbachisaurus tessonei* sp. nov., a new Sauropoda from the Albian-Cenomanian of Argentina: new evidence on the origin of the Diplodocidae. GAIA 11:13–33

Calvo JO, Rubilar-Rogers D, Moreno K (2004) A new Abelisauridae (Dinosauria: Theropoda) from northwest Patagonia. Ameghiniana 41:555–563

Calvo JO, Porfiri J, Pol D, González Riga BJ, de la Fuente M, Rougier GW (2011) Vertebrados continentales mesozoicos. In: Leanza HA, Arregui C, Carbone O, Danieli JC, Vallés JM (eds) Geología y recursos naturales de la Provincia del Neuquén, 18° Congreso Geológico Argentino, Neuquén, Relatorio, pp 539–556

Campbell KE (ed) (2004) The Paleogene mammalian fauna of Santa Rosa, Amazonian Peru. Nat Hist Mus Los Angeles County, Sci Ser

Campbell KE, Frailey CD, Romero-Pittman L (2004) The Paleogene Santa Rosa local fauna of Amazonian Peru: Geographic and geologic setting. In: Campbell KE (ed) The Paleogene mammalian fauna of Santa Rosa, Amazonian Peru. Nat Hist Mus Los Angeles County, Sci Ser, vol 40, pp 3–14

Campos DA, Suarez JM, Riff D, Kellner AWA (2001) Short note on a new Baurusuchidae (Crocodyliformes, Metasuchia) from Late Cretaceous of Brazil. Bol do Mus Nac 57:1–7

Campos DA, Oliveira GR, Figueiredo RG, Riff D, Azevedo SAK, Carvalho LB, Kellner AWA (2011) On a new peirosaurid crocodyliform from the Upper Cretaceous, Bauru Group, southeastern Brazil. An Acad Bras Ciênc 83:317–327

Candeiro CRA, Martinelli AG, Avilla LS, Rich TH (2006) Tetrapods from the Upper Cretaceous (Turonian–Maastrichtian) Bauru Group of Brazil: a reappraisal. Cretac Res 27:923–946

Candeiro CRA, Nava W, Martinelli AG, Forasiepi AM, Scanferla CA, Muzzopappa P (2009) New lizard record (Diapsida; Lepidosauria) from the Late Cretaceous Adamantina Formation, Brazil. Bull Geosci 84:573–576

Candeiro CRA, Fanti F, Cau A, Nava WR, Novas FE (2012) First evidence of an unenlagiid (Dinosauria, Theropoda, Maniraptora) from the Bauru Group, Brazil. Cretac Res 32:223–226

Candela AM, Rasia LL, Pérez ME (2012) Paleobiology of Santacrucian caviomorph rodents: A morphofunctional approach. In: Vizcaíno SF, Kay RF, Bargo MS (eds) Early Miocene paleobiology in Patagonia. Cambridge University Press, Cambridge, pp 287–305

Candia Halupczok DJ, Sánchez ML, Veiga GD, Apesteguía S (2018) Dinosaur tracks in the Kokorkom Desert, Candeleros Formation (Cenomanian, Upper Cretaceous), Patagonia Argentina: Implications for deformation structures in dune fields. Cretac Res 83:194–206

Carvalho AB, Zaher H, Nava WR (2003) A new anuran (Lissamphibia: Tetrapoda) from the continental Late Cretaceous Bauru Basin, State of São Paulo. 18° Congresso Brasileiro de Paleontologia, Brasília, Abstracts:188

Carvalho IS, Bertini RJ (1999) *Mariliasuchus*, um novo Crocodylomorpha (Notosuchia) do Cretáceo da Bacia Bauru, Brasil. Geol Colomb 24:83–105

Carvalho IS, Arruda-Campos AC, Nobre PH (2005) *Baurusuchus salgadoensis*, a new Crocodylomorpha from the Bauru Basin (Cretaceous), Brazil. Gondwana Res 8:11–30

Carvalho IS, Vasconcellos FM, Tavares SAS (2007) *Montealtosuchus arrudacamposi*, a new peirosaurid crocodile (Mesoeucrocodylia) from the Late Cretaceous Adamantina Formation of Brazil. Zootaxa 1607:35–46

Carvalho IS, Gasparini ZB, Salgado L, Vasconcellos FM, Marinho TS (2010) Climate's role in the distribution of the Cretaceous terrestrial Crocodyliformes throughout Gondwana. Palaeogeogr Palaeoclimatol Palaeoecol 297:252–262

Carvalho IS, Teixeira VPA, Ferraz MLF, Ribeiro LCB, Martinelli AG, Neto FM, Sertich JJ, Cunha GC, Cunha IC, Ferraz PF (2011) *Campinasuchus dinizi* gen. et sp. nov., a new Late Cretaceous baurusuchid (Crocodyliformes) from the Bauru Basin, Brazil. Zootaxa 2871:19–42

Casadío S (1998) Las ostras del límite Cretácico-Paleógeno de la cuenca Neuquina (Argentina). Su importancia bioestratigráfica y Paleobiogeográfica. Ameghiniana 35:449–471

Casamiquela RM (1961) Sobre la presencia de un mamífero en el primer elenco (icnológico) de vertebrados del Jurásico de la Patagonia. Physis 22:225–233

Casamiquela RM (1964a) Estudios icnológicos. Problemas y métodos de la icnología con aplicación al estudio de pisadas mesozoicas (Reptilia, Mammalia) de la Patagonia. Colegio Industrial Pío IX, Buenos Aires

Casamiquela RM (1964b) Sobre un dinosaurio hadrosáurido de la Argentina. Ameghiniana 3:285–312

Casamiquela RM (1967) Un nuevo dinosaurio ornitisquio Triásico (*Pisanosaurus mertii*; Ornithopoda) de la Formación Ischigualasto, Argentina. Ameghiniana 4:47–64

Casamiquela RM (1969a) Historia geológica del valle de Huahuel Niyeu, área extraandina del suroeste de la Provincia de Río Negro, República Argentina (con especial énfasis en el Pleistoceno). Rev Asoc Geol Argent 24:287–329

Casamiquela RM (1969b) La presencia en Chile de *Aristonectes* Cabrera (Plesiosauria) del Maestrichtense del Chubut, Argentina. Edad y carácter de la transgresión "Rocanense". 4° Jornadas Geológicas Argentinas, Mendoza, Actas, vol 1, pp 199–213

Casamiquela RM (1978) La zona litoral de la transgresión Maástrichtense en el norte de Patagonia. Aspectos ecológicos. Ameghiniana 15:137–148

Casamiquela RM (1980) Nota sobre restos de un reptil aetosauroideo (Thecodontia, Aetosauria) de Quimal, Cordillera de Domeyko, Antofagasta. Prueba de la existencia del Neotriásico continental en los Andes del Norte de Chile. 2° Congreso Argentino de Paleontología y Bioestratigrafía y 1° Congreso Latinoamericano de Paleontología, Buenos Aires, Actas, vol 1, pp 135–142

Case JA, Woodburne MO, Chaney DS (1988) A new genus and species of polydolopid marsupial from the La Meseta Formation, Late Eocene, Seymour Island, Antarctic Peninsula. In: Feldmann RM, Woodburne MO (eds) Geology and paleontology of Seymour Island, Geol Soc Am Mem 169:505–521

Case JA, Goin FJ, Woodburne MO (2005) "South American" marsupials from the Late Cretaceous of North America and the origin of marsupial cohorts. J Mammal Evol 12:461–494

Cassini GH, Cerdeño E, Villafañe AL, Muñoz NA (2012) Paleobiology of Santacrucian native ungulates (Meridiungulata: Astrapotheria, Litopterna and Notoungulata). In: Vizcaíno SF, Kay RF, Bargo MS (eds) Early Miocene paleobiology in Patagonia. Cambridge University Press, Cambridge, pp 243–286

Castro MC, Goin FJ, Ortiz-Jaureguizar E, Vieytes EC, Tsukui K, Ramezani J, Batezelli A, Marsola JCA, Langer MC (2018) A Late Cretaceous mammal from Brazil and the first radioisotopic age for the Bauru Group. R Soc Open Sci 5:180482

Chacaltana C, Benites Cañote AL, Valdivia W, Chumpitaz M (2015) La Formación Chota en la Cuenca Bagua: evaluación e implicancias del criterio de clasificación litoestratigráfico. Bol Soc Geol Peru 110:63–66

Chiappe LM (1988) A new trematochampsid crocodile from the Early Cretaceous of north-western Patagonia, Argentina and its palaeobiogeographical and phylogenetic implications. Cretac Res 9:379–389

Chiappe LM, Coria RA (2004) Auca Mahuevo, un extraordinario sitio de nidificación de dinosaurios saurópodos del Cretácico Tardío, Neuquén, Argentina. Ameghiniana 41:591–596

Chiappe LM, Dingus L (2001) Walking on eggs. Scribner, New York

Chiappe LM, Coria RA, Dingus L, Jackson FD, Chinsamy A, Fox M (1998) Sauropod dinosaur embryos from Late Cretaceous of Patagonia. Nature 396:258–261

Chiappe LM, Salgado ML, Coria RA (2001) Embryonic skulls of titanosaur sauropod dinosaurs. Science 293:2444–2446

Chiappe LM, Nava W, Martinelli AG, Tucker R, Alvarenga H (2018) A remarkable assemblage of enantiornithine birds from the Late Cretaceous of southeastern Brazil. 78° Annual Meeting, Society of Vertebrate Paleontology, New Mexico, Abstract:105

Chimento NR, Agnolín FL, Novas FE (2012) The Patagonian fossil mammal *Necrolestes*: A Neogene survivor of Dryolestoidea. Rev Mus Argent Cienc Nat "B Rivadavia", ns 14:261–306

Chimento NR, Agnolín FL, Novas F (2015) The bizarre 'metatherians' *Groeberia* and *Patagonia*, late surviving members of gondwanatherian mammals. Hist Biol 27:603–623

Chimento N, Agnolín FL, Martinelli AG (2016) Mesozoic mammals from South America: implications for understanding early mammalian faunas from Gondwana. Mus Argent Cienc Nat "B Rivadavia". Contribuciones 6:199–209

Chimento NR, Agnolín FL, Tsuihiji T, Manabe M, Novas FE (2020) New record of a Mesozoic gondwanatherian mammaliaform from Southern Patagonia. Sci Nat 107:49

Chornogubsky L, Gelfo JJ (2011) Un nuevo mesungulátido (Mammalia, Dryolestoidea) del Cretácico Tardío de Argentina. Ameghiniana 48(Suppl):9R

Chornogubsky L, Goin FJ, Reguero MA (2009) A reassessment of Antarctic polydolopid marsupials (Middle Eocene, La Meseta Formation). Antarct Sci 21:285–297

Ciancio MR, Carlini AA, Campbell KE, Scillato-Yané GJ (2013) New Palaeogene cingulates (Mammalia, Xenarthra) from Santa Rosa, Peru and their importance in the context of South American faunas. J Syst Palaeontol 11:727–741

Cione AL (1987) Part II -The fishes. In: Bonaparte JF (ed) The Late Cretaceous fauna of Los Alamitos, Patagonia, Argentina. Rev Mus Argent Cienc Nat "B Rivadavia", Paleontología, vol 3, pp 111–120

Cione AL, Gourie S, Goin FJ, Poiré D (2007) *Atlantoceratodus*, a new genus of lungfish from Upper Cretaceous of South America and Africa. Rev Mus La Plata, Paleontología 10(62):1–12

Cione AL, Gasparini G, Soibelzon E, Soibelzon LH, Tonni EP (2015) The Great American Biotic Interchange. A South American perspective. Springer, Dordrecht

Cisneros JC, Schultz CL (2003) *Soturnia caliodon* n. g. n. sp., a procolophonid reptile from the upper Triassic of Southern Brazil. Neues Jahrb Geol Paläontol 227:365–380

Clarke JA, Chiappe LM (2001) A new carinate bird from the Late Cretaceous of Patagonia (Argentina). Am Mus Novit 3323:1–23

Clyde CW, Wilf P, Iglesias A, Slingerland RL, BarnumT Bij PK, Bralower TJ, Brinkhuis H, Comer EE, Huber BT, Ibañez-Mejia I, Jicha BR, Krause JM, Schueth JD, Singer BS, Raigemborn MS, Schmitz MD, Sluijs A, Zamaloa MC (2014) New age constraints for the Salamanca Formation and lower Río Chico Group in the western San Jorge Basin, Patagonia, Argentina: Implications for Cretaceous-Paleogene extinction recovery and land mammal age correlations. Geol Soc Am Bull 126:289–306

Coates AG, Stallard RF (2013) How old is the Isthmus of Panama? Bull Mar Sci 89:801–813

Coates AG, Stallard RF (2015) Historia Natural del Istmo de Panamá. In: Rodríguez Mejía F, O'Dea A (eds) Historia Natural del Istmo de Panamá. Toppan Lee Fung, Panamá, pp 17–27

Codorniú L, Carabajal AP, Pol D, Unwin D, Rauhut OWM (2016) A Jurassic pterosaur from Patagonia and the origin of the pterodactyloid neurocranium. PeerJ 4:e2311

Coira B (1979) Descripción geológica de la Hoja 40 d, Ingeniero Jacobacci, Provincia de Río Negro. Dirección Nacional del Servicio Geológico, Buenos Aires, Bol 168:1–104

Colombi CE, Santi-Malnis P, Correa GA, Martínez RN, Fernández E, Abelín DO, Praderio A, Apaldetti C, Alcober O, Drovandi J (2015) La Formación Balde de Leyes, una nueva unidad estratigráfica de la Cuenca Triásica de Marayes-El Carrizal, San Juan. Rev Asoc Geol Argent 72:456–469

Colombi CE, Limarino CQ, Alcober OA (2017) Allogenic controls on the fluvial architecture and fossil preservation of the Upper Triassic Ischigualasto Formation, NW Argentina. Sediment Geol 362:1–16

Comer EE, Slingerland RL, Krause JM, Iglesias A, Clyde WC, Raigemborn MS, Wilf P (2015) Sedimentary facies and depositional environments of diverse Early Paleocene floras, north-central San Jorge Basin, Patagonia, Argentina. Palaios 30:553–573

Cope ED (1882) Mammalia in the Laramie Formation. Am Nat 16:830–831

Coria RA (2001) A new theropod from the Late Cretaceous of Patagonia. In: Tanke DH, Carpenter K (eds) Mesozoic vertebrate life. Indiana University Press, Bloomington, pp 3–9

Coria RA (2016) A review of the ornithischian dinosaurs from Argentina. In: Agnolín FA, Lio GL, Brissón Egli F, Chimento NR, Novas FE (eds) Historia evolutiva y paleobiogeográfica de los vertebrados de América del Sur. Mus Argent Cienc Nat "B Rivadavia", Contribuciones, vol 6, pp 109–117

Coria RA, Salgado L (1995) A new giant carnivorous dinosaur from the Cretaceous of Patagonia. Nature 377:224–226

Coria RA, Salgado L (1996a) "*Loncosaurus argentinus*" Ameghino, 1899 (Ornithischia, Ornithopoda): A revised description with comments on its phylogenetic relationships. Ameghiniana 33:373–376

Coria RA, Salgado L (1996b) A basal iguanodontian (Ornithischia, Ornithopoda) from the Late Cretaceous of South America. J Vertebr Paleontol 16:445–457

Coria RA, Chiappe LM, Dingus L (2002a) A new close relative of *Carnotaurus sastrei* Bonaparte, 1985 (Theropoda: Abelisauridae) from the Late Cretaceous of Patagonia. J J Vertebr Paleontol 22:460–465

Coria RA, Currie PJ, Eberth D, Garrido AC (2002b) Bird footprints from the Anacleto Formation (Late Cretaceous), Neuquén, Argentina. Ameghiniana 39:453–464

Coria RA, Cambiaso AV, Salgado L (2007) New records of basal ornithopod dinosaurs in the Cretaceous of North Patagonia. Ameghiniana 44:473–477

Coria RA, González Riga B, Casadío S (2012) Un nuevo hadrosáurido (Dinosauria, Ornithopoda) de la Formación Allen, Provincia de La Pampa, Argentina. Ameghiniana 49:552–572

Coria RA, Filippi LS, Chiappe LM, García R, Arcucci AB (2013) *Overosaurus paradasorum* gen. et sp. nov., a new sauropod dinosaur (Titanosauria: Lithostrotia) from the Late Cretaceous of Neuquén, Patagonia, Argentina. Zootaxa 3683:357–376

Cortés JM, Baldoni A (1984) Plantas fósiles jurásicas del sur del Río Chubut medio. 9° Congreso Geológico Argentino, Buenos Aires, Actas, vol 4, pp 432–443

Cox CB (1962) Preliminary diagnosis of *Ischigualastia*, a new genus of dicynodont from Argentina. Breviora 156:8–9

Croft DA, Gelfo JN, López GM (2020) Splendid innovation: the extinct South American native ungulates. Annu Rev Earth Planet Sci 48:259–290

Crompton AW, Musinsky C, Rougier GW, Bhullar BAS, Miyamae JA (2018) Origin of the lateral wall of the mammalian skull: fossils, monotremes and therians revisited. J Mammal Evol 25:301–313

Cuitiño JI, Fernicola JC, Kohn MJ, Trayler R, Naipauer M, Bargo MS, Kay RF, Vizcaíno SR (2016) U-Pb geochronology of the Santa Cruz Formation (Early Miocene) at the Río Bote and Río Santa Cruz (southernmost Patagonia, Argentina): Implications for the correlation of fossil vertebrate localities. J S Am Earth Sci 70:198–210

Cuitiño JI, Varela AN, Ghiglione MC, Richiano S, Poiré DG (2019) The Austral-Magallanes Basin (southern Patagonia): a synthesis of its stratigraphy and evolution. Lat Am J Sedimentol Basin Anal 26:155–166

Cúneo NR, Panza JL (2008) El Bosque Petrificado de Madre e Hija: El Parque Jurásico de la Patagonia. In: CSIGA (ed) Sitios de interés geológico de la República Argentina. Instituto de Geología y Recursos Minerales y Servicio Geológico Minero Argentino, Buenos Aires, pp 781–789

Cúneo R, Ramezani J, Scasso R, Pol D, Escapa IH, Zavattieri AM, Bowring SA (2013) High-precision U-Pb geochronology and a new chronostratigraphy for the Cañadón Asfalto Basin, Chubut, central Patagonia: Implications for terrestrial faunal and floral evolution in Jurassic. Gondwana Res 24:1267–1275

Cúneo NR, Gandolfo MA, Zamaloa MC, Hermsen E (2014) Late Cretaceous aquatic plant world in Patagonia, Argentina. PLoS ONE 9(8):e104749

Czaplewski NJ (2010) Bats (Mammalia: Chiroptera) from Gran Barranca (Early Miocene, Colhuehuapian), Chubut Province, Argentina. In: Madden RH, Carlini AA, Vucetich MG, Kay RF (eds) The paleontology of Gran Barranca: evolution and environmental change through the Middle Cenozoic of Patagonia. Cambridge University Press, Cambridge, pp 240–252

Czaplewski NJ, Campbell KE (2004) A possible bat (Mammalia: Chiroptera) from the ?Eocene of Amazonian Peru. In: Campbell KE (ed) The Paleogene mammalian fauna of Santa Rosa, Amazonian Peru. Nat Hist Mus Los Angeles County, Sci Ser, vol 40, pp 141–144

Danderfer JC, Vera P (1992) Carta Geológica de la Provincia del Neuquén, Departamento Confluencia (1:200.000). Dirección Provincial de Minería, Servicio Geológico Neuquino, Zapala

de Barrio RE (1993) El volcanismo ácido Jurásico en el Noroeste de Santa Cruz, Argentina. 12° Congreso Geológico Argentino, Mendoza, Actas, vol 4, pp 189–198

de Barrio RE, Panza JL, Nullo FE (1999) Jurásico y Cretácico del Macizo del Deseado, Provincia de Santa Cruz. In: Caminos R (ed) Geología Argentina. Servicio Geológico Minero Argentino, Buenos Aires, pp 511–527

de Broin F (1987) Part IV -Chelonia. In: Bonaparte JF (ed) The Late Cretaceous fauna of Los Alamitos, Patagonia, Argentina. Rev Mus Argent Cienc Nat "B Rivadavia", Paleontología vol 3, pp 131–139

de Broin F, de la Fuente MS (1993) Les tortues fossiles d'Argentine: Synthèse. Ann Paléont (Vert) 79:169–232

Defler T (2019) History of terrestrial mammals in South America. How South American mammalian fauna changed from the Mesozoic to Recent times. Springer, Dordrecht

Degrange FJ, Noriega JI, Areta JI (2012) Diversity and paleobiology of the Santacrucian birds. In: Vizcaíno SF, Kay RF, Bargo MS (eds) Early Miocene paleobiology in Patagonia. Cambridge University Press, Cambridge, pp 138–155

de la Fuente MS, Lapparent de Broin F, Manera de Bianco T (2001) The oldest and first nearly complete skeleton of a chelid, of the *Hydromedusa* sub-group (Chelidae, Pleurodira), from the Upper Cretaceous of Patagonia. Bull Soc Géol Fr 172:237–244

de la Fuente M, Sterli J, Maniel I (2014) Origin, evolution and biogeographic history of South American turtles. Springer, Dordrecht

de la Fuente MS, Maniel IJ, Jannello JM, Filippi LS, Cerda I (2015) Long-necked chelid turtles from the Campanian of northwestern Patagonia with comments on K/P survivorship of the genus *Yaminuechelys*. CR Palevol 14:563–576

Delcourt R, Iori FV (2018) A new Abelisauridae (Dinosauria: Theropoda) from São José do Rio Preto Formation, Upper Cretaceous of Brazil and comments on the Bauru Group fauna. Hist Biol. https://doi.org/10.1080/08912963.2018.1546700

Del Fueyo GM (1998) Coniferous woods from the Upper Cretaceous of Patagonia, Argentina. Rev Esp Paleontol 13:43–50

Desojo AB (2003) Redescripción del aetosaurio *Chilenosuchus forttae* Casamiquela (Diapsida: Arcosauria): Presencia de Triásico continental en el norte de Chile. Rev Geol Chile 30:53–63

Desojo JB, Ezcurra MD, Kischlat EE (2012) A new aetosaur genus (Archosauria: Pseudosuchia) from the Early Late Triassic of Southern Brazil. Zootaxa 3166:1–33

Desojo JB, Fiorelli LE, Ezcurra MD, Martinelli AG, Ramezani J, Da Rosa ÁAS, von Baczko MB, Trotteyn MJ, Montefeltro FC, Ezpeleta M, Langer MC (2020) The Late Triassic Ischigualasto Formation at cerro Las Lajas (La Rioja, Argentina): fossil tetrapods, high-resolution chronostratigraphy, and faunal correlations. Sci Rep 10:12782

de Valais S (2009) Ichnotaxonomic review of *Ameghinichnus*, a mammal ichnogenus from the La Matilde Formation (Middle Jurassic), Santa Cruz Province, Argentina. Zootaxa 2203:1–21

de Valais S (2011) Revision of dinosaur ichnotaxa from the La Matilde Formation (Middle Jurassic), Santa Cruz Province, Argentina. Ameghiniana 48:28–42

Dias-Brito D, Musacchio EA, Castro JC, Maranhão MSAS, Suarez JM, Rodrigues R (2001) Grupo Bauru: Uma unidade continental do Cretáceo no Brasil–concepções baseadas em dados micropaleontológicos, isotópicos e estratigráficos. Rev Paléobiol 20:245–304

Dias-da-Silva S, Cabreira SF, Roberto-da-Silva LR (2011) Occurrence of giant stereospondyl remains in the Santa Maria Formation (Middle Triassic of southern Brazil). Alcheringa 35:11–19

Domnanovich NS, Marsicano C (2012) The Triassic dicynodont *Vinceria* (Therapsida, Anomodontia) from Argentina and a discussion on basal Kannemeyeriiformes. Geobios 45:173–186

D'Orazi Porchetti S, Bertini RJ, Langer MC (2018) Proposal for ichnotaxonomic allocation of therapsid footprints from the Botucatu Formation (Brazil). Ichnos 25:192–207

Echeveste H, Fernández R, Bellieni G, Tessone M, Llambías E, Schlamuk I, Piccirillo E, De Min A (2001) Relaciones entre las Formaciones Bajo Pobre y Chon Aike (Jurásico Medio a Superior) en el área Estancia El Fénix-Cerro Huemul, zona centro-occidental del Macizo del Deseado, Provincia de Santa Cruz. Rev Asoc Geol Argent 56:548–558

Escapa IH, Sterli J, Pol D, Nicoli L (2008a) Jurassic tetrapods and flora of Cañadón Asfalto Formation in Cerro Cóndor area, Chubut Province. Rev Asoc Geol Argent 63:613–624

Escapa IH, Cúneo NR, Axsmith B (2008b) A new genus of the Cupressaceae (*sensu lato*) from the Jurassic of Patagonia; implications for conifer megasporangiate cone homologies. Rev Palaeobot Palynol 151:110–122

Ezcurra MD (2017) A new early coelophysoid neotheropod from the Late Triassic of northwestern Argentina. Ameghiniana 54:506–538

Ezcurra MD, Agnolín FL (2012) A new global palaeobiogeographical model for the late Mesozoic and early Tertiary. Syst Biol 61:553–566

Ezcurra MD, Leucona A, Martinelli AG (2010a) A new basal archosauriform diapsid from the Early Triassic of Argentina. J Vertebr Paleontol 30:1433–1450

Ezcurra MD, Agnolín FL, Novas FE (2010b) An abelisauroid dinosaur with a non-atrophied manus from the Late Cretaceous Pari Aike Formation of southern Patagonia. Zootaxa 2450:1–25

Ezcurra MD, Fiorelli LE, Martinelli AG, Rocher S, von Baczko MB, Ezpeleta M, Taborda JRA, Hechenleitner EM, Trotteyn MJ, Desojo JB (2017) Deep faunistic changes preceded the rise of dinosaurs in southwestern Pangaea. Nat Ecol Evol 1:1477–1483

Fachini TS, Hsiou AS (2011) Presence of an anilioid snake from the Late Cretaceous of Adamantina Formation, Brazil. Ameghiniana 48(Suppl):164 R

Falaschi P, Grosfeld J, Zamuner AB, Foix N, Rivera SM (2011) Growth architecture and silhouette of Jurassic conifers from La Matilde Formation, Patagonia, Argentina. Palaeogeogr Palaeoclimatol Palaeoecol 302:122–141

Fernandes ACS, Netto RG, Carvalho IS (1988) O icnogênero *Taenidium* na Formação Botucatu. An Acad Bras Ciênc 60:493

Fernandes ACS, Carvalho IS, Netto RG (1990) Icnofósseis de invertebrados da Formação Botucatu, São Paulo (Brasil). An Acad Bras Ciênc 62:45–49

Fernandes LA, Basilici G (2009) Transition of ephemeral palustrine to aeolian deposits in arid-semiarid environment of the intracratonic Bauru Basin (Upper Cretaceous, Brazil). Cretac Res 30:605–614

Fernandes LA, Coimbra AM (1992) A cobertura Cretácea suprabasáltica no Estado do Paraná e Pontal do Paranapanema (SP): Os grupos Bauru e Caiuá. 37° Congresso Brasileiro de Geologia, São Paulo. Bol 2:506–508

Fernandes LA, Coimbra AM (1996) A Bacia Bauru (Cretáceo Superior, Brasil). An Acad Bras Ciênc 68:195–205

Fernandes LA, Coimbra AM (2000) Revisão estratigráfica da parte oriental da Bacia Bauru (Neocretáceo). Rev Bras Geociênc 30:717–728

Fernandes MA, Carvalho IS (2007) Pegadas fósseis da Formação Botucatu (Jurássico Superior–Cretáceo Inferior): O registro de um grande dinossauro Ornithopoda na Bacia do Paraná. In: Carvalho IS, Cassab RCT, Schwanke C, Carvalho MA, Fernandes ACS, Carvalho MSS, Arai M, Oliveira MEQ (eds) Paleontologia: Cenários da vida. Interciência, Rio de Janeiro, pp 425–432

Fernandes MA, Carvalho IS (2008) Revisão diagnóstica para a icnoespécie de tetrápode mesozóico *Brasilichnium elusivum* (Leonardi, 1981) (Mammalia) da Formação Botucatu, Bacia do Paraná, Brasil. Ameghiniana 45:167–173

Fernandes MA, Fernandes LBR, Souto PRF (2004) Occurrence of urolites related to dinosaurs in the Lower Cretaceous of the Botucatu Formation, Paraná Basin, São Paulo State, Brazil. Rev Bras Paleontol 7:263–268

Fernández MS, Gasparini Z (2012) Campanian and Maastrichtian mosasaurs from Antarctic Peninsula and Patagonia, Argentina. Bull Soc Géol Fr 183:93–102

Fernícola JC, Albino AM (2012) Amphibians and squamate reptiles from the Santa Cruz Formation (late Early Miocene), Santa Cruz Province, Argentina: paleoenvironmental and paleobiological considerations. In: Vizcaíno SF, Kay RF, Bargo MS (eds) Early Miocene paleobiology in Patagonia. Cambridge University Press, Cambridge, pp 129–137

Ferreira GS, Iori FV, Hermanson G, Langer MC (2018) New turtle remains from the Late Cretaceous of Monte Alto-SP, Brazil, including cranial osteology, neuroanatomy and phylogenetic position of a new taxon. PalZ 92:481–498

Feruglio E (1938) El Cretácico superior del Lago San Martín (Patagonia) y de las regiones adyacentes. Physis 12:293–342

Feruglio E (1944) Estudios geológicos y glaciológicos en la región del Lago Argentino (Patagonia). Acad Nac Cienc Córdoba, Bol 37:3–255

Feruglio E (1949) Descripción geológica de la Patagonia. Dirección General Yacimientos Petroliferos Fiscales, Buenos Aires

Figari EG, Scasso RA, Cúneo RN, Escapa IH (2015) Estratigrafía y evolución geológica de la Cuenca de Cañadón Asfalto, Provincia del Chubut, Argentina. Lat Am J Sedimentol Basin Anal 22:135–169

Filippi LS, Garrido AC (2008) *Pitekunsaurus macayai* gen. et sp. nov., nuevo titanosaurio (Saurischia, Sauropoda) del Cretácico Superior de la Cuenca Neuquina, Argentina. Ameghiniana 45:575–590

Filippi LS, García RA, Garrido AC (2011) A new titanosaur sauropod dinosaurs from the Upper Cretaceous of North Patagonia, Argentina. Acta Paleontol Pol 56:505–520

Fiorelli LE, Rocher S, Martinelli AG, Ezcurra MD, Hechenleitner EM, Ezpeleta M (2018) Tetrapod burrows from the Middle–Upper Triassic Chañares Formation (La Rioja, Argentina) and its palaeoecological implications. Palaeogeogr Palaeoclimatol Palaeoecol 496:85–102

Flynn JJ, Parrish JM, Rakotosamimanana B, Simpson WF, Wyss AR (1999) A Middle Jurassic mammal from Madagascar. Nature 401:57–60

Forasiepi AM, Lopez Armengol MF (1999) *Potamolithus windhauseni* (Parodiz, 1961) (Gastropoda-Hidrobiidae) in Los Alamitos Formation (Campanian–Maastrichtian) from Río Negro Province, Argentina. 7° International Symposium on Mesozoic Terrestrial Ecosystems, Buenos Aires, Abstracts:25–26

Forasiepi AM, Martinelli AG (2003) A femur of a monotreme (Mammalia, Monotremata) from the Early Paleocene Salamanca Formation of Patagonia, Argentina. Ameghiniana 40:625–630

Forasiepi AM, Rougier GW (2009) Additional data on Early Paleocene metatherians (Mammalia) from Punta Peligro (Salamanca Formation, Argentina): Comments based on the morphology of the petrosal bone. J Zool Syst Evol Res 47:391–398

Forasiepi AM, Coria RA, Hurum J, Currie PJ (2012) First dryolestoid (Mammalia, Dryolestida) dentary from the Coniacian of Patagonia, Argentina. Ameghiniana 49:497–504

Fosdick JC, Schwartz TM, Romans B, Ali R, Leonard JS, Bostelmann JE, Ugalde Peralta R, Bernhard A, Graham AS (2016) Cenozoic evolution of the Magallanes-Austral Basin and Patagonian Fold-Thrust Belt: A tale of inheritance and sediment recycling. American Association of Petroleum Geologists Annual Convention & Exhibition, Calgary, Abstract

Foulger GR (2018) Origin of the South Atlantic igneous province. J Volcanol Geotherm Res 355:2–20

Fox RC (1985) Upper molar structure in the Late Cretaceous symmetrodont *Symmetrodontoides* Fox, and a classification of the Symmetrodonta (Mammalia). J Paleontol 59:21–26

Frailey CD, Campbell KE (2004) Paleogene rodents from Amazonian Peru: The Santa Rosa local fauna. In: Campbell KE (ed) The Paleogene mammalian fauna of Santa Rosa, Amazonian Peru. Nat Hist Mus Los Angeles County, Sci Ser, vol 40, pp 71–130

Franchi M, Ardolino AA, Remesal M (2001) Hoja Geológica N°4166- III, Cona Niyeu. Provincia de Río Negro. Instituto de Geología y Recursos Naturales, SEGEMAR, Buenos Aires, Bol 262:1–114

Francischini H, Dentzien-Dias PC, Fernandes MA, Schultz CL (2015) Dinosaur ichnofauna of the Upper Jurassic/Lower Cretaceous of the Paraná Basin (Brazil and Uruguay). J S Am Earth Sci 63:180–190

Francischini H, Dentzien-Dias PC, Gobbi V, Fernandes MA (2018) As lendas e a ciência por trás dos répteis gigantes de Araraquara. Rev Biol 18:31–36

Frenguelli J (1948) Estratigrafía y edad del llamado "Retico" en la Argentina. An Soc Arg Est Geograf 8:159–309

Frenguelli J (1949) Los estratos con "*Estheria*" en el Chubut (Patagonia). Rev Asoc Geol Argent 4:11–24

Gaetano LC, Rougier GW (2011) New materials of *Argentoconodon fariasorum* (Mammaliaformes, Triconodontidae) from the Jurassic of Argentina and its bearing on triconodont phylogeny. J Vertebr Paleontol 31:829–843

Gaetano LC, Rougier GW (2012) First amphilestid from South America: a molariform from the Jurassic Cañadón Asfalto Formation, Patagonia, Argentina. J Mammal Evol 19:235–248

Gaetano LC, Marsicano CA, Rougier GW (2013) A revision of the putative Late Cretaceous triconodonts from South America. Cretac Res 46:90–100

Gallego OF (1994) Conchóstracos Jurásicos de Santa Cruz y Chubut, Argentina. Ameghiniana 31:333–345

Gallego OF, Shen Y-B (2004) Revision of a conchostracan form from the La Amarga Formation (Lower Cretaceous), Neuquen Basin, Argentina. Rev Bras Paleontol 7:5–12

Gallego OF, Shen YB, Cabaleri NG, Hernández M (2010) The genus *Congestheriella* Kobayashi, 1954 (Conchostraca, Afrograptioidea): Redescription and new combination to *Isaura olsoni* Bock from Venezuela and a new species from Argentina (Upper Jurassic). Alavesia 3:11–24

Gallego OF, Cabaleri NG, Armella C, Volkheimer W, Ballent SC, Martínez S, Monferran MD, Silva Nieto DG, Páez MA (2011) Paleontology, sedimentology and paleoenvironment of a new fossiliferous locality of the Jurassic Cañadón Asfalto Formation, Chubut Province, Argentina. J S Am Earth Sci 31:54–68

Gallina PA, Otero A (2015) Reassessment of *Laplatasaurus araukanicus* (Sauropoda: Titanosauria) from the Upper Cretaceous of Patagonia, Argentina. Ameghiniana 52:487–501

Galton PM, Carpenter K (2016) The plated dinosaur *Stegosaurus longispinus* Gilmore, 1914 (Dinosauria: Ornithischia; Upper Jurassic, western USA), type species of *Alcovasaurus* n. gen. Neues Jahrb Geol Paläontol 279:185–208

Gandolfo MA, Cúneo NR (2005) Fossil Nelumbonaceae from the La Colonia Formation (Campanian–Maastrichtian, Upper Cretaceous), Chubut, Patagonia, Argentina. Rev Palaeobot Palynol 133:169–178

Gandolfo MA, Marenssi SA, Santillana SN (1998a) Flora y paleoclima de la Formación La Meseta (Eoceno Medio), Isla Marambio (Seymour), Antártida. In: Casadio S (ed) Paleógeno de América del Sur y de la Península Antártica. Asociación Paleontológica Argentina, Buenos Aires, Publicación Especial, vol 5, pp 155–162

Gandolfo MA, Hoc P, Santillana SN, Marenssi S (1998b) Una flor fósil morfológicamente afín a las Grossulariaceae (Orden Rosales) de la Formación La Meseta (Eoceno Medio), Isla Marambio, Antártida. In: Casadio S (ed) Paleógeno de América del Sur y de la Península Antártica. Asociación Paleontológica Argentina, Buenos Aires, Publicación Especial, vol 5, pp 147–153

Garrido AC (2010) Estratigrafía del Grupo Neuquén, Cretácico Superior de la Cuenca Neuquina (República Argentina): Nueva propuesta de ordenamiento litoestratigráfico. Rev Mus Argent Cienc Nat "B Rivadavia", ns 12:121–177

Garrido AC (2011) Grupo Neuquén (Cenomaniano–Campaniano). In: Leanza HA, Arregui C, Carbone O, Danieli JC, Vallés JM (eds) Geología y recursos naturales de la Provincia del Neuquén, 18° Congreso Geológico Argentino. Relatorio, Neuquén, pp 231–244

Gasparini Z, de la Fuente M (2000) Tortugas y plesiosaurios de la Formación La Colonia (Cretácico Superior) de Patagonia, Argentina. Rev Esp Paleontol 15:23–35

Gasparini Z, Goñi R (1985) Los plesiosaurios cretácicos de América del Sur y del continente antárctico. 8° Congresso Brasileiro de Paleontología, Brasilia, Coletanea de Trabalhos Paleontológicos, Série Geología 27:55–63

Gasparini Z, Spalletti LA (1990) Un nuevo cocodrilo en los depósitos mareales maastrichtianos de la Patagonia noroccidental. Ameghiniana 27:141–150

Gasparini Z, Sterli J, Parras A, O'Gorman J, Salgado L, Varela J, Pol D (2015) Late Cretaceous reptilian biota of the La Colonia Formation, central Patagonia, Argentina: occurrences, preservation and paleoenvironments. Cretac Res 54:154–168

Gayet M, Brito PM (1989) Ichthyofaune nouvelle du Crétacé Supérieur du Groupe Baurú (Etats de São Pãulo et Minas Gerais, Brésil). Geobios 22:841–847

Gayet M, Marshall LG, Sempere T (1991) The Mesozoic and Paleocene vertebrates of Bolivia and their stratigraphic context: a review. In: Suarez-Soruco R (ed) Fósiles y facies de Bolivia. Vertebrados. Rev Tec YPFB, vol 12, pp 393–433

Gayet M, Sempere T, Cappetta H, Jaillard E, Lévy A (1993) La presence de fossils marins dans le Crétacé terminal des Andes Centrales et ses conséquences paléogéographiques. Palaeogeogr Palaeoclimatol Palaeoecol 102:283–319

Gayet M, Marshall LG, Sempere T, Meunier FJ, Cappetta H, Rage J-C (2001) Middle Maastrichtian vertebrates (fishes, amphibians, dinosaurs and other reptiles, mammals) from Pajcha Pata (Bolivia). Biostratigraphic, palaeoecologic and palaeobiogeographic implications. Palaeogeogr Palaeoclimatol Palaeoecol 169:39–68

Gelfo JN (2007) The condylarth *Raulvaccia peligrensis* (Mammalia: Didolodontidae) from the Paleocene of Patagonia, Argentina. J Vertebr Paleontol 27:651–660

Gelfo JN, Pascual R (2001) *Peligrotherium tropicalis* (Mammalia, Dryolestida) from the Early Paleocene of Patagonia, a survival from a Mesozoic Gondwanan radiation. Geodiversitas 23:369–379

Gelfo JN, Ortiz-Jaureguizar E, Rougier GW (2007) New remains and species of the "condylarth" genus *Escribania* (Mammalia: Didolodontidae) from the Early Paleocene of Patagonia, Argentina. Earth Env Sci T R So 98:127–138

Gelfo JN, Goin FJ, Woodburne MO, de Muizon C (2009) Biochronological relationships of the earliest South American Paleogene mammalian faunas. Palaeontology 52:251–269

Genise JF, Gallego OF, Melchor RN, de Valais S (2002) Capullos fósiles de tricópteros (*Conchindusia* isp.) (Insecta) construidos con valvas de conchóstracos del Jurásico de la Patagonia, Argentina. 8° Congreso Argentino de Paleontología y Bioestratigrafía. Corrientes, Abstracts:93

Ghidella ME, Lawver LA, Gahagan LM (2007) Break-up of Gondwana and opening of the South Atlantic: Review of existing plate tectonic models. USGS, Short Research Paper 055. https://doi.org/10.3133/of2007-1047.srp055

Gianni GM, Dávila FM, Echaurren A, Fennell L, Tobal J, Navarrete C, Quezada P, Folguera A, Giménez M (2018) A geodynamic model linking Cretaceous orogeny, arc migration, foreland dynamic subsidence and marine ingression in southern South America. Earth-Sci Rev 185:437–462

Gobbo-Rodrigues SR, Petri S, Bertini RJ (1999) Ocorrências de ostrácodes na Formação Adamantina do Grupo Bauru, Cretáceo Superior da Bacia do Paraná e possibilidades de correlação com depósitos isócronos argentinos. Parte I - Família Ilyocyprididae. Acta Geol Leopoldina 23:3–13

Godoy PL, Montefeltro FC, Norell MA, Langer MC (2014) An additional baurusuchid from the Cretaceous of Brazil with evidence of interspecific predation among Crocodyliformes. PLoS ONE 9(5):e97138

Goin FJ, Abello MA (2013) Los Metatheria sudamericanos de comienzos del Neógeno (Mioceno temprano, edad mamífero Colhuehuapense): Microbiotheria y Polydolopimorphia. Ameghiniana 50:51–78

Goin FJ, Candela AM (2004) New Paleogene marsupials from the Amazon Basin of eastern Peru. In: Campbell KE (ed) The Paleogene mammalian fauna of Santa Rosa, Amazonian Peru. Nat Hist Mus Los Angeles County, Sci Ser, vol 40, pp 15–60

Goin FJ, Reguero MA (1993) Un “enigmático insectívoro” del Eoceno de Antártida. Ameghiniana 30(Suppl):108R

Goin FJ, Carlini AA, Pascual R (1986) Un probable marsupial del Cretácico tardío del Norte de Patagonia, Argentina. 4° Congreso Argentino de Paleontología y Bioestratigrafía, Mendoza, Actas, vol 2, pp 43–47

Goin FJ, Case JA, Woodburne MO, Vizcaíno SF, Reguero MA (1999) New discoveries of “opossum-like” marsupials from Antarctica (Seymour Island, Medial Eocene). J Mammal Evol 6:335–365

Goin FJ, Forasiepi AM, Candela AM, Ortiz Jaureguizar E, Pascual R, Archer M, Godthelp H, Muirhead J, Augee M, Hand S, Wroe S (2002) Earliest Paleocene marsupials from Patagonia. 1° International Palaeontological Congress, Sydney, Abstracts:68

Goin FJ, Vieytes EC, Vucetich MG, Carlini AA, Bond M (2004a) Enigmatic mammal from the Paleogene of Peru. In: Campbell KE (ed) The Paleogene mammalian fauna of Santa Rosa, Amazonian Peru. Nat Hist Mus Los Angeles County, Sci Ser, vol 40, pp 145–153

Goin FJ, Ortiz Jaureguizar E, Pascual R, Rougier GW, Cladera G (2004b) El primer registro de un Bonapartheriidae (Marsupialia, Polydolopimorphia) en el Paleoceno Inferior (Formación Salamanca, “Edad-Mamífero” Peligrense) de América del Sur. Ameghiniana 41(Suppl):49R

Goin FJ, Pascual R, Tejedor MF, Gelfo JN, Woodburne MO, Case JA, Reguero MA, Bond M, López GM, Cione AL, Udrizar Sauthier D, Balarino L, Scasso RA, Medina FA, Ubaldón MC (2006a) The earliest Tertiary therian mammal from South America. J Vertebr Paleontol 26:505–510

Goin FJ, Reguero MA, Pascual R, Koenigswald W von, Woodburne MO, Case JA, Vieytes C, Marenssi SA, Vizcaíno SF (2006b) First gondwanatherian mammal from Antarctica. In: Francis JE, Pirrie D, Crame JA (eds) Cretaceous-Tertiary high-latitude palaeoenvironments, James Ross Basin, Antarctica. Geological Society of London, Special Publications, vol 258, pp 135–144

Goin FJ, Zimicz N, Reguero MA, Santillana SN, Marenssi SA, Moly JJ (2007a) New mammal from the Eocene of Antarctica, and the origins of the Microbiotheria. Rev Asoc Geol Argent 62:597–603

Goin FJ, Abello A, Bellosi E, Kay R, Madden R, Carlini AA (2007b) Los Metatheria sudamericanos de comienzos del Neógeno (Mioceno Temprano, Edad-Mamífero Colhuehuapense). Parte 1: Introducción, Didelphimorphia y Sparassodonta. Ameghiniana 44:29–71

Goin FJ, Tejedor MF, Chornogubsky L, López GM, Gelfo JN, Bond M, Woodburne MO, Gurovich Y, Reguero M (2012) Persistence of a Mesozoic, non-therian mammalian lineage (Gondwanatheria) in the mid-Paleogene of Patagonia. Naturwissenschaften 99:449–463

Goin FJ, Woodburne MO, Zimicz AN, Martin GM, Chornogubsky L (2016) A brief history of South American metatherians: evolutionary contexts and intercontinental dispersals. Springer, Dordrecht

Goin FJ, Martinelli AG, Soto-Acuña S, Vieytes EC, Manríquez LME, Fernández RA, Pino JP, Trevisa C, Kaluza J, Reguero MA, Leppe M, Ortiz H, Rubilar-Rogers D, Vargas AO (2020) First Mesozoic mammal from Chile: the southernmost record of a Late Cretaceous gondwanatherian. Bol Mus Nac Hist Nat Chile 69:5–31

Goldberg K, García AJV (2000) Palaeobiogeography of the Bauru Group, a dinosaur-bearing Cretaceous unit, northeastern Paraná Basin, Brazil. Cretac Res 21:241–254

Gómez RO, Báez AM, Rougier GW (2008) An anilioid snake from the Upper Cretaceous of northern Patagonia. Cretac Res 29:481–488

González Díaz E (1964) Rasgos geológicos y evolución geomorfológico de la Hoja 27-D (San Rafael) y zona occidental vecina (Provincia de Mendoza). Rev Asoc Geol Arg 19:151–188

Gregory WK (1910) The orders of mammals. Bull Am Mus Nat Hist 27:1–524

Griffin M, Parras A (2012) Oysters from the base of Santa Cruz Formation (late Early Miocene) of Patagonia. In: Vizcaíno SF, Kay RF, Bargo MS (eds) Early Miocene paleobiology in Patagonia. Cambridge University Press, Cambridge, pp 83–90

Griffin M, Varela AN (2012) Systematic palaeontology and taphonomic significance of the mollusc fauna from the Mata Amarilla Formation (lower Upper Cretaceous), southern Patagonia, Argentina. Cretac Res 37:164–176

Groeber P, Stipanicic PH (1953) Triásico. In: Groeber P (ed) Geografía de la República Argentina. Sociedad Argentina de Estudios Geográficos, Buenos Aires, pp 9–141

Guler MV, Borel CM, Brinkhuis H, Navarro E, Astini R (2014) Brackish to freshwater dinoflagellate cyst assemblages from the La Colonia Formation (Paleocene?), northeastern Patagonia, Argentina. Ameghiniana 51:141–153

Gurovich Y, Beck R (2009) The phylogenetic affinities of the enigmatic mammalian clade Gondwanatheria. J Mammal Evol 16:25–49

Gutiérrez NM, Le Roux JP, Vásquez A, Carreño C, Pedroza V, Araos J, Oyarzún JL, Pino PJ, Rivera HA, Hinojosa LF (2017) Tectonic events reflected by palaeocurrents, zircon geochronology, and palaeobotany in the Sierra Baguales of Chilean Patagonia. Tectonophysics 695:76–99

Haluza A, Canale JI (2013) A partial pterosaur ulna from the Candeleros Formation (Neuquén Group), Late Cretaceous of Argentina. Hist Nat 3:5–12

Harper T, Parras A, Rougier GW (2019) *Reigitherium* (Meridiolestida, Mesungulatoidea), an enigmatic Late Cretaceous mammal from Patagonia, Argentina: morphology, affinities, and dental evolution. J Mammal Evol 26:447–478

Hatcher JB (1903) Narrative of the expeditions. Geography of Southern Patagonia. In: Scott WB (ed) Reports of the Princeton Expeditions to Patagonia, 1896–1899: J.B. Hatcher in charge. The University, Sttutgart, xvi+, pp 1–224

Heredia S, Salgado L (1999) Posición estratigráfica de los estratos supracretácicos portadores de dinosaurios en el Lago Pellegrini, Patagonia septentrional, Argentina. Ameghiniana 36:229–234

Hopson JA, Rougier GW (1993) Braincase structures in the oldest known skull of a therian mammal: Implications for mammalian systematics and cranial evolution. Am J Sci 293:268–299

Horn BLD, Melo TM, Schutlz CL, Philipp RP, Kloss HP, Goldberg K (2014) A new third-order sequence stratigraphic framework applied to the Triassic of the Paraná Basin, Rio Grande do Sul, Brazil, based on structural, stratigraphic and paleontological data. J S Am Earth Sci 55:123–132

Howell JA, Schwarz E, Spalletti LA, Veiga GD (2005) The Neuquén Basin: an overview. Geol Soc Spec Pub 252:1–14

Hsiou AS, Nydam R, Simões T, Pretto F, Onary S, Martinelli AG, Liparini A, Romo de Vivar Martínez P, Soares MB, Schultz CL, Caldwell M (2019)A new clevosaurid from the Triassic (Carnian) of Brazil and the rise of sphenodontians in Gondwana. Sci Rep 9:11821

von Huene F (1929) Los saurísquios y ornitísquios del Cretáceo Argentino. An Mus La Plata, Paleontología 3:1–196

von Huene F (1931) Die fossilen Fährten im Rhät von Ischigualasto in Nordwest-Argentinien. Palaeobiol 4:99–112

von Huene F (1935) Lieferung 1. Anomodontia. Die fossilen Reptilien des Südamerikanischen Gondwanalandes. Ergebnisse der Sauriergrabungen in Südbrasilien 1928/29. C. H. Beck Verlagsbuchhandlung, München, pp 1–82

von Huene F (1942) Lieferungen 3/4. Pseudosuchia, Saurischia, Rhynchosauridae und Schlussabschnitt. Die fossilen Reptilien des Südamerikanischen Gondwanalandes. Ergebnisse der Sauriergrabungen in Südbrasilien 1928/29. C. H. Beck Verlagsbuchhandlung, München, pp 161–332

Hugo CA, Leanza HA (2001a) Hoja Geológica 3969-IV, General Roca, provincias del Neuquén y Río Negro, Instituto de Geología y Recursos Naturales, SEGEMAR, Buenos Aires. Bol 308:1–71

Hugo CA, Leanza HA (2001b) Hoja Geológica 3966-III, Villa Regina, Provincia Río Negro, Instituto de Geología y Recursos Naturales, SEGEMAR, Buenos Aires. Bol 309:1–53

Iglesias A, Zamuner AB, Poiré DG, Larriestra F (2007a) Diversity, taphonomy and palaeoecology of an angiosperms flora from Cretaceous (Cenomanian–Coniacian) in Southern Patagonia, Argentina. Palaeontology 50:445–466

Iglesias A, Wilf P, Johnson KR, Zamuner AB, Cúneo NR, Matheos SD, Singer BS (2007b) A Paleocene lowland macroflora from Patagonia reveals significantly greater richness than North American analogs. Geology 35:947–950

Iglesias A, Artabe AE, Morel EM (2011) The evolution of Patagonian climate and vegetation from the Mesozoic to the present. Biol J Linn Soc 103:409–422

Iori FV, Carvalho IS (2009) *Morrinhosuchus luziae*, um novo Crocodylomorpha Notosuchia da Bacia Bauru, Brasil. Rev Bras Geociênc 39:717–725

Iori FV, Carvalho IS (2011) *Caipirasuchus paulistanus*, a new sphagesaurid (Crocodylomorpha, Mesoeucrocodylia) from the Adamantina Formation (Upper Cretaceous, Turonian–Santonian), Bauru Basin, Brazil. J Vertebr Paleontol 31:1255–1264

Iori FV, García KL (2012) *Barreirosuchus franciscoi*, um novo Crocodylomorpha Trematochampsidae da Bacia Bauru, Brasil. Rev Bras Geociênc 42:397–410

Iori FV, Marinho TS, Carvalho IS, Campos ACDA (2013) Taxonomic reappraisal of the sphagesaurid crocodyliform *Sphagesaurus montealtensis* from the Late Cretaceous Adamantina Formation of São Paulo State, Brazil. Zootaxa 3686:183–200

Ivany LC, Lohmann KC, Blake DB, Hasiuk F, Aronson RB, Glass A, Moody R (2008) Eocene climate record of a high southern latitude continental shelf: Seymour Island, Antarctica. Geol Soc Am Bull 120:659–678

Jensen JA (2001) The road to Chilecito. Queen Victoria Museum and Art Gallery, Launceston

Juárez Valieri RD, Fiorelli LE, Cruz LE (2007) *Quilmesaurus curriei* Coria, 2001. Su validez taxonómica y relaciones filogenéticas. Rev Mus Argent Cienc Nat "B Rivadavia", ns 9:59–66

Juárez Valieri RD, Haro JA, Fiorelli LE, Calvo JO (2010). A new hadrosauroid (Dinosauria: Ornithopoda) from the Allen Formation (Late Cretaceous) of Patagonia, Argentina. Rev Mus Argent Cienc Nat "B Rivadavia", ns 11:217–231

Jud NA, Gandolfo MA, Iglesias A, Wilf P (2018) Fossil flowers from the Early Palaeocene of Patagonia, Argentina, with affinity to Schizomerieae (Cunoniaceae). Ann Bot 121:431–442

Katz H (1963) Revision of Cretaceous stratigraphy in Patagonian Cordillera of Última Esperanza, Magallanes Province, Chile. AAPG Bull 47:506–524

Kay RF (2010) A new primate from the Early Miocene of Gran Barranca, Chubut Province, Argentina: paleoecological implications. In: Madden RH, Carlini AA, Vucetich MG, Kay RF (eds) The paleontology of Gran Barranca: evolution and environmental change through the Middle Cenozoic of Patagonia. Cambridge University Press, Cambridge, pp 220–240

Kay RF, Madden RM, Vucetich MG, Carlini AA, Mazzoni MM, Ré GH, Heizler M, Sandeman H (1999) Revised age of the Casamayoran South American Land Mammal "Age"climatic and biotic implications. Proc Natl Acad Sci USA 96:13235–13240

Kay RF, Fleagle JG, Mitchell TRT, Colbert M, Bown T, Powers DW (2008) The anatomy of *Dolichocebus gaimanensis*, a stem platyrrhine monkey from Argentina. J Hum Evol 54:323–382

Kay RF, Perry JMG, Malinzak M, Allen KL, Kirk EC, Plavcan JM, Fleagle JG (2012) Paleobiology of Santacrucian primates. In: Vizcaíno SF, Kay RF, Bargo MS (eds) Early Miocene paleobiology in Patagonia. Cambridge University Press, Cambridge, pp 306–330

Kellner AWA, Azevedo SAK (1999) A new sauropod dinosaur (Titanosauria) from the Late Cretaceous of Brazil. Nat Sci Mus Monogr 15:111–142

Kellner AWA, Campos DA, Azevedo SAK, Trotta MNF, Henriques DDR, Craik MMT, Silva HP (2006) On a new titanosaur sauropod from the Bauru Group, Late Cretaceous of Brazil. Bol do Mus Nac 74:1–31

Kellner AWA, Campos DA, Riff D, Andrade MB (2011) A new crocodylomorph (Sphagesauridae, Notosuchia) with horn-like tubercles from Brazil. Zool J Linn Soc 163:57–65

Kielan-Jaworowska Z, Cifelli RL, Luo Z-X (2004) Mammals from the age of dinosaurs. Origins, evolution, and structure. Columbia University Press, New York

Kielan-Jaworowska Z, Ortiz-Jaureguizar E, Vieytes EC, Pascual R, Goin F (2007) First ?cimolodontan multituberculate mammal from South America. Acta Paleontol Pol 52:257–262

Kloster AC, Gnaedinger SC (2018) Coniferous wood of *Agathoxylon* from the La Matilde Formation, (Middle Jurassic), Santa Cruz, Argentina. J Paleontol 92:546–567

Kokogian D, Spalletti LA, Morel EM, Artabe AE, Martínez RN, Alcober OA, Milana JP, Zavattieri AM (2001) Estratigrafía del Triásico argentino. In: Artabe AE, Morel EM, Zamuner AB (eds) El Sistema Triásico en la Argentina. Fundación Museo de La Plata Francisco Pascasio Moreno, La Plata, pp 23–54

Krapovickas V, Garrido AC (2006) Sobre una huella de Theropoda de la Formación Anacleto (Campaniano Temprano), Provincia del Neuquén, Argentina. Ameghiniana 43(Suppl):43R

Krause DW, Bonaparte JF (1993) Superfamily Gondwanatherioidea: A previously unrecognized radiation of multituberculate mammals in South America. Proc Natl Acad Sci USA 90:9379–9383

Krause DW, Hoffmann S, Wible JR, Kirk EC, Schultz JA, von Koenigswald W, Groenke JR, Rossie JB, O'Connor PM, Seiffert ER, Dumont ER, Holloway WL, Rogers RR, Rahantarisoa RJ, Kemp AD, Andriamialison H (2014) First cranial remains of a gondwanatherian mammal reveal remarkable mosaicism. Nature 515:512–517

Krause DW, Hoffmann S, Hu Y, Wible JR, Rougier GW, Kirk EC, Groenke JR, Rogers RR, Rossie JB, Schultz JA, Evans AR, von Koenigswald W, Rahantarisoa LJ (2020) Skeleton of Cretaceous mammal from Madagascar reflects long-term insularity. Nature 581:421–427

Kurtz F (1921) Atlas de plantas fósiles de la República Argentina. Acad Nac Cienc Córdoba, Actas 7:129–153

Lacerda MB, Schultz CL, Bertoni-Machado C (2015) First 'rauisuchian' archosaur (Pseudosuchia, Loricata) for the Middle Triassic *Santacruzodon* Assemblage Zone (Santa Maria Supersequence), Rio Grande do Sul State, Brazil. PLoS ONE 10(2):e0118563

Lacerda MB, Mastrantonio BM, Fortier DC, Schultz CL (2016) New insights on *Prestosuchus chiniquensis* Huene, 1942 (Pseudosuchia, Loricata) based on new specimens from the "Tree Sanga" Outcrop, Chiniquá Region, Rio Grande do Sul, Brazil. PeerJ 4:e1622

Ladevèze S, Asher RJ, Sánchez-Villagra MR (2008) Petrosal anatomy in the fossil mammal *Necrolestes*: Evidence for metatherian affinities and comparisons with the extant marsupial mole. J Anat 213:686–697

La Nacion (2004) Paisaje lunar, fósiles y huellas del tiempo en Ischigualasto. La Nacion August 29th 2004

Langer MC, Schultz CL (2000) A new species of the Late Triassic rhynchosaur *Hyperodapedon* from the Santa Maria Formation of south Brazil. Palaeontology 43:633–652

Langer MC, Abdala NF, Richter M, Benton M (1999) A sauropodomorph dinosaur from the Upper Triassic (Carnian) of southern Brazil. C R Acad Sci 329:511–517

Langer MC, Ribeiro AM, Schultz CL, Ferigolo J (2007) The continental tetrapod bearing Triassic of south Brazil. Bull N M Mus Nat Hist Sci 41:201–218

Langer MC, Ramezani J, Da-Rosa ÁAS (2018) U-Pb age constraints on dinosaur rise from south Brazil. Gondwana Res 57:133–140

Lapparent de Broin F, de la Fuente MS (2001) Oldest world Chelidae (Chelonia, Pleurodira), from the Cretaceous of Patagonia, Argentina. C R Acad Sci 333:463–470

Laza JH, Reguero MA (1993) Extensión faunística de la antigua Región Neotropical en la Península Antártica durante el Eoceno. Ameghiniana 26(Suppl):245R

Leanza HA, Hugo CA (1997) Hoja Geológica 3969-III—Picún Leufú, provincias del Neuquén y Río Negro. Instituto de Geología y Recursos Naturales, SEGEMAR, Buenos Aires. Bol 218:1–135

Leanza HA, Apesteguía S, Novas FE, de la Fuente MS (2004) Cretaceous terrestrial beds from the Neuquén Basin (Argentina) and their tetrapod assemblages. Cretac Res 25:61–87

Leche W (1907) Zur Entwicklungsgeschichte des Zahn-Systems der Saügetiere, zugleich ein Beitrag zur Stammesgeschichte dieser Tiergruppe. Zweite Teil: Phylogenie. Zweites Heft: Die Familien der Centetidae, Solenodontidae und Chrysochloridae. Zoologica 20:1–158

Leonardi G (1980) On the discovery of an abundant ichno-fauna (vertebrates and invertebrates) in the Botucatu Formation s.s. in Araraquara, São Paulo. Brazil. An Acad Bras Ciênc 52:559–567

Leonardi G (1981) Novo icnogênero de tetrápode Mesozóico da Formação Botucatu, Araraquara, SP. An Acad Bras Ciênc 53:793–805

Leonardi G (1992) Sulle prime piste fossili di tetrapodi del Paraguay. Paleocronache 1992:66–67

Leonardi G (1994) Annotated atlas of South America tetrapod footprints (Devonian to Holocene) with an appendix on Mexico and Central America. Companhia de Pesquisa de Recursos Minerais, Brasília, pp 1–248

Leonardi G, Carvalho IS (2002) Jazigo Icnofossilífero do Ouro-Araraquara (SP). In: Schobbenhaus C, Campos DA, Queiroz ET, Winge M, Berbert-Born M (eds) Sítios geológicos e paleontológicos do Brasil. Departamento Nacional de Producción Mineral, Brasília, pp 39–48

Leonardi G, Carvalho IS, Fernandes MA (2007) The desert ichnofauna from Botucatu Formation (Upper Jurassic-Lower Cretaceous), Brazil. In: Carvalho IS, Cassab RCT, Schwanke C, Carvalho MA, Fernandes ACS, Carvalho MSS, Arai M, Oliveira MEQ (eds) Paleontologia: Cenários da vida. Interciência, Rio de Janeiro, pp 379–391

Lesta P, Ferello R (1972) Region extraandina de Chubut y norte de Santa Cruz. In: Leanza H (ed) Geología regional Argentina. Acad Nac Cienc Córdoba, Córdoba, pp 601–653

Loope D, Schmitt J, Jackson F (2000) Thunder platters: Thin discontinuous limestones in Late Cretaceous continental mudstones of Argentina may be chemically infilled sauropod tracks. Geological Society of America, Abstracts with Programs:A450

Lorente M, Chornogubsky L, Goin FJ (2016) On the existence of non-microbiotherian Australidelphian marsupials (Diprotodontia) in the Eocene of Patagonia. Palaeontology 59:533–547

Luo Z-X, Kielan-Jaworowska Z, Cifelli RL (2002) In quest for a phylogeny of Mesozoic Mammals. Acta Palaeontol Pol 47:1–78

Lydekker R (1893) Contributions to the study of the fossil vertebrates of Argentina. I. The dinosaurs of Patagonia. An Mus La Plata, Paleontología 2:1–14

Macellari CE, Barrio CA, Manassero M (1989) Upper Cretaceous to Paleocene depositional sequences and sandstone petrography of southwestern Patagonia (Argentina and Chile). J S Am Earth Sci 2:223–239

MacFadden BJ (1990) Chronology of Cenozoic primate localities in South America. J Hum Evol 19:7–21

Machado EB, Avilla LDS, Nava WR, Campos DDA, Kellner AW (2013) A new titanosaur sauropod from the Late Cretaceous of Brazil. Zootaxa 3701:301–321

MacPhee R, Reguero MA, Strgnac P, Nishida M, Jacobs P (2008) Out of Antarctica: Paleontological reconnaissance of Livingston island (South Shetlands) and Seymour island (James Ross Group). J Vertebr Paleontol 28(Suppl):110A

Macrini TE, Rougier GW, Rowe T (2007) Description of a cranial endocast from the fossil mammal *Vincelestes neuquenianus* (Theriiformes) and its relevance to the evolution of endocranial characters in therians. Anat Rec 290:875–892

Madden RH, Carlini AA, Vucetich MG, Kay RF (eds) (2010) The paleontology of Gran Barranca: evolution and environmental change through the Middle Cenozoic of Patagonia. Cambridge University Press, Cambridge

Makovicky PJ, Apesteguía S, Agnolín FL (2005) The earliest dromaeosaurid theropod from South America. Nature 437:1007–1011

Makovicky PJ, Apesteguía S, Gianechini FA (2012) A new coelurosaurian theropod from the La Buitrera fossil locality of Río Negro, Argentina. Fieldiana 5:90–98

Malkowski MA, Schwartz TM, Sharman GR, Sickmann ZT, Graham SA (2016) Stratigraphic and provenance variations in the early evolution of the Magallanes-Austral foreland basin: Implications for the role of longitudinal versus transverse sediment dispersal during arc-continent collision. Geol Soc Am Bull 129:349–371

Malumián N, Náñez C (2011) The Late Cretaceous-Cenozoic transgressions in Patagonia and the Fuegian Andes: Foraminifera, palaeoecology, and palaeogeography. Biol J Linn Soc 103:269–288

Mancuso AC, Caselli AT (2012) Paleolimnology evolution in rift basins: The Ischigualasto-Villa Unión Basin (Central-Western Argentina) during the Triassic. Sediment Geol 275–276:38–54

Mancuso AC, Gaetano LC, Leardi JM, Abdala NF, Arcucci AB (2014) The Chañares Formation: a window to a Middle Triassic tetrapod community. Lethaia 47:244–265

Manríquez LM, Lavina EL, Fernández RA, Trevisan C, Leppe MA (2019) Campanian-Maastrichtian and Eocene stratigraphic architecture, facies analysis, and paleoenvironmental evolution of the northern Magallanes Basin (Chilean Patagonia). J S Am Earth Sci 93:102–118

Marenssi SA, Santillana SN, Rinaldi CA (1998) Stratigraphy of the La Meseta Formation (Eocene), Marambio (Seymour) Island, Antarctica. In: Casadio S (ed) Paleógeno de América del Sur y de la

Península Antártica. Asociación Paleontológica Argentina, Buenos Aires, Publicación Especial, vol 5, pp 137–146

Marinho TS, Carvalho IS (2009) An armadillo-like sphagesaurid crocodyliform from the Late Cretaceous of Brazil. J S Am Earth Sci 27:36–41

Marinho TS, Iori FV, Carvalho IS, Vasconcellos FM (2013) *Gondwanasuchus scabrosus* gen. et sp. nov., a new terrestrial predatory crocodyliform (Mesoeucrocodylia: Baurusuchidae) from the Late Cretaceous Bauru Basin of Brazil. Cretac Res 44:104–111

Marsh OC (1889) Discovery of Cretaceous Mammalia. Am J Sci 38:81–92

Marshall LG (1976) Fossil localities for Santacrucian (Early Miocene) mammals, Santa Cruz Province, Southern Patagonia, Argentina. J Paleontol 50:1129–1142

Marsola JSA, Grellet-Tinner G, Montefeltro FC, Sayão JM, Hsiou AS, Langer MC (2014) The first fossil avian egg from Brazil. Alcheringa 38:563–567

Martin T, Rauhut OWM (2005) Mandible and dentition of *Asfaltomylos patagonicus* (Australosphenida, Mammalia) and the evolution of tribosphenic teeth. J Vertebr Paleontol 25:414–425

Martin T, Goin FJ, Chornogubsky L, Gelfo JN, Schultz J (2013) Early Late Cretaceous (Cenomanian) mammals and other vertebrates from the Mata Amarilla Formation of Southern Patagonia (Argentina). J Vertebr Paleontol 33(Suppl):172A

Martinelli AG, Forasiepi AM (2004) Late Cretaceous vertebrates from Bajo de Santa Rosa (Allen Formation), Río Negro Province, Argentina, with the description of a new sauropod dinosaur (Titanosauridae). Rev Mus Argent Cienc Nat "B Rivadavia", ns 6:257–305

Martinelli AG, Rougier GW (2007) On *Chaliminia musteloides* (Eucynodontia: Tritheledontidae) from the Late Triassic of Argentina and a phylogeny of Ictidosauria. J Vertebr Paleontol 27:442–460

Martinelli AG, Soares MB (2016) Evolution of South American cynodonts. Mus Argent Cienc Nat "B Rivadavia". Contribuciones 6:183–197

Martinelli AG, Teixeira VPA (2015) The Late Cretaceous vertebrate record from the Bauru Group at the Triângulo Mineiro, southeastern Brazil. Bol Geol Min España 126:129–158

Martinelli AG, Bonaparte JF, Schultz CB (2005) A new tritheledontid (Therapsida, Eucynodontia) from the Late Triassic of Rio Grande do Sul (Brazil) and its phylogenetic relationships among carnivorous non-mammalian eucynodonts. Ameghiniana 42:191–208

Martinelli AG, de la Fuente M, Abdala NF (2009) *Diademodon tetragonus* Seeley, 1894 (Therapsida: Cynodontia) in the Triassic of South America and its biostratigraphic implications. J Vertebr Paleontol 29:852–862

Martinelli AG, Chornogubsky L, Abello A, Goin FJ, Reguero M (2014) The first non-therian dryolestoid from Antarctica. 2014 SCAR Open Science Conference, Auckland, Abstracts:432

Martinelli AG, Soares MB, Schwanke C (2016) Two new cynodonts (Therapsida) from the Middle–early Late Triassic of Brazil and comments on South American probainognathians. PLoS ONE 11(10):e0162945

Martinelli AG, Kammerer CF, Melo TP, Paes Neto VD, Ribeiro AM, Da-Rosa ÁAS, Schultz CL, Soares MB (2017a) The African cynodont *Aleodon* (Cynodontia, Probainognathia) in the Triassic of southern Brazil and its biostratigraphic significance. PLoS ONE 12(6):e0177948

Martinelli AG, Eltink E, Da-Rosa ÁAS, Langer MC (2017b) A new cynodont (Therapsida) from the *Hyperodapedon* Assemblage Zone (upper Carnian–Norian) of southern Brazil improves the Late Triassic probainognathian diversity. Pap Palaeontol 3:401–423

Martinelli AG, Marinho T, Iori F, Ribeiro LCB (2018) The first *Caipirasuchus* (Mesoeucrocodylia, Notosuchia) from the Late Cretaceous of Minas Gerais, Brazil: New insights on sphagesaurid anatomy and taxonomy. PeerJ 6:e5594

Martinelli AG, Escobar JA, Francischini H, Kerber L, Müller RT, Rubert R, Schultz CL, Da-Rosa ÁAS (2020) New record of a stahleckeriid dicynodont (Therapsida, Dicynodontia) from the Late Triassic of southern Brazil and biostratigraphic remarks on the *Riograndia* Assemblage Zone. Hist Biol. https://doi.org/10.1080/08912963.2020.1850715

Martínez LCA, Iglesias A, Artabe AE, Varela AN, Apesteguía S (2017) A new Encephalarteae trunk (Cycadales) from the Cretaceous of Patagonia (Mata Amarilla Formation, Austral Basin), Argentina. Cretac Res 72:81–94

Martínez RN, Apaldetti C (2017) A late Norian-Rhaetian coelophysid neotheropod (Dinosauria, Saurischia) from the Quebrada del Barro Formation, northwestern Argentina. Ameghiniana 54:488–505

Martínez RN, Sereno PC, Alcober OA, Colombi CE, Renne PR, Montañez IP, Currie BS (2011) A basal dinosaur from the Dawn of the Dinosaur era in Southwestern Pangaea. Science 331:206–210

Martínez RN, Apaldetti C, Alcober OA, Colombi CE, Sereno PC, Fernandez E, Santi Malnis P, Correa GA, Abelín D (2013a) Vertebrate succession in the Ischigualasto Formation. J Vertebr Paleontol 32(Suppl):10–30A

Martínez RN, Apaldetti C, Colombi CE, Praderio A, Fernández E, Malnis PS, Correa GA, Abelin D, Alcober O (2013b) A new sphenodontian (Lepidosauria: Rhynchocephalia) from the Late Triassic of Argentina and the early origin of the herbivore opisthodontians. Proc R Soc Lond B 280:20132057

Martínez RN, Apaldetti C, Correa G, Colombi CE, Fernández E, Santi Malnis P, Praderio A, Abelín D, Benegas LG, Aguilar-Cameo A, Alcober AO (2015) A new Late Triassic vertebrate assemblage from northwestern Argentina. Ameghiniana 52:379–390

Martínez RN, Cecilia Apaldetti C, Correa GA, Abelín D (2016) A Norian lagerpetid dinosauromorph from the Quebrada del Barro Formation, northwestern Argentina. Ameghiniana 53:1–13

Mazzoni MM, Spalletti LA, Iñíguez MA, Teruggi ME (1981) El Grupo Bahía Laura en el Gran Bajo de San Julián, Provincia de Santa Cruz. 8° Congreso Geológico Argentino. San Luis, Actas 3:485–507

Melchor RN, de Valais S, Genise JF (2004) Middle Jurassic mammalian and dinosaur footprints and petrified forests from the volcaniclastic La Matilde Formation. In: Bellosi ES, Melchor RN (eds) Fieldtrip guidebook. 1° International Congress on Ichnology, Trelew, pp 47–63

Melo TP, Abdala NF, Soares MB (2015) The Malagasy cynodont *Menadon besairiei* (Cynodontia; Traversodontidae) in the Upper Triassic of Brazil. J Vertebr Paleontol 35(6):e1002562

Méndez AH, Novas F, Iori F (2012) First record of Megaraptora (Theropoda, Neovenatoridae) from Brazil. CR Palevol 11:251–256

Menegazzo MC, Catuneanu O, Chang HK (2016) The South American retroarc foreland system: The development of the Bauru Basin in the back-bulge province. Mar Petrol Geol 73:131–156

Meng J, Hu Y-M, Wang Y-Q, Wang X-L, Li C-K (2006) A Mesozoic gliding mammal from northeastern China. Nature 444:889–893

Meyer CA, Hippler D, Lockley MG (2001) The Late Cretaceous vertebrate ichnofacies of Bolivia — facts and implications. Asociación Paleontológica Argentina, Buenos Aires, Publicación Especial 7:133–138

Meyer CA, Marty D, Belvedere M (2018) Titanosaur trackways from the Late Cretaceous El Molino Formation of Bolivia (Cal Orcko, Sucre). Ann Soc Geol Pol 88:223 – 241

Milani EJ, França AB, Schneider RL (1994) Bacia do Paraná. Bol Geociênc Petrobras 8:69–82

Milani EJ, Faccini UF, Scherer CMS, Araújo LM, Cupertino LM (1998) Sequences and stratigraphic hierarchy of the Paraná Basin (Ordovician to Cretaceous), southern Brazil. Bol Inst Geociência USP, Sêrie Científica 29:125–173

Milani EJ, Melo JHG, Souza PA, Fernandes LA, França AB (2007) Bacia do Paraná. Bol. Geociência, Petrobras 15:265–287

Montanelli SB (1987) Presencia de Pterosauria (Reptilia) en la Formación La Amarga (Hauteriviano–Barremiano), Neuquén, Argentina. Ameghiniana 24:109–113

Montefeltro FC, Langer MC, Schultz CL (2010) Cranial anatomy of a new genus of hyperodapedontine rhynchosaur (Diapsida, Archosauromorpha) from the Upper Triassic of southern Brazil. Earth Environ Sci Trans R Soc Edinb 101:27–52

Montefeltro FC, Larsson HCE, Langer MC (2011) A new baurusuchid (Crocodyliformes, Mesoeucrocodylia) from the Late Cretaceous of Brazil and the phylogeny of Baurusuchidae. PLoS ONE 6(7):e21916

Monti M, Franzese JR (2016) Análisis tectonoestratigráfico del Grupo Puesto Viejo (San Rafael, Argentina): Evolución de un rift continental triásico. Latin Am J Sedim Basin Analy 23:1–33

Moreno FP, Mercerat A (1891) Catálogo de los pájaros fósiles de la República Argenitna conservados en el Museo de La Plata. An Mus La Plata, Paleontología 1:1–71

Mörs T, Reguero M, Vasilyan D (2020) first fossil frog from Antarctica: Implications for Eocene high latitude climate conditions and Gondwanan cosmopolitanism of Australobatrachia. Sci Rep 10:5051

Mourier T, Jaillard É, Laubacher G, Noblet C, Pardo A, Sigé B, Taquet P (1986) Découverte de de restes dinosauriens et mammalien d'âge Crétacé Supérieur à la base des couches rouges du synclinal de Bagua (Andes nord-péruviennes): Aspects stratigraphiques, sédimentologiques et paléogéographiques concernant la régression fini-Crétacée. Bull Soc Géol France 2:171–175

Mourier T, Bengtson P, Bonhomme M, Buge E, Cappetta H, Crochet J-Y, Feist M, Hirsch KF, Jaillard E, Laubacher G, Lefrano JP, Moullade M, Noblet C, Pons D, Rey J, Sigé B, Tambareau Y, Taquet P (1988) The Upper Cretaoeous-Lower Tertiary marine to continental transition in the Bagua Basin, Northern Peru: paleontology, biostratigraphy, radiometry, correlations. Newsl Stratigr 19:143–177

Mpodozis C, Ramos V (2008) Jurassic tectonics in Argentina and Chile: extension, oblique subduction, rifting, drift and collisions? Rev Asoc Geol Argent 63:481–497

Mudelsee M, Bickert T, Lear CH, Lohmann G (2014) Cenozoic climate changes: a review based on time series analysis of marine benthic δ18 O records. Rev Geophys 52:333–374

Müller RT, Langer MC, Dias-da-Silva S (2018) An exceptionally preserved association of complete dinosaur skeletons reveals the oldest long-necked sauropodomorphs. Biol Lett 14:20180633

Musacchio EA (1970) Ostrácodos de las Superfamilias Cytheracea y Darwinulacea de la Formación La Amarga (Cretácico Inferior), Provincia de Neuquén, Argentina. Ameghiniana 7:301–316

Musacchio EA (1971a) Hallazgo del género *Cypridea* en Argentina y consideraciones estratigráficas sobre la Formación La Amarga (Cretácico Inferior), en la Provincia del Neuquén, Argentina. Ameghiniana 8:105–125

Musacchio EA (1971b) Carophytas de la Formación La Amarga (Cretácico Inferior), Provincia del Neuquén. Rev Mus La Plata, ns, Paleontología 6:19–38

Musacchio EA (1973) Charophytas y ostrácodos no marinos del Grupo Neuquén (Cretácico Superior) en algunos afloramientos de las provincias de Río Negro y Neuquén, República Argentina. Rev Mus La Plata, ns, Paleontología 48:1–32

Musacchio EA (1995) Estratigrafía y micropalentología del Jurásico y el Cretácico en la comarca del Valle Medio del Río Chubut, Argentina. 6° Congreso Argentino de Paleontología y Bioestratigrafía, Trelew, Actas:179–187

Musacchio EA, Beros C, Pujana I (1990) Microfósiles continentales del Jurásico y el Cretácico en Chubut y su contribución a la bioestratigrafía de la Cuenca del Golfo de San Jorge, Argentina. In: Volkheimer W (ed) Bioestratigrafía de los sistemas regionales del Jurásico y Cretácico de América del Sur. Comité Sudamericano del Jurásico–Cretácico, vol 2, pp 355–383

Musser AM (2003) Review of the monotreme fossil record and comparison of palaeontological and molecular data. Comp Biochem Physiol, Part A Mol Integr Physiol 136:927–942

Muzzopappa P, Báez AM (2013) Presence of a calyptocephallelid frog in the Paleocene of the San Jorge Gulf Basin, Chubut, Patagonia, Argentina. Reunión de Comunicaciones de la APA, Córdoba, Abstracts:62–63

Muzzopappa P, Martinelli AG, Gardenes JP, Rougier GW (2020) Exceptional avian pellet from the Paleocene of Patagonia and description of its content: a new sp. of calyptocephalellid (Neobatrachia) anuran. Pap Palaeontol. https://doi.org/10.1002/spp2.1333

Naeser CW, Crochet JY, Jaillard E, Laubacher G, Mourier T, Sigé B (1991) Tertiary fission-track ages from the Bagua syncline (northern Peru): stratigraphic and tectonic implications. J S Am Earth Sci 4:61–71

Nascimento PM, Zaher H (2010) A new species of *Baurusuchus* (Crocodyliformes, Mesoeucrocodylia) from the Upper Cretaceous of Brazil, with the first complete postcranial skeleton described for the family Baurusuchidae. Pap Avulsos Zool 50:323–361

Nava WR, Martinelli AG (2011) A new squamate lizard from the Upper Cretaceous Adamantina Formation (Bauru Group), São Paulo State, Brazil. An Acad Bras Ciênc 83:291–299
Nobre PH, Carvalho IS (2006) *Adamantinasuchus navae*: a new Gondwanan Crocodylomorpha (Mesoeucrocodylia) from the Late Cretaceous of Brazil. Gondwana Res 10:370–378
Novas FE (2009) The age of dinosaurs in South America. Indiana University Press, Bloomington
Novas FE, Cambiaso AV, Ambrosio A (2004) A new basal iguanodontian (Dinosauria, Ornithischia) from the Upper Cretaceous of Patagonia. Ameghiniana 41:75–82
Novas FE, Salgado L, Calvo JO, Agnolín FL (2005) Giant titanosaur (Dinosauria, Sauropoda) from the Late Cretaceous of Patagonia. Rev Mus Argent Cienc Nat "B Rivadavia", ns 7:37–41
Novas FE, Ezcurra MA, Leucona A (2008) *Orkoraptor burkei* nov. gen. et sp., a large theropod from the Maastrichtian Pari Aike Formation, southern Patagonia, Argentina. Cretac Res 29:468–480
Novas FE, Pol D, Canale JI, Porfiri JD, Calvo JO (2009) A bizarre Cretaceous theropod dinosaur from Patagonia and the evolution of Gondwanan dromaeosaurids. Proc R Soc Lond B 276:1101–1107
Novas FE, Ezcurra MD, Agnolín FL, Pol D, Ortíz R (2012a) New Patagonian Cretaceous theropod sheds light about the early radiation of Coelurosauria. Rev Mus Argent Cienc Nat "B Rivadavia", ns 14:57–81
Novas FE, Kundrat M, Agnolín FL, Ezcurra MND, Ahlberg PE, Isasi MP, Arriagada A, Chafrat P (2012b) A new large pterosaur from the Late Cretaceous of Patagonia. J Vertebr Paleontol 32:1447–1452
Novas FE, Agnolín FL, Rozadilla S, Aranciaga-Rolando AM, Brisson-Egli F, Motta MJ, Cerroni M, Ezcurra MD, Martinelli AG, D'Angeo JS, Alvarez-Herrera G, Gentil AR, Bogan S, Chimento NR, García-Marsà JA, Lo Coco G, Miquel SE, Brito FF, Vera EI, Perez Loinaze VS, Fernández MS, Salgado L (2019) Paleontological discoveries in the Chorrillo Formation (upper Campanian–lower Maastrichtian, Upper Cretaceous), Santa Cruz Province, Patagonia, Argentina. Rev Mus Argent Cienc Nat "B Rivadavia", ns 21:217–293
Nullo F, Blasco G, Risso C, Combina A, Otamendi J (2006) Hoja Geológica 5172-I y 5175-II, El Calafate, Provincia de Santa Cruz. Instituto de Geología y Recursos Minerales, SEGEMAR, Buenos Aires, Bol 396:1–86
O'Gorman JP (2016) A small body sized non-aristonectine elasmosaurid (Sauropterygia, Plesiosauria) from the Late Cretaceous of Patagonia with comments on the relationships of the Patagonian and Antarctic elasmosaurids. Ameghiniana 53:245–268
O'Gorman JP, Varela AN (2010) The oldest lower Upper Cretaceous plesiosaurs (Reptilia, Sauropterygia) from southern Patagonia, Argentina. Ameghiniana 47:447–459
Olivera DE, Zavattieri AM, Quattrocchio ME (2015) The palynology of the Cañadón Asfalto Formation (Jurassic), Cerro Cóndor depocentre, Cañadón Asfalto Basin, Patagonia, Argentina: Palaeoecology and palaeoclimate based on ecogroup analysis. Palynology 39:362–386
O'Meara RN, Thompson RS (2014) Were there Miocene meridiolestidans? Assessing the phylogenetic placement of *Necrolestes patagonensis* and the presence of a 40 million year merdiolestidan ghost lineage. J Mammal Evol 21:271–284
Ortega F, Gasparini Z, Buscalioni A, Calvo JO (2000) A new *Araripesuchus* (Crocodylomorpha, Lower Cretaceous) from northwestern Patagonia. J Vertebr Paleontol 20:57–76
Ortiz-Jaureguizar E (1996) Paleobiogeografía y paleoecología de los mamíferos continentales de América del Sur durante el Cretácico Tardío–Paleoceno: Una revisión. Estud Geol 52:83–94
Osborn HF (1887) On the structure and classification of the Mesozoic Mammalia. Proc Acad Nat Sci Phila 38:282–292
Ottone EG (2009) La flora cretácica de Cuenca Neuquina, su significado paleoambiental y paleoclimático. Rev Asoc Geol Argent 65:373–386
Ottone EG, Monti M, Marsicano CA, de la Fuente MS, Naipauer M, Armstrong R, Mancuso AC (2014) A new Late Triassic age for the Puesto Viejo Group (San Rafael depocenter, Argentina): SHRIMP U-Pb zircon dating and biostratigraphic correlations across southern Gondwana. J S Am Earth Sci 56:186–199

Owen R (1871) Monograph of the fossil Mammalia of the Mesozoic formations. Monogr Palaeontol Soc 33:1–115
Pacheco C, Müller R, Langer M, Pretto F, Kerber L, Dias da Silva S (2019) *Gnathovorax cabreirai*: a new early dinosaur and the origin and initial radiation of predatory dinosaurs. PeerJ 7:e7963
Pacheco JA (1913) Notas sobre a geologia do vale do Rio Grande a partir da foz do Rio Pardo até sua confluência com o Rio Paranaíba. Comissão Geográfica Geológica, São Paulo 1:33–38
Páez Arango N (2008) Dental and craniomandibular anatomy of *Peligrotherium tropicalis*: the evolutionary radiation of South American dryolestoid mammals. Unpublished Ms thesis, University of Louisville
Page R, Ardolino A, de Barrio RE, Franchi M, Lizuain A, Page S, Silva Nieto D (1999) Estratigrafía del Jurásico y Cretácico del Macizo de Somún Curá, provincias de Río Negro y Chubut. In: Caminos R (ed) Geología Argentina. Servicio Geológico Minero Argentino, Buenos Aires, pp 460–488
Panza JL, Genini A (1998) Hoja Geológica 4769-IV, Monumento Natural Bosques Petrificados, escala 1:250000, Provincia de Santa Cruz. Dirección Nacional del Servicio Geológico, Buenos Aires, Bol 258:1–163
Papú OH, Sepúlveda EG (1995) Datos palinológicos de la Formación Los Alamitos en la localidad de Montoniló, Departamento 25 de Mayo, Río Negro, Argentina. Sus relaciones con unidades colindantes coetáneas. 6° Congreso Argentino de Paleontología y Bioestratigrafía, Trelew, Actas:195–200
Parras A, Manera de Bianco T, Montalvo C (2004) Campanian–Maastrichtian coprolites from northern Patagonia: Paleoenvironmental and paleobiological significance. 1° International Congress on Ichnology, Ichnia, Abstracs:6
Pascual R (1992) Unpublished letter from R. Pascual to H de Macedo, La Plata
Pascual R (2006) Evolution and geography: the biogeographic history of South American land mammals. Ann Missouri Bot Gard 93:209–230
Pascual R, Bondesio P (1976) Notas sobre vertebrados de la frontera Cretácico–Terciaria. III: Ceratodontidae (Peces Osteichtyes, Dipnoi) de la Formación Coli Toro y de otras unidades del Cretácico Tardío de Patagonia y sur de Mendoza. Sus implicancias paleobiogeográficas. 6° Congreso Geológico Argentino, Bahía Blanca, Actas 1:565–577
Pascual R, Carlini AA (1987) A new superfamily in the extensive radiation of South American Paleogene marsupials. Field Zool 39:99–110
Pascual R, Ortiz-Jaureguizar E (2007) The Gondwanan and South American episodes: two major moments in the history of South American mammals. J Mammal Evol 14:75–137
Pascual R, Archer M, Ortiz-Jaureguizar E, Prado JL, Godthelp H, Hand SH (1992) First discovery of monotremes in South America. Nature 356:704–705
Pascual R, Ortiz-Jaureguizar E, Prado JL (1996) Land mammals: Paradigm for Cenozoic South American geobiotic evolution. In: Arratia G (ed) Contributions of southern South America to vertebrate paleontology. Münch Geowiss Abh A Paläontol, vol 30, pp 265–319
Pascual R, Goin FJ, Krause DW, Ortiz-Jaureguizar E, Carlini AA (1999) The first gnathic remains of *Sudamerica*: Implications for gondwanathere relationships. J Vertebr Paleontol 19:373–382
Pascual R, González P, Ardolino A, Puerta PF (2000) A highly derived docodont from the Patagonian Late Cretaceous: Evolutionary implications for Gondwanan mammals. Geodiversitas 22:395–414
Pascual R, Balarino ML, Udrizar Sauthier DE (2001) The K/T and Tertiary–Pleistocene South American mammalian turnovers. Similar phenomena? 7° International Symposium on Mesozoic Terrestrial Ecosystems. Asociación Paleontológica Argentina, Buenos Aires, Publicación Especial 7:151–156
Pascual R, Goin FJ, Balarino L, Udrizar Sauthier DE (2002) New data on the Paleocene monotreme *Monotrematum sudamericanum*, and the convergent evolution of triangulate molars. Acta Palaeontol Pol 47:487–492
Patterson B (1958) Affinities of the Patagonian fossil mammal *Necrolestes*. Breviora 94:1–14

Perea D, Soto M, Veroslavsky G, Matínez S, Ubilla M (2009) A Late Jurassic fossil assemblage in Gondwana: Biostratigraphy and correlations of the Tacuarembó Formation, Paranó Basin, Uruguay. J S Am Earth Sci 28:168–179

Pereda-Suberbiola X, Galton PM, Mallison H, Novas FE (2012) A plated dinosaur (Ornithischia, Stegosauria) from the Early Cretaceous of Argentina, South America: an evaluation. Alcheringa 37:1–14

Perkins ME, Fleagle JG, Heizler MT, Nash B, Bown TM, Tauber AA, Dozo MT (2012) Tephrochronology of the Miocene Santa Cruz and Pinturas formations, Argentina. In: Vizcaíno SF, Kay RF, Bargo MS (eds) Early Miocene paleobiology in Patagonia. Cambridge University Press, Cambridge, pp 23–40

Petriella B (1972) Estudio de maderas petrificadas del Terciario inferior del area central de Chubut (Cerro Bororó). Rev Mus La Plata, ns, Paleontología 6:159–254

Petriella B, Archangelsky S (1975) Vegetación y ambiente en el Paleoceno de Chubut. 1° Congreso Argentino de Paleontología y Bioestratigrafía, Tucumán, Actas:257–270

Pfuhl HA, McCave IN (2005) Evidence for Late Oligocene establishment of the Antarctic Circumpolar Current. Earth Planet Sci Lett 235:715–728

Philipp RP, Schultz CL, Kloss HP, Horn BLD, Soares MB, Basei MAS (2018) Middle Triassic SW Gondwana paleogeography and sedimentary dispersal revealed by integration of stratigraphy and U-Pb zircon analysis: the Santa Cruz Sequence, Paraná Basin, Brazil. J S Am Earth Sci 88:216–237

Piatnitzky A (1936) Estudio geológico de la región del río Chubut y del río Genua. Bol Inf Petrol 137:83–118

Pinheiro AEP, Pereira PVLGdC, de Souza RG, Brum AS, Lopes RT, Machado AS, Bergqvist LP, Simbras FM (2018) Reassessment of the enigmatic crocodyliform "*Goniopholis*" *paulistanus* Roxo, 1936: Historical approach, systematic, and description by new materials. PLoS ONE 13(8):e0199984

Pires EF, Guerra-Sommer M, Scherer CMS, Santos AR, Cardoso E (2011) Early Cretaceous coniferous woods from a paleoerg (Paraná Basin, Brazil). J S Am Earth Sci 32:96–109

Pol D (2003) New remains of *Sphagesaurus huenei* (Crocodylomorpha: Mesoeucrocodylia) from the Late Cretaceous of Brazil. J Vertebr Paleontol 23:817–831

Pol D, Apesteguía S (2005) New *Araripesuchus* remains from the early Late Cretaceous (Cenomanian–Turonian) of Patagonia. Am Mus Novit 3490:1–38

Pol D, Rauhut OWM (2012) A Middle Jurassic abelisaurid from Patagonia and the early diversification of theropod dinosaurs. Proc R Soc Lond B 279:3170–3175

Pol D, Rauhut O, Carballido JL (2009) Skull anatomy of a new basal eusauropod from the Middle Jurassic of Patagonia. J Vertebr Paleontol 29(Suppl):165A

Pol D, Rauhut OWM, Becerra M (2011) A Middle Jurassic heterodontosaurid dinosaur from Patagonia and the evolution of heterodontosaurids. Naturwissenschaften 98:369–379

Pol D, Nascimento PM, Carvalho AB, Riccomini C, Pires-Domingues RA, Zaher H (2014) A new notosuchian from the Late Cretaceous of Brazil and the phylogeny of advanced notosuchians. PLoS ONE 9(4):e93105

Pöthe de Baldis ED (1984) Microfloras cenozoicas. In: Ramos VA (ed) Geología y recursos naturales de la Provincia de Río Negro. 9° Congreso Geológico Argentino, San Carlos de Bariloche, Relatorio 2:393–411

Powell JE (1986) Revisión de los titanosáuridos de América del Sur. Unpublished PhD thesis, Universidad Nacional de Tucumán

Powell JE (1987) Part VI-The titanosaurids. In: Bonaparte JF (ed) The Late Cretaceous fauna of Los Alamitos, Patagonia, Argentina. Rev Mus Argent Cienc Nat "B Rivadavia", Paleontología 3:147–153

Powell JE (2003) Revision of South American titanosaurid dinosaurs: palaeobiological, palaeobiogeographical and phylogenetic aspects. Rec Queen Victoria Mus 111:1–173

Prámparo MB, Volkheimer W (2002) Nuevos hallazgos de palinomorfos en la Formación La Amarga, Miembro Bañados de Caichigüe, Cuenca Neuquina sudoriental, Argentina. Ameghiniana 39:395–404

Prevosti FJ, Forasiepi AM, Ercoli MD, Turazzini GF (2012) Paleoecology of the mammalian carnivores (Metatheria, Sparassodonta) of the Santa Cruz Formation (late Early Miocene). In: Vizcaíno SF, Kay RF, Bargo MS (eds) Early Miocene paleobiology in Patagonia. Cambridge University Press, Cambridge, pp 173–193

Price LI (1945) A new reptile from the Late Cretaceous of Brazil. Serviço Geológico e Mineralogía 25:1–8

Price LI (1950) On a new crocodilian, *Sphagesaurus*, from the Cretaceous of the state of São Paulo, Brazil. An Acad Bras Ciênc 22:77–85

Price LI (1953) Quelônios da Formação Baúru, Cretáceo terrestre do Brasil Meridional. Divisão de Geología e Mineralogía, Bol 147:1–39

Prieto-Marquez A, Salinas GC (2010) A re-evaluation of *Secernosaurus koerneri* and *Kritosaurus australis* (Dinosauria, Hadrosauridae) from the Late Cretaceous of Argentina. J Vertebr Paleontol 30:813–837

Quattrocchio ME, Volkheimer W (2000) Danian microfloral provinces in Argentina. Rev Esp Paleontol 15:3–11

Ramos VA (2009) Anatomy and global context of the Andes: main geologic features and the Andean orogenic cycle. Geol Soc Am Mem 204:31–65

Ramos VA, Ghiglione MC (2008) Tectonic evolution of the patagonian Andes. Dev Quat Sci 11:57–71

Rauhut OWM (2005) Osteology and relationships of a new theropod dinosaur from the Middle Jurassic of Patagonia. Palaeontology 48:87–110

Rauhut OWM (2007) A fragmentary theropod skull from the Middle Jurassic of Patagonia. Ameghiniana 44:479–483

Rauhut OWN, Pol D (2019) Probable basal allosauroid from the early Middle Jurassic Cañadón Asfalto Formation of Argentina highlights phylogenetic uncertainty in tetanuran theropod dinosaurs. Sci Rep 9:18826

Rauhut OWM, Martin T, Ortiz-Jaureguizar E, Puerta P (2002) A Jurassic mammal from South America. Nature 416:165–168

Ré G, Bellosi ES, Heizler M, Vilas J, Madden R, Carlini AA, Kay R, Vucetich MG (2010) Geochronology for the Sarmiento Formation at Gran Barranca. In: Madden RH, Carlini AA, Vucetich MG, Kay RF (eds) The paleontology of Gran Barranca: evolution and environmental change through the Middle Cenozoic of Patagonia. Cambridge University Press, Cambridge, pp 46–59

Reguero MA (2016) Evolution of the Cenozoic mammals from Antarctica. In: Agnolín FA, Lio GL, Brissón Egli F, Chimento NR, Novas FE (eds) Historia evolutiva y paleobiogeográfica de los vertebrados de América del Sur. Mus Argent Cienc Nat "B Rivadavia", Contribuciones 6:403–412

Reguero MA, Goin FJ, Hospitaleche CA, Dutra T, Marenssi SA (2013) Late Cretaceous/Paleogene west Antarctica terrestrial biota and its intercontinental affinities. Springer, Dordrecht

Reig OA (1963) La presencia de dinosaurios saurisquios en los "Estratos de Ischigualasto" (Mesotriásico Superior) de las provincias de San Juan y La Rioja (República Argentina). Ameghiniana 3:3–20

Riccardi AC (1974) *Eubaculites* Spath (Ammonoidea) del Cretácico superior de Argentina. Ameghiniana 11:379–399

Riccardi AC (1980) Presencia de epizoos en un amonoideo heteromorfo del Maastrichtiano de Argentina: Una inferencia paleoautoecológica. Ameghiniana 17:11–14

Rich TH, Vickers-Rich P, Constanine A, Flannery TF, Kool L, van Klaveren NA (1997) A tribosphenic mammal from the Mesozoic of Australia. Science 278:1438–1442

Rich TH, Vickers-Rich P, Constanine A, Flannery TF, Kool L, van Klaveren NA (1999) Early Cretaceous mammals from Flat Rocks, Victoria, Australia. Rec Queen Vic Mus 106:1–35

Rich TH, Flannery TF, Trusler P, Kool L, van Klaveren NA, Vickers-Rich P (2001a) A second tribosphenic mammal from the Mesozoic of Australia. Rec Queen Vic Mus 110:1–10

Rich TH, Vickers-Rich P, Trusler P, Flannery TF, Cifelli R, Constantine A, Kool L, van Klaveren N (2001b) Monotreme nature of the Australian Early Cretaceous mammal *Teinolophos.* Acta Paleontol Pol 46:113–118

Rich TH, Flannery TF, Trusler P, Kool L, van Klaveren NA, Vickers-Rich P (2002) Evidence that monotremes and ausktribosphenids are not sister groups. J Vertebr Paleontol 22:466–469

Rich TH, Hopson JA, Gill PG, Trusler P, Rogers-Davidson S, Morton S, Cifelli RL, Pickering D, Kool L, Siu K, Burgmann FA, Senden T, Evans AR, Wagstaff BE, Seegets-Villiers D, Corfe IJ, Flannery TF, Walker K, Musser AM, Archer M, Pian R, Vickers-Rich P (2016) The mandible and dentition of the Early Cretaceous monotreme *Teinolophos trusleri.* Alcheringa 40:475–501

Riff D, Kellner AWA (2001) On the dentition of *Baurusuchus pachecoi* Price (Crocodyliformes, Metasuchia) from the Late Cretaceous of Brazil. Bol do Mus Nac 59:1–15

Rivera HA, Le Roux JP, Farías M, Gutiérrez NM, Sánchez A, Palma-Heldt S (2020) Tectonic controls on the Maastrichtian-Danian transgression in the Magallanes-Austral foreland basin (Chile): implications for the growth of the Southern Patagonian Andes. Sediment Geol 403:105645

Rodríguez M, Getino P, Sepúlveda E (1995) Elementos faunísticos del Cretácico–Terciario marino en el Rincón de Colitoro Grande, Provincia de Río Negro. 6º Congreso Argentino de Paleontología y Bioestratigrafía, Trelew, Actas:239–246

Rogers RR, Swisher CC III, Sereno PC, Monetta AM, Forster CA (1993) The Ischigualasto tetrapod assemblage, Late Triassic, Argentina, and 40Ar/39Ar dating of dinosaurs origins. Science 260:794–797

Rogers RR, Arcucci AB, Sereno PC, Forster CA, May CL (2001) Paleoenvironment and taphonomy of the Chanãres Formation tetrapod assemblage (Middle Triassic), northwestern Argentina: Spectacular preservation in volcanogenic concretions. Palaios 16:61–481

Rojas Vera EA, Orts DL, Folguera A, Zamora Valcarce G, Bottesi G, Fennell L, Chiachiarelli F, Ramos VA (2016) The transitional zone between the Southern Central and Northern Patagonian Andes (36–39°S). In: Folguera A, Naipauer M, Sagripanti L, Ghiglione MC, Orts DL, Giambiagi L (eds) Growth of the Southern Andes. Springer, Dordrecht, pp 99–114

Romer AS (1945) Vertebrate paleontology. University of Chicago Press, Chicago

Romer AS (1966) The Chañares (Argentina) Triassic Reptile Fauna. I. Introduction. Breviora 247:1–14

Romer AS (1973) The Chañares (Argentina) Triassic Reptile Fauna. XX. Summary. Breviora 413:1–20

Romero EJ (1968) *Palmoxylon patagonicum* n. sp. del Terciario Inferior de la Provincia de Chubut, Argentina. Ameghiniana 5:417–432

Romo de Vivar PR, Martinelli AG, Hsiou AS, Soares MB (2020) A new rhynchocephalian from the Late Triassic of southern Brazil enhances eusphenodontian diversity. J Syst Palaeontol 18:1103–1126

Rossetti L, Lima EF, Waichel BL, Hole MJ, Simões MS, Scherer CMS (2018) Lithostratigraphy and volcanology of the Serra Geral Group, Paraná-Etendeka Igneous Province in Southern Brazil: towards a formal stratigraphical framework. J Volcanol Geotherm Res 355:98–114

Rougier GW (1993) *Vincelestes neuquenianus* Bonaparte (Mammalia, Theria), un primitivo mamífero del Cretácico Inferior de la Cuenca Neuquina. Unpublished PhD thesis, University of Buenos Aires

Rougier GW, Wible JR (2006) Major changes in the ear region and basicranium of early mammals. In: Carrano MT, Gaudin TJ, Blob RW, Wible JR (eds) Amniote paleobiology: perspectives on the evolution of mammals, birds, and reptiles. University of Chicago Press, Chicago, pp 269–311

Rougier GW, Wible JR, Hopson JA (1992) Reconstruction of the cranial vessels in the Early Cretaceous mammal *Vincelestes neuquenianus*: implications for the evolution of the mammalian cranial vascular system. J Vertebr Paleontol 12:188–216

Rougier GW, de La Fuente MS, Arcuchi A (1995) Late Triassic turtles from South America. Science 268:844–858

Rougier GW, Novacek MJ, Pascual R, Gelfo JN, Cladera G (2000) New Late Cretaceous mammals from Argentina and the survival of Mesozoic lineages in the Patagonian Early Tertiary. J Vertebr Paleontol 20:65A

Rougier GW, Martinelli AG, Forasiepi AM, Novacek MJ (2007a) New Jurassic mammals from Patagonia, Argentina: a reappraisal of australosphenidan morphology and interrelationships. Am Mus Novit 3566:1–54

Rougier GW, Garrido AC, Gaetano L, Puerta P, Corbitt C, Novacek MJ (2007b) First Jurassic triconodont from South America. Am Mus Novit 3580:1–17

Rougier GW, Chornogubsky L, Casadio S, Paéz Arango N, Giallombardo A (2009a) Mammals from the Allen Formation, Late Cretaceous, Argentina. Cretac Res 30:223–238

Rougier GW, Forasiepi AM, Hill RV, Novacek MJ (2009b) New mammalian remains from the Late Cretaceous La Colonia Formation, Patagonia, Argentina. Acta Palaeontol Pol 54:195–212

Rougier GW, Gaetano L, Drury BR, Colella R, Gómez RO, Páez Arango N (2011a) A review of the Mesozoic mammalian record of South America. In: Calvo J, Porfiri J, González Riga B, Dos Santos D (eds) Dinosaurios y paleontología desde América Latina. Editorial de la Universidad Nacional de Cuyo, Mendoza, pp 195–214

Rougier GW, Apesteguía S, Gaetano LC (2011b) Highly specialized mammalian skulls from the Late Cretaceous of South America. Nature 479:98–102

Rougier GW, Wible JR, Beck RMD, Apesteguía S (2012) The Miocene mammal *Necrolestes* demonstrates the survival of a Mesozoic nontherian lineage into the late Cenozoic of South America. Proc Natl Acad Sci USA 109:20053–20058

Rowe TB, Rich TH, Vickers-Rich P, Springer M, Woodburne MO (2008) The oldest platypus and its bearing on divergence timing of the platypus and echidna clades. Proc Natl Acad Sci USA 105:1238–1242

Rubilar D, Chatterjee S, Vargas A (2002) A crocodylomorph from the Late Triassic of Chile. 1° Congreso Latinoamericano de Paleontología de Vertebrados, Santiago de Chile, Actas 1:47

Saban R (1954) Phylogénie des insectivores. Bull Mus Natl Hist Nat, Ser 2, 26:419–432

Salamuni R, Bigarella JJ (1967) Botucatu Formation. In: Bigarella JJ, Becker RD, Pinto ID (eds) Problems in Brazilian Gondwana Geology. 1° International Symposium on the Gondwana Stratigraphy and Paleontology, Mar de la Plata, Abstracts:197–206

Salgado L (1996) *Pellegrinisaurus powelli* nov. gen. et sp. (Sauropoda, Titanosauridae) from the Upper Cretaceous of Lago Pellegrini, northwestern Patagonia, Argentina. Ameghiniana 33:355–365

Salgado L, Azpilicueta C (2000) Un nuevo saltasaurino (Sauropoda, Titanosauridae) de la Provincia de Río Negro (Formación Allen, Cretácico Superior), Patagonia, Argentina. Ameghiniana 37:259–264

Salgado L, Bonaparte JF (1991) Un nuevo saurópodo Dicraeosauridae, *Amargasaurus cazaui* gen. et sp. nov., de la Provincia del Neuquén, Argentina. Ameghiniana 28:333–346

Salgado L, Coria RA (1993) El género *Aeolosaurus* (Sauropoda-Titanosauridae) en la Formación Allen (Campaniano–Maastrichtiano) de la Provincia de Río Negro, Argentina. Ameghiniana 30:119–128

Salgado L, Coria RA (1996) First evidence of an ankylosaur (Dinosauria, Ornithischia) in South America. Ameghiniana 33:367–371

Salgado L, Coria RA (2009) *Barrosasaurus casamiquelai* gen. et sp. nov., a new titanosaur (Dinosauria, Sauropoda) from the Anacleto Formation (Late Cretaceous: early Campanian) of Sierra Barrosa (Neuquén, Argentina). Zootaxa 2222:1–16

Salgado L, Coria R, Heredia S (1997a) New materials of *Gasparinisaura cincosaltensis* (Ornithischia, Ornithopoda) from the Upper Cretaceous of Argentina. J Paleontol 71:933–940

Salgado L, Coria RA, Calvo JO (1997b) Presencia del género *Aeolosaurus* (Sauropoda, Titanosauridae) en la Formación Los Alamitos, Cretácico Superior de la Provincia de Río Negro, Argentina. Geociencias 2:44–49

Salgado L, Garrido AC, Cocca S, Cocca R (2004) Lower Cretaceous rebbachisaurid sauropods from Cerro Aguada del León (Lohan Cura Formation), Neuquén Province, northwestern Patagonia, Argentina. J Vertebr Paleontol 24:903–912

Salgado L, Apesteguía S, Heredia SE (2005) A new specimen of *Neuquensaurus australis*, a Late Cretaceous saltasaurine titanosaur from North Patagonia. J Vertebr Paleontol 25:623–634

Salgado L, de Souza Carvalho I, Garrido AC (2006) *Zapalasaurus bonapartei*, un nuevo dinosaurio saurópodo de la Formación La Amarga (Cretácico Inferior), noroeste de Patagonia, Provincia de Neuquén, Argentina. Geobios 39:695–707

Salgado L, Coria RA, Magalhaes Ribeiro CM, Garrido AC, Rogers R, Simon ME, Arcucci AB, Curry-Rogers K, Paulina Carabajal A, Apesteguía S, Fernandez M, Talevi M (2007) Upper Cretaceous dinosaur nesting site of Río Negro (Salitral Ojo de Agua and Salinas de Trapalcó-Salitral de Santa Rosa), northern Patagonia, Argentina. Cretac Res 28:392–404

Sanford RM, Lange FW (1960) Basin study approach for oil evaluation of Paraná miogeosyncline of South Brazil. Am Assoc Pet Geol Bull 44:1316–1370

Santamarina PE, Barreda VD, Iglesias A, Varela NA (2018) Salvinialean megaspores in the Late Cretaceous of southern Patagonia, Argentina. Acta Paleontol Pol 63:607–616

Santucci RM, Arruda-Campos AC (2011) A new sauropod (Macronaria, Titanosauria) from the Adamantina Formation, Bauru Group, Upper Cretaceous of Brazil and the phylogenetic relationships of Aeolosaurini. Zootaxa 3085:1–33

Santucci RM, Bertini RJ (2006) A new titanosaur from western São Paulo state, Upper Cretaceous Bauru Group, southeast Brazil. Palaeontology 49:59–66

Scanferla CA, Canale JI (2007) The youngest record of the Cretaceous snake genus *Dinilysia* (Squamata, Serpentes). S Am J Herpetol 2:76–81

Schultz CL, Scherer CM, Barberena MC (2000) Biostratigraphy of southern Brazilian Middle-Upper Triassic. Rev Bras Geociênc 30:491–494

Schultz CL, Martinelli AG, Soares MB, Pinheiro FL, Kerber L, Horn BLD, Pretto FA, Muller RT, Melo TP (2020) Triassic faunal successions of the Paraná Basin, southern Brazil. J S Am Earth Sci 104:102846

Scillato-Yané GJ (1986) Los Xenarthra fósiles de Argentina (Mammalia, Edentata). 4° Congreso Argentino de Paleontología y Bioestratigrafía, Mendoza, Actas 2:151–155

Scillato-Yané GJ, Pascual R (1985) Un peculiar Xenarthra del Paleoceno medio de Patagonia (Argentina). Su importancia en la sistemática de los Paratheria. Ameghiniana 21:316–318

Scotese CR (2004) A continental drift flipbook. J Geol 112:729–741

Scotese CR (2013a) Map Folio 44, Late Triassic (210 Ma, Norian), PALEOMAP PaleoAtlas for ArcGIS, vol 3, Jurassic and Triassic. PALEOMAP Project, Evanston, IL

Scotese CR (2013b) Map Folio 37 Middle Jurassic (Bajocian and Bathonian, 169.7 Ma), PALEOMAP PaleoAtlas for ArcGIS, vol 3, Triassic and Jurassic Paleogeographic, Paleoclimatic and Plate Tectonic Reconstructions. PALEOMAP Project, Evanston, IL

Scotese CR (2013c) Map Folio 17, Late Cretaceous, (Maastrichtian, 68 Ma), PALEOMAP PaleoAtlas for ArcGIS, vol 2, Cretaceous Paleogeographic, Paleoclimatic and Plate Tectonic Reconstructions. PALEOMAP Project, Evanston, IL

Scotese CR (2013d) Map Folio 5, Middle/Late Miocene, (Serravallian & Tortonian, 10.5 Ma), PALEOMAP PaleoAtlas for ArcGIS, vol 1, Cenozoic. PALEOMAP Project, Evanston, IL

Scott WB (1903–1904) Mammalia of the Santa Cruz Beds. I. Edentata. In: Scott WB (ed) Reports of the Princeton University Expedition to Patagonia, 1896–1899, vol 5, Paleontology II. Princeton University Press, Princeton, pp 1–364

Scott WB (1905). Mammalia of the Santa Cruz beds. Part II. Insectivora. In: Scott WB (ed) Reports of the Princeton University Expedition to Patagonia, 1896–1899. Vol. 5, Paleontology II. Princeton University Press, Princeton, pp 365–383

Scott WB (1910). Mammalia of the Santa Cruz beds. Part I. Litopterna. In: Scott WB (ed) Reports of the Princeton University Expedition to Patagonia, 1896–1899, vol 7, Paleontology IV. Princeton University Press, Princeton, pp 1–156

Scott WB (1912). Mammalia of the Santa Cruz beds. Part II. Toxodonta. Part III. Entelonychia. In: Scott WB (ed) Reports of the Princeton University Expedition to Patagonia, 1896–1899, vol 6, Paleontology III. Princeton University Press, Princeton, pp 111–300

Scott WB (1913) A history of land mammals in the Western Hemisphere. MacMillan, New York

Scott WB (1928a). Mammalia of the Santa Cruz beds. Part IV. Astrapotheria. In: Scott WB (ed) Reports of the Princeton University Expedition to Patagonia, 1896–1899, vol 6, Paleontology III. Princeton University Press, Princeton, pp 301–341

Scott WB (1928b). Mammalia of the Santa Cruz beds. Part IV: Primates. I In: Scott WB (ed) Reports of the Princeton University Expedition to Patagonia, 1896–1899, vol 6, Paleontology III. Princeton University Press, Princeton, pp 342–351

Scott WB (1937) The Astrapotheria. Proc Am Philos Soc 77:309–393

Sereno PC (2013) Preface. Vertebrate succession in the Ischigualasto Formation. J Vertebr Paleontol 32 (Suppl):1–9

Sereno PC, Novas FE (1992) The complete skull and skeleton of an early dinosaur. Science 258:1137–1140

Sereno PC, Forster CA, Rogers RR, Monetta AM (1993) Primitive dinosaur skeleton from Argentina and the early evolution of Dinosauria. Nature 361:64–66

Sereno PC, Martínez R, Wilson JA, Varricchio DJ, Alcober OA, Larson HC (2008) Evidence for avian intrathoracic air sacs in a new predatory dinosaur from Argentina. PLoS ONE 3(9):e3303

Shockey BJ, Hitz R, Bond M (2004) Paleogene notoungulates from the Amazon Basin of Peru. In: Campbell KE (ed) The Paleogene mammalian fauna of Santa Rosa, Amazonian Peru. Nat Hist Mus Los Angeles County, Sci Ser 40:61–70

Sigogneau-Russell D (1989) Découverte du premier Symmétrodonte (Mammalia) du continent africain. C R Acad Sci 309:921–926

Sigogneau-Russell D (1991) Découverte du premier mammifère tribosphénique de Mésozoïque africain. C R Acad Sci 313:1635–1640

Sigogneau-Russell D (1995) Two possibly aquatic triconodont mammals from the Early Cretaceous of Morocco. Acta Palaeontol Pol 40:149–162

Sigogneau-Russell D, Ensom P (1998) *Thereuodon* (Theria, Symmetrodonta) from the Lower Cretaceous of North Africa and Europe, and a brief review of symmetrodonts. Cretac Res 19:445–470

Sigogneau-Russell D, Monbaron M, de Kaenel E (1990) Nouvelle données sur le gisement á mammiféres mésozoïques du Haut-Atlas Marocain. Geobios 23:461–483

Silva Nieto DG, Cabaleri N, Salani F, Coluccia A (2002) Cañadón Asfalto, una cuenca de tipo "Pull Apart" en el área de Cerro Cóndor, Provincia del Chubut. 15° Congreso Geológico Argentino, El Calafate, Actas 1:238–243

Simões TR, Caldwell MW, Weinschütz LC, Wilner E, Kellner AWA (2017) Mesozoic lizards from Brazil and their role in early squamate evolution in South America. J Herpetol 51:307–315

Simón ME, Calvo JO (2002) Un primitivo titanosaurio (Sauropoda) del Chocón, Formación Candeleros, (Cenomaniano Temprano), Neuquén, Argentina. Ameghiniana 39(Suppl):17R

Simón ME, Kellner AWA (2003) New sphenodontid (Lepidosauria, Rhynchocephalia, Eilenodontinae) from the Candeleros Formation, Cenomanian of Patagonia, Argentina. Bol Mus Nac 68:1–12

Simpson GG (1937) An ancient eusuchian crocodile from Patagonia. Am Mus Novit 965:1–20

Simpson GG (1945) The principles of classification and a classification of mammals. Bull Am Mus Nat Hist 85:1–350

Simpson GG (1980) Splendid Isolation: the curious history of South American mammals. Yale University Press, New Haven

Sinclair WJ (1906) Mammalia of the Santa Cruz beds. Part III Marsupialia. In: Scott WB (ed) Reports of the Princeton University Expedition to Patagonia, 1896–1899, vol 4, Paleontology I. Princeton University Press, Princeton, pp 333–460

Smith AG, Smith DG, Funnell BM (1994) Atlas of Mesozoic and Cenozoic coastlines. Cambridge University Press, Cambridge

Soares PC, Landim PMB, Fúlfaro VJ, Sobreiro Neto AF (1980) Ensaio de caracterização do Cretáceo no Estado de São Paulo: Grupo Bauru. Rev Bras Geociênc 10:177–185

Soares MVT, Basilici G, Marinho TS, Martinelli AG, Marconato A, Mountney NP, Colombera L, Mesquita AF, Vasques JT. Abrantes Junior FR, Ribeiro LCB (2020) Sedimentology of a distributive fluvial system (Upper Cretaceous, Bauru Basin, Brazil): The Serra da Galga Formation, a new lithostratigraphic unit. Geol J. https://doi.org/10.1002/gj.3987

Spalletti LA, Franzese JR (2007) Mesozoic paleogeography and paleoenvironmental evolution of Patagonia (southern South America). In: Gasparini Z, Salgado L, Coria RA (eds) Patagonian mesozoic reptiles. Indiana University Press, Bloomington, pp 29–49

Spalletti LA, Mazzoni MM (1979) Estratigrafía de la Formación Sarmiento en la barranca sur del lago Colhue-Huapi, Provincia del Chubut. Rev Asoc Geol Argent 34:271–281

Spalletti LA, Iñiguez Rodriguez M, Mazzoni M (1982) Edades radiométricas de piroclastitas y volcanitas del Grupo Bahia Laura, Gran Bajo de San Julián, Santa Cruz. Rev Asoc Geol Argent 37:483–485

Spalletti LA, Morel E, Artabe A, Ganuza D, Bodnar J (2011) Sedimentología de la Formación Carrizal (Triásico) en el depocentro Marayes-El Carrizal, Provincia de San Juan, República Argentina. Lat Am J Sedimentol Basin Anal 18:89–104

Sterli J (2008) A new, nearly complete stem turtle from the Jurassic of South America with implications for turtle evolution. Biol Lett 4:286–289

Sterli J, de la Fuente MS (2010) Anatomy of *Condorchelys antiqua* Sterli, 2008 and the origin of the modern jaw closure mechanism in turtles. J Vertebr Paleontol 30:351–366

Sterli J, de la Fuente MS (2011) A new turtle from the La Colonia Formation (Campanian–Maastrichtian), Patagonia, Argentina, with remarks on the evolution of the vertebral column in turtles. Palaeontology 54:63–78

Sterli J, de la Fuente MS (2013) New evidence from the Palaeocene of Patagonia (Argentina) on the evolution and palaeo-biogeography of Meiolaniformes (Testudinata, new taxon name). J Syst Palaeontol 11:835–852

Sterli J, Rougier GW, de la Fuente MS (2007) Anatomy and relationships of *Palaeochersis talampayensis*, a Late Triassic turtle from Argentina. Palaeontographica Abt A 281:1–61

Sterli J, de la Fuente MS, Rougier GW (2019) New remains of *Condorchelys antiqua* (Testudinata) from the Early-Middle Jurassic of Patagonia: Anatomy, phylogeny, and paedomorphosis in the early evolution of turtles. J Vertebr Paleontol 34:e1480112

Stilwell JD, Zinsmeister WJ (1992) Molluscan systematics and biostratigraphy: lower Tertiary La Meseta Formation, Seymour Island, Antarctic Peninsula. Antarct Res Ser 55:1–152

Stipanicic PN (1979) El Triásico de valle del río de Los Patos (provincia de San Juan). In: Turner JCM (ed) 2° Simposio de Geología Regional Argentina. Acad Nac Cienc Córdoba, Córdoba, vol 1, pp 695–744

Stipanicic PN, Bonaparte JF (1979) Cuenca triásica de Ischigualasto-Villa Unión (Provincias de San Juan y La Rioja). In: Turner JCM (ed) 2° Simposio de Geología Regional Argentina. Acad Nac Cienc Córdoba, Córdoba, vol 1, pp 507–536

Stipanicic PN, Bonetti MIR (1970) Posiciones estratigráficas y edades de las principales floras jurásicas argentinas II. Floras doggerianas y málmicas. Ameghiniana 7:101–118

Stipanicic PN, Reig OA (1955) Breve noticia sobre el hallazgo de anuros en el denominado "complejo porfírico de la Patagonia Extraandina", con consideraciones acerca de la composición geológica del mismo. Rev Asoc Geol Argent 10:215–233

Stipanicic PN, Reig OA (1957) El "complejo porfírico de la Patagonia Extraandina" y su fauna de anuros. Acta Geol Lill 1:185–298

Stipanicic PN, Rodrigo F, Baulíes OL, Martínez CG (1968) Las formaciones presenonianas en el denominado Macizo Nordpatagónico y regiones adyacentes. Rev Asoc Geol Argent 23:67–98

Stipanicic PN, González Díaz E, Zavattieri AM (2007) Grupo Puesto Viejo nom. transl. por Formación Puesto Viejo González Díaz, 1964, 1967: nuevas interpretaciones paleontológicas, estratigráficas y cronológicas. Ameghiniana 44:759–761

Stollhofen H (1999) Karoo Synrift—Sedimentation und ihre tektonische Kontrolle am entstehenden Kontinentalrand Namibias. Z Dtsch Geol Ges 149:519–623

Suárez JM (1969) Um quelônio da Formação Bauru. 12° Congresso Brasileiro de Paleontologia, São Paulo, Actas:167–176

Suguio K, Coimbra AM (1972) Madeira fóssil silicíficada na Formação Botucatu. Ciencia Cult 24:1049–1055

Szalay FS (1994) Evolutionary history of the marsupials and an analysis of osteological characters. Cambridge University Press, Cambridge

Tasch P, Volkheimer W (1970) Jurassic conchostracans from Patagonia. University of Kansas Paleontological Contributions, Paper 50:1–23

Tauber AA, Luna CA, Palacios ME (2005) El registro de *Necrolestes patagonensis* Ameghino, 1891 (Mammalia) en La Formación Santa Cruz (Mioceno), Patagonia Austral. In: Cabaleri N, Cingolani CA, Linares E, López de Luchi MG, Ostera HA, Panarello HO (eds). 16° Congreso Geológico Argentino, La Plata, Actas 5:157–164

Tejedor MF, Goin FJ, Gelfo JN, López GM, Bond M, Carlini AA, Scillato-Yané GJ, Woodburne MO, Chornogubsky L, Aragón E, Reguero MA, Czaplewski NJ, Vincon S, Martin GM, Ciancio MR (2009) New Early Eocene mammalian fauna from western Patagonia, Argentina. Am Mus Novit 3638:1–43

Tettamanti C, Moyano Paz D, Varela AN, Tineo DE, Gómez-Peral LE, Poiré DG, Cereceda A, Odino Barreto AL (2018) Sedimentology and fluvial styles of the uppermost Cretaceous continental deposits of the Austral-Magallanes Basin, Patagonia, Argentina. Lat Am J Sedimentol Basin Anal 25:149–168

Torres T, Marenssi SA, Santillana SN (1994) Maderas fósiles de la isla Seymour, Formación La Meseta, Antártica. Serie Científica INACH 44:17–38

Trevisan C, Dutra T, Wielberger T, Leppe M, Manríquez L (2020) An austral fern assemblage from the Upper Cretaceous (Campanian) beds of Cerro Guido, Magallanes Basin, Chilean Patagonia. Cretac Res 106:104215

Turner AH, Calvo JO (2005) A new sebecosuchian crocodyliform from the Late Cretaceous of Patagonia. J Vertebr Paleontol 25:87–98

Turner S, Regelous M, Hawkesworth C, Montovani M (1994) Magmatism and continental breakup in the South Atlantic: High precision $^{40}$Ar-$^{49}$Ar geochronology. Earth Planet Sci Lett 121:333–348

Ulansky RE (2014) Dinosaurs classification. Basal Thyreophora & Stegosauria. Dinologia:1–8

Uliana MA, Biddle KT (1988) Mesozoic-Cenozoic paleogeographic and geodynamic evolution of southern South America. Rev Bras Geoc 18:182–190

Uliana MA, Dellapé DA (1981) Estratigraía y evolución paleoambiental de la sucesión Maastrichtiano–Eoterciaria del engolfamiento neuquino (Patagonia septentrional). 8° Congreso Geológico Argentino, San Luis, Actas 3:673–711

Vallatti P (1986) Conchostracos jurásicos de la Provincia del Chubut, Argentina. 4° Congreso Argentino de Paleontología y Bioestratigrafía, Mendoza, Actas 4:29–38

Van Valen LM (1988) Faunas of a southern world. Nature 333:113

Varela AN, Poiré DG (2008) Paleogeografía de la Formación Mata Amarilla, Cuenca Austral, Patagonia, Argentina. 12° Reunión Argentina de Sedimentología, Buenos Aires, Actas:183

Varela AN, Poiré DG, Martin T, Gerdes A, Goin FJ, Gelfo JN, Hoffmann S (2012) U-Pb zircon constraints on the age of the Cretaceous Mata Amarilla Formation, Southern Patagonia, Argentina: its relationship with the evolution of the Austral Basin. Andean Geol 39:359–379

Varela AN, Raigemborn MS, Richiano S, White T, Poiré DG, Lizzoli S (2018) Late Cretaceous paleosols as paleoclimate proxies of high-latitude Southern Hemisphere: Mata Amarilla Formation, Patagonia, Argentina. Sediment Geol 363:83–95

Vildoso-Morales CA (1991) Sobre mamíferos del Cretácico terminal del noroeste peruano. 8° Jornadas Argentinas de Paleontología de Vertebrados, La Rioja, Abstracts:10

Vizcaíno SF, Kay RF, Bargo MS (eds) (2012a) Early Miocene paleobiology in Patagonia. Cambridge University Press, Cambridge

Vizcaíno SF, Fernícola JC, Bargo MS (2012b) Paleobiology of Santacrucian glyptodonts and armadillos (Xenarthra, Cingulata). In: Vizcaíno SF, Kay RF, Bargo MS (eds) Early Miocene paleobiology in Patagonia. Cambridge University Press, Cambridge, pp 194–215

de Vleeschouwer D, Vahlenkamp M, Crucifix M, Pälike H (2017) Alternating Southern and Northern Hemisphere climate response to astronomical forcing during the past 35 m.y. Geology 45:375–378

Vogt M, Leppe M, Stinnesbeck W, Jujihara T, Mansilla M, Ortiz H, Manríquez L, González E (2014) Depositional environment of Maastrichtian (Late Cretaceous) dinosaur-bearing deltaic deposits of the Dorotea Formation, Magallanes Basin, southern Chile. 23° Latin American Colloquium on Earth Sciences, Heidelberg, Abstracts:156

Volkheimer W (1973) Observaciones geológicas en el área de Ingeniero Jacobacci y adyacencias (Provincia de Río Negro). Rev Asoc Geol Argent 18:13–16

Volkheimer W (1978) Microfloras fósiles. In: Rolleri EO (ed) Geología y recursos naturales de la provincia del Neuquén, 7° Congreso Geológico Argentino, Neuquén, Actas:147–161

Volkheimer W, Quattrocchio M, Cabaleri N, García V (2008) Palynology and paleoenvironment of the Jurassic lacustrine Cañadón Asfalto Formation at Cañadón Lahuincó locality, Chubut Province, Central Patagonia, Argentina. Rev Esp Micropaleontol 40:77–96

Volkheimer W, Gallego OF, Cabaleri NG, Armella C, Narvaez PL, Silva Nieto DG, Paez MA (2009) Stratigraphy, palynology, and conchostracans of a Lower Cretaceous sequence at the Cañadón Calcareo locality, extra-Andean central Patagonia; age and palaeoenvironmental significance. Cretac Res 30:270–282

Volkheimer W, Quattrocchio ME, Cabaleri NG, Narváez PL, Rosenfeld U, Scafati L, Melendi DL (2015) Environmental and climatic proxies for the Cañadón Asfalto and Neuquén Basins (Patagonia, Argentina): review of Middle to Upper Jurassic continental and near coastal sequences. Rev Bras Paleontol 18:71–82

von Baczko MB, Desojo JB (2016) Cranial anatomy and palaeoneurology of the archosaur *Riojasuchus tenuisceps* from the Los Colorados Formation, La Rioja, Argentina. PLoS ONE 11(2):e0148575

von Staesche K (1937) *Podocnemis brasiliensis* n. sp. aus der oberen Kreide Brasiliens. Neues Jahrb Geol Paläontol 77:291–309

Vucetich MG, Kramarz A, Candela MA (2010) Colhuehuapian rodents from Gran Barranca and other Patagonian localities: the state of the art. In: Madden RH, Carlini AA, Vucetich MG, Kay RF (eds) The paleontology of Gran Barranca: evolution and environmental change through the Middle Cenozoic of Patagonia. Cambridge University Press, Cambridge, pp 206–219

Wible JR, Rougier GW (2017) Craniomandibular anatomy of the subterranean Meridolestidan *Necrolestes patagonensis* Ameghino, 1891 (Mammalia, Cladotheria) from the Early Miocene of Patagonia. Ann Carnegie Mus 84:183–251

Wichmann R (1924) Nuevas observaciones geológicas en la parte oriental del Neuquén y en el Territorio del Río Negro. Ministerio de Agricultura, Sección Geología 2:1–22

Wichmann R (1927) Sobre la facies lacustre Senomaniana de los estratos con dinosaurios y su fauna. Acad Nac Cienc Córdoba, Bol 30:383–405

Wilf P, Cúneo R, Escapa IH, Pol D, Woodburne MO (2013) Splendid and seldom isolated: the paleobiogeography of Patagonia. Annu Rev Earth Planet Sci 41:561–603

Wilson GP, Arens NC (2001) The evolutionary impact of an epeiric seaway on Late Cretaceous and Paleocene palynofloras of South America. 7° International Symposium on Mesozoic Terrestrial Ecosystems. Asociación Paleontológica Argentina, Buenos Aires, Publicación Especial 7:185–189

Winge H (1941) The interrelationships of the mammalian genera. Volume I. Monotremata, Marsupialia, Insectivora, Chiroptera, Edentata. C.A. Reitzels Forlag, København

Woodburne MO (2003) Monotremes as pretribosphenic mammals. J Mammal Evol 10:195–248

Woodburne MO (2010) The Great American Biotic Interchange: dispersals, tectonics, climate, sea level and holding pens. J Mammal Evol 17:245–264

Woodburne MO, Zinsmeister WJ (1984) The first land mammal from Antarctica and its biogeographic implications. J Paleontol 54:913–948

Woodburne MO, Rich TH, Springer MS (2003) The evolution of tribospheny and the antiquity of mammalian clades. Mol Phylogenet Evol 28:360–385

Woodburne MO, Goin FJ, Bond M, Carlini AA, Gelfo JN, López GM, Iglesias A, Zimicz AN (2014a) Paleogene land mammal faunas of South America: a response to global climatic changes and indigenous floral diversity. J Mammal Evol 21:1–73

Woodburne MO, Goin FJ, Raigemborn MS, Heizler M, Gelfo JN, Oliveira EV (2014b) Revised timing of the South American Early Paleogene land mammal ages. J S Am Earth Sci 54:109–119

Yeo D, Cumberlidge N, Klaus S (2014) Advances in freshwater decapod systematics and biology. Crustaceana Monographs 19. Brill, Leiden

Zachos J, Pagani M, Sloan L, Thomas E, Billups K (2001) Trends, global rhythms, aberrations in global climate 65 Ma to Present. Science 292:686–693

Zaher HD, Langer MC, Fara E, Carvalho IS, Arruda JT (2003) A mais antiga serpente (Anilioidea) brasileira: Cretáceo Superior do Grupo Bauru, General Salgado, SP. Paleontol Destaque 44:50–51

Zambrano JJ (1987) Las cuencas sedimentarias de América del Sur durante el Jurásico y Cretácico: Su relación con la actividad tectónica y magmática. In: Volkheimer W (ed) Bioestratigrafía de los sistemas regionales del Jurásico y Cretácico de América del Sur, Mendoza, Comité Sudamericano del Jurásico–Cretácico 1:1–48

Zamuner AB, Falaschi P, Bamford M, Iglesias A, Poiré DG, Varela AN, Larriestra F (2006) Anatomía y paleocología de dos bosques in situ de la Zona de Tres Lagos, Formación Mata Amarilla, Cretácico Superior, Patagonia, Argentina. 13° Simposio Argentino de Paleobotánica y Palinología, Bahía Blanca, Actas:55

Zavattieri AM, Arcucci AB (2007) Edad y posición estratigráfica de los tetrápodos del Cerro Bayo de Potrerillos (Triásico), Mendoza, Argentina. Ameghiniana 44:133–142

Zavattieri AM, Papú OH (1993) Microfloras mesozoicas. In: Ramos V (ed) Geología y Recursos Naturales de Mendoza. 12° Congreso Geológico Argentino y 2° Congreso de Exploración de Hidrocarburos, Mendoza, Relatorio Capítulo II-9:309–316

Zerfass H, Lavina EL, Schultz CL, García AJV, Faccini UF, Chemale F (2003) Sequence stratigraphy of continental Triassic strata of southernmost Brazil: a contribution to southwestwern Gondwana palaeogeography and palaeoclimate. Sediment Geol 161:85–105

Zerfass H, Chemale-Jr F, Schultz CL, Lavina EL (2004) Tectonics and sedimentation in southern South America during Triassic. Sediment Geol 166:265–292

# Chapter 3
# The Origin and the Radiation of Early Mammals: A Southern Perspective

*Perhaps no other major adaptive shift between widely different grades of vertebrate organization is so well documented by fossils as is the reptile-mammal transition. This has allowed an interpretation of the adaptive basis for much of the morphological transformation.*
James A. Hopson and Leonard B. Radinsky
*Vertebrate paleontology: New approaches and new insights*, 1980

**Abstract** Non-mammaliaform cynodonts, formerly called "mammal-like reptiles," illustrate earlier states of the morphological architecture in the mammalian lineage. These mammalian forerunners show unique character combinations without direct counterparts among living vertebrates reflecting adaptations long lost along the millions of years of cynodont history. The fossil record from South America, originating mostly from the Middle to Late Triassic of Argentina and Brazil, is one of the most prolific worldwide. SA non-mammalian cynodonts are systematically diverse, including approximately 40 species that present great morphological disparity in skull shape, tooth morphology, pattern of tooth replacement, masticatory mechanisms, and locomotory architectures. In this chapter, we summarize the record of SA non-mammaliaform cynodonts.

**Keywords** Synapsida · Therapsida · Cynodontia · Triassic · Argentina · Brazil

## 3.1 Introduction

Mammals are a surviving group of synapsids, a branch of amniotes that diverged from reptiles about 320 Ma (Reisz 1972, 1986). The earliest synapsids date to the beginning of the Carboniferous. They were numerous and diverse, occupying different roles in the ancient terrestrial ecosystems as do mammals, their modern-day descendants (Reisz 1972, 1986; Kemp 2005; Luo 2007; O'Leary et al. 2013). Throughout their history, hundreds of groups with disparate morphologies appeared and became extinct. Since at least the Paleogene, mammals (Mammalia) are the sole survivors of the group (O'Leary et al. 2013), with more than 6,400 extant species (e.g., Burgin et al. 2018).



G. W. Rougier et al., *Mesozoic Mammals from South America and Their Forerunners*, Springer Earth System Sciences, https://doi.org/10.1007/978-3-030-63862-7_3

Cynodontia is a monophyletic group of synapsids, which includes Mammalia and their closest relatives. They are first recorded in the Late Permian, survived major mass extinction events (e.g., Permian/Triassic, Triassic/Jurassic, and Cretaceous/Paleogene), and radiated throughout the Mesozoic and Cenozoic. Cynodontia is cladistically defined as the most inclusive clade of synapsids, including Mammalia and excluding *Bauria* (Hopson and Kitching 2001). The paraphyletic group "non-mammaliaform cynodonts" is commonly used in the literature as simply "cynodonts" in a practical sense, in order to exclude Mammaliaformes, in which the crown group Mammalia is nested (Rowe 1988; see also Chap. 1).

In this chapter, we focus on SA non-mammaliaform cynodonts, highlighting their diversity and bearing on the origin of mammals. We summarize findings and major anatomical characteristics, providing a backdrop to the rise of many features we commonly associate with mammals, like complex mastication and sophistication in the sensory apparatuses, which in fact have a long history preceding Mammalia. Many SA findings have a profound impact on our understanding and current interpretations of the origin and evolution of the mammalian bauplan.

## 3.2 Non-mammaliaform Cynodonts

Non-mammaliaform cynodonts (hereafter cynodonts) were highly diverse, with global distribution from the Late Permian to the Early Cretaceous (e.g., Kemp 2005; Abdala and Ribeiro 2010; Matsuoka et al. 2016). Cynodonts are claimed to extend into the Early Paleocene (Fox et al. 1992), however, the arguments to support the taxonomic referral of this fossil are questionable and further studies are needed. The oldest and basalmost cynodonts are from the beginning of the Late Permian (early Lopingian) of the *Tropidostoma* Assemblage Zone (AZ) of the Karoo Basin, Western Cape Province, South Africa. The group is represented by a few specimens of *Charassognathus gracilis* (Botha et al. 2007) and *Abdalodon diastematicus* (Kammerer 2016). By the end of the Permian, cynodonts were more abundant and taxonomically diverse, especially in the *Dicynodon* AZ of the Karoo Basin (late Lopingian) and in other chrono-correlated assemblages of Eastern Europe (e.g., Abdala et al. 2019). Later and just before the Permian/Triassic mass extinction, species of the genus *Procynosuchus,* as well as *Dvinia, Nanictosaurus, Uralocynodon,* and *Nanocynodon* were the dominant taxa with a broad distribution in Africa and Eastern Europe (Kemp 2005; Ruta et al. 2013). These early cynodonts have large postdentary bones, numerous, fairly homogeneous postcanine teeth, and non-fully ossified secondary palate. Permian cynodonts are to date unknown in SA, but the discovery of diverse Late Permian therapsid associations in Southern Brazil (e.g., Cisneros et al. 2011) has opened new possibilities for fossil findings on the continent.

Basal epicynodonts, Thrinaxodontidae and Galesauridae, appeared in the Late Permian and became abundant during the Early Triassic. The monophyly of these groups, however, is still controversial (Sidor and Smith 2004; Abdala 2007; Van den Brandt and Abdala 2018). In particular, *Thrinaxodon* is one of the best studied and most abundant basal epicynodonts, so far recovered in the Early Triassic of South Africa and Antarctica (Colbert and Kitching 1977; Botha and Chinsamy 2005; Kemp 2005; Abdala et al. 2013; Jasinoski et al. 2015). In contrast, basal epicynodonts are poorly represented in SA. The species *Prozostrodon brasiliensis* from the Late Triassic of Brazil was originally erected with the name *Thrinaxodon brasiliensis* (Barberena et al. 1987) because of resemblances with African species of this genus in the lower postcanine teeth, with triconodont-like crowns, enlarging the temporal and spatial distribution of the genus. However, a later revision of the holotype material suggested that it belongs to a different genus, *Prozostrodon* (Bonaparte and Barberena 2001), which possess several derived probainognathian features (e.g., Bonaparte and Barberena 2001; Liu and Olsen 2010; Martinelli et al. 2017a, c; Guignard et al. 2018; Pacheco et al. 2018). In addition, a few postcranial remains referred to indeterminate Cynodontia were reported from the Early Triassic (?Olenekian), *Procolophon* AZ (Sanga do Cabral Supersequence) from Rio Grande do Sul (Abdala et al. 2002b; Da-Rosa et al. 2009). The material resembles the postcranium of *Galesaurus* and *Thrinaxodon*; however, the taxonomy of these specimens is problematic due to the poor condition of the fossils (there are isolated fragments of humeri and femora but no cranial remains). In fact, some of these specimens were considered non-cynodont synapsids (Piñeiro et al. 2015), and it is also possible that some specimens could belong to dicynodont therapsids or even procolophonian parareptiles (Martinelli et al. 2016b; Dias-da-Silva et al. 2017).

The Eucynodontia clade quickly radiated during the end of the Early Triassic in different ecological forms, exhibiting unique cranial and dental morphologies (e.g., Hopson and Kitching 2001; Kemp 2005; Ruta et al. 2013). Their body sizes varied from tiny opossum- to large tapir-like forms, and specialized into carnivory, herbivory, and insectivory. With the exception of Trirachodontidae (Gomphodontia) and Tritylodontidae (see below; Luo 1994; Hopson and Kitching 2001), other eucynodontian groups are well-documented in Argentina and Brazil. A few postcranial remains from the Late Triassic of the Los Colorados Formation were tentatively assigned to Tritylodontidae by Bonaparte (1971), but the presence of this group in SA is not confirmed. Instead, the specimens can be referred to the tritheledontid *Chaliminia* (Martinelli and Soares 2016), a cynodont not infrequent in the Los Colorados Formation (Bonaparte 1980; Martinelli and Rougier 2007), or even to an indeterminate cynodont (Gaetano et al. 2017).

Recent phylogenetic analyses recognized two major eucynodontian clades: Cynognathia and Probainognathia, the latter including crown Mammalia (e.g., Hopson and Kitching 2001; Abdala 2007; Bonaparte et al. 2005; Liu and Olsen 2010; Ruta et al. 2013; Martinelli et al. 2016a, 2017a, b; Wallace et al. 2019) (Fig. 3.1). One of the major conflicts in cynodont phylogeny is the placement of the highly specialized herbivorous clade Tritylodontidae, which originally was considered as a terminal group of Traversodontidae due to morphological features associated with herbivory observed in the zygoma and dentition (e.g., Crompton and Ellenberger 1957; Hopson and Barghusen 1986; Sues 1986; Hopson and Kitching 2001). However, other

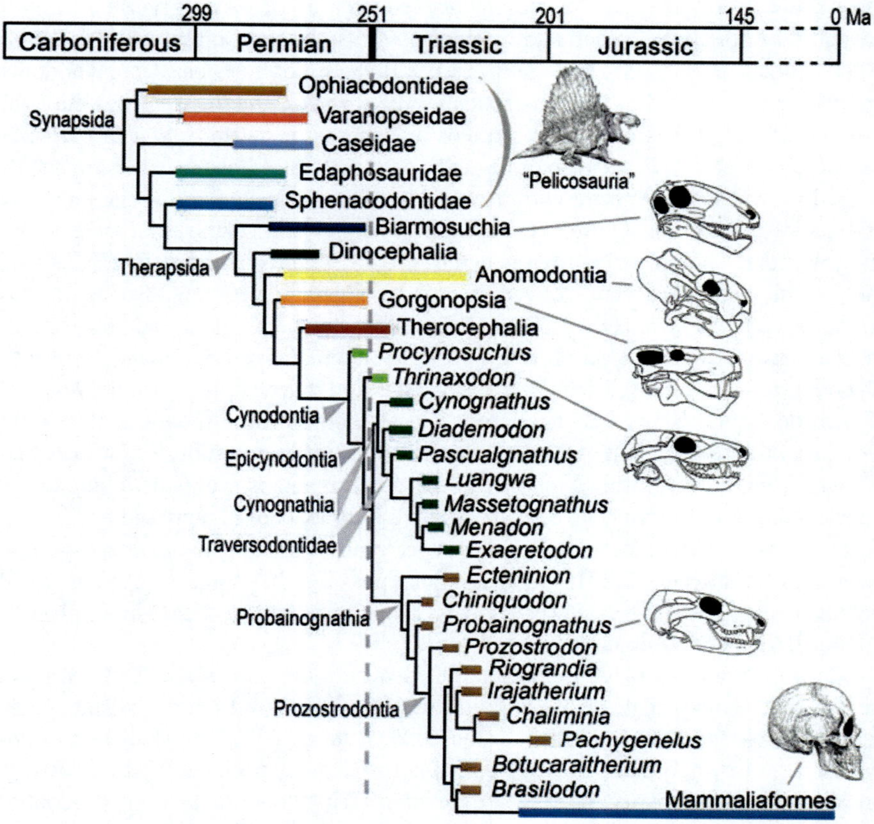


**Fig. 3.1** Phylogenetic relationships of Synapsida, focused on Cynodontia (based on several contributions; see text)

hypotheses place this clade within probainognathians, close to Mammaliaformes (Rowe 1988; Luo 1994; Abdala 2007; Liu and Olsen 2010; Wallace et al. 2019), due to the well-ossified medial wall of the orbit (with a large contribution of the orbitosphenoid and ascending process of the palatine to the orbital wall), quadrate with stapedial process, jugular foramen separated from fenestra cochleae (=fenestra rotunda) with the latter in life covered by the secondary tympanic membrane, and completely divided tooth roots, among other features. The phylogenetic position and interspecific relationships of the taxa included among tritylodontids, gomphodonts, and probainognathians are still unresolved issues because all current hypotheses involve many homoplastic traits (e.g., Luo 1994). Findings of more basal tritylodontids or "more transitional" taxa (either traversodontids or probainognathians) are needed to produce better-supported phylogenies. A new traversodontid from the Upper Ntawere Formation of Zambia with a rare mix of traits eventually could add support to the hypothesis of tritylodontids as gomphodonts, accordingly to Sidor et al. (2016).

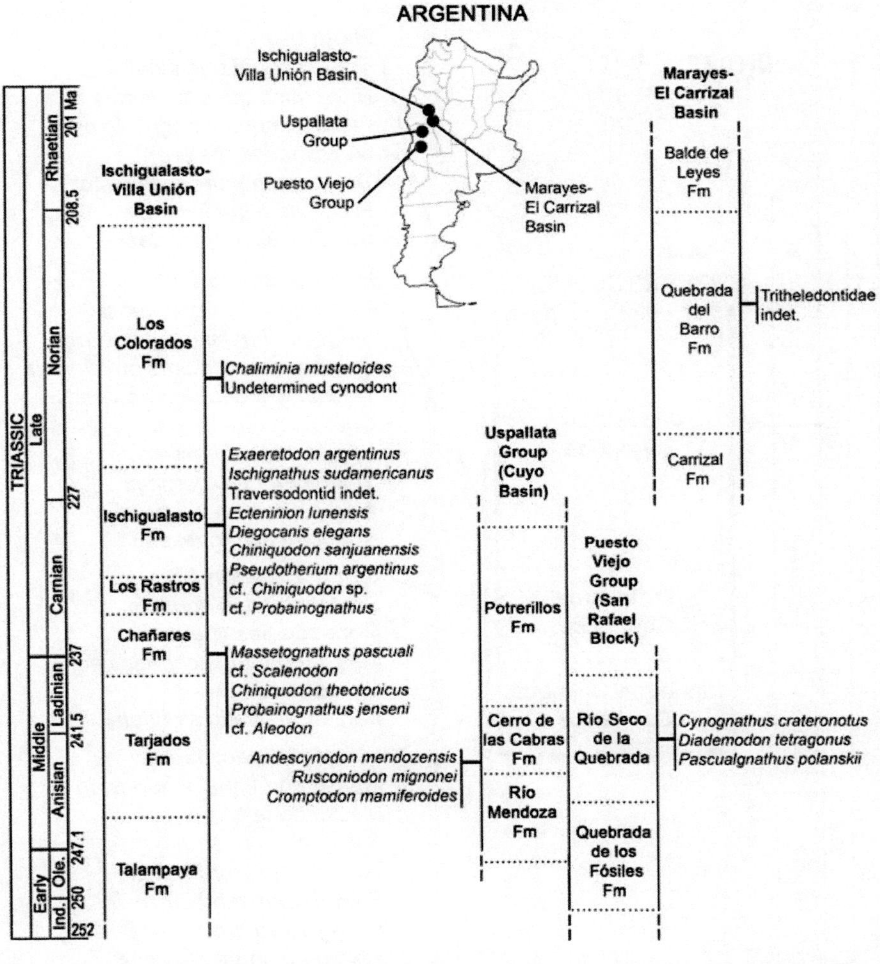


**Fig. 3.2** Geographic and stratigraphic occurrence of known species of non-mammaliaform cynodonts in Argentina. Abbreviation: Fm, Formation; Ind, Induan; Ole, Olenekian

As mentioned in Chap. 2, the cynodonts from SA come from the Triassic rocks of Argentina (Fig. 3.2) and Brazil (Fig. 3.3), comprising 38 genera and 41 species, plus several fossil material pending studies.

### 3.2.1 *Cynognathia*

Cynognathia is a heterogeneous group recovered from Middle to Late Triassic beds, excluding tritylodontids, which are found throughout Pangea. *Cynognathus crateronotus* represents the sister taxon of three clades nested within Gomphodontia:

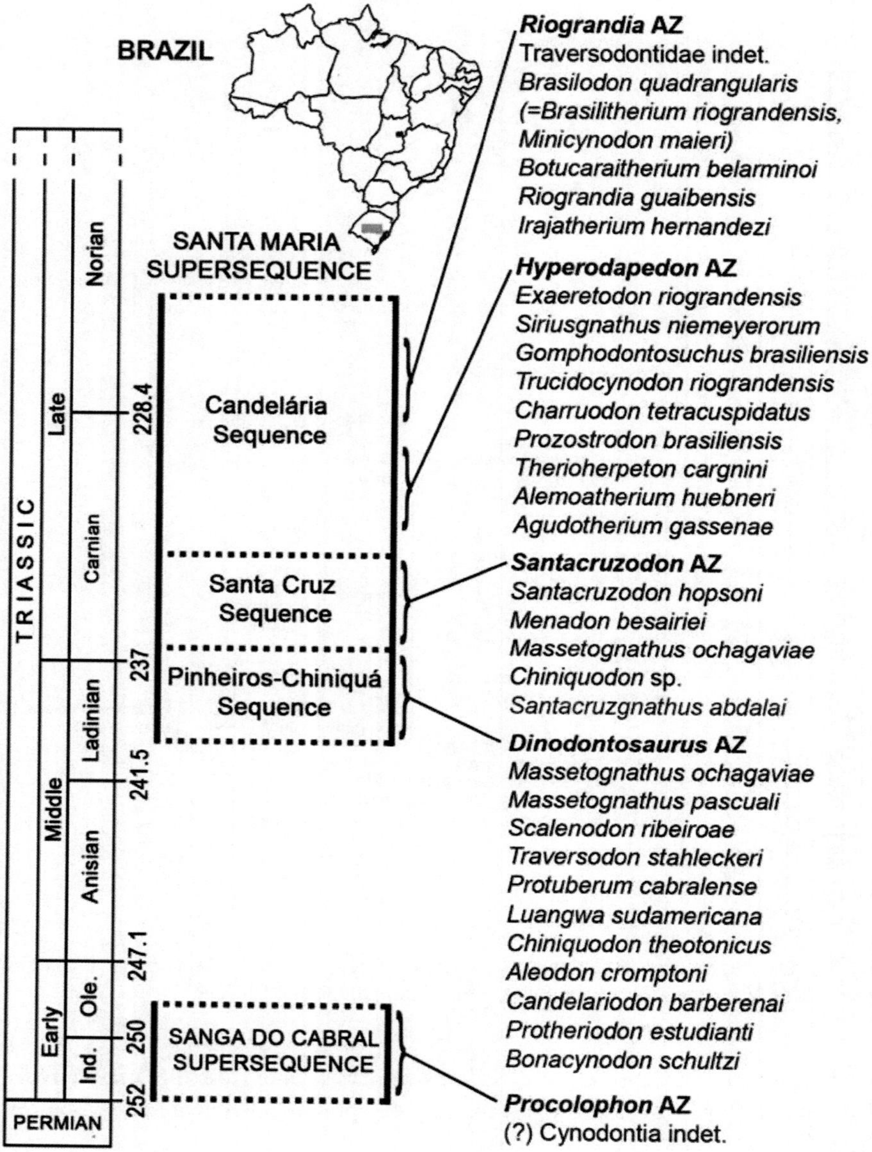


**Fig. 3.3** Geographic and stratigraphic occurrence of known species of non-mammaliaform cynodonts in Brazil. Abbreviations: AZ, Assemblage Zone; Ind, Induan; Ole, Olenekian

Diademodontidae, Trirachodontidae, and Traversodontidae. At present, trirachodontids have not been found in SA.

*Cynognathus crateronotus* is the most common taxon of the ?Olenekian–Anisian *Cynognathus* AZ (Karoo Basin) of South Africa (Kitching 1977), with broad distribution throughout several localities in Africa, Antarctica, and Argentina (Bonaparte 1969; Hammer 1995; Abdala 1996). It is a highly specialized taxon with a robustly built skull, long snout, dorsoventrally deep zygomatic arch, and a conspicuous carnivorous dentition, including large canines and sectorial postcanine teeth with recurved cusps (Broili and Schröder 1934) (Fig. 3.4a). Its sister-taxon relationship with the herbivorous/omnivorous gomphodont clade is mainly based on the features of the zygomatic arch (Hopson and Kitching 2001), which is dorsoventrally tall with a large participation of the squamosal and a well-developed suborbital process at the rostral base of the zygoma. In SA, *Cynognathus* is represented by a single specimen, including a skull with articulated jaws (Fig. 3.3a), a dorsal vertebra, and right humerus, from the Río Seco de la Quebrada Formation, Puesto Viejo Group, Mendoza, Argentina (Bonaparte 1969; Abdala 1999; Fig. 3.5). The specimen was originally assigned to a distinctive species, *C. minor*, by Bonaparte (1969), but later synonymized with the broadly distributed species, *C. crateronotus* (Abdala 1996).

Diademodontidae includes herbivorous to omnivorous cynodonts with both sectorial as well as gomphodont functional postcanines, and with a complex pattern of tooth replacement (Hopson 1971). Several genera and species have been included within this group (see historical review in Martinelli et al. 2009), but most of them were based on poorly preserved specimens and only two taxa are currently recognized: *Titanogomphodon crassus*, restricted to Namibia (Bradu and Grine 1979), and *Diademodon tetragonus*, found in South Africa, Zambia, Tanzania, Namibia, Antarctica, and Argentina (Hammer 1995; Abdala et al. 2005; Martinelli et al. 2009). *Diademodon tetragonus* is very abundant in the *Cynognathus* AZ of South Africa (Hancox et al. 1995; Abdala et al. 2005). *Diademodon* has a narrow and elongated snout and a broad temporal region, the zygomatic arch is robust and high, and the postcanines exhibit conspicuous heterodonty including mesial teeth with a circular outline, ovoid gomphodont teeth with multiple crests at the middle, and sectorial to sub-sectorial teeth at the back of the tooth series (Fig. 3.4b, c). In SA, *Diademodon* was found at the Río Seco de la Quebrada Formation, Puesto Viejo Group, Mendoza (Argentina), based on partially preserved skull and jaws (Martinelli and de la Fuente 2008; Martinelli et al. 2009; Fig. 3.4b, c).

Traversodontidae is the most diverse clade of gomphodonts, and both are temporally and geographically broadly distributed (Abdala and Ribeiro 2010; Liu and Abdala 2014; Hendrickx et al. 2020). The members of this clade exhibited the most specialized dentition among cynodonts, together with that of tritylodontids. The postcanines are labiolingually wide and basined with a transverse crest in upper and lower teeth that would allow crushing during mastication (e.g., Crompton 1972a; Hendrickx et al. 2020) (Figs. 3.6, 3.7 and 3.8). Considering the attachment areas seen in the skull and jaws, the adductor jaw musculature was well-developed. Among the clade, sequential tooth replacement likely allowed precise occlusion, analogous to that present in several mammalian herbivores today. The tooth replacement patterns

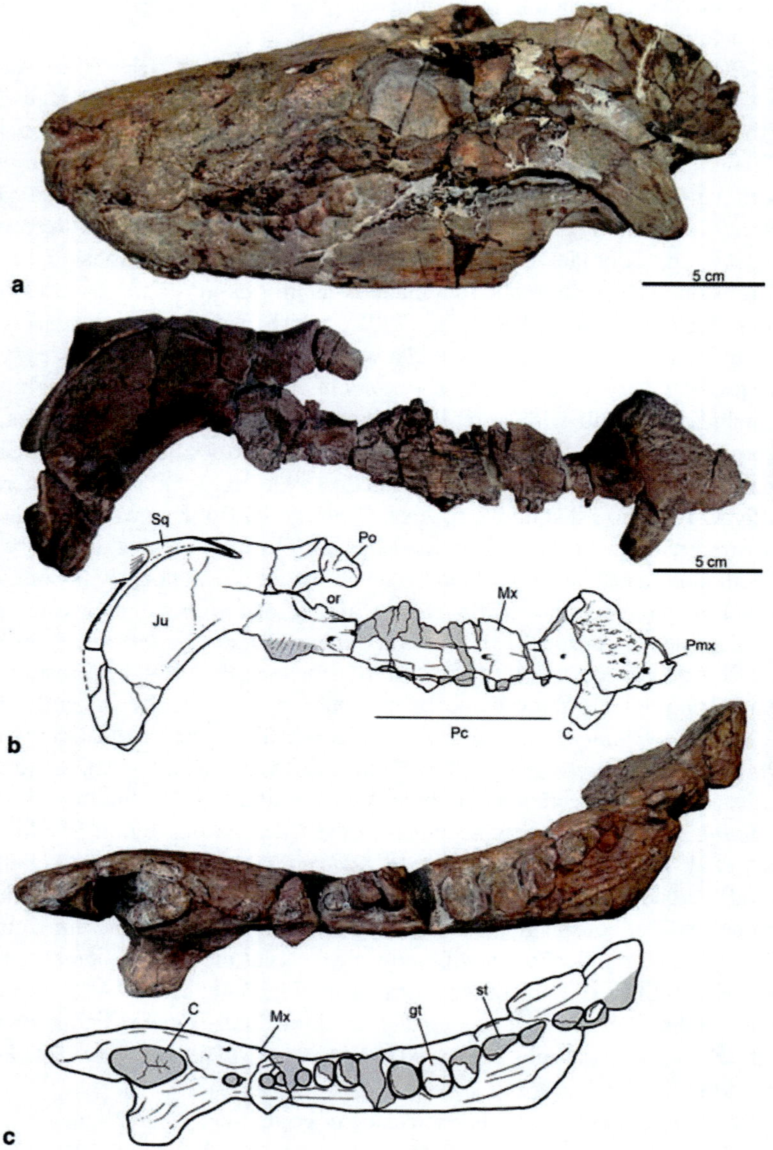


**Fig. 3.4** Cynognathids and diademodontids from South America. *Cynognathus crateronotus*, skull (PVL 3859) in lateral view (**a**). *Diademodon tetragonus*, fragmentary skull (MHNSR-PV 357) in lateral view, with accompanying line drawing (**b**), and detail of the left maxilla in occlusal view, with accompanying line drawing (**c**). Abbreviations: C, upper canine; gt, gomphodont tooth; Ju, jugal; Mx, maxilla; or, orbit; Pc, upper postcanine teeth; Pmx, premaxilla; Po, postorbital; Sq, squamosal; st, sectorial tooth

**Fig. 3.5** *Cynognathus crateronotus*, artistic reconstruction by Jorge L. Blanco

recognized among members of this clade are also varied. For example, the basal *Andescynodon* replaced posterior sectorial postcanines in juveniles by gomphodont teeth, which remained as permanent as in adults (Goñi 1986). The mechanism of tooth replacement in *Massetognathus* consists of the addition of the distal teeth at the end of the tooth series during ontogeny. Consequently, during growth the number of teeth increases: the mesial postcanines are usually strongly worn out while the distal ones have less worn crowns (Romer 1967; Abdala and Giannini 2000). In other taxa, such as *Exaeretodon*, mesial postcanines are lost during ontogeny and new teeth are added at the distal end of the tooth series, with a forward movement of the elements (Goñi and Goin 1990; Abdala et al. 2002a; O'Meara et al. 2018). One interesting condition was recently discovered in the postcanine teeth of *Menadon besairiei*. The postcanines of this species are long and columnar, with open roots. They were not replaced even among the oldest individuals and remained functional after the complete wear of the crown enamel (Melo et al. 2019). The continuous wear was compensated by the prolonged growth of each postcanine, resulting in dentine hypsodont teeth, similar to the condition of extant xenarthran mammals (Melo et al. 2019).

The morphology of the postcanines also exhibits different specializations along the clade. For example, some traversodontids (e.g., *Massetognathus* and *Exaeretodon*; Figs. 3.7 and 3.8) have a mechanism of "shouldering" between each postcanine tooth, forming a continuous dental battery to process food. Also, the morphology of the incisors differs considerably in traversodontids. Some taxa have enlarged incisors, likely procumbent, such as in *Exaeretodon*, whereas others have leaf-shaped crowns, with several tiny accessory cusps on the cutting edge, as seen in *Massetognathus* and *Santacruzodon* (Martinelli et al. 2014).

Traversodontids are better documented in SA and Africa than in other regions worldwide. These areas have provided the best fossil record for the group by the Middle to Late Triassic, with at least 16 species across Brazil (10) and Argentina (6) *versus* eight in Tanzania (*Scalenodon angustifrons, Mandagomphodon attridge,* and *Man. hirschsoni*), Zambia (*Luangwa drysdalli*), Namibia (*Luangwa* sp. and *Etjoia dentitransitus*), Lesotho (*Scalenodontoides macrodontes*), and Madagascar (*Dadadon isaloi* and *Menadon besairiei*).

Argentinean traversodontids (Fig. 3.2) were recovered from the Cerro de Las Cabras (Uspallata Group), Río Seco de la Quebrada (Puesto Viejo Group),

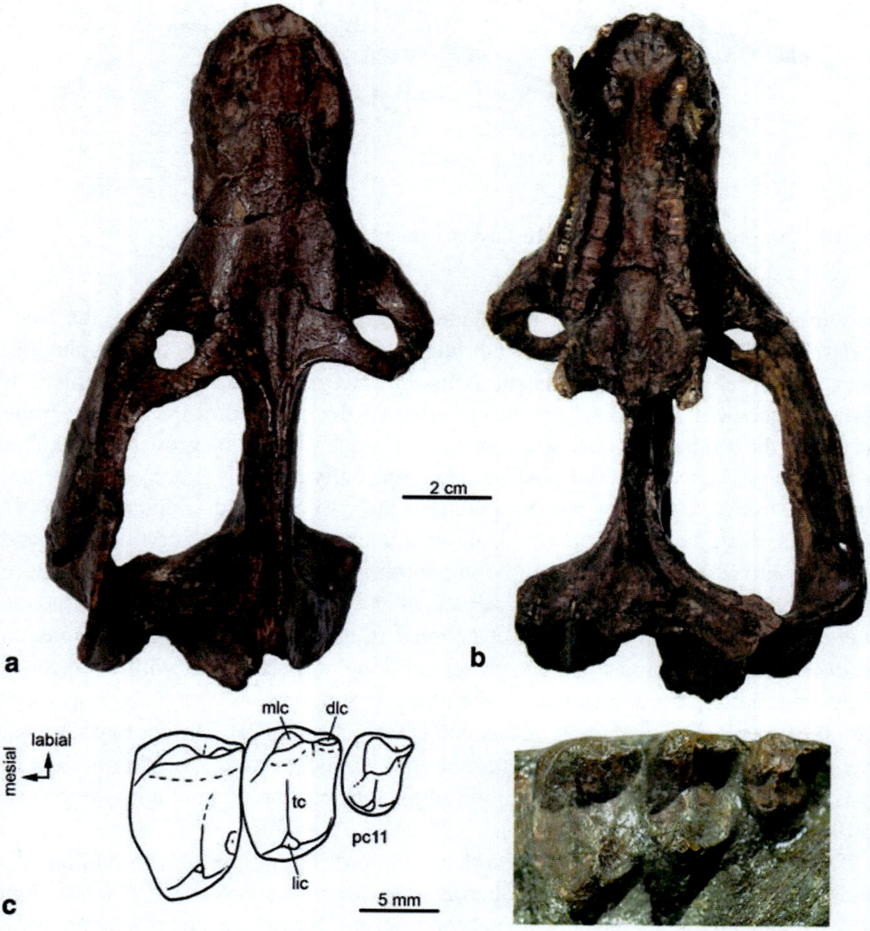


**Fig. 3.6** The traversodontid *Pascualgnathus polanskii* from the Middle–early Late Triassic of Argentina. Skull (holotype, MLP 65-VI-18–1) in dorsal (**a**) and ventral (**b**) views. Detail of the last three left upper postcanine teeth (PVL 4416) in occlusal view, with accompanying line drawing (**c**). Abbreviations: dlc, distal labial cusp; lic, lingual cusp; mlc, main labial cusp; tc, transverse crest

Chañares, and Ischigualasto (Agua de la Peña Group) Formations. The first litostratigraphic unit also includes *Andescynodon mendozensis* and *Rusconiodon mignonei* (Bonaparte 1970; *contra* Liu and Powell 2009); the Río Seco de la Quebrada Formation includes *Pascualgnathus polanskii* (Bonaparte 1966; Martinelli 2010) (Fig. 3.6); the Chañares Formation includes *Massetognathus pascuali* (Fig. 3.7) and cf. *Scalenodon* sp. (Romer 1967, 1972; Abdala and Giannini 2000; Ezcurra et al. 2017), and the Ischigualasto Formation includes *Exaeretodon argentinus* and *Ischignathus sudamericanus* (Cabrera 1943; Bonaparte 1962, 1963a; *contra* Liu 2007). Other named traversodontids from Argentina, such as *Colbertosaurus muralis* (?Potrerillos Formation; Minoprio 1954, 1957) and *Proexaeretodon vincei* (Ischigualasto Formation; Bonaparte 1963b), were considered *nomen dubium* or synonymized with other taxa but further studies are needed, in particular, for the latter taxon.

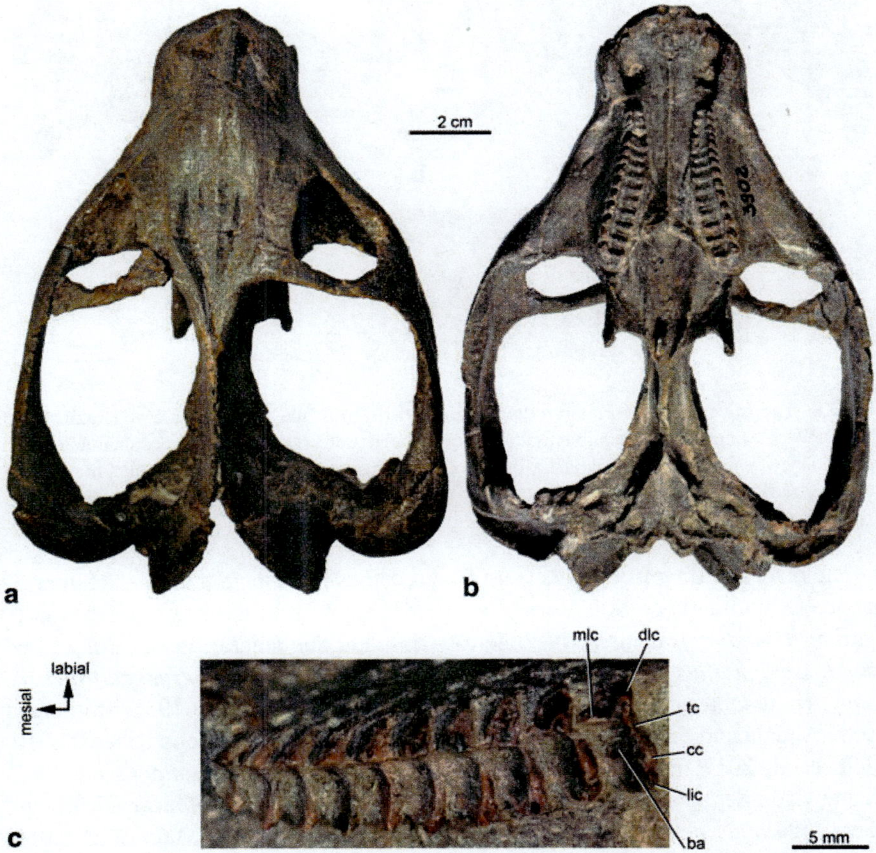


**Fig. 3.7** The traversodontid *Massetognathus pascuali* from the early Late Triassic of Argentina. Skull (PVL 3902) in dorsal (**a**) and ventral (**b**) views. Detail of the left upper postcanine teeth (PVL4729) in occlusal view (**c**). Abbreviations: ba, basin; cc, central cusp; dlc, distal labial cusp; lic, lingual cusp; mlc, main labial cusp; tc, transverse crest

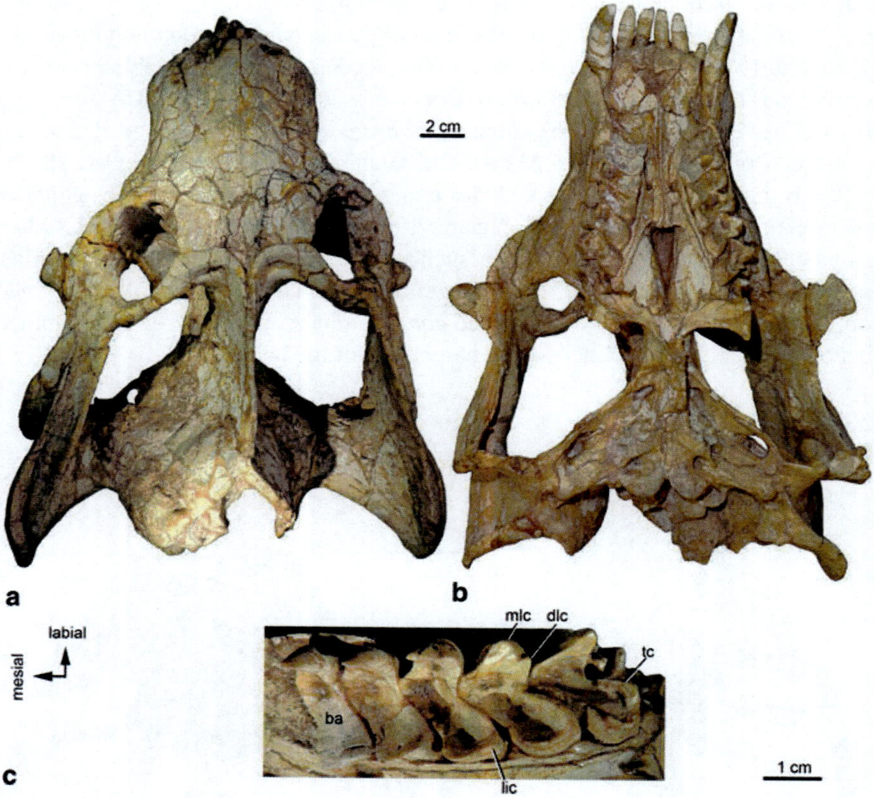


**Fig. 3.8** The traversodontid *Exaeretodon riograndensis* from the Late Triassic of Brazil. Skull (MCT-PV1522) in dorsal (**a**) and ventral (**b**) views, and detail of the left upper postcanine teeth in occlusal view (**c**). Abbreviations: ba, basin; dlc, distal labial cusp; lic, lingual cusp; mlc, main labial cusp; tc, transverse crest

The record of traversodontids from Brazil comes from the Santa Maria Supersequence (Rio Grande do Sul; Horn et al. 2014; Fig. 3.3). In the late Ladinian–early Carnian *Dinodontosaurus* AZ (Pinheiros-Chiniquá Sequence), the taxa described are *Massetognathus ochagaviae*, *Traversodon stahleckeri*, *Protuberum cabralensis*, *Luangwa sudamericana*, and *Scalenodon ribeiroae* (e.g., Huene 1936; Barberena 1981; Abdala and Sá-Teixeira 2004; Reichel et al. 2009; Abdala and Ribeiro 2010; Melo et al. 2017; Pavanatto et al. 2020). The early Carnian *Santacruzodon* AZ (Santa Cruz Sequence) includes *Santacruzodon hopsoni*, *Menadon besairiei*, and *Massetognathus ochagaviae* (Abdala and Ribeiro 2003, 2010; Melo et al. 2015; Schmitt et al. 2019), and the late Carnian *Hyperodapedon* AZ (Candelária Sequence) includes *Exaeretodon riograndensis* (Figs. 3.8, 3.9), *Siriusgnathus niemeyerorum*, and *Gomphodontosuchus brasiliensis* (Abdala et al. 2002a; Abdala and Ribeiro 2010; Pavonatto et al. 2018). A conspicuous record of traversodontids was briefly reported

**Fig. 3.9** *Exaeretodon riograndensis,* artistic reconstruction by Jorge L. Blanco

by Ribeiro et al. (2011) for the *Riograndia* AZ (Candelária Sequence), which represents the youngest occurrence for the Late Triassic (early Norian; Soares et al. 2011a; Langer et al. 2018) of SA.

The fossils from SA trace much of the evolution of traversodontids, which exemplified the main morphological changes within the group. Basal taxa are represented by *Pascualgnathus* (Fig. 3.6) and *Andescynodon*, with generalized skull morphology similar to that of the Middle Triassic trirachodontids, bearing simple incisors, large canines, and gomphodont postcanines with oval to subrectangular occlusal outlines (Fig. 3.6c). *Massetognathus* (Fig. 3.7) and *Santacruzodon* have an intermediate morphology between basal traversodontids and gomphodontosuchines (e.g., *Gomphodontosuchus*, *Exaeretodon*, and *Menadon*). In these taxa, the zygomatic arch becomes dorsoventrally tall, the incisors are leaf-shaped with multiple cusps, the canines are reduced, and there is an incipient shouldering between postcanines. Gomphodontosuchines show the most conspicuous modifications in the incisor and postcanine dentition reaching the largest body mass among the family, at about 30 kg. *Exaeretodon* is one of the best-known taxa and fairly abundant (Bonaparte 1962; Abdala et al. 2002a; Figs. 3.8, 3.9). The incisors are enlarged and procumbent with a thick layer of enamel covering the labial surface (Fig. 3.8b). The postcanine tooth row is tightly packaged and each tooth has a transversely broad occlusal basin (Fig. 3.8c). The skull and lower jaws accommodated large packages of musculature along the zygomatic arch, temporal fenestra (Fig. 3.8), and coronoid process of the dentary, which may have resulted in an effective grinding mechanism.

In SA, the fossil record of traversodontids is richer and taxonomically more diverse in Brazil than Argentina, especially when the coeval *Dinodontosaurus* AZ and the Chañares Formation are compared. It is likely that taphonomic biases are affecting the different sedimentary basins, and further studies in the Argentinean fossil localities are required.

### *3.2.2 Probainognathia*

Middle to Late Triassic probainognathians from Argentina and Brazil form the basis for understanding most craniodental transformations that occurred prior to the establishment of mammaliaform features (e.g., Quiroga 1980; Luo and Crompton 1994; Hopson and Kitching 2001; Bonaparte et al. 2003, 2005; Kielan-Jaworowska et al. 2004; Martinelli and Rougier 2007; Oliveira et al. 2010; Ruta et al. 2013). Within probainognathians, Brazilian taxa grouped into the Brasilodontidae (sensu Bonaparte et al. 2005) were postulated as the sister-group of mammaliaforms (Bonaparte et al. 2003, 2005, 2012; Luo 2007; Liu and Olsen 2010), providing fresh data to the classic disputed sister taxon hypotheses of the Tritheledontidae *versus* the Tritylodontidae (Luo 1994).

There are about 20 probainognathian species in Argentina and Brazil (Figs. 3.2 and 3.3), several of them based on well-known specimens. The oldest SA probainognathian is *Cromptodon mamiferoides* from the Cerro de Las Cabras Formation (Bonaparte 1972; Zavattieri and Arcucci 2007; originally referred in the literature to the Río Mendoza Formation), Uspallata Group, Mendoza, Argentina. Its phylogenetic position is still controversial (Abdala and Giannini 2002). It was first classified as a Galesauridae (Bonaparte 1972) but later related to Chiniquodontidae (Hopson 1991). The postcanine dentition of *Cromptodon* is very interesting because has a complexity similar to *Aleodon* and some prozostrodontians (Martinelli 2017). Further material is needed to evaluate its phylogenetic relationships.

The Ischigualasto-Villa Unión Basin includes *Chiniquodon theotonicus* and *Probainognathus jenseni* (Romer 1969, 1970, 1973; Abdala and Giannini 2002) from the lower Carnian Chañares Formation; *Ecteninion lunensis* (Martínez et al. 1996), *Diegocanis elegans* (Martínez et al. 2013a), *Chiniquodon sanjuanensis*, cf. *Chiniquodon* sp. (Martínez et al. 2013b), cf. *Probainognathus* sp. (Bonaparte and Crompton 1994; but see Fernández et al. 2011), and *Pseudotherium argentinus* (Wallace et al. 2019) from the upper Carnian Ischigualasto Formation; and *Chaliminia musteloides* (Bonaparte 1980; Martinelli and Rougier 2007) from the Norian Los Colorados Formation (Fig. 3.2). Recently, new cynodont specimens from the Los Colorados Formation have been informally mentioned, but those are still under study (Gaetano et al. 2018).

A few years ago, Martínez et al. (2015) mentioned the finding of a new faunal association at the Upper Triassic Quebrada del Barro Formation, Marayes-El Carrizal Basin that includes tritheledontids. This finding has the potential to considerably enrich the knowledge of the group.

In Brazil, probainognathians are documented all along the Santa Maria Supersequence (Fig. 3.3). The Ladinian–early Carnian *Dinodontosaurus* AZ includes *Chiniquodon theotonicus* (Abdala and Giannini 2002), *Candelariodon barberenai* (Oliveira et al. 2011b; Martinelli et al. 2017c), *Protheriodon estudianti* (Bonaparte et al. 2006; Martinelli et al. 2016a), and *Aleodon cromptoni* (Martinelli et al. 2017b). The early Carnian *Santacruzodon* AZ includes *Santacruzgnathus abdalai* (Soares et al. 2011b; Martinelli et al. 2015) and *Chiniquodon* sp. (Abdala et al. 2001), the latter possibly representing a new species (Bertoni et al. 2016). The late Carnian *Hyperodapedon* AZ includes *Trucidocynodon riograndensis* (Oliveira et al. 2010), *Prozostrodon brasiliensis* (Bonaparte and Barberena 2001; Guignard et al. 2018; Pacheco et al. 2018), *Therioherpeton cargnini* (Bonaparte and Barberena 1975, 2001; Martinelli et al. 2017a), *Charruodon tetracuspidatus* (Abdala and Ribeiro 2010; Martinelli et al. 2017a), *Agudotherium gassenae* (Stefanello et al. 2020), and *Alemoatherium huebneri* (Martinelli et al. 2017a). The *Riograndia* AZ includes *Brasilodon quadrangularis*, *Brasilitherium riograndensis* (Bonaparte et al. 2003, 2005; Martinelli and Bonaparte 2011; Martinelli 2017; Guignard et al. 2019b), *Minicynodon maieri* (Bonaparte et al. 2010), *Botucaraitherium belarminoi* (Soares et al. 2014), *Riograndia guaibensis* (Bonaparte et al. 2001; Soares et al. 2014; Guignard et al. 2019a), and *Irajatherium hernandezi* (Martinelli et al. 2005; Oliveira et al. 2011a). The three former species of the *Riograndia* AZ (*Brasilodon quadrangularis*, *Brasilitherium riograndensis*, and *Minicynodon maieri*) were recently considered synonyms (Liu and Olsen 2010; Martinelli 2017), with priority in *B. quadrangularis*, but an in-depth study is in process (Martinelli 2017). All the mentioned species can be clustered in distinctive groups among probainognathians (see below).

The Ecteniniidae (e.g., *Trucidocynodon riograndensis*; Fig. 3.10a, b) and Chiniquodontidae (e.g., *Chiniquodon theotonicus*; Fig. 3.10c, d) include basal clades with specialized dentition for carnivory, with the exception of the species of the genus *Aleodon*, which developed a wide lingual cingulum consistent with a more omnivorous diet (Crompton 1955; Martinelli et al. 2017b). Carnivorous forms of these families are characterized by having sectorial postcanines with three to four aligned and strongly curved cusps (Fig. 3.10). Particularly in ecteniniids, the margins of each cusp are finely serrated and the last postcanine teeth are overlapped (Martínez et al. 2013a; Oliveira et al. 2010; Stefanello et al. 2018; Fig. 3.10b). Chiniquodontids have a longer secondary palate than ecteniniids, and the pterygoid flange is greatly elongated, ending in a thin projection (Abdala and Giannini 2002). In ecteniniids, the snout length is equal to the length of the temporal region, the zygomatic arch is more slender than in chiniquodontids, and the orbitosphenoid in the orbital wall is well-ossified. The specialized carnivorous dentition is present in basal probainognathians and it is to some degree similar to the one present in *Cynognathus crateronotus*. In more inclusive clades of probainognathian cynodonts, the dentitions reveal features consistent with omnivorous to insectivorous diets.

In *Probainognathus* and especially in prozostrodontians (with the exception of some tritylodontids—e.g., *Kayentatherium*, if this group is included within probainognathians), there is a reduction in body size (e.g., Lautenschlager et al. 2019)

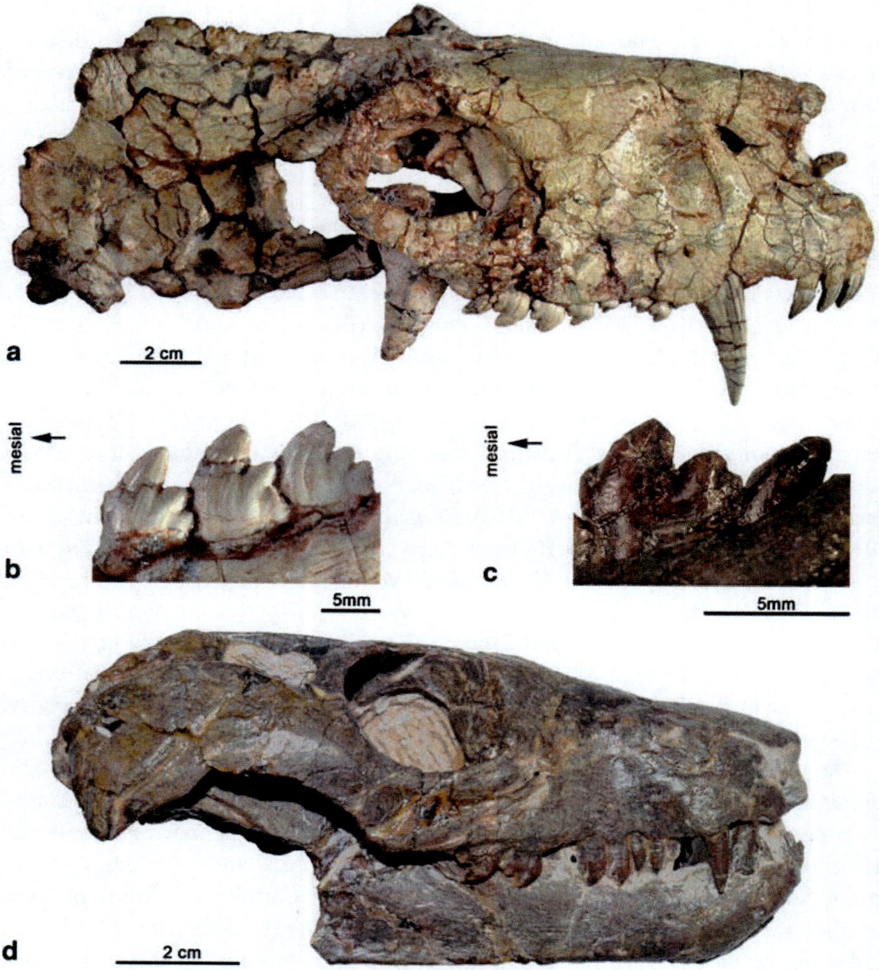


**Fig. 3.10** Selected non-mammaliaform probainognathians from the Triassic of South America. Skull of the ecteniniid *Trucidocynodon riograndensis* (holotype, UFRGS-PV-1051-T) in lateral view (**a**) and detail of left middle lower postcanine teeth in labial view (**b**). Detail of left distal postcanine teeth of the chiniquodontid *Chiniquodon theotonicus* (PVL 4444) in labial view (**c**), and skull and jaw in lateral view (**d**)

and a great complexity in the postcanine tooth morphology, coupled with conspicuous modifications in the skull and jaws (e.g., Romer 1970; Rougier et al. 1992; Martinelli and Rougier 2007; Rodrigues et al. 2014; Martinelli et al. 2015; Wallace et al. 2019). Particularly, *Probainognathus* has a skull with a broad temporal region, which is longer than the snout (Abdala and Giannini 2002; Fig. 3.11a). The sectorial upper and lower postcanines have tall crowns with a main central cusp, and well-developed accessory cusps, one mesial and two distal (Fig. 3.11b). *Probainognathus*

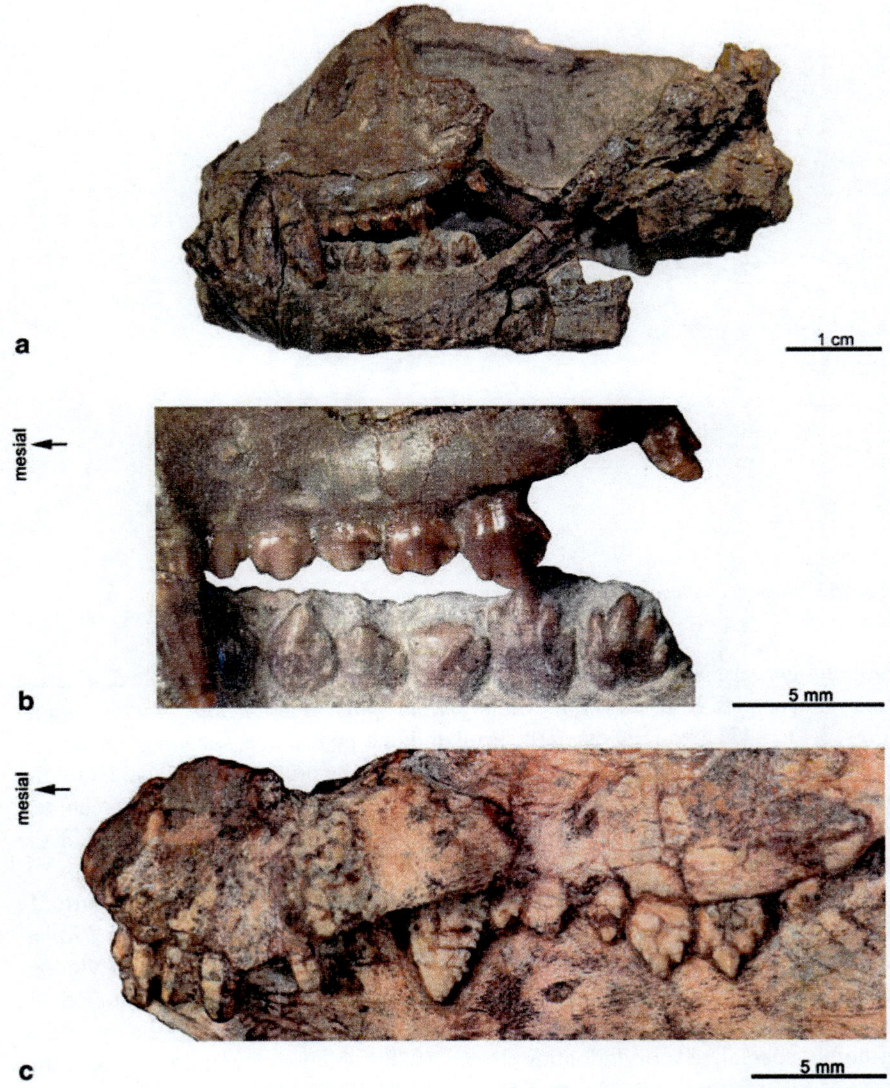


**Fig. 3.11** Selected non-mammaliaform probainognathians from the Triassic of South America. Skull and jaw of the probainognathid *Probainognathus jenseni* (PVL 4673) in lateral view (**a**), and detail of left upper and lower dentition in labial view (**b**). Detail of the snout with the dentition of the probainognathid *Bonacynodon schultzi* (MCT-1716-R) in lateral view (right side, inverted) (**c**)

has an additional craniomandibular articulation, in which not only the dentary but also the surangular contact the squamosal (Crompton 1972b). This trait together with the mammal-like appearance of the skull and dentition was strong evidence during the 70's to consider *Probainognathus* the sister-taxon of mammaliaforms (Romer 1970). The assignment of specimens from the Ischigualasto Formation to cf.

**Fig. 3.12** *Bonacynodon schultzi*, artistic reconstruction by Jorge L. Blanco

*Probainognathus* sp. is controversial and the material possibly represents a distinctive taxon (Fernández et al. 2011). *Probainognathus jenseni* and *Bonacynodon schultzi* (Figs. 3.11, 3.12) are members of the Probainognathidae clade, showing "intermediate" features between basal probainognathians and the prozostrodontians, including incipient cingular cusps in postcanines, less frequent replacement of postcanines, incipient glenoid fossa in the squamosal for surangular bone, and absence of the entepicondylar foramen in the humerus.

Prozostrodontia is a clade (Liu and Olsen 2010) that includes some partially known taxa such as *Prozostrodon* and *Therioherpeton* (Fig. 3.13), perhaps *Protheriodon* (Fig. 3.13), as well as *Pseudotherium* (Fig. 3.14), tritheledontids, tritylodontids, brasilodontids (*Brasilodon* and *Botucaraitherium*), and mammaliaforms. In this clade, many mammaliaform apomorphies are already present. With the exception of highly specialized tritylodontids (Rowe 1988; which some authors considering them as relative to gomphodonts; Hopson and Kitching 2001), most prozostrodontians are small-sized animals with insectivorous to omnivorous dietary habits. The zygomatic arch is slender (Figs. 3.13, 3.14 and 3.15). The ossified postorbital bar is absent, with the loss or reduction of the prefrontal and postorbital bones. The horizontal ramus of the dentary is usually low (e.g., brasilodontids) and the symphysis is unfused. The postcanine morphology is quite disparate among prozostrodontians. Some taxa (e.g., *Prozostrodon*, *Botucaraitherium*, and *Brasilodon*; Fig. 3.15) have postcanine teeth very similar to early mammaliaforms (Bonaparte et al. 2012), bearing a "triconodont-like" pattern with cuspidate cingula and constricted root (Fig. 3.15).

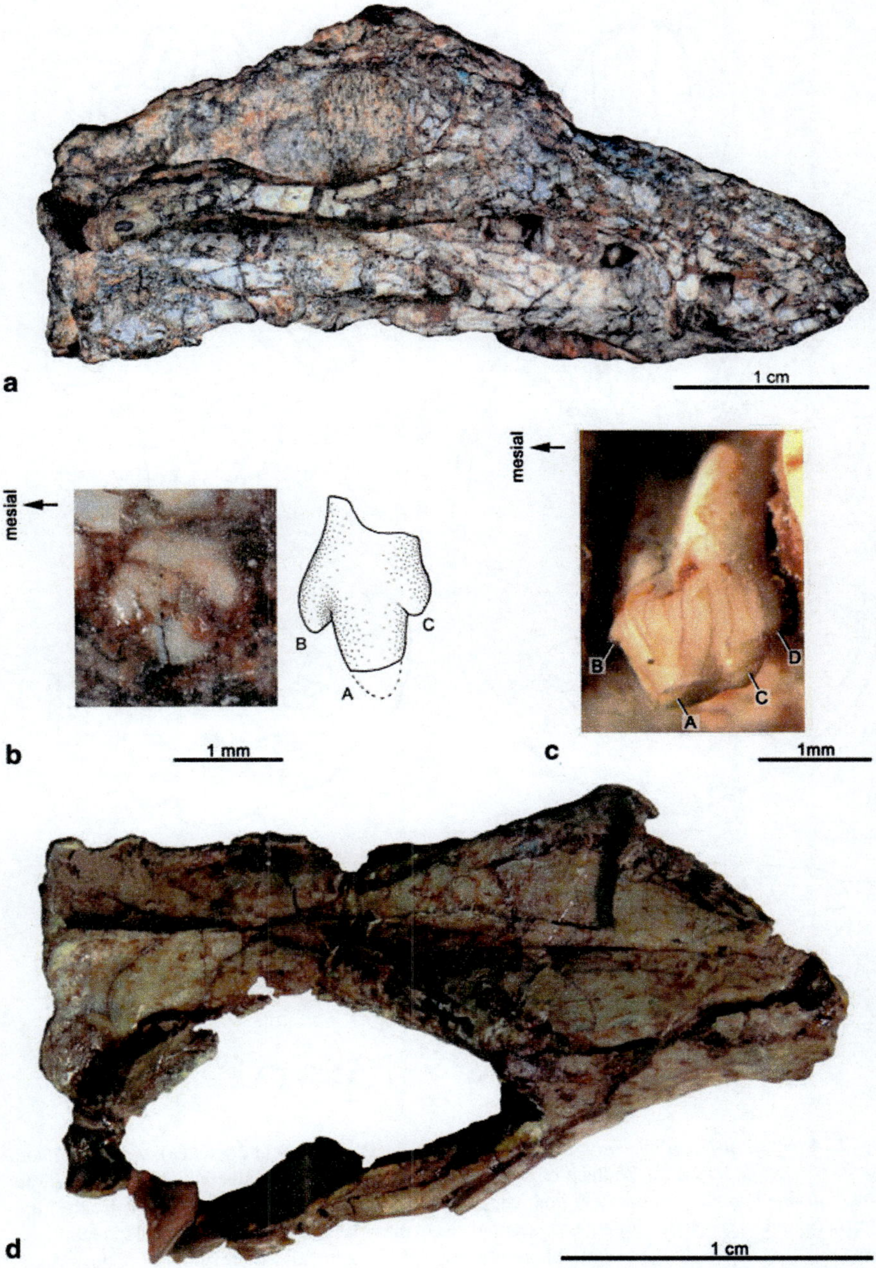


**Fig. 3.13** Selected non-mammaliaform probainognathians from the Triassic of South America. Skull and jaw of *Protheriodon estudianti* (holotype, UFRGS-PV-0962-T) in lateral view (**a**) and detail of left middle upper postcanine tooth in labial view (**b**). Detail of right (inverted) fourth upper postcanine of *Therioherpeton cargnini* (holotype, MVP 05.22.04) in labial view (**c**), and skull in dorsal view (**d**). Abbreviations: A, B, C, D refer to names of upper cusps of postcanine teeth

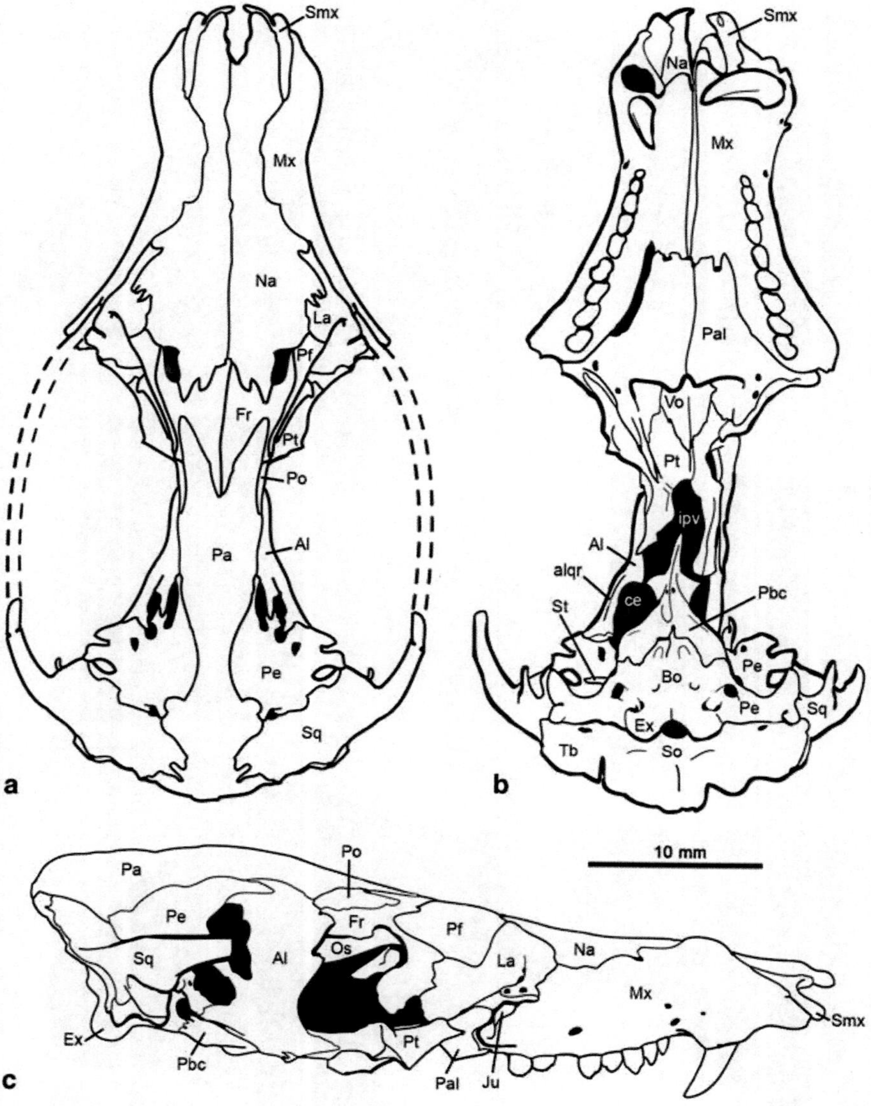


**Fig. 3.14** Skull of *Pseudotherium argentinus* (holotype, PVSJ 882) in dorsal (**a**), ventral (**b**), and lateral (**c**) views, based on Wallace et al. (2019). Abbreviations: Al, alisphenoid; alqr, quadrate ramus of alisphenoid; Bo, basioccipital; ce, cavum epiptericum; Ex, exoccipital; Fr, frontal; ipv, interpterygoid vacuity; Ju, jugal; La, lacrimal; Mx, maxilla; Na, nasal; Os, orbitosphenoid; Pa, parietal; Pal, palatine; Pbc, parabasisphenoid complex; Pe, petrosal; Pf, prefrontal; Po, postorbital; Pt, pterygoid; Smx, septomaxilla; So, supraoccipital; Sq, squamosal; St, stapes; Tb, tabular; Vo, vomer

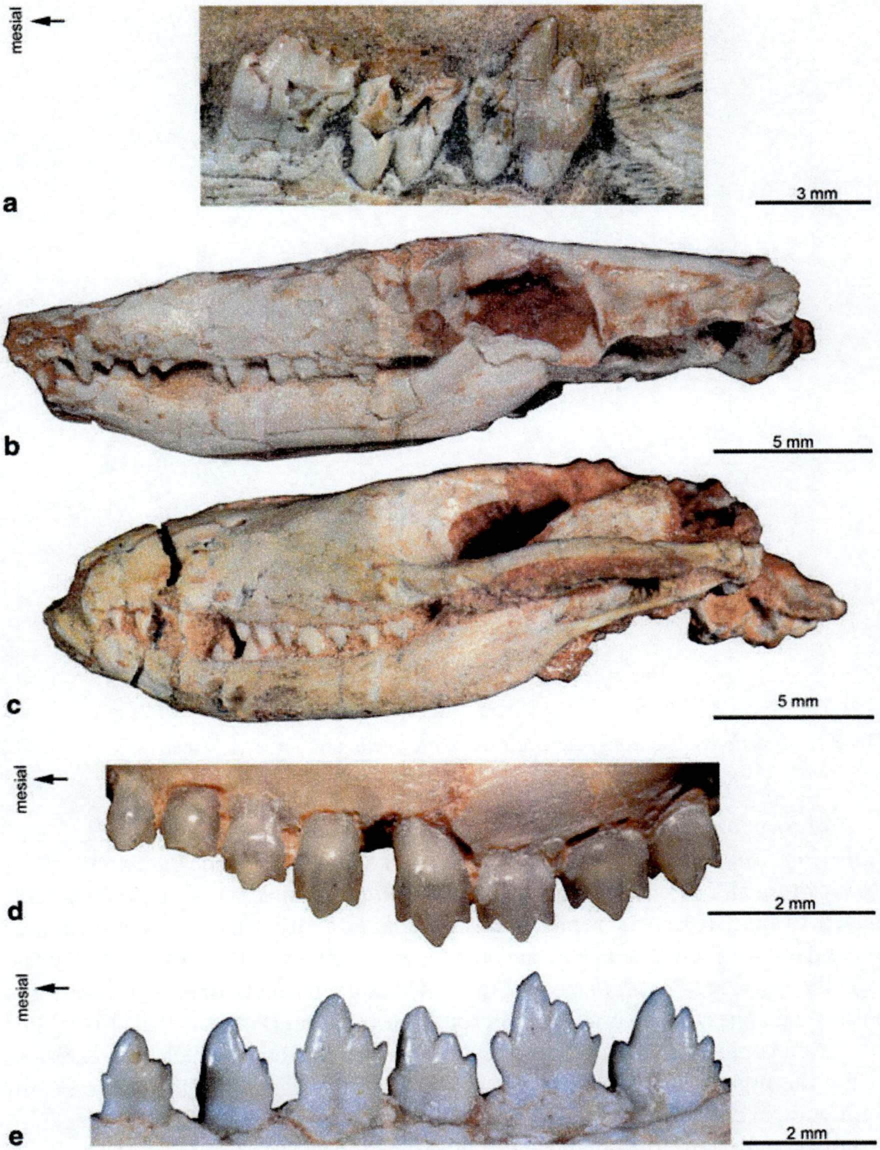


**Fig. 3.15** Selected non-mammaliaform probainognathians from the Triassic of South America. Left (inverted) distal lower postcanines of *Botucaraitherium belarminoi* (holotype, MMACR-PV-003-T) in lingual view (**a**). *Brasilodon quadrangularis*, skull and jaw of specimens UFRGS-PV-929-T (formerly *Brasilitherium riograndensis*) (**b**) and UFRGS-PV-1030-T (formerly holotype of *Minicynodon maieri*) in lateral view (**c**), and detail of right (inverted) upper postcanine teeth (holotype, UFRGS-PV-0611-T) (**d**) and left lower postcanine teeth (UFRGS-PV-0603 T, formerly *B. riograndensis*) in labial view (**e**)

**Fig. 3.16** *Brasilodon quadrangularis*, artistic reconstruction by Jorge L. Blanco

*Prozostrodon*, *Therioherpeton*, *Protheriodon*, possibly *Candelariodon*, *Pseudotherium*, *Botucaraitherium*, and *Brasilodon* (Bonaparte and Barberena 2001; Bonaparte et al. 2003, 2005; Martinelli et al. 2017a, b; Wallace et al. 2019; Figs. 3.13, 3.14, 3.15 and 3.16) have several cranial, mandibular, and dental patterns more similar to basal mammaliaforms than to any other cynodont (Figs. 3.13, 3.14 and 3.15). The recently described *Pseudotherium* (Fig. 3.14) has a reduced prefrontal and vestigial postorbital bones, but an ossified postorbital bar is absent (Wallace et al. 2019). The prootic and ophistotic are fused, constituting the petrosal bone, which have enclosed in life the organs of hearing and balance (Fig. 3.14); there is a bulgy promontorium with an elongated but uncoiled cochlea, such as in *Brasilodon.* The dentition is only partially preserved in this taxon, with upper postcanine teeth similar to those described for *Prozostrodon* (Bonaparte and Barberena 2001).

In particular, *Brasilodon* (including its alleged junior synonyms *Brasilitherium* and *Minicynodon*) (Figs. 3.15 and 3.16) summarizes most mammal-like features achieved in probainognathians, including the lack of prefrontal and postorbital bones, slender and low zygomatic arch, extremely low horizontal ramus of the dentary, fused ophistotic and prootic (i.e., presence of petrosal bone) with a promontorium, quadrate with a well-developed stapedial process, brain and olfactory bulb size slightly larger than in other cynodonts, partially ossified turbinals (but see Crompton et al. 2017), caudal process of dentary reaching a more caudal position than the surangular, lower middle postcanines with mesiolingual cusp e and lingual cusp g, presence of a tongue (by cusp d) and groove-like (by cusps b/e) mesiodistal interlocking system among middle postcanine teeth, and well-constricted roots with two separated innervation canals (Bonaparte et al. 2003, 2005, 2012; Martinelli and Bonaparte 2011; Rodrigues et al. 2013, 2014; Ruff et al. 2014; Martinelli 2017; Martinelli et al. 2017c, 2019) (Fig. 3.15).

Within Prozostrodontia, the Ictidosauria clade (sensu Martinelli and Rougier 2007) is well represented in SA by *Riograndia*, *Irajatherium*, and *Chaliminia* (Figs. 3.17 and 3.18). *Riograndia* is a basal form (perhaps it represents a still poorly known lineage; Fig. 3.17a, b) related to Tritheledontidae based on the number and shape of the incisors. *Riograndia* has flat, non-cingulated upper and lower postcanines, with up to nine small cusps along the cutting edges of the crown (Bonaparte et al. 2001; Soares et al. 2011a; Fig. 3.17a, b). In contrast, the basal tritheledontid *Irajatherium* and *Chaliminia* (Fig. 3.17c–e) have postcanines with morphology closer to *Pachygenelus* than to *Riograndia* (Martinelli et al. 2005; Martinelli and Rougier 2007), bearing fewer cusps of larger size on bulbous crowns.

In recent years, the discovery of small-sized probainognathians in the Late Triassic of Southern Brazil (Bonaparte et al. 2003, 2005, 2012; Soares et al. 2014; Martinelli et al. 2017a, b) and Argentina (Wallace et al. 2019) have considerably improved the knowledge on prozostrodontians and their impact on the understanding of the origin of mammals (Bonaparte 2012; Rodrigues et al. 2013, 2014, 2019; Martinelli et al. 2017a, 2017c; Botha-Brink et al. 2018; Guignard et al. 2019b; Wallace et al. 2019). These and new specimens recently unearthed are the basis for future studies.

## 3.3 Concluding Remarks

Eucynodonts are conspicuous elements of the Middle to Late Triassic tetrapod faunal associations of SA (Abdala and Ribeiro 2010; Martinelli and Soares 2016; Martinelli et al. 2018; Abdala et al. 2020). The diversity and abundance of traversodontids and probainognathians illustrate the major evolutionary radiation of cynodonts at the beginning of the Mesozoic. Although collections are relatively abundant, the material of high quality, and they are been actively investigated, issues related to the systematics, phylogeny, paleobiology, and biostratigraphy remain unclear and new efforts are required to address them properly.

In some cases, cynodont species have been used as index fossils playing an important role in biostratigraphic correlations (Fig. 3.19; see also Martinelli et al. 2017b; Abdala et al. 2020). The presence of *Cynognathus* and *Diademodon* in western Argentina supports the correlation of the Río Seco de la Quebrada Formation with the ?Olenekian–Anisian *Cynognathus* AZ of South Africa (Bonaparte 1982; Martinelli et al. 2009). The presence of *Massetognathus* and *Chiniquodon* in the *Dinodontosaurus* AZ and the Chañares Formation, and *Exaeretodon* in the *Hyperodapedon* AZ and the Ischigualasto Formation has helped to correlate Brazilian and Argentinean

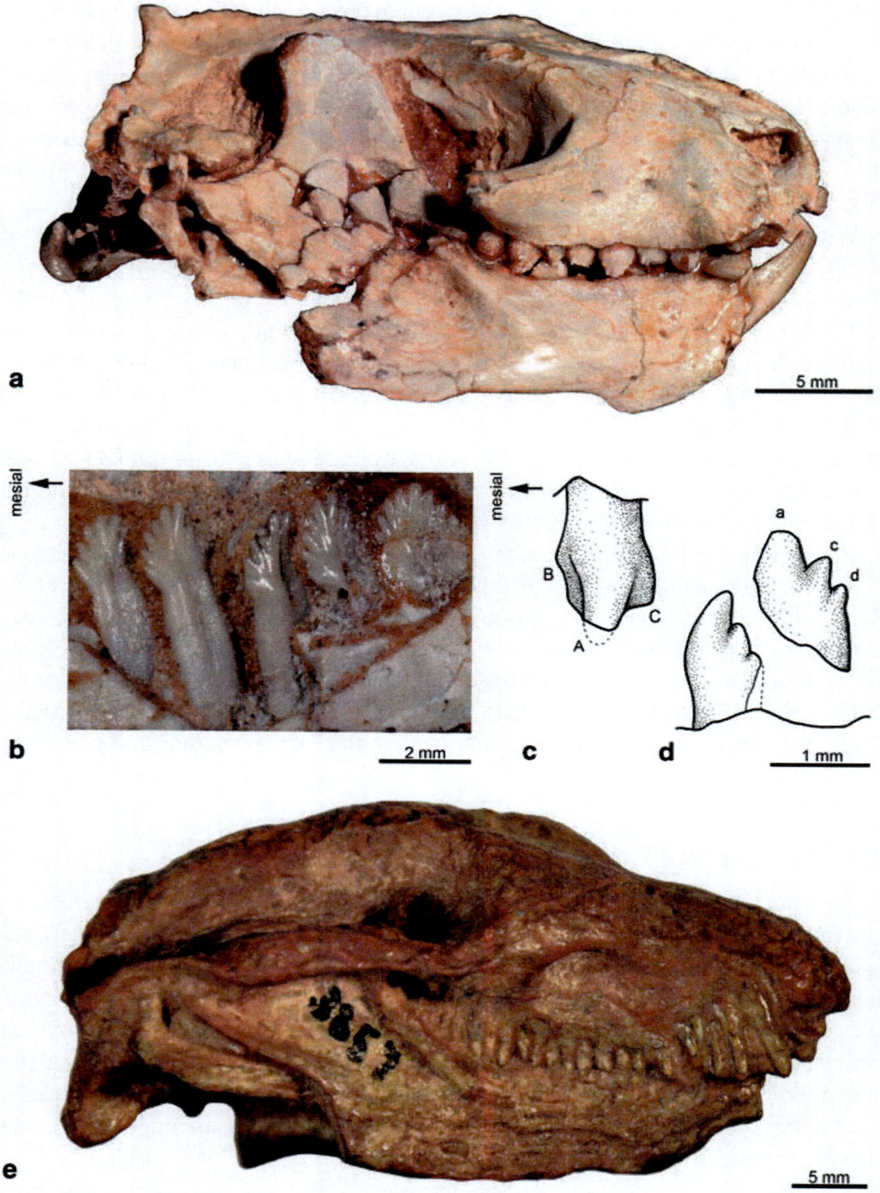


**Fig. 3.17** Selected non-mammaliaform probainognathians from the Triassic of South America. Skull and jaw of *Riograndia guaibensis* (UFRGS-PV-0596-T) in lateral view (**a**) and detail of left (inverted) mesial/middle postcanines (UFRGS-PV 0833-T) in lingual view (**b**). Detail of left sixth upper (**c**) and left second and third lower (**d**) postcanine teeth of *Chaliminia musteloides* (PULR 081), and articulated skull and jaw (holotype, PVL 3857) in lateral view (**e**). Abbreviations: A/a, B, C/c, d refer to names of upper/lower cusps of postcanine teeth

**Fig. 3.18** *Riograndia guaibensis*, artistic reconstruction by Jorge L. Blanco

faunal assemblages (e.g., Langer et al. 2007; Abdala and Ribeiro 2010; Martinelli et al. 2017b). The occurrence of *Menadon* in the *Santacruzodon* AZ and in the basal "Isalo II" (Morondava Basin) supports the correlation between Brazil and Madagascar (Melo et al. 2015) (Fig. 3.19).

In recent years, radiometric dating in SA Triassic units from the Ischigualasto-Villa Unión Basin (e.g., Rogers et al. 1993; Santi Malnis et al. 2011; Martínez et al. 2013b; Marsicano et al. 2016; Ezcurra et al. 2017), Uspallata Group (e.g., Spalletti et al. 2008), Santa Maria Supersequence (Langer et al. 2018; Philipp et al. 2018), and Puesto Viejo Group (Ottone et al. 2014) has sometimes challenged the scheme constructed by biostratigraphic correlations. In both Brazil and Argentina, the radiometric dating and faunal compositions (mainly synapsids and archosauromorphs) are largely congruent (see Langer 2005; Langer et al. 2007; Ezcurra et al. 2017; Martinelli et al. 2017b; Schmitt et al. 2019). Nonetheless, a putative date from the Puesto Viejo Group (235.8 $\pm$ 2.0 Ma; Ottone et al. 2014) indicates that the cynodonts (*Cynognathus*, *Diademodon*) from the Río Seco de la Quebrada Formation are lower Carnian and, consequently, are similar in age to the base of the Chañares Formation (Ischigualasto-Villa Unión Basin) and the base of the Santa Maria Supersequence, yet the faunal composition is radically different (see Martinelli et al. 2017b; Schmitt et al. 2019). Formerly, the cynodonts from the Río Seco de la Quebrada Formation were considered indicative of a lower Middle Triassic age, due to correlation with the *Cynognathus* AZ of the Karoo Basin (South Africa). Regardless of the correlation between Argentina and South Africa by two specific taxa (see Martinelli et al. 2009), the lack of these cynodonts in the well-documented Ischigualasto-Villa Unión Basin (Argentina) and Santa Maria Supersequence (Brazil) suggests caution regarding the inferred age for the San Rafael Group. A similar age for the Río Seco de la Quebrada

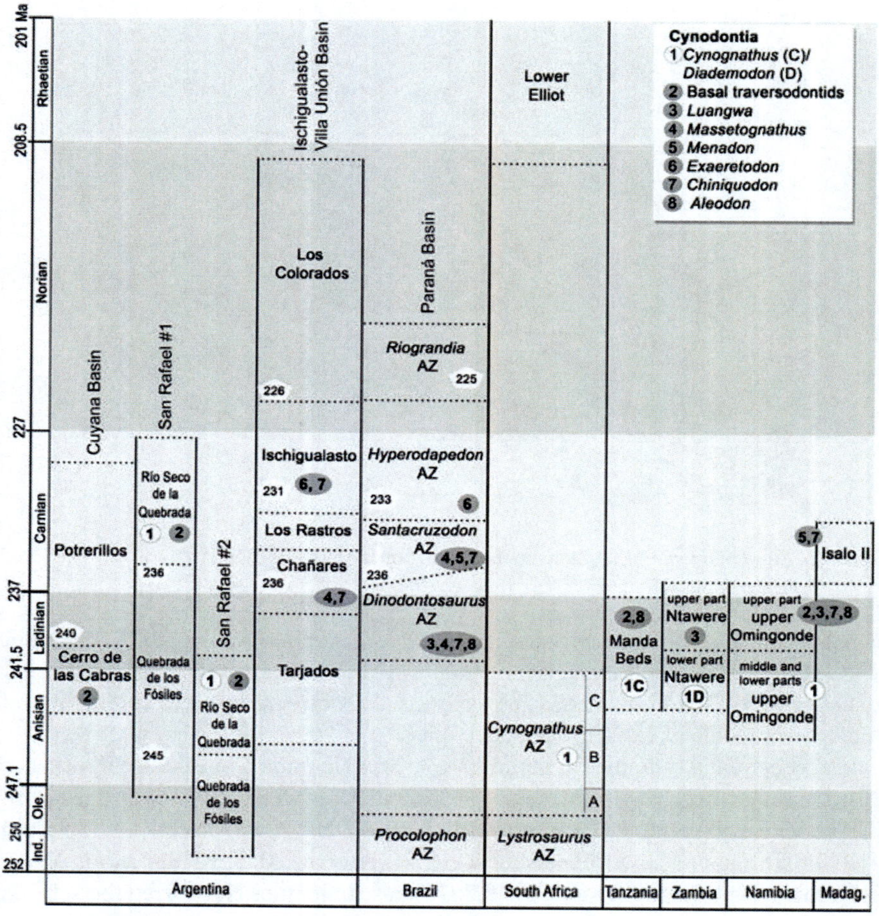


**Fig. 3.19** Main Gondwanan Triassic units with selected cynodonts used for biostratigraphic correlations and radiometric data (modified from Martinelli et al. 2017b). San Rafael #1 considers radiometric data over faunas and San Rafael #2 considers fauna over radiometric data. Abbreviations: AZ, Assemblage Zone; Ind, Induan; Madag, Madagascar; Ole, Olenekian

and Chañares Formations (Ottone et al. 2014; Marsicano et al. 2016) raises the question of why the faunal associations of two nearby areas are quite distinctive in composition. This lack of faunal congruence calls into question the accuracy of the radiometric dating obtained for the Puesto Viejo Group (see also discussion in Sues 2016).

The numerous non-mammaliaform cynodonts recently discovered in Southern Brazil and the classic specimens collected by Alfred S. Romer from western Argentina in the '60 s and '70 s represent the core evidence on the major morphological transformations diagnosing the rise of mammaliaforms. Mammaliaform research agenda mirrored closely the pre-cladistic concept of Mammalia, when arguments centered around what characters "made" a taxon a mammal, instead of the current approach in which a taxon is phylogenetically defined and the characters diagnose the group. Mammaliaforms achieve many complex morphologies considered hallmarks we regularly associate with mammals; these morphological transformations involve a wide variety of character systems including changes in tooth patterns and replacement, lower jaw suspensory, middle and inner ear, palate, braincase, chewing musculature, and appendicular skeleton, to name a few (e.g., Romer 1970; Crompton 1972b; Romer and Lewis 1973; Quiroga 1980, 1984; Rougier et al. 1992; Bonaparte and Barberena 2001; Abdala et al. 2002a, b; Bonaparte et al. 2003, 2005, 2012; Martinelli et al. 2005, 2017a, c, 2019; Luo 2007; Martinelli and Rougier 2007; Martinelli and Bonaparte 2011; Soares et al. 2011a; Rodrigues et al. 2013, 2014, 2019; Martinelli 2017; Guignard et al. 2019b; Wallace et al. 2019). The rise of these characters involving every major area of the body is one of the most momentous episodes in the history of the mammalian lineage. The fossil material from SA is uniquely pertinent in providing the evidence to determine how and when these complex character systems arose, in which sequence with regard to each other, and in inferring the biology of these long-extinct forms. Those numerous and well-preserved specimens form an evolutionary sequence of ~ 25 Ma, from the upper Middle to Upper Triassic (*Dinodontosaurus* AZ/Chañares Formation to *Riograndia* AZ/Los Colorados Formation in Brazil/Argentina, respectively). For this crucial span of time, there is no close parallel anywhere in the world; our understanding of the mammalian precursors rests heavily on the reading of these well-preserved and relatively abundant cynodont communities. Several taxa, such as the traversodontids *Massetoganthus* and *Exaeretodon* and the probainognathians *Chiniquodon, Probainognathus, Riograndia, Pseudotherium,* and *Brasilodon*, are iconic and an obligate reference for any research program focusing on the origin and early transformations of Mammalia and its closest relatives.

The non-mammalian cynodont record in SA dwindles and disappears by the end of the Triassic; *Chaliminia musteloides* (Bonaparte 1980; Martinelli and Rougier 2007) from the Los Colorados Formation (~227–213 Ma) is the youngest known species. This is clearly a truncated and incomplete story. There is a priori no reason to suspect an abrupt discontinuity between the latest Triassic–earliest Jurassic faunas in SA. Unfortunately, the earliest Jurassic sediments are not abundant and they are yet to yield mammaliaforms. Certainly, the final chapters of SA non-mammalian cynodonts are waiting for the enterprising paleontologist on some desolate slope of Patagonia, the craggy mountains of the cordillera, or the forested hills of Brazil.

## References

Abdala NF (1996) Redescripción del cráneo y reconsideración de la validez de *Cynognathus minor* (Eucynodontia-Cynodonthidae) del Triásico Inferior de Mendoza. Ameghiniana 33:115–126

Abdala NF (1999) Elementos postcraneanos de *Cynognathus* (Synapsida-Cynodontia) del Triásico Inferior de la Provincia de Mendoza, Argentina. Rev Española Paleont 14:13–24

Abdala NF (2007) Redescription of *Platycraniellus elegans* (Therapsida, Cynodontia) from the Lower Triassic of South Africa, and the cladistic relationships of eutheriodonts. Palaeontology 50:591–618

Abdala NF, Giannini NP (2000) Gomphodont cynodonts of the Chañares Formation: The analysis of an ontogenetic sequence. J Vertebr Paleontol 20:501–506

Abdala NF, Giannini NP (2002) Chiniquodontid cynodonts: Systematic and morphometric considerations. Palaeontology 45:1151–1170

Abdala F, Ribeiro AM (2003) A new traversodontid cynodont from the Santa Maria Formation (Ladinian-Carnian) of southern Brazil, with a phylogenetic analysis of Gondwanan traversodontids. Zool J Linn Soc 139:529–545

Abdala NF, Sá-Teixeira AM (2004) A traversodontid cynodont of African affinity in the South American Triassic. Palaeontol Africana 40:11–22

Abdala NF, Ribeiro AM (2010) Distribution and diversity patterns of Triassic cynodonts (Therapsida, Cynodontia) in Gondwana. Palaeogeogr Palaeoclimatol Palaeoecol 286:202–217

Abdala NF, Ribeiro AM, Schultz CL (2001) A rich cynodont fauna of Santa Cruz do Sul, Santa Maria Formation (Middle–Late Triassic), southern Brazil. Neues Jahrb Geol Paläontol 2001:669–687

Abdala NF, Barberena MC, Dornelles J (2002a) A new species of the traversodontid cynodont *Exaeretodon* from the Santa Maria Formation (Middle/Late Triassic) of southern Brazil. J Vertebr Paleontol 22:313–325

Abdala NF, Dias-da-Silva S, Cisneros JC (2002b) First record of nonmammalian cynodonts (Therapsida) in the Sanga do Cabral Formation (Early Triassic) of southern Brazil. Palaeontol Afr 38:93–98

Abdala NF, Hancox PJ, Neveling J (2005) Cynodonts from the uppermost Burgersdorp Formation, South Africa, and their bearing on the biostratigraphy and correlation of the Triassic *Cynognathus* Assemblage Zone. J Vertebr Paleontol 25:192–199

Abdala NF, Jasinoski SC, Fernandez V (2013) Ontogeny of the Early Triassic cynodont *Thrinaxodon liorhinus* (Therapsida): Dental morphology and replacement. J Vertebr Paleontol 33:1408–1431

Abdala NF, Gaetano LC, Smith RMH, Rubidge BS (2019) A new large cynodont from the Late Permian (Lopingian) of the South African Karoo Basin and its phylogenetic significance. Zool J Linn Soc 186:983–1005

Abdala F, Gaetano LC, Martinelli AG, Soares MB, Hancox JP, Rubidge BS (2020) Non-mammaliaform cynodonts from western Gondwana and the significance of Argentinean forms in enhancing understanding of the group. J S Am Earth Sci 104:102884

Barberena MC (1981) Uma nova espécie de *Massetognathus* (*Massetognathus ochagaviae*, sp. nov.) da Formação Santa Maria, Triássico do Rio Grande do Sul. Pesquisas 14:181–195

Barberena MC, Bonaparte JF, Sá-Teixeira AM (1987) *Thrinaxodon brasiliensis* sp. nov., a primeira ocorrência de cinodontes galessauros para o Triássico do Rio Grande do Sul. 10° Congresso Brasileiro de Paleontologia, Rio de Janeiro, Anais:67–76

Bertoni R, Martinelli AG, Ribeiro AM (2016) Novas informações sobre *Chiniquodon* sp. da Zona de Assembleia de *Santacruzodon* (Sequência Santa Cruz, Supersequência Santa Maria), Triássico Médio–Superior do Brasil. 10° Simpósio Brasileiro de Paleontologia de Vertebrados, Rio de Janeiro, Bol:42

Bonaparte JF (1962) Descripción del cráneo y mandíbula de *Exaeretodon frenguellii* Cabrera y su comparación con Diademodontidae, Tritylodontidae y los cinodontes sudamericanos. Pub Mus Munic Cienc Nat Mar Del Plata 1:135–202

Bonaparte JF (1963a) Descripción de *Ischignathus sudamericanus* n. gen. n. sp., nuevo cindonte gonfodonte del Triásico Medio Superior de San Juan, Argentina (Cynodontia-Traversodontidae). Acta Geol Lill 4:111–128

Bonaparte JF (1963b) Un nuevo cinodonte gonfodonte del Triásico Medio Superior de San Juan, *Proexaeretodon vincei* n. gen., n. sp (Cynodontia-Traversodontidae). Acta Geol Lill 4:129–133

Bonaparte JF (1966) Una nueva "fauna" Triásica de Argentina. (Therapsida: Cynodontia-Dicynodontia). Consideraciones filogenéticas y paleobiogeográficas. Ameghiniana 4:243–296

Bonaparte JF (1969) *Cynognathus minor* n. sp. (Therapsida-Cynodontia). Nueva evidencia de vinculación faunística Afro-Sudamericana a principios del Triásico. 1° International Symposium on Gondwana Stratigraphy and Palaeontology, Mar del Plata 1967, Gondwana Stratigraphy:273–281

Bonaparte JF (1970) Annotated list of the South American Triassic tetrapods. In: Haughton SH (ed) 2° Gondwana Symposium. CSIR Proc Pap, Pretoria, pp 665–682

Bonaparte JF (1971) Los tetrápodos del sector superior de la Formación Los Colorados, La Rioja, Argentina (Triásico Superior). Op Lilloana 22:1–183

Bonaparte JF (1972) *Cromptodon mamiferoides*, Galesauridae de la Formación Río Mendoza, Mendoza, Argentina (Therapsida-Cynodontia). Ameghiniana 9:343–353

Bonaparte JF (1980) El primer ictidosaurio (Reptilia-Therapsida) de América del Sur, *Chaliminia musteloides*, del Triásico Superior de La Rioja, Argentina. 2° Congreso Argentino de Paleontología y Bioestratigrafía y 1° Congreso Latinoamericano de Paleontología. Buenos Aires Actas 1:123–133

Bonaparte JF (1982) Faunal replacement in the Triassic of South America. J Vertebr Paleontol 2:362–371

Bonaparte JF (2012) Evolution of the Brasilodontidae (Cynodontia-Eucynodontia). Hist Biol 25:643–653

Bonaparte JF, Barbarena MC (1975) A possible mammalian ancestor from the Middle Triassic of Brazil (Therapsida-Cynodontia). J Paleontol 49:931–936

Bonaparte JF, Barberena MC (2001) On two advanced carnivorous cynodonts from the Late Triassic of Southern Brazil. Bull Mus Comp Zool 156:59–80

Bonaparte JF, Crompton AW (1994) A juvenile probainognathid cynodont skull from the Ischigualasto Formation and the origin of mammals. Rev Mus Arg Cienc Nat "B Rivadavia" 5:1–12

Bonaparte JF, Ferigolo J, Ribeiro AM (2001) A primitive Late Triassic "ictidosaur" from Rio Grande do Sul, Brazil. Palaeontology 44:623–635

Bonaparte JF, Martinelli AG, Schultz CL, Rubert R (2003) The sister group of mammals: Small cynodonts from the Late Triassic of southern Brazil. Rev Bras Paleontol 5:5–27

Bonaparte JF, Martinelli AG, Schultz CL (2005) New information on *Brasilodon* and *Brasilitherium* (Cynodontia, Probainognathia) from the Late Triassic of southern Brazil. Rev Bras Paleontol 8:25–46

Bonaparte JF, Soares MB, Schultz CL (2006) A new non-mammalian cynodont from the Middle Triassic of southern Brazil and its implications for the ancestry of mammals. Bull N M Mus Nat Hist Sci 37:599–607

Bonaparte JF, Schultz CL, Soares MB, Martinelli AG (2010) La fauna local de Faxinal do Soturno, Triásico Tardío de Rio Grande do Sul, Brasil. Rev Bras Paleontol 13:233–246

Bonaparte JF, Soares MB, Martinelli AG (2012) Discoveries in the Late Triassic of Brazil improve knowledge on the origin of mammals. Hist Nat Ser 3:5–30

Botha J, Chinsamy A (2005) Growth patterns of *Thrinaxodon liorhinus*, a non-mammalian cynodont from the Lower Triassic of South Africa. Palaeontology 48:385–394

Botha J, Abdala NF, Smith R (2007) The oldest cynodont: New clues on the origin and early diversification of the Cynodontia. Zool J Linn Soc 149:477–492

Botha-Brink J, Soares MB, Martinelli AG (2018) Osteohistology of Late Triassic prozostrodontian cynodonts from Brazil. PeerJ 6:e5029

Bradu D, Grine FE (1979) Multivariate analysis of *Diademodon* crania from South Africa and Zambia. S African J Sci 75:441–448

Broili F, Schröder J (1934) Zur Osteologie des Kopfes von *Cynognathus*. Sitz Bayerischen Akad Wiss 1934:95–128

Burgin CJ, Colella JP, Kahn PL, Upham NS (2018) How many species of mammals are there? J Mammal 99:1–14

Cabrera A (1943) El primer hallazgo de terápsidos en la Argentina. Notas Mus La Plata, Paleont 8:317–331

Cisneros JC, Abdala NF, Rubidge BS, Dentzien-Dias PC, Bueno AO (2011) Dental occlusion in a 260-million-year-old therapsid with saber canines from the Premian of Brazil. Science 331:1603–1605

Colbert EH, Kitching JW (1977) Triassic cynodont reptiles from Antarctica. Am Mus Novit 2611:1–30

Crompton AW (1955) On some Triassic cynodonts from Tanganyika. Proc Zool Soc Lond 1125:617–669

Crompton AW (1972a) Postcanine occlusion in cynodonts and tritylodonts. Bull Br Mus Nat Hist 21:27–71

Crompton AW (1972b) Evolution of the jaw articulation in cynodonts. In: Joysey KA, Kemp TS (eds) Studies in vertebrate evolution. Oliver & Boyd, Edinburgh, pp 231–253

Crompton AW, Ellenberger F (1957) On a new cynodont from the Molteno Beds and the origin of the tritylodontids. Ann S Afr Mus 44:1–13

Crompton AW, Owerkowicz T, Bhullar B-A, Musinsky C (2017) Structure of the nasal region of non-mammalian cynodonts and mammaliaforms: Speculations on the evolution of mammalian endothermy. J Vertebr Paleontol 37:e1269116

Da-Rosa AAS, Piñeiro G, Dias-da-Silva S, Cisneros JC, Feltrin FF, Neto LW (2009) Bica São Tomé, um novo sítio fossilífero para o Triássico Inferior do sul do Brasil. Rev Bras Paleontol 12:67–76

Dias-da-Silva S, Pinheiro F, Da-Rosa ÁAS, Martinelli AG, Schultz CL, Silva Neves E, Modesto SP (2017) Biostratigraphic review of the Lower Triassic Sanga do Cabral Supersequence from South America, with a description of new material attributable to the parareptile genus *Procolophon*. J S Am Earth Sci 79:281–296

Ezcurra MD, Fiorelli LE, Martinelli AG, Rocher S, von Baczko MB, Ezpeleta M, Taborda JRA, Hechenleitner EM, Trotteyn MJ, Desojo JB (2017) Deep faunistic turnovers preceded the rise of dinosaurs in southwestern Pangaea. Nat Ecol Evol 1:1477–1483

Fernández E, Martinelli AG, Martínez RN, Abdala NF (2011) Redescription and taxonomic reinterpretation of cf. *Probainognathus* (Cynodontia, Probainognathia) from the Upper Triassic Ischigualasto Formation (San Juan, Argentina). Ameghiniana 48(Suppl):107R

Fox RC, Youzwyshyn GP, Krause DW (1992) Post-Jurassic mammal-like reptile from the Paleocene. Nature 358:233–235

Gaetano LC, Abdala NF, Govender R (2017) The postcranial skeleton of the Lower Jurassic *Tritylodon longaevus* from Southern Africa. Ameghiniana 54:1–35

Gaetano LC, Alonso Muruaga P, Krapovickas V, Leardi JM, Otero A, Apaldetti C (2018) New cynodont specimens from Los Colorados Formation. Reunión de Comunicaciones de la APA, Puerto Madryn, Abstracts:62

Goñi R (1986) Reemplazo de dientes postcaninos en *Andescynodon mendozensis* Bonaparte (Cynodontia, Traversodontidae). 4°Congreso Argentino de Paleontología y Bioestratigrafía. Mendoza, Actas 2:7–14

Goñi R, Goin FJ (1990) Morfología dentaria y biomecánica masticatoria de los cinodontes (Reptilia, Therapsida) del Triásico argentino: II. *Exaeretodon frenguellii* Cabrera (Traversodontidae). Ameghiniana 27:327–336

Guignard M, Martinelli AG, Soares MB (2018) Reassessment of the anatomy of *Prozostrodon brasiliensis* (Eucynodontia: Probainognathia) and implications for postural evolution of non-mammaliaform cynodonts. J Vertebr Paleontol 38:e1511570

Guignard ML, Martinelli AG, Soares MB (2019a) Postcranial anatomy of *Riograndia guaibensis* (Cynodontia: Ictidosauria). Geobios 53:9–21

Guignard ML, Martinelli AG, Soares MB (2019b) The postcranial anatomy of *Brasilodon quadrangularis* and the acquisition of mammaliaform traits among non-mammaliaform cynodonts. PLoS ONE 14(5):e0216672

Hammer WR (1995) New therapsids from the upper Fremouw Formation (Triassic) of Antarctica. J Vertebr Paleontol 15:105–112

Hancox PJ, Shishkin MA, Rubidge BS, Kitching JW (1995) A threefold subdivision of the *Cynognathus* Assemblage Zone (Beaufort Group, South Africa) and its palaeogeographic implications. S African J Sci 91:143–144

Hendrickx C, Gaetano LC, Choiniere JN, Mocke H, Abdala F (2020) A new traversodontid cynodont with a peculiar postcanine dentition from the Middle/Late Triassic of Namibia and dental evolution in basal gomphodonts. J Syst Palaeont 18:1669–1706

Hopson JA (1971) Postcanine replacement in the gomphodont cynodont *Diademodon*. In: Kermack DM, Kermack KA (eds) Early mammals. Zool J Linn Soc 50, suppl 1:1–21

Hopson JA (1991) Systematics of the non-mammalian Synapsida and implications for patterns of evolution in synapsids. In: Schultze H-P, Trueb L (eds) Origins of the higher groups of tetrapods. Comstock Publishing Associates, Ithaca, Controversy and consensus, pp 635–693

Hopson JA, Barghusen HR (1986) An analysis of therapsid relationships. In: Hotton N, MacLean PD, Roth JJ, Roth EC (eds) The ecology and biology of mammal-like reptiles. Smithsonian Institution Press, Washington, pp 83–106

Hopson JA, Kitching JW (2001) A probainognathian cynodont from South Africa and the phylogeny of nonmammalian cynodonts. Bull Mus Comp Zool 156:5–35

Horn BLD, Melo TM, Schultz CL, Philipp RP, Kloss HP, Goldberg K (2014) A new third-order sequence stratigraphic framework applied to the Triassic of the Paraná Basin, Rio Grande do Sul, Brazil, based on structural, stratigraphic and paleontological data. J S Am Earth Sci 55:123–132

Huene F von (1936) Lieferung 2. Cynodontia. Die Fossilen Reptilien des Südamerikanischen Gondwanalandes. Ergebnisse der Sauriergrabungen in Südbrasilien 1928/29. C. H. Beck'sche Verlagsbuchhandlung, München, pp 83–160

Jasinoski SC, Abdala NF, Fernandez V (2015) Ontogeny of the Early Triassic cynodont *Thrinaxodon liorhinus* (Therapsida): Cranial morphology. Anat Record 298:1440–1464

Kammerer CF (2016) A new taxon of cynodont from the *Tropidostoma* Assemblage Zone (Upper Permian) of South Africa, and the early evolution of Cynodontia. Pap Palaeontol 2:387–397

Kemp TS (2005) The origin and evolution of mammals. Oxford University Press, Oxford

Kielan-Jaworowska Z, Cifelli RL, Luo Z-X (2004) Mammals from the age of dinosaurs. Origins, evolution, and structure. Columbia University Press, New York

Kitching JW (1977) The distribution of the Karoo vertebrate fauna. Bernard Price Inst Palaeontol Res Mem 1:1–131

Langer MC (2005) Studies on continental Late Triassic tetrapod biochronology. I. The type locality of *Saturnalia tupiniquim* and the faunal succession in south Brazil. J S Am Earth Sci 19:205–218

Langer MC, Ribeiro AM, Schultz CL, Ferigolo J (2007) The continental tetrapod-bearing Triassic of south Brazil. Bull N M Mus Nat Hist Sci 41:201–218

Langer MC, Ramezani J, Da-Rosa ÁAA (2018) U-Pb age constraints on dinosaur rise from south Brazil. Gondwana Res 57:133–140

Lautenschlager S, Gill P, Luo Z-X, Fagan MJ, Rayfield EJ (2019) The role of miniaturization in the evolution of the mammalian jaw and middle ear. Nature 561:533–537

Liu J (2007) The taxonomy of the traversodontid cynodonts *Exaeretodon* and *Ischignathus*. Rev Bras Paleontol 10:133–136

Liu J, Abdala NF (2014) Phylogeny and taxonomy of the Traversodontidae. In: Kammerer CF, Angielczyk KD, Fröbisch J (eds) Early evolutionary history of the Synapsida. Springer, Dordrecht, pp 255–279

Liu J, Olsen PE (2010) The phylogenetic relationships of Eucynodontia (Amniota, Synapsida). J Mammal Evol 17:151–176

Liu J, Powell J (2009) Osteology of *Andescynodon* (Cynodontia, Traversodontidae) from the Middle Triassic of Argentina. Am Mus Novit 3674:1–19

Luo Z-X (1994) Sister-group relationships of mammals and transformations of diagnostic mammalian characters. In: Fraser NC, Sues H-D (eds) In the shadow of the dinosaurs: early Mesozoic tetrapods. Cambridge University Press, Cambridge, pp 98–128

Luo Z-X (2007) Transformation and diversification in the early mammalian evolution. Nature 450:1011–1019

Luo Z-X, Crompton AW (1994) Transformation of the quadrate (incus) through the transition from non-mammalian cynodonts to mammals. J Vertebr Paleontol 14:341–374

Marsicano CA, Irmis RB, Mancuso AC, Mundile R, Chemale F (2016) The precise temporal calibration of dinosaur origins. Proc Natl Acad Sci USA 113:509–513

Martinelli AG (2010) On the postcanine dentition of *Pascualgnathus polanskii* Bonaparte (Cynodotia: Traversodontidae) from the Middle Triassic of Argentina. Geobios 43:629–638

Martinelli AG (2017) Contribuição ao conhecimento dos cinodontes probainognátios (Therapsida, Cynodontia, Probainognathia) do Triássico da América do Sul e seu impacto na origem dos Mammaliaformes. Unpublished Phd thesis, Universidade Federal do Rio Grande do Sul

Martinelli AG, Bonaparte JF (2011) Postcanine replacement in *Brasilodon* and *Brasilitherium* (Cynodontia, Probainognathia) and its bearing in cynodont evolution. In: Calvo J, Porfiri J, González Riga B, Dos Santos D (eds) Dinosaurios y paleontología desde América Latina. Editorial de la Universidad Nacional de Cuyo, Mendoza, pp 179–186

Martinelli AG and de la Fuente M (2008) Los cinodontes no-mamaliaformes de la Formación Puesto Viejo, San Rafael, Mendoza. 3° Encuentro Científico ICES, Malargüe, Actas:67–74

Martinelli AG, Rougier GW (2007) On *Chaliminia musteloides* (Eucynodontia: Tritheledontidae) from the Late Triassic of Argentina and a phylogeny of Ictidosauria. J Vertebr Paleontol 27:442–460

Martinelli AG, Soares MB (2016) Evolution of South American cynodonts. Mus Argent Cienc Nat "B Rivadavia." Contribuciones 6:183–197

Martinelli AG, Bonaparte JF, Schultz CB (2005) A new tritheledontid (Therapsida, Eucynodontia) from the Late Triassic of Rio Grande do Sul (Brazil) and its phylogenetic relationships among carnivorous non-mammalian eucynodonts. Ameghiniana 42:191–208

Martinelli AG, de la Fuente M, Abdala NF (2009) *Diademodon tetragonus* Seeley, 1894 (Therapsida: Cynodontia) in the Triassic of South America and its biostratigraphic implications. J Vertebr Paleontol 29:852–862

Martinelli AG, Ezcurra MD, Melo TP, Gaetano LC, Soares MB, Fiorelli LE, Desojo JB (2014) *Massetognathus pascuali* and *Santacruzodon hopsoni* (Cynodontia, Traversodontidae) from the late Middle—early Late Triassic of South America: New insights on their premaxillary dentition. 9° Simpósio Brasileiro de Paleontologia de Vertebrados, Vitoria, Abstracts:80

Martinelli AG, Soares MB, Schwanke C (2015) Nuevos aportes al conocimiento de los cinodontes probainognatios (Therapsida) del Triásico Medio (Zona de Asociación de *Dinodontosaurus*) del Sur de Brasil. 5° Congreso Latinoamericano de Paleontología de Vertebrados, Colonia del Sacramento, Abstracts:64

Martinelli AG, Soares MB, Schwanke C (2016a) Two new cynodonts (Therapsida) from the Middle-early Late Triassic of Brazil and comments on South American probainognathians. PLoS ONE 11(10):e0162945

Martinelli AG, Dias-Da-Silva S, Pinheiro FL, Da-Rosa ÁAS, Schultz CL, Modesto SP, Soares MB (2016b) Cynodonts from the Sanga do Cabral Supersequence revisited: are they really present in the Early Triassic of Brazil? Reunião Anual da Sociedade Brasileira de Paleontologia, Santa Maria. Abstracts:68–69

Martinelli AG, Eltink E, Da-Rosa ÁAS, Langer MC (2017a) A new cynodont (Therapsida) from the *Hyperodapedon* Assemblage Zone (upper Carnian–Norian) of southern Brazil improves the Late Triassic probainognathian diversity. Pap Palaeontol 3:401–423

Martinelli AG, Kammerer CF, Melo TP, Paes Neto VD, Ribeiro AM, Da-Rosa ÁAS, Schultz CL, Soares MB (2017b) The African cynodont *Aleodon* (Cynodontia, Probainognathia) in the Triassic of southern Brazil and its biostratigraphic significance. PLoS ONE 12(6):e0177948

Martinelli AG, Soares MB, Oliveira TV, Rodrigues PG, Schultz CL (2017c) The Triassic eucynodont *Candelariodon barberenai* revisited and the early diversity of stem prozostrodontians. Acta Palaeontol Pol 62:527–542

Martinelli AG, Soares MB, Forasiepi AM (2018) Cinodontes triásicos y mamíferos del tiempo de los dinosaurios. Ciencia Hoy 27:52–58

Martinelli AG, Gill PG, Corfe IJ, Rodrigues PG, Fonseca PH, Schultz CL, Soares MB, Rayfield EJ (2019) Ontogeny and tooth replacement in the Brazilian cynodonts *Brasilodon quadrangularis, Brasilitherium riograndensis* and *Minicynodon maieri*. Reunión de Comunicaciones de la APA, La Plata, Abstracts:62–63

Martínez RN, May CL, Forster CA (1996) A new carnivorous cynodont from the Ischigualasto Formation (Late Triassic, Argentina), with comments on eucynodont phylogeny. J Vertebr Paleontol 16:271–284

Martínez RN, Fernández E, Alcober OA (2013a) A new non-mammaliaform eucynodont from the Carnian-Norian Ischigualasto Formation, northwestern Argentina. Rev Bras Paleontol 16:61–76

Martínez RN, Apaldetti C, Alcober OA, Colombi CE, Sereno PC, Fernandez E, Santi Malnis P, Correa GA, Abelín D (2013b) Vertebrate succession in the Ischigualasto Formation. J Vertebr Paleontol 32(Memmoir 12):10–30

Martínez RN, Apaldetti C, Correa G, Colombi CE, Fernández E, Malnis PS, Praderio A, Abelín D, Benegas LG, Aguilar-Cameo A, Alcober OA (2015) A new Late Triassic vertebrate assemblage from northwestern Argentina. Ameghiniana 52:379–390

Matsuoka H, Kusuhashi N, Corfe IJ (2016) A new Early Cretaceous tritylodontid (Synapsida, Cynodontia, Mammaliamorpha) from the Kuwajima Formation (Tetori Group) of central Japan. J Vertebr Paleont 36:e1112289

Melo TP, Abdala NF, Soares MB (2015) The Malagasy cynodont *Menadon besairiei* (Cynodontia; Traversodontidae) in the Upper Triassic of Brazil. J Vertebr Paleontol 35(6):e1002562

Melo TP, Martinelli AG, Soares MB (2017) A new gomphodont cynodont (Traversodontidae) from the Middle-Late Triassic *Dinodontosaurus* Assemblage Zone of the Santa Maria Supersequence, Brazil. Palaeontology 60:571–582

Melo TP, Ribeiro AM, Martinelli AG, Soares MB (2019) Earliest evidence of molariform hypsodonty in a Triassic stem-mammal. Nat Commun 10:2841

Minoprio JL (1954) Theriodonte en el Triásico de Mendoza. an Soc Cient Arg 157:31–37

Minoprio JL (1957) Nota aclaratoria sobre *Colbertia muralis* (Cambio de denominación). Ameghiniana 1:113

O'Leary MA, Bloch JI, Flynn JJ, Gaudin TJ, Giallombardo A, Giannini NP et al (2013) The placental mammal ancestor and the post-K-Pg radiation of placentals. Science 339:662–667

Oliveira TV, Soares MB, Schultz CL (2010) *Trucidocynodon riograndensis* gen. nov. et sp. nov. (Eucynodontia), a new cynodont from the Brazilian Upper Triassic (Santa Maria Formation). Zootaxa 2382:1–71

Oliveira TV, Martinelli AG, Soares MB (2011) New material of *Irajatherium hernandezi* Martinelli, Bonaparte, Schultz & Rubert 2005 (Eucynodontia, Tritheledontidae) from the Upper Triassic (Caturrita Formation, Paraná Basin) of Brazil. Paläontol Z 85:67–82

Oliveira TV, Soares MB, Schultz CL, Rodrigues CN (2011) A new carnivorous cynodont (Synapsida, Therapsida) from the Brazilian Middle Triassic (Santa Maria Formation): *Candelariodon barberenai* gen. et sp. nov. Zootaxa 3027:19–28

O'Meara R, Dirks W, Martinelli AG (2018) Enamel formation and growth in non-mammalian cynodonts. R Soc Open Sci 5:172293

Ottone EG, Monti M, Marsicano CA, de la Fuente MS, Naipauer M, Armstrong R, Mancuso AC (2014) A new Late Triassic age for the Puesto Viejo Group (San Rafael depocenter, Argentina): SHRIMP U-Pb zircon dating and biostratigraphic correlations across southern Gondwana. J S Am Earth Sci 56:186–199

Pacheco CP, Martinelli AG, Pavanatto AEB, Soares MB, Dias-da-Silva S (2018) *Prozostrodon brasiliensis*, a probainognathian cynodont from the Late Triassic of Brazil: second record and improvements on its dental anatomy. Hist Biol 30:475–485

Pavanatto AEB, Pretto FA, Kerber L, Müller RT, Da-Rosa AAS, Dias-da-Silva S (2018) A new Upper Triassic cynodont-bearing fossiliferous site from southern Brazil, with taphonomic remarks and description of a new traversodontid taxon. J S Am Earth Sci 88:179–196

Pavanatto AEB, Da-Rosa AAS, Temp-Müller R, Roberto-da-Silva L, Ribeiro AM, Martinelli AG, Dias-da-Silva S (2020) Bortolin site, a new fossiliferous locality in the Triassic (Ladinian/Carnian) of southern Brazil. Rev Bras Paleontol 23:123–137

Philipp RP, Schultz CL, Kloss HP, Horn BLD, Soares MB, Basei MAS (2018) Middle Triassic SW Gondwana paleogeography and sedimentary dispersal revealed by integration of stratigraphy and U-Pb zircon analysis: the Santa Cruz Sequence, Paraná Basin, Brazil. J S Am Earth Sci 88:216–237

Piñeiro G, Ferigolo J, Ribeiro AM, Velozo P (2015) Reassessing the affinities of vertebral remains from Permo-Triassic beds of Gondwana. CR Palevol 14:387–401

Quiroga JC (1980) The brain of the mammal-like reptile *Probainognathus jenseni* (Therapsida, Cynodontia). A correlative paleo-neoneurological approach to the neocortex at the reptilemammal transition. J Hirnforsch 21:299–336

Quiroga JC (1984) The endocranial cast of the advanced mammal-like reptile *Therioherpeton cargnini* (Therapsida-Cynodontia) form the Middle Triassic of Brazil. J Hirnforsch 25:285–290

Reichel M, Schultz CL, Soares MB (2009) A new traversodontid cynodont (Therapsida, Eucynodontia) from the Middle Triassic Santa Maria Formation of Rio Grande do Sul. Palaeontology 52:229–250

Reisz RR (1972) Pelycosaurian reptiles from the middle Pennsylvanian of North America. Bull Mus Comp Zool 144:27–60

Reisz RR (1986) Pelycosauria, encyclopedia of paleoherpetology, 17A. Gustav Fischer Verlag, Stuttgart

Ribeiro AM, Abdala NF, Bertoni RS (2011) Traversodontid cynodonts (Therapsida-Eucynodontia) from two Upper Triassic localities of the Parana Basin, southern Brazil. Ameghiniana 48(Suppl):111R

Rodrigues PG, Ruf I, Schultz CL (2013) Digital Reconstruction of the otic region and inner ear of the non-mammalian cynodont *Brasilitherium riograndensis* (Late Triassic, Brazil) and its relevance to the evolution of the mammalian ear. J Mammal Evol 20:291–307

Rodrigues PG, Ruf I, Schultz CL (2014) Study of a digital cranial endocast of the non-mammaliaform cynodont *Brasilitherium riograndensis* (Later Triassic, Brazil) and its relevance to the evolution of the mammalian brain. Paläontol Z 88:329–352

Rodrigues PG, Martinelli AG, Schultz CL, Corfe IJ, Gill PG, Soares MB, Rayfield EJ (2019) Digital endocast of *Riograndia guaibensis* (Late Triassic, Brazil) and the evolution of the brain in non-mammalian cynodonts. Hist Biol 31:1195–1212

Rogers RR, Swisher CC III, Sereno PC, Monetta AM, Forster CA, Martínez RN (1993) The Ischigualasto tetrapod assemblage, Late Triassic, Argentina, and 40Ar/39Ar dating of dinosaur origins. Science 260:794–797

Romer AS (1967) The Chañares (Argentina) Triassic reptile fauna. III. Two new Gomphodonts *Massetognathus pascuali* and *M. teruggii*. Breviora 264:1–25

Romer AS (1969) The Chañares (Argentina) Triassic reptile fauna. V. A new chiniquodontid cynodont, *Probelesodon lewisi* —cynodont ancestry. Breviora 333:1–24

Romer AS (1970) The Chañares (Argentina) Triassic reptile fauna. VI. A chiniquodontid cynodont with incipient squamosal-dentary jaw articulation. Breviora 344:1–18

Romer AS (1972) The Chañares (Argentina) Triassic reptile fauna. XVII. The Chañares gomphodonts. Breviora 396:1–9

Romer AS (1973) The Chañares (Argentina) Triassic reptile fauna. XVIII. *Probelesodon minor*, a new species of carnivorous cynodont—family Probainognathidae nov. Breviora 401:1–4

Romer AS, Lewis AD (1973) The Chañares (Argentina) Triassic reptile fauna. XIX. Postcranial materials of the cynodonts *Probelesodon* and *Probainognathus*. Breviora 407:1–26

Rougier GW, Wible JR, Hopson JA (1992) Reconstruction of the cranial vessels in the Early Cretaceous mammal *Vincelestes neuquenianus*: Implications for the evolution of the mammalian cranial vascular system. J Vertebr Paleontol 12:188–216

Rowe TB (1988) Definition, diagnosis, and origin of Mammalia. J Vertebr Paleontol 8:241–264

Ruff I, Maier W, Rodrigues PG, Schultz CL (2014) Nasal anatomy of the non-mammaliaform cynodont *Brasilitherium riograndensis* (Eucynodontia, Therapsida) reveals new insight into mammalian evolution. Anat Record 297:2018–2030

Ruta M, Botha-Brink J, Mitchell SA, Benton MJ (2013) The radiation of cynodonts and the ground plan of mammalian morphological diversity. Proc R Soc Lond B 280:20131865

Santi Malnis P, Kent DV, Colombi CE, Geuna SE (2011) Quebrada de la Sal magnetostratigraphic section, Los Colorados Formation, Upper Triassic Ischigualasto-Villa Unión Basin, Argentina. Latinmag Lett 1:1–7

Schmitt MR, Martinelli AG, Melo TP, Soares MB (2019) On the occurrence of the traversodontid *Massetognathus ochagaviae* (Synapsida, Cynodontia) in the early Late Triassic *Santacruzodon* Assemblage Zone (Santa Maria Supersequence, southern Brazil): Taxonomic and biostratigraphic implications. J S Am Earth Sci 93:36–50

Sidor CA, Smith RMH (2004) A new galesaurid (Therapsida: Cynodontia) from the Lower Triassic of South Africa. Palaeontology 47:535–556

Sidor CA, Hopson JA, Angielczyk KD, Nesbitt SJ, Peecook BR, Smith RMH, Steyer JS, Tabor NJ, Tolan S (2016) A new species of traversodont cynodont with tritylodont-like features and possible arboreal adaptations from the upper Ntawere Formation, northeastern Zambia. 76° Annual Meeting, Society of Vertebrate Paleontology, Salt Lake City, Abstracts: 224

Soares MB, Schultz CL, Horn BLD (2011) New information on *Riograndia guaibensis* Bonaparte, Ferigolo & Ribeiro, 2001 (Eucynodontia, Tritheledontidae) from the Late Triassic of southern Brazil: Anatomical and biostratigraphic implications. An Acad Brasil Ciênc 83:329–354

Soares MB, Abdala NF, Bertoni CM (2011) A sectorial toothed cynodont from the Triassic Santa Cruz do Sul fauna, Santa Maria Formation, Southern Brazil. Geodiversitas 33:265–278

Soares MB, Martinelli AG, Oliveira TV (2014) A new prozostrodontian cynodont (Therapsida) from the Late Triassic *Riograndia* Assemblage Zone (Santa Maria Supersequence) of Southern Brazil. An Acad Brasil Ciênc 86:1673–1691

Spalletti LA, Fanning CM, Rapela CW (2008) Dating the Triassic continental rift in the southern Andes: the Potrerillos Formation, Cuyo Basin, Argentina. Geol Acta 6:267–283

Stefanello M, Müller RT, Kerber L, Martínez RN, Dias-da-Silva S (2018) Skull anatomy and phylogenetic assessment of a large specimen of Ecteniniidae (Eucynodontia: Probainognathia) from the Upper Triassic of southern Brazil. Zootaxa 4457:351–378

Stefanello M, Kerber L, Martinelli AG, Dias-da-Silva S (2020) A new prozostrodontian cynodont (Eucynodontia, Probainognathia) from the Upper Triassic of Southern Brazil. J Vertebr Paleontol 40(3):e1782415

Sues H-D (1986) The skull and dentition of two tritylodontid synapsids from the Lower Jurassic of western North America. Bull Mus Comp Zool 151:217–268

Sues H-D (2016) Dating the origin of dinosaurs. Proc Natl Acad Sci USA 113:480–481

Van den Brandt M, Abdala NF (2018) Cranial morphology and phylogenetic analysis of *Cynosaurus suppostus* (Therapsida, Cynodontia) from the Upper Permian of the Karoo Basin, South Africa. Palaeontol Africana 52:201–221

Wallace RVS, Martínez R, Rowe T (2019) First record of a basal mammaliamorph from the early Late Triassic Ischigualasto Formation of Argentina. PLoS ONE 14(8):e0218791

Zavattieri AM, Arcucci AB (2007) Edad y posición estratigráfica de los tetrápodos del Cerro Bayo de Potrerillos (Triásico), Mendoza, Argentina. Ameghiniana 44:133–142

# Chapter 4
# Australosphenidans

*Since the first Mesozoic mammal was discovered to the present day, there has never been any real agreement among students as to the relationships of these various orders among themselves or to later mammals, or even as to the real existence of the various groups as such. This lack of any approach toward unanimity, even as regards the very fundamentals of the whole problem, has been partly due to the faulty nature of the known materials. The inadequacy of knowledge has been so deeply felt that, in general, students of mammals, aside from two or three specialists, have not even felt it necessary to attempt to grasp the true scope and meaning of what has been known. Attention has chiefly been focused on the last third of mammalian history without troubling the obscurity of the first two thirds.*
George Gaylord Simpson
*American Mesozoic Mammalia*, 1929

**Abstract** As presently understood Australosphenida is a clade of Gondwanan taxa, including an array of Jurassic and Cretaceous extinct forms and the extant monotremes as the sole survivors. Mesozoic Australosphenida show plesiomorphic features in the lower jaw and derived tribosphenic, or tribosphenic-like, dentition, possibly acquired independently from boreal tribosphenic taxa, namely, therians and their immediate relatives. However, non-monotreme australosphenidans are known by rare and incomplete material making alternative hypotheses of dental homologies viable. Fossil and extant monotremes are hard to relate to Mesozoic non-monotreme australosphenidans, but a few dental and mandibular characters support them as members of the group. South American australosphenidans are the oldest currently known undisputed mammals, highlighting the early acquisition of complex tribosphenic dentitions and the importance of the fossil record from southern continents in the history of early mammals.

**Keywords** Henosferidae · Ornithorhynchidae · Monotremata · Tribosphenic molars · Talonid

 
G. W. Rougier et al., *Mesozoic Mammals from South America and Their Forerunners*, Springer Earth System Sciences, https://doi.org/10.1007/978-3-030-63862-7_4

## 4.1 Introduction

Australosphenida was defined by Luo et al. (2001) based on discoveries from the last few decades in Gondwana. The group composition has been somewhat stable over the years, and includes *Ambondro* (Flynn et al. 1999), Ausktribosphenida (Rich et al. 1997, 2001a; Rich and Vickers-Rich 2004), Henosferidae (Rauhut et al. 2002; Rougier et al. 2007), and Monotremata (e.g., Archer et al. 1985, 1992, 1993; Pascual et al. 1992a, b; Flannery et al. 1995; Rich et al. 1999, 2001b; Rich et al. 2016). Together with *Ambondro* from the Middle Jurassic (Bathonian) of Madagascar (Flynn et al. 1999), the SA australosphenidans, *Asfaltomylos patagonicus* and *Henosferus molus* from the Early–Middle Jurassic (Toarcian–Bajocian?) of Patagonia, Argentina (Cúneo et al. 2013) suggest that the group had already diversified and spread over much of what would become Gondwana by the mid-Mesozoic (Rauhut et al. 2002; Rougier et al. 2007). Given the antiquity of henosferids, the evolution of australosphenidans can be accommodated equally well in a Pangeic or Gondwanan setting.

The stem therians *Brancatherulum tendagurense* (Dietrich 1927; Simpson 1928) and *Tendagurutherium dietrichi* (Heinrich 1998; Heinrich et al. 2011) from the Jurassic Tendaguru beds of Tanzania exhibit some resemblances with henosferids. *Brancatherulum tendagurense* is known by a partial edentulous lower jaw (Dietrich 1927; Simpson 1928) and *Tendagurutherium dietrichi* by a posterior portion of the right lower jaw with the last molar (Heinrich 1998; Heinrich et al. 2011). Both species show a long and relatively low horizontal ramus of the dentary, with a well-marked Meckelian groove, postdentary trough, and spoon-like angular process similar to that of the SA australosphenidan *Henosferus* (Rougier et al. 2007). The only known molar of *Tendagurutherium* is poorly preserved, and it was interpreted as peramurid-grade, lacking a basined talonid (Heinrich 1998). However, most of the talonid is broken off and it is not possible to verify, if instead, it was basined as in henosferids. Its taxonomic position is consequently uncertain (Kielan-Jaworoska et al. 2004); however, it is tempting to consider that the Tendaguru species are close phylogenetically to the Jurassic radiation of mammals with complex (tribosphenic-like) molars and primitive lower jaws (Chimento et al. 2016), as is the case of *Ambondro* and henosferids. This could be a tantalizing option; however, the material is too incomplete to allow their inclusion in a strict phylogenetic analysis and we regard their relationships as unresolved.

Australosphenidans show a striking combination of derived tooth morphology and primitive lower jaw. The lower teeth have a sharp trigonid and a basined talonid that was capable of grinding; all the hallmark features of tribosphenic molars as defined by Simpson (1936) can be recognized in these molars. There is a continuous, shelf-like mesial cingulid wrapping around the mesiolingual corner of the trigonid, extending to the lingual side of the tooth, the interlocking mechanism between molars is absent, and the hypoconulid is procumbent. The dentary (at least in Ausktribosphenida and Henosferidae) has a prominent postdentary trough that likely housed postdentary bones, and the angular process is elevated, rather than downturned as in tribosphenidans (Luo et al. 2001, 2002; Kielan-Jaworowska et al. 2004). This character combination and phylogenetic interpretation (Fig. 4.1a) implies that the tribosphenic-like molar pattern evolved independently in two distinctive lineages: the Gondwanan Australosphenida and the Laurasian Tribosphenida (= Boreosphenida sensu Luo et al. 2001, 2002; see also Chap. 1). The Australosphenida would have acquired independently a molar pattern functionally similar or equivalent to the tribosphenic molars that characterize marsupials, placentals (Simpson 1936), and their immediate stems (Tribotheria in the sense of Kielan-Jaworowska et al. 2004). This complex molar pattern in australosphenidans predates that in tribosphenidans by more than 30 Ma (Rauhut et al. 2002) if the holartic *Juramaia sinensis* is indeed

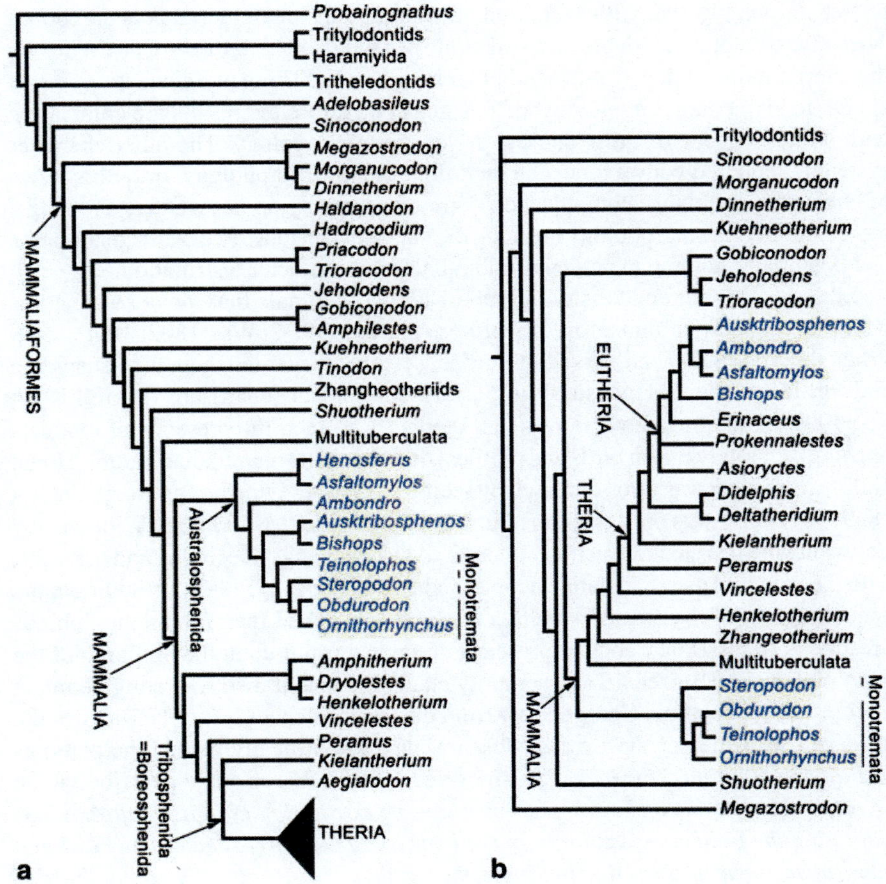


**Fig. 4.1** Phylogenetic trees with alternative hypotheses on the relationships of the Australosphenida. Luo et al. (2001, 2002), Rauhut et al. (2002), Rougier et al. (2007), with Australosphenida (including Monotremata) as stem Theria (**a**). Rich et al. (2002), Woodburne (2003), Woodburne et al. (2003), with Australosphenida (excluding Monotremata) as part of Eutheria (**b**)

Jurassic (Luo et al. 2011), or even longer if *Juramaia* is instead of an Early Cretaceous age (King and Beck 2020).

Alternative phylogenetic hypotheses have suggested that australosphenidans (excluding Monotremata) are a monophyletic group, placed at the root of Eutheria (e.g., Rich et al. 2002; Woodburne 2003) (Fig. 4.1b), or even sister taxon to *Vincelestes* plus therians (Krause et al. 2014). The first alternative phylogenetic hypothesis suggests that tribosphenic grinding molars evolved once but that the therian (i.e., placentals and marsupials) middle ear with the three suspended auditory ossicles (malleus, incus, and stapes that transmit vibrations from the tympanic membrane to the receptors of the inner ear) evolved twice and independently.

Both hypotheses about the phylogenetic position of the Australosphenida (monophyletic group together with Monotremata or nested inside Eutheria, or at the root of Theria) are theoretical viable and still controversial. The bulk of evidence relies on the interpretation of dental and mandibular morphology. These two character systems appear to give conflicting phylogenetic signals, therefore the results are compatible with whichever one of those subsets drives the tree topology. The other character system is rendered convergent. The dentition, the jaw morphology, or both subsets of characters are highly homoplastic. There are currently no known taxa showing a more intermediate morphology bridging the gap between these two character systems and potentially helping to achieve a more balanced character distribution.

Likewise, Monotremata is the group of living mammals that includes creatures with the strangest combination of features (Shaw 1799; Collins 1802; Hall 1999). When the first specimens arrived in Europe, scientists regarded them with suspicion and considered them as monstrous impostures made by artful artisans (Burrell 1927; Eco 1999). The absence of nipples and uterus (or in fact, the presence of oviducts opening separately like in birds and reptiles) questioned its mammalian nature (Home 1802), a topic that was also addressed by seeing platypuses lying on eggs (e.g., Lesson 1839—oviparity was only confirmed in 1884 by the Scottish zoologist William Hay Caldwell who traveled to Australia in order to discover eggs of echidna and platypus, with the help of Australian native people; Caldwell 1885; Hall 1999). In addition, the general aspect of the creatures, with a duck beak, webbed feet, fur, an amphibious lifestyle (Collins 1802), and the presence of a poison spur upon the hindlimb of the male platypuses (Burrell 1927), certainly challenged naturalists and enlightenment intellectuals of the time. George Shaw from the Department of Natural History of the Modern Curiosities of the British Museum, who scientifically named the platypus as *Platypus anatinus* (meaning "flat-foot duck-like"), marvelously described it as: *Of all the Mammalia yet known it seems the most extraordinary in its conformation; exhibiting the perfect resemblance of the beak of a Duck engrafted on the head of a quadruped. So accurate is the similitude, that, at first view, it naturally excites the idea of some deceptive preparation by artificial means; the very epidermis, proportions, serratures, manner of opening, and other particulars is the beak of a shoveler, or other broad-billed species of duck, presenting themselves to the view; nor is it without the most minute and rigid examination that we can persuade ourselves of its being the real beak or snout of a quadruped* (Shaw 1799). In brief, Thomas Bewick indicated that platypuses *possess a threefold nature, that of a fish, a bird, and a quadruped, and* [are] *related to nothing that we have hitherto seen* (Bewick 1800). More than two centuries later, monotremes still awake the interest of researchers worldwide.

Living monotremes are known by two families and five species: Ornithorhynchidae, represented by the Australian platypus *Ornithorhynchus anatinus*, and Tachyglossidae, represented by the small short-beaked echidna *Tachyglossus aculeatus*, and the larger long-beaked echidnas *Zaglossus* that includes the species *Z. bruijnii* (western New Guinea), *Z. bartoni* (central and eastern New Guinea), and *Z. attenboroughi* (restricted to the Cyclops Mountains in north New Guinea) (e.g., Flannery and Groves 1998; Musser 2003; Helgen et al. 2012; Bino et al. 2019).

The monotreme fossil record is not abundant but it has considerably increased in the last few decades, in particular, that from the Australian fossil sites (e.g., Musser 2003, 2013). Known fossil echidnas include *Megalibgwilia* ("*Zaglossus*") *robustus* from a Middle Miocene (13–14 Ma) lead golden mine from New South Wales and the Pleistocene *Megalibgwilia oweni*, "*Zaglossus*" *hacketti*, and fossil material of the current species *Zaglossus* cf. *Z. bruijnii* and *Tachyglossus aculeatus* (e.g., Griffiths et al. 1991; Musser 2003; Helgen et al. 2012). An outstanding fossil monotremes is the exquisitely preserved skull, in addition to isolated dentition and fragmentary lower jaws, of the ornithorhynchid *Obdurodon dicksoni* collected from the Riversleigh World Heritage Area in Queensland, from Early–Middle Miocene (20–15 Ma) deposits (Archer et al. 1992, 1993; Musser and Archer 1998). The skull morphology of *O. dicksoni* immediately recalls that of a living platypus, with a large bill, but with permanent teeth in the adults. This locality also produced the similar *Obdurodon tharalkooschild* based on an isolated lower molariform. This later species represents the largest ornithorhynchid known so far (Pian et al. 2013). Isolated teeth from older Oligocene deposits (~25 Ma) at the Lake Eyre region of central Australia have also provided material corresponding to *Obdurodon insignis* and a second unnamed species (*Obdurodon* sp. A) (Woodburne and Tedford 1975; Archer et al. 1978; Woodburne et al. 1993).

The Mesozoic record is scarcer but provides significant data on earlier monotremes (or perhaps alternatively, from stem monotreme taxa; Fig. 4.1a). A conclusive phylogenetic position for the putative Mesozoic monotremes is still pending and it is bounded to the as yet unresolved relationships between platypuses and echidnas (e.g., Rowe et al. 2008; Phillips et al. 2009a, b; Pian et al. 2016).

The record of the putative Mesozoic monotremes includes three taxa from the Early Cretaceous of Australia: *Teinolophos trusleri*, *Steropodon galmani*, and possibly *Kollikodon ritchiei* (but see Musser 2005; Musser et al. 2019). *Teinolophos trusleri* was discovered in Aptian (~115 Ma) deposits from the Flat Rocks locality, Victoria (Rich et al. 2001b, 2016). This is the same locality producing the ausktribosphenians *Ausktribosphenos nyktos* and *Bishops whitmorei* (Rich et al. 1997, 1999, 2001a). *Teinolophos trusleri* was originally based on a lower jaw with a single molar, but later, 14 partial lower jaws and a single isolated upper premolar were referred to this taxon (Rich et al. 2016). *Teinolophos* was initially described as a eupantothere (Rich et al. 1999), but later it was re-interpreted as a monotreme (Rich et al. 2001b, 2016). The dentary of *Teinolophos* is long and deep, but its mesial region is elongated and thin as it is in extant monotremes. However, it differs from them in having a deep dentary with seven antemolar teeth, which are separated by discrete diastemata. Based on known specimens, the dental formula is I?/i2, C?/c1, P?/p4, M?/m5 (Rich et al. 2016). The only known upper tooth has long roots, suggesting a deep vertical snout, different from the flat-bill Cenozoic and living ornithorhynchids (Rich et al. 2016). The most recent review of the *Teinolophos* jaw morphology (Rich et al. 2016) concluded that a rod of postdentary bones attached to the jaw was lacking in the adult of this species, as suggested earlier by Bever et al. (2005), Rougier et al. (2005), and Rowe et al. (2008) (*contra* Rich et al. 2005a, b). The authors inferred that *Teinolophos* would have an intermediate condition, in which the middle ear ossicles retained a connection to a persisting Meckel's cartilage although not to the dentary (Rich et al. 2016). This intermediate condition is similar to what can be seen in developing monotremes (Zeller 1989).

*Steropodon galmani* comes from the Aptian–Albian (~110 Ma) deposits at Lightning Ridge, New South Wales (Archer et al. 1985), known by a fragmentary lower jaw with dentition. The material of *Steropodon* was the first record of Mesozoic mammals in Australia. The teeth of *Steropodon* recall not only those of the ornithorhynchid *Obdurodon* but also that of *Teinolophos* and to some degree those of tribosphenic therians (Luo et al. 2001, 2002; Musser 2003), which suggest that resemblances are based on plesiomorphies. The similarities of the dentitions of fossil monotremes with those of therians raise the possibility that a tribosphenic molar pattern, or something morphologically similar, could be present in the last common ancestor of mammals and constitute the primitive morphology from which the disparate distinctive morphologies of the major groups of Mesozoic mammals derive. A few current tree topologies make this idea viable, but most do not optimize a "tribosphenic" molar as ancestral to the successive taxa along the therian stem lineage.

*Kollikodon ritchiei*, another peculiar mammal from Lightning Ridge, was originally described as a monotreme (Flannery et al. 1995). The species is currently known by a partial right dentary with three bizarre molars and a fragment of right maxilla with the last premolariform and four bunodont molariforms (Flannery et al. 1995; Pian et al. 2016). Inclusion of this taxon in a broad-sampled data matrix by Pian et al. (2016) placed *Kollikodon* as a successive sister taxon to *Teinolophos*, *Steropodon*, plus platypus and echidnas, or alternatively, in an unresolved trichotomy with *Teinolophos* and other monotremes. These results suggest that the monotreme nature of *Kollikodon* (and the other putative Mesozoic monotremes) is at least ambiguous. Furthermore, less than 6% of the characters were scored for *Kollikodon* in that study and the primary homologies of several dental structures require assumptions difficult to corroborate. As such, the monotreme affinities for *Kollikodon* are in our view controversial, and alternative hypotheses (e.g., Musser 2005; Musser et al. 2019 who suggest allotherian affinities) should be entertained as viable.

Monotremata were thought to be restricted to the Australian continent, an iconic product of island isolation. However, this traditional view was challenged in the eighties by the finding of *Monotrematum sudamericanum* in the Paleocene Punta Peligro fossil site, eastern Patagonia (Pascual et al. 1992a, b, 2002). The Patagonian taxon is the sole monotreme recovered outside Australia. The finding opened alternatives for the biogeography of the group. Were Paleocene monotremes from Patagonia immigrants from Australia/Antarctica? Was *Monotrematum* a surviving member of a Mesozoic stock present in Patagonia? Were monotremes Gondwanan in distribution during the Mesozoic? The presence of henosferids in the Jurassic of Patagonia lends support to an older history of the group in the continent, predicting a missing record of monotremes and their stem lineages in SA.

## 4.2 Systematics

Mammalia Linnaeus 1758
Australosphenida Luo et al. 2001
Henosferidae Rougier et al. 2007
*Asfaltomylos* Rauhut et al. 2002

Type species: *Asfaltomylos patagonicus* Rauhut et al. 2002.

Included species: The type only.

*Asfaltomylos patagonicus* Rauhut et al. 2002
(Fig. 4.2)

Holotype: MPEF-PV 1671, left dentary with roots and crown fragments of the last three premolars and m1–m3 (Fig. 4.2).

Locality and horizon: Queso Rallado locality, about 3 miles northwest of the Cerro Cóndor village, Chubut Province, Patagonia, Argentina. Cañadón Asfalto Formation, Toarcian–Bajocian? (Lower to Middle Jurassic).

Diagnosis (from Rauhut et al. 2002; Martin and Rauhut 2005): The lower dental formula is I?/i?, C?/c?, P?/p?5, M?/m3. All preserved teeth are double-rooted. Molars have fully basined talonids, with hypoconid and hypoconulid (broken on M1). Trigonids are lingually open: angle of trigonid is obtuse in m1 (~130°) and acute in m2–m3 (~80°). Molars have faint lingual cingulids at the base of the paraconid. There is no distal metacristid on talonids. The dentary is slender with a gently rising coronoid process. The angular process is posteroventrally positioned and the dental foramen is placed anterior to the origin of the coronoid process. There is no well-delimited pterygoid fossa. A distinct postdentary trough is present, subdivided by longitudinal striations. Meckel's groove is not well-preserved in the holotype, but there is a prominent postdentary through. *Asfaltomylos* differs from *Peramus* and other pretribosphenic stem therians by the presence of a broad, fully basined talonid with wear on the hypoconid and hypoconulid, lack of distal metacristid, anterior position of the mandibular foramen, presence of postdentary trough, and lack of pterygoid fossa. Differs from Theria, and *Holoclemensia* and *Pappotherium*, by the presence of a lingually open trigonid. Important characters shared with australosphenidans are the presence of a lingual cingulid at the base of the paraconid (the wrapping cingulid of Luo et al. 2001) and talonids which are wider than long. Differs from *Ambondro*, from

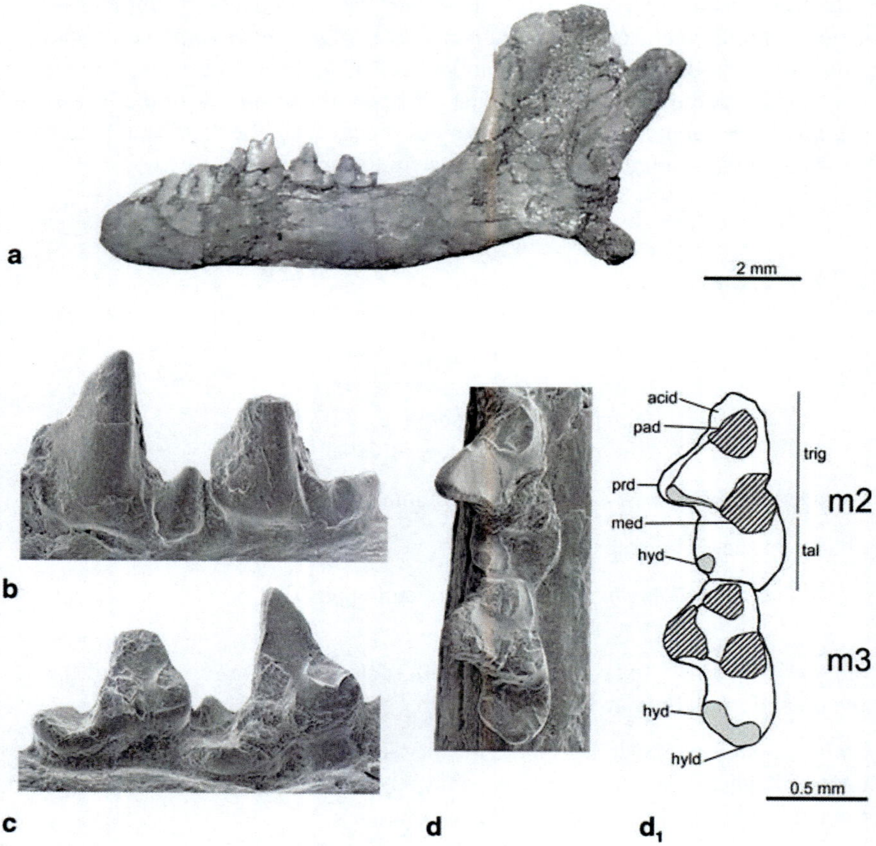


**Fig. 4.2** *Asfaltomylos patagonicus* from Queso Rallado locality, Chubut, Argentina; Cañadón Asfalto Formation, Toarcian–Bajocian? (Lower to Middle Jurassic). Holotype (MPEF-PV 1671), left dentary in labial view (**a**); detail of the last molars in labial (**b**), lingual (**c**), and occlusal (**d**) views, and accompanying line drawing (**d**$_1$). Abbreviations: acid, anterior cingulid; hyd, hypoconid; hyld, hypoconulid; m2–m3, lower molars; med, metaconid; pad, paraconid; prd, protoconid; tal, talonid; trig, trigonid. Grey pattern indicates wear facets following Martin and Rauhut (2005), oblique lines indicate breakage. **b–d** modified from Martin and Rauhut (2005)

the Middle Jurassic of Madagascar, by the lack of a distal metacristid and a weaker lingual cingulid, which does not wrap around the mesial side. The Early Cretaceous australosphenidan *Ausktribosphenos*, from Australia, is characterized by a number of distinctive autapomorphic features, such as additional cristids on the metaconid and talonid, which are not present in *Asfaltomylos* or *Henosferus*. Differs from the Early Cretaceous *Bishops* through the lack of additional cristids in the talonid, a weaker lingual cingulid, and a more slender protoconid. *Asfaltomylos* is about twice smaller than *Henosferus*, with a proportionately taller horizontal ramus of the dentary and coronoid process, less-defined medial ridge on the dentary, longer retromolar space, smaller diastemata between premolars, and stronger lingual cingulid that extends from the paraconid to the metaconid in the molars.

Comments: The upper molars of *Asfaltomylos patagonicus* are still unknown. In order to infer the morphology of the upper elements and how they work, Martin and Rauhut (2005) studied the wear in the lower molars of *Asfaltomylos*. Wear covers the apices of hypoconid and hypoconulid, the hypocristid, and the distal part of the cristid obliqua but no wear facets were detected on the talonid basin. The pattern of *Asfaltomylos* resembles other australosphenidans (*Ambondro* and *Ausktribosphenos*), including toothed monotremes (*Obdurodon*). A similar conclusion was reached by Davis (2011) arguing for the absence of a fully functional protocone occluding in the talonid basin. Standard wear facets of tribosphenic molars, as defined by Crompton (1971) and refined by Davis (2011), are absent. The authors concluded that in addition to a possible protocone, other structures in the upper molar, such as ridges or crests in the middle or lingual part of the molar, could be present to produce the wear on apices of the hypoconid, hypoconulid, and hypocristid (Martin and Rauhut 2005). We, however, regard more likely that a basined talonid as the one in both *Asfaltomylos* and *Henosferus* was functionally partially or fully tribosphenic (see below).

*Henosferus* Rougier et al. 2007

Type species: *Henosferus molus* Rougier et al. 2007.

Included species: The type only.

*Henosferus molus* Rougier et al. 2007
(Figs. 4.3, 4.4, 4.5 and 4.6)

Holotype: MPEF 2353: Right lower jaw with well-preserved dentary, bearing fragmentary p1, p2, and almost complete m1 (Fig. 4.3a).

Locality and horizon: Queso Rallado locality, about 3 miles northwest of the Cerro Cóndor village, Chubut Province, Patagonia, Argentina. Cañadón Asfalto Formation, Toarcian–Bajocian? (Lower to Middle Jurassic).

Diagnosis (from Rougier et al. 2007): The lower dental formula is I?/i4, C?/c1, P?/p5, M?/m3. Diastemata between most premolars variously developed. Molars with an obtuse to straight trigonid and a basined talonid with two well-developed cusps and a ridge-like structure in the position of entoconid/entocristid; paraconid procumbent and placed in a more labial position than metaconid; talonid slightly wider and much lower than trigonid; talonid wider than long; blunt prominent hypoconid not fully differentiated from broad, bulbous hypoconulid connected by a broad, low crest; talonid lingually closed by a strong rounded entocristid, well-developed lingually. Slender lower jaw having a Meckelian groove, prominent medial flange associated with a lateral ridge, and deep postdentary trough; and prominent, transversely wide, spoon-like angular process with a medial crest occupying a position homologous to the pterygoid crest of other mammals.

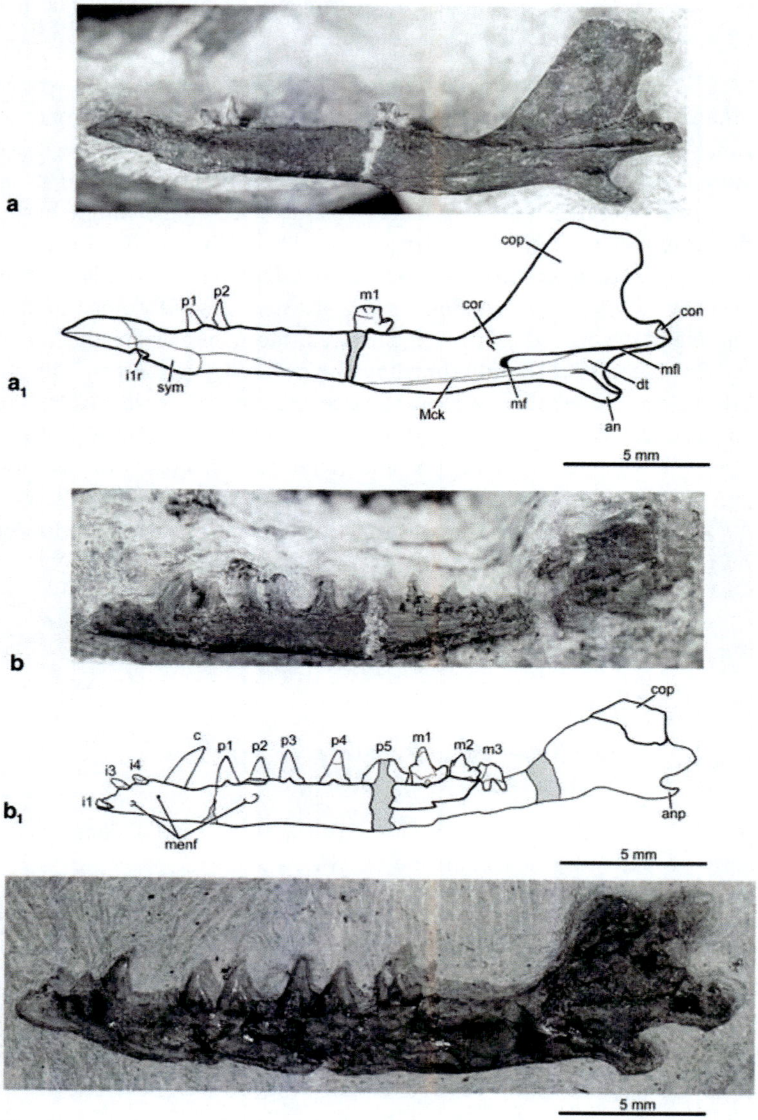


**Fig. 4.3** *Henosferus molus* from Queso Rallado locality, Chubut, Argentina; Cañadón Asfalto Formation, Toarcian–Bajocian? (Lower to Middle Jurassic). Holotype (MPEF-PV 2353), right dentary in lingual view (**a**) and accompanying line drawing (**a**$_1$); MPEF-PV 2354, left dentary in labial view (**b**) and accompanying line drawing (**b**$_1$); MPEF-PV 2357, left dentary in labial view (**c**). Abbreviations: anp, angular process; c, lower canine; con, condyle; cop, coronoid process; cor, scar for the paradentary coronoid bone; dt, postdentary trough; i1–i3, lower incisors; i1r, root of the lower first incisor; m1–m3, lower molars; Mck, Meckelian groove; menf, mental foramina; mf, mandibular foramen; mfl, medial flange; p1–p5, lower premolars; sym, mandibular symphysis. Grey pattern indicates broken bone and matrix. Modified from Rougier et al. (2007)

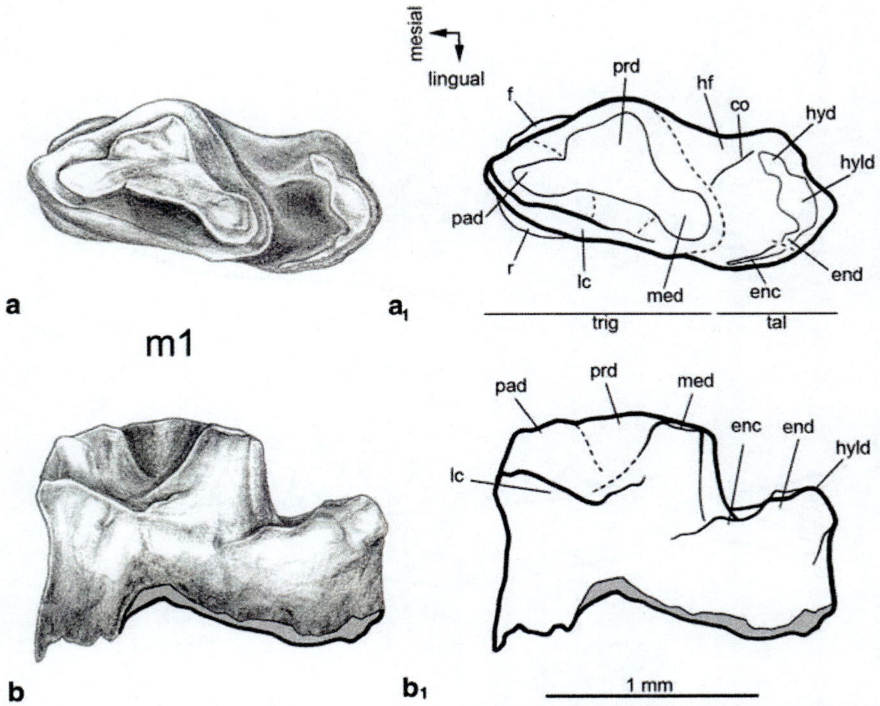


**Fig. 4.4** *Henosferus molus* from Queso Rallado locality, Chubut, Argentina; Cañadón Asfalto Formation, Toarcian–Bajocian? (Lower to Middle Jurassic). First molar (m1) (MEFP 2353) in occlusal (**a**) and lingual (**b**) views, with accompanying line drawings ($\mathbf{a}_1$, $\mathbf{b}_1$). Abbreviations: co, cristid obliqua; enc, entocristid; end, entoconid; f, cingular cuspule f; hf, hypoflexid; hyd, hypoconid; hyld, hypoconulid; lc, lingual cingulid; med, metaconid; pad, paraconid; prd, protoconid; r, root; tal, talonid; trig, trigonid

Comments: The well-preserved specimens of *Henosferus molus* show more evidence than other ausktribosphenidans in the plesiomorphic features of the lower jaw (Fig. 4.3). It has a Meckelian groove, a prominent medial flange associated with a lateral ridge of the dentary, and a deep postdentary trough (Fig. 4.3a). It was interpreted that *Henosferus* retained a full complement of postdentary bones, in a generalized mammaliaform arrangement, but reduced from the condition in *Morganucodon* (e.g., Kermack et al. 1973) and other early mammaliaforms (Fig. 4.5). The ear structures in both henosferids were still essentially mandibular in nature (Rougier et al. 2007) and as long as they are recovered as mammals, the free middle ear ossicles of monotremes, marsupials, and placentals will optimize as convergent at least once in any possible tree topology. The m1 of the holotype shows a pattern with a trigonid and a fully basined talonid to which tribosphenic terminology can be applied (Fig. 4.4).

*Henosferus* is about twice the size of *Asfaltomylos*, with large diastemata between premolars, low horizontal ramus of dentary, well-defined medial ridge on dentary, low coronoid process with the anterior edge less vertical, lower condyle, shorter

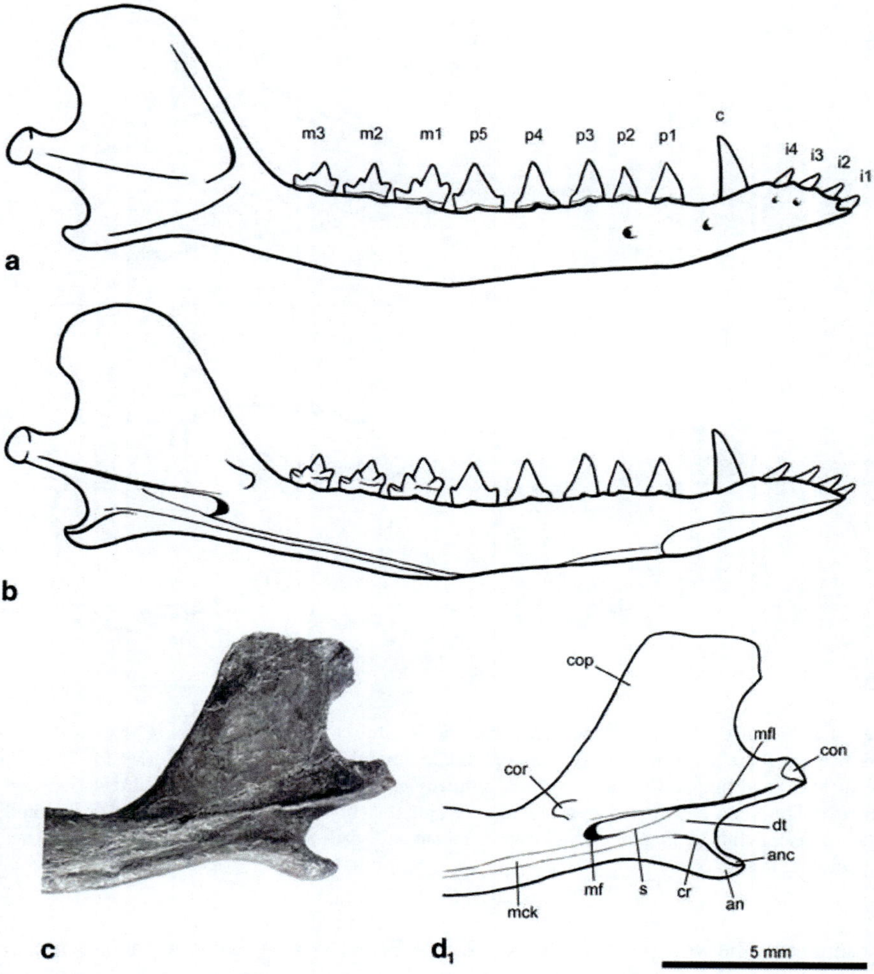


**Fig. 4.5** *Henosferus molus* from Queso Rallado locality, Chubut, Argentina; Cañadón Asfalto Formation, Toarcian–Bajocian? (Lower to Middle Jurassic). Reconstruction of the right lower jaw in labial (**a**) and lingual (**b**) (inverted) views. Holotype (MPEF-PV 2353), detail of posterior portion of the right lower jaw in lingual (**c**) view and accompanying line drawing ($\mathbf{c}_1$). Abbreviations: an, angular process; anc, concave surface of the angular process; con, condyle; cop, coronoid process; cor, scar for the paradentary coronoid bone; cr, medial crest of the angular process; dt, dentary trough; i1–i4, lower incisors; m1–m3, lower molars; Mck, Meckelian groove; mf, mandibular foramen; mfl, medial flange; p1–p5, lower premolars; s, step may be homologous to the "diagonal ridge" of *Morganucodon* (Kermack et al. 1973)

**Fig. 4.6** *Henosferus molus*, artistic reconstruction by Jorge L. Blanco

retromolar space, distal metacristid on m1–m2 very weak in *Henosferus* and absent in *Asfaltomylos*, and weaker lingual cingulid than in *Asfaltomylos* (Rougier et al. 2007) (Fig. 4.5). *Henosferus* (Fig. 4.6) is more abundant than *Asfaltomylos* in the Jurassic Cañadón Asfalto Formation.

Monotremata Bonaparte 1837
Ornithorhynchidae Gray 1825
*Monotrematum* Pascual et al. 1992b

Type species: *Monotrematum sudamericanum* Pascual et al. 1992b.

Included species: The type only.

*Monotrematum sudamericanum* Pascual et al. 1992b
(Figs. 4.7 and 4.8)

Holotype: MLP 91-I-1-1, isolated right M2 (Fig. 4.7b).

Locality and horizon: Punta Peligro, southeastern Chubut Province (central Patagonia), Argentina; Salamanca Formation; Hansen Member ("Banco Negro Inferior"), Lower Paleocene (Peligran SALMA, Danian).

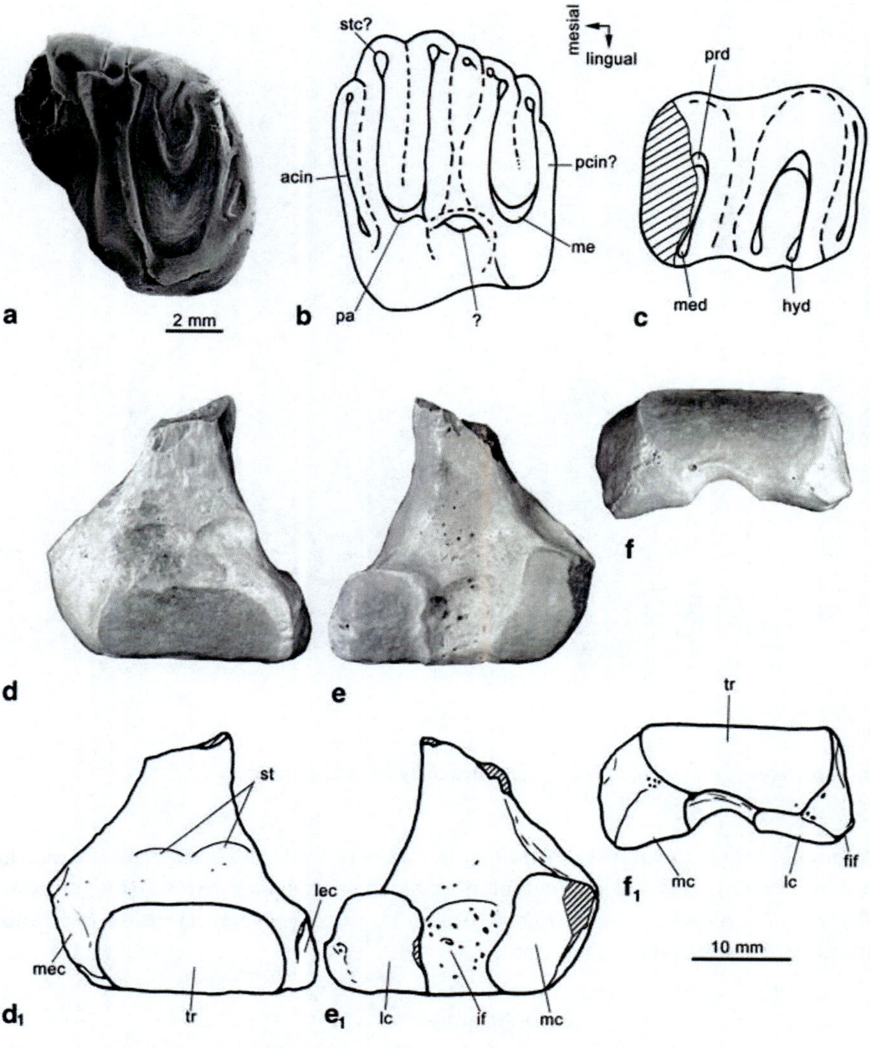


**Fig. 4.7** *Monotrematum sudamericanum* from Punta Peligro locality, Chubut, Argentina; Salamanca Formation, Danian, Lower Paleocene. MPEF-PV 2282, isolated probable left M2 (**a**); line drawing reconstruction of the holotype (MLP 91-I-1-1), isolated right M2 (based on Pascual et al. 1992b), reversed (**b**); line drawing reconstruction of MPEF-PV 1635, isolated right m1 (based on Pascual et al. 2002), reversed (**c**); MACN-CH 1888, fragmentary left femur in anterior (**d**), posterior (**e**), and distal (**f**) views, and accompanying line drawings (**d**$_1$–**f**$_1$), based on Forasiepi and Martinelli (2003). Abbreviations: acin, anterior cingulum (= precingulum); fif, articular facet for the fibula; hyd, hypoconid; if, intercondylar fossa; lc, lateral condyle; lec, lateral epicondyle; mc, medial condyle; me, metacone; mec, medial epicondyle; med, metaconid; pa, paracone; pcin?, posterior cingulum (= postcingulum); prd, protoconid; st, supratrochlear depression; stc?, stylocone; tr, trochlea;?, cusp of uncertain homology. Cusp homologies based on Luo et al. (2002); in particular, cusp indicated with "?" in (**a**) was referred to as the "protocone", but this homology is uncertain (see text). Oblique lines indicate breakage

**Fig. 4.8** *Monotrematum sudamericanum*, artistic reconstruction by Jorge L. Blanco

Diagnosis (from Pascual et al. 1992b, 2002): Cheek tooth pattern similar to that of *Obdurodon*, the most closely similar ornithorhynchid, but double in size. It differs from *Steropodon galmani*, in being much larger in all comparative dimensions. It differs from *Obdurodon insignis* and *O. dicksoni* in having a more bulbous mesiolingual cusp (paracone) on M2. The posterior lobe of m1 is single-rooted, but preserving a vertical anterior sulcus that apparently divided two original roots, which remain well-separated in *Obdurodon* spp. It differs from *Ornithorhynchus anatinus* in having well-developed teeth and structures of crown morphology, as the molars of *O. anatinus* are vestigial.

Comments: Before the finding of *Monotrematum sudamericanum* in the Paleocene outcrops of Patagonia (Pascual et al. 1992a), Monotremata were thought to be restricted to the Australian continent (Pascual et al. 1992b).

The known material of *Monotrematum* provides evidence on the evolution of the dentition in ornithorhynchids, otherwise ontogenetically lost in (adult) living platypuses (e.g., Simpson 1929; Luckett and Zeller 1989). The molar of *Monotrematum* is morphologically very similar to that of *Obdurodon* spp. from the late Oligocene–middle Miocene of Australia (Woodburne and Tedford 1975; Archer et al. 1978, 1992, 1993; Musser and Archer 1998; Pian et al. 2013). In occlusal view, both upper and lower molars have lingual and buccal cusp rows; cusps are more numerous in the buccal row of the upper molars and in the lingual row of the lowers, and a transverse subparallel blade system connects some of the labial and lingual cusps (Archer et al. 1992, 1993; Pascual et al. 1992b, 2002). On the basis of tooth similarities with *Obdurodon*, several authors have argued that *Monotrematum*, together with *Obdurodon*, are archaic toothed platypuses (Ornithorhynchidae) (Pascual et al. 1992b; Musser 1999); molecular evidence suggests a much more recent split for the last common ancestor of living monotremes (Phillips et al. 2009a, b; see also Camens 2010).

The homology of the main molariform cusps in monotremes and by extension in some australosphenidans is problematic. Initial assessment of cusp homologies of the highly modified deciduous molariforms of platypuses with those of later therians was only partially helpful (Simpson 1929). Nonetheless, Simpson's detailed and masterful comparison of *Ornithorhynchus* with most major dental types known at the time contains some shrewd observations: (1) there is evidence of a regular crown pattern with a small number of cusps; (2) the upper and lower teeth are of a similar pattern but reversed; (3) the outer side of the lower teeth and the inner side of the upper teeth are more prominent; (4) the teeth interlock and overlap in occlusion. The extension of these conclusions into cusp homologies was hampered by Simpson's attachment to the then traditional view that the primitive upper molariform cusp was a homologue of the protocone (Osborn 1897, 1907; Simpson 1961) instead of the paracone (Butler 1939; Patterson 1956), an idea Simpson would only consider possible very late in his career (Simpson 1971). It was not until much later with the recognition of toothed monotremes (Woodburne and Tedford 1975; Pascual et al. 1992a, b, c; Archer et al. 1993) that a pattern of reversed triangles was broadly accepted for monotreme dentition. The homology of the lower molariform mesial triangle is fairly universally accepted and formed by a buccal protoconid, a mesiolingual paraconid, and a distolingual metaconid. In their influential review of Mesozoic mammals, Luo et al. (2002) further recognized the distal half of the molariform as having a bicuspid talonid formed by the hypocone and hypoconulid. In their view, the upper molar has all three major tribosphenic cusps, robust paracone, metacone, and a small, but functional, protocone. We agree with the interpretation of the trigonid and regard the talonid in monotremes as formed by the primitive talonid cusp in a distolingual position, the remaining structures of the talonid being neomorphs (Fig. 4.7c). The suggestion of Luo et al. (2002) of a protocone in the upper molariforms was quickly challenged (Woodbourne 2003; Woodburne et al. 2003) with the purported protocone regarded as a cingular cusp, a view we share but that is often left as an unresolved matter and scored as "?" in phylogenetic matrices (see Krause et al. 2014, 2020) or considered to be absent (Rougier et al. 2011, 2012; Averianov and Lopatin 2014), as we are inclined to do here. Therefore, in our view the upper molar of the monotremes is highly derived, with prominent paracone and metacone, a distinct stylocone, and stylar cusps. The stylar shelf is divided into a mesial and a distal portion by sharp crests stemming from the paracone and metacone. The protocone is absent and there are well-developed mesial and distal cingula (Fig. 4.7a, b). The lower molars show a primitive trigonid with protoconid, paraconid, and metaconid; in some teeth, the paraconid is missing (Fig. 4.7c). The talonid is not basined and the primitive talonid cusp is of distolingual position (Fig. 4.7c).

*Monotrematum* has relatively thick prismatic enamel, with hexagonal packing, substantial interprismatic area, and a wide outer aprismatic zone (Wood and Rougier 2005). The prism shape and packing suggest a plesiomorphic condition widely present among Mesozoic groups (Wood and Rougier 2005). The reduction of mastication in the toothed monotremes seems to be mirrored by the simplification and eventual loss of prismatic structure in the enamel of younger platypuses (Wood and Rougier 2005).

A small facial portion of a skull from the BNI, Punta Peligro, was tentatively assigned to *Monotrematum sudamericanum*. The material was presented in abstract format only (Carlini et al. 2002), pending formal publication. It consists of a fragmentary maxilla with four roots of the postcanine dentition. The presence in the maxilla of a large infraorbital foramen and multiple apertures suggest hypertrophy of the trigeminal system which permits the existence of an incipient system of electric receptors, perhaps similar to Ornithorhynchidae (Carlini et al. 2002), which may be ancestral to monotremes as a whole (Musser 2013; Rich et al. 2016).

Additionally, two fragmentary distal femora have been assigned to *Monotrematum sudamericanum* (Fig. 4.7d–f) (Forasiepi and Martinelli 2003). Measurements compared to *Ornithorhynchus anatinus* suggested a total length of the animal of ~700 mm. Based on the molar size, the species *Obdurodon tharalkooschild* from the Middle Miocene of Riversleigh is the largest Ornithorhynchidae, slightly larger than *Monotrematum* (Pian et al. 2013). However, the tachyglossid *Zaglossus hacketti* is the largest Monotremata so far (Forasiepi and Martinelli 2003).

At present, it is uncertain if *Monotrematum* is a post-K/Pg immigrant from Australia-Antarctica, a Cretaceous immigrant from Australia-Antarctica, or an example of a surviving local Mesozoic lineage. Paleocene immigration into South America seems unlikely given that nothing substantive has changed in the paleogeography of the Southern Atlantic (Rabinowitz and LaBrecque 1979; Setoyama and Kanungo 2020). It is to be expected that monotremes were part of the Mesozoic communities of SA, but if their origin is related to henosferids, either an Australian, Antarctic, or South American origin is viable.

## 4.3 Concluding Remarks

South American australosphenidans are quite dissimilar from each other in morphology, size, and likely ecology; they range in time from early in the Jurassic to the Paleocene. On the one hand, the odd henosferids show a surprising combination of derived dentition and plesiomorphic looking jaws (Rauhut et al. 2002; Martin and Rauhut 2005; Rougier et al. 2007), on the other, the few remains of Paleocene monotremes appear to be morphologically similar to toothed Miocene platypuses but retaining a primitive enamel microstructure in their molars (Pascual et al. 1992a, b, 2002; Forasiepi and Martinelli 2003; Wood and Rougier 2005).

The henosferids are currently the oldest undisputed members of Mammalia with an age expected to be around 176 Ma (Toarcian), according to the youngest date for the Cañadón Asfalto Formation (Cúneo et al. 2013). The basal mammalian position for henosferids reflects a tree topology dominated by the primitive mandibular characters (Rauhut et al. 2002; Rougier et al. 2007; Huttenlocker et al. 2018) and by inference a primitive middle ear not too dissimilar to that of basal mammaliaforms

(Rougier et al. 2007; Meng et al. 2016, 2020; Wang et al. 2019). However, the alternative position of australosphenidans at the base of Theria is a rearrangement not too costly from a parsimony point of view in most analyses (Rougier et al. 2007) and recovered in others (Krause et al. 2014 and studies deriving from it). In fact, Flynn et al. (1999) recovered the australosphenidan *Ambondro*, from the Middle Jurassic of Madagascar, as an unresolved generalized tribosphenic mammal on the stem therian lineage. Implications for the origin of the three major groups of living mammals rest on the resolution of the seeming conflict between the phylogenetic signal of dentition, on the one hand, and lower jaw/middle ear morphology on the other. If henosferids and *Ambondro* are part of the stem lineage of Monotremata, reverse triangular molars are likely basal for mammals; even if the tribosphenic nature of the teeth is questioned, the homology of the trigonid and its likely upper opposite triangle can be seen as foundational features of Mammalia. A corollary of this view, dependent in part on the tree topology defended, is that some Mesozoic groups with molars with cusps in line, forming an obtuse angle, or with multicuspidated teeth, likely derive from ancestors where the main molar cusps form a well-developed triangle. Multituberculates, euharamiyidans, "triconodonts", and other mammals with obtuse angle molariforms would potentially derive from a configuration similar to henosferids. Teeth buccolingually broad engage in a wide range of mediolateral translation and flatter masticatory orbit (Crompton 1971; Davis 2011; Schultz and Martin 2014; Schultz et al. 2018). The increased medial component on these dentitions suggests the presence of sophisticated pterygoid and masseteric musculature, similar to that present in therians (Crompton and Hiiemae 1970; Crompton and Sita-Lumsden 1970; Lautenschlager et al. 2017; Bhullar et al. 2019).

The triangular arrangement of molariform cusps is present in several non-mammalian groups such as docodonts (Butler 1988; Luo and Martin 2008; Schultz et al. 2019), *Kuehneotherium,* and allies, among others (Kermack et al. 1968; Newham et al. 2020), even though a non-mammalian physiology has been argued (Newham et al. 2020). The homology of the three main cusps in the lower molar, the protoconid, paraconid, and metaconid of a therian molar, can be traced beyond Mammalia (Patterson 1956). It appears likely that not only these cusps, but also some degree of triangulation are primitive for the last common ancestor of mammals. Henosferids would further elaborate on that ancestral morphology.

The alternative view is that the last common ancestor of therians, monotremes, and henosferids had cusps in lines, and the detailed similarities between toothed monotremes, henosferids, and *bona fide* tribosphenic mammals are all convergent. We recognize current phylogenies render the talonid of henosferids and monotremes by and large as not homologous with the therian counterpart. The homology of the trigonid, however, depends on character optimization dictated by the arrangement of taxa like "triconodonts" and allotherians; an ancestral triangular arrangement of the main cusps is viable for them under certain tree topologies. During proof reading, two new australosphenidans from the Early Cretaceous of Australia have been described, including a toothed-monotreme (Rich et al. 2020a) and an ausktribosphenid (Rich et al. 2020b). The first was named *Stirtodon elizabethae*, based on an isolated upper premolar that represents the largest known Mesozoic monotreme (i.e., premolar about twice the size of the P2 of the Cenozoic *Obdurodon dicksoni*; Rich et al. 2020a). The ausktribosphenid was named *Kryoparvus gerriti* on the basis of a right dentary with partial dentition, which is about two-third smaller (Rich et al. 2020b) than the other Australian ausktribosphenids (*Ausktribosphenos nyktos* and *Bishops whitmorei*; Rich et al. 1997, 2001a).

## References

Archer M, Plane MD, Pledge NS (1978) Additional evidence for interpreting the Miocene *Obdurodon insignis* Woodburne and Tedford, 1975, to be a fossil platypus (Ornithorhynchidae: Monotremata) and a reconsideration of the status of *Ornithorhynchus agilis* De Vis, 1885. Aust Zool 20:9–27

Archer M, Flannery TF, Ritchie A, Molnar RE (1985) First Mesozoic mammal from Australia, an Early Cretaceous monotreme. Nature 318:363–366

Archer M, Jenkins FA, Hand SJ, Murray P, Godthelp H (1992) Description of the skull and non-vestigial dentition of a Miocene platypus (*Obdurodon dicksoni* n sp.) from Riversleigh. In: Augee ML (ed) Platypus and echidnas. The Royal Society of New South Wales, Sydney, pp 15–27

Archer M, Murray P, Hand SJ, Godthelp H (1993) Reconsideration of monotreme relationships based on the skull and dentition of the Miocene *Obdurodon dicksoni*. In: Szalay FS, Novacek MJ, McKenna MC (eds) Mammal phylogeny: Mesozoic differentiation, multituberculates, monotremes, early therians, and marsupials. Springer-Verlag, New York, pp 30–44

Averianov AO, Lopatin AV (2014) On the phylogenetic position of monotremes (Mammalia, Monotremata). Paleontol J 48:426–446

Bever GS, Rowe T, Ekdale EG, Macrini TE, Colbert MW, Balanoff AM (2005) Comment on independent origins of middle ear bones in monotremes and therians. Science 309:1432a

Bewick T (1800) A general history of quadrupeds. Edward Walker, Newcastle-upon-Tyne

Bhullar B-AS, Manafzadeh AR, Miyamae JA, Hoffman EA, Brainerd EL, Musinsky C, Crompton AW (2019) Rolling of the jaw is essential for mammalian chewing and tribosphenic molar function. Nature 566:528–532

Bino G, Kingsford RT, Archer M, Connolly JH, Day J, Dias K, Goldney D, Gongora J, Grant T, Griffiths J, Hawke T, Klamt M, Lunney D, Mijangos L, Munks S, Sherwin W, Serena M, Temple-Smith P, Thomas J, Williams G, Whittington C (2019) The platypus: evolutionary history, biology, and an uncertain future. J Mammal 100:308–327

Bonaparte CL (1837) New systematic arrangement of vertebrated animals. Trans Linn Soc London 18:247–304

Burrell H (1927) The Platypus. Its discovery, zoological position, form and characteristics, habits, life history, etc. Angus & Robertson Limited, Sydney

Butler PM (1939) The teeth of the Jurassic mammals. Proc Zool Soc Lond B 109:329–356

Butler PM (1988) Docodont molars as tribosphenic analogues (Mammalia, Jurassic). In: Russell DE, Santoro JP, Sigogneau-Russell D (eds) Teeth Revisited: Proceedings of the VII International Symposium on Dental Morphology, Paris, Mem Mus Nat Hist Nat Paris, pp 329–340

Caldwell WH (1885) On the development of the monotremes and *Ceratodus*. J R Soc New South Wales 18:117–122

Camens AB (2010) Were Early Tertiary monotremes really all aquatic? Inferring paleobiology and phylogeny from a depauperate fossil record. Proc Natl Acad Sci USA 107:E12

Carlini AA, Ortiz Jaureguizar E, Rougier GW (2002) El primer resto craneano de Monotremata del Paleoceno temprano de Patagonia (Argentina). 1° Congreso Latinoamericano de Paleontología de Vertebrados, Santiago de Chile, Abstracts

Chimento N, Agnolín FL, Martinelli AG (2016) Mesozoic mammals from South America: implications for understanding early mammalian faunas from Gondwana. Mus Argent Cienc Nat B Rivadavia Contribuciones 6:199–209

Collins D (1802) An account of the english colony in New South Wales. A Strahan, Printers Street, London

Crompton AW (1971) The origin of the tribosphenic molar. In: Kermack DM, Kermack KA (eds) Early mammals. Zool J Linn Soc 50:65–87

Crompton AW, Hiiemae KM (1970) Molar occlusion and mandibular movements during occlusion in the American opossum, *Didelphis marsupialis*. Zool J Linn Soc 49:21–47

Crompton AW, Sita-Lumsden A (1970) Functional significance of the therian molar pattern. Nature 227:197–199

Cúneo R, Ramezani J, Scasso R, Pol D, Escapa IH, Zavattieri AM, Bowring SA (2013) High-precision U-Pb geochronology and a new chronostratigraphy for the Cañadón Asfalto Basin, Chubut, central Patagonia: Implications for terrestrial faunal and floral evolution in Jurassic. Gondwana Res 24:1267–1275

Davis B (2011) Evolution of the tribosphenic molar pattern in early mammals, with comments on the dual-origin hypothesis. J Mammal Evol 18:224–227

Dietrich WO (1927) *Brancatherulum* n. g., ein Proplacentalier aus dem obersten Jura des Tendaguru in Deutsch-Ostafrika. Zentralbl Mineral, Teil 4. Hist Paläontol 1927:423–426

Eco U (1999) Kant and the platypus: essays on language and cognition. Alastair McEwen Books, New York

Flannery TF, Groves CP (1998) A revision of the genus *Zaglossus* (Monotremata, Tachyglossidae), with description of new species and subspecies. Mammalia 62:367–396

Flannery TF, Archer M, Rich TH, Jones R (1995) A new family of monotremes from the Cretaceous of Australia. Nature 377:418–420

Flynn JJ, Parrish JM, Rakotosamimanana B, Simpson WF, Wyss AR (1999) A Middle Jurassic mammal from Madagascar. Nature 401:57–60

Forasiepi AM, Martinelli AG (2003) Femur of a monotreme (Mammalia, Monotremata) from the early Paleocene Salamanca Formation of Patagonia, Argentina. Ameghiniana 40:625–630

Gray JE (1825) Outline of an attempt at the disposition of the mammalia into tribes and families with the list of the genera apparently appertaining to each tribe. Ann Philos 10:336–344

Griffiths M, Wells RT, Barrie DJ (1991) Observations on the skulls of fossil and extant echidnas (Monotremata: Tachyglossidae). Aust Mammal 14:87–101

Hall BK (1999) The paradoxical platypus. Bioscience 49:211–218

Heinrich W-D (1998) Late Jurassic mammals from Tendaguru, Tanzania, East Africa. J Mammal Evol 5:269–290

Heinrich W-D, Bussert R, Aberhan M (2011) A blast from the past: the lost world of dinosaurs at Tendaguru, East Africa. Geol Today 27:101–106

Helgen KH, Miguez RP, Kohen JL, Helgen LE (2012) Twentieth century occurrence of the Long-beaked echidna *Zaglossus bruijnii* in the Kimberley region of Australia. ZooKeys 255:103–132

Home E (1802) A description of the anatomy of the *Ornithorhynchus paradoxus*. Philos Trans R Soc B 1802:67–84

Huttenlocker AK, Grossnickle DM, Kirkland JI, Schultz JA, Luo Z-X (2018) Late-surviving stem mammal links the lowermost cretaceous of North America and Gondwana. Nature 558:108–112

Kermack DM, Kermack KA, Musset F (1968) The Welsh pantothere *Kuehneotherium praecursoris*. Zool J Linn Soc 47:407–423

Kermack KA, Mussett F, Rigney HW (1973) The lower jaw of *Morganucodon*. Zool J Linn Soc 53:87–175

Kielan-Jaworowska Z, Cifelli RL, Luo Z-X (2004) Mammals from the age of dinosaurs. Origins, evolution, and structure. Columbia University Press, New York

King B, Beck RMD (2020) Tip dating supports novel resolutions of controversial relationships among early mammals. Proc R Soc Lond B 287:20200943

Krause DW, Hoffmann S, Wible JR, Kirk EC, Schultz JA, von Koenigswald W, Groenke JR, Rossie JB, O'Connor PM, Seiffert ER, Dumont ER, Holloway WL, Rogers RR, Rahantarisoa RJ, Kemp AD, Andriamialison H (2014) First cranial remains of a gondwanatherian mammal reveal remarkable mosaicism. Nature 515:512–517

Krause DW, Hoffmann S, Hu Y, Wible JR, Rougier GW, Kirk EC, Groenke JR, Rogers RR, Rossie JB, Schultz JA, Evans AR, von Koenigswald W, Rahantarisoa LJ (2020) Skeleton of Cretaceous mammal from Madagascar reflects long-term insularity. Nature 581:421–427

Lautenschlager S, Gill P, Luo Z-X, Fagan MJ, Rayfield EJ (2017) Morphological evolution of the mammalian jaw adductor complex. Biol Rev 92:1910–1940

Lesson RP (1839) Voyage autour du monde. Entrepris par ordre du gouvernement sur la Corvette La Coquille, Pourrat Frères Editeurs, Paris

Linnaeus C (1758) Systema naturae per regna tria naturae, secundum classes, ordines, genera, species, cum characteribus, differentiis, synonymis, locis. vol. 1: Regnum animale. Editio decima, reformata. Laurentii Salvii, Stockholm

Luckett WP, Zeller U (1989) Developmental evidence for dental homologies in the monotreme *Ornithorhynchus* and its systematic implications. Z Sdugetierkunde 54:193–204

Luo Z-X, Martin T (2008) Analysis of molar structure and phylogeny of docodont genera. Bull Carnegie Mus Nat Hist 39:27–47

Luo Z-X, Cifelli RL, Kielan-Jaworowska Z (2001) Dual origin of tribosphenic mammals. Nature 409:53–57

Luo Z-X, Kielan-Jaworowska Z, Cifelli RL (2002) In quest for a phylogeny of Mesozoic mammals. Acta Palaeontol Pol 47:1–78

Luo Z-X, Yuan C-X, Meng Q-J, Ji Q (2011) A Jurassic eutherian mammal and divergence of marsupials and placentals. Nature 476:442–445

Martin T, Rauhut OWM (2005) Mandible and dentition of *Asfaltomylos patagonicus* (Australosphenida, Mammalia) and the evolution of tribosphenic teeth. J Vertebr Paleontol 25:414–425

Meng J, Bi S, Zheng X, Wang X (2016) Ear ossicle morphology of the Jurassic euharamiyidan *Arboroharamiya* and evolution of mammalian middle ear. J Morph 279:441–457

Meng J, Mao F, Han G, Zheng X-T, Wang X-L, Wang Y (2020) A comparative study on auditory and hyoid bones of Jurassic euharamiyidans and contrasting evidence for mammalian middle ear evolution. J Anat 236:50–71

Musser AM (1999) Diversity and relationships of living and extinct monotremes. Aust Mammal 21:8–9

Musser AM (2003) Review of the monotreme fossil record and comparison of palaeontological and molecular data. Comp Biochem Physiol A 136:927–942

Musser AM (2005) Investigations into the evolution of Australian mammals with a focus on monotremata. Unpublished PhD thesis, University of New South Wales

Musser AM (2013) Classification and evolution of the monotremes. In: Ashwell K (ed) Neurobiology of monotremes: Brain evolution in our distant mammalian cousins. CSIRO Publishing, Collingwood, Victoria, pp 1–16

Musser AM, Archer M (1998) New information about the skull and dentary of the Miocene platypus *Obdurodon dicksoni* and a discussion of ornithorhynchid relationships. Philos Trans R Soc B 353:1063–1079

Musser AM, Lamanna MC, Martinelli AG, Salisbury SW, Ahyong S, Jones R (2019) The first non-mammalian cynodonts from Australia and the unusual nature of Australian Cretaceous continental tetrapod faunas. 79th Annual Meeting, Society of Vertebrate Paleontology, Brisbane. (Abstract:157)

Newham E, Gill PG, Brewer P, Benton MJ, Fernandez V, Gostling NJ, Haberthür D, Jernvall J, Kankanpää T, Kallonen A, Navarro C, Pacureanu A, Richards K, Brown KB, Schneider P, Suhonen H, Tafforeau P, Williams KA, Zeller-Plumhoff B, Corfe IJ (2020) Reptile-like physiology in Early Jurassic stem-mammals. Nat Commun 11:5121

Osborn HF (1897) Trituberculy: a review dedicated to the late professor Cope. Am Nat 31:993–1016

Osborn HF (1907) Evolution of mammalian molar teeth. MacMillan and Company, New York
Pascual R, Archer M, Ortiz-Jaureguizar E, Prado JL, Godthelp H, Hand SJ (1992a) First discovery of monotremes in South America. Nature 356:704–705
Pascual R, Archer M, Ortiz-Jaureguizar E, Prado JL, Godthelp H, Hand SJ (1992b) The first non-Australian monotreme: an early Paleocene South American platypus (Monotremata, Ornithorhynchidae). In: Augee ML (ed) Platypus and echidnas. The Royal Society of New South Wales, Sydney, pp 1–14
Pascual R, Ortiz-Jaureguizar E (1992) Evolutionary pattern of land mammal faunas during the Late Cretaceous and Paleocene in South America: a comparison with the North American pattern. Ann Zool Fennici 28:245–252
Pascual R, Goin FJ, Balarino L, Udrizar Sauthier DE (2002) New data on the Paleocene monotreme *Monotrematum sudamericanum*, and the convergent evolution of triangulate molars. Acta Palaeontol Pol 47:487–492
Patterson B (1956) Early Cretaceous mammals and the evolution of mammalian molar teeth. Fieldiana Geol 13:1–105
Phillips MJ, Bennett TH, Lee MS (2009a) Molecules, morphology, and ecology indicate a recent, amphibious ancestry for echidnas. Proc Natl Acad Sci USA 106:17089–17094
Phillips MJ, Bennett TH, Lee MS (2009b) Reply to camens: how recently did modern monotremes diversify? Proc Natl Acad Sci USA 107:E12
Pian R, Archer M, Hand SJ (2013) A new, giant platypus, *Obdurodon tharalkooschild* sp. nov. (Monotremata, Ornithorhynchidae), from the Riversleigh World Heritage Area Australia. J Vertebr Paleontol 33:1255–1259
Pian R, Archer M, Hand SJ, Beck RMD, Cody A (2016) The upper dentition and relationships of the enigmatic Australian Cretaceous mammal *Kollikodon ritchiei*. Mem Mus Vic 74:97–105
Rabinowitz PD, LaBrecque J (1979) The Mesozoic South Atlantic Ocean and evolution of its continental margins. J Geophys Res Solid Earth 84:5973–6002
Rauhut OWM, Martin T, Ortiz-Jaureguizar E, Puerta P (2002) The first Jurassic mammal from South America. Nature 416:165–168
Rich TH, Vickers-Rich P (2004) Diversity of Early Cretaceous mammals from Victoria, Australia. Bull Am Mus Nat Hist 285:36–53
Rich TH, Vickers-Rich P, Constanine A, Flannery TF, Kool L, van Klaveren NA (1997) A tribosphenic mammal from the Mesozoic of Australia. Science 278:1438–1442
Rich TH, Vickers-Rich P, Constanine A, Flannery TF, Kool L, van Klaveren NA (1999) Early Cretaceous mammals from Flat Rocks, Victoria, Australia. Rec Queen Vic Mus 106:1–35
Rich TH, Flannery TF, Trusler P, Kool L, van Klaveren NA, Vickers-Rich P (2001a) A second tribosphenic mammal from the Mesozoic of Australia. Rec Queen Vic Mus 110:1–10
Rich TH, Vickers-Rich P, Trusler P, Flannery TF, Cifelli RL, Constanine A, Kool L, van Klaveren NA (2001b) Monotreme nature of the Australian Early Cretaceous mammal *Teinolophos trusleri*. Acta Palaeontol Pol 46:113–118
Rich TH, Flannery TF, Trusler P, Kool L, van Klaveren NA, Vickers-Rich P (2002) Evidence that monotremes and ausktribosphenids are not sister groups. J Vertebr Paleontol 22:466–469
Rich TH, Hopson JA, Musser AM, Flannery TF, Vickers-Rich P (2005a) Independent origins of middle ear bones in monotremes and therians. Science 307:910–914
Rich TH, Hopson JA, Musser AM, Flannery TF, Vickers-Rich P (2005b) Response to comments on independent origins of middle ear bones in monotremes and therians. Science 309:1492
Rich TH, Hopson JA, Gill PG, Trusler P, Rogers-Davidson S, Morton S, Cifelli RL, Pickering D, Kool L, Siu K, Burgmann FA, Senden T, Evans AR, Wagstaff BE, Seegets-Villiers D, Corfe IJ, Flannery TF, Walker K, Musser AM, Archer M, Pian R, Vickers-Rich P (2016) The mandible and dentition of the Early Cretaceous monotreme *Teinolophos trusleri*. Alcheringa 40:475–501
Rich TH, Flannery TF, Vickers-Rich P (2020a) Evidence for a remarkably large toothed-monotreme from the Early Cretaceous of Lightning Ridge, NSW, Australia. In: Prasad GVR, Patnaik R (eds), Biological consequences of plate tectonics: New perspectives on post-Gondwana break-up—A

Tribute to Ashok Sahni. Springer Vertebrate Paleobiology and Paleoanthropology Series, pp 77–81

Rich TH, Trusler P, Kool L, David Pickering D, Evans A, Siu K, Maksimenko A, Kundrat M, Gostling NJ, Morton S, Vickers-Rich P (2020b) A third, remarkably small, tribosphenic mammal from the Mesozoic of Australia. In: Prasad GVR, Patnaik R (eds), Biological consequences of plate tectonics: New perspectives on post-Gondwana break-up—A Tribute to Ashok Sahni. Springer Vertebrate Paleobiology and Paleoanthropology Series, pp 67–75

Rougier GW, Forasiepi AM, Martinelli AG (2005) Comment on independent origins of middle ear bones in monotremes and therians II. Science 309:1432b

Rougier GW, Martinelli AG, Forasiepi AM, Novacek MJ (2007) New Jurassic mammals from Patagonia, Argentina: A reappraisal of australosphenidan morphology and interrelationships. Am Mus Novit 3566:1–54

Rougier GW, Apesteguía S, Gaetano LC (2011) Highly specialized mammalian skulls from the Late Cretaceous of South America. Nature 479:98–102

Rougier GW, Wible JR, Beck RMD, Apesteguía S (2012) The Miocene mammal *Necrolestes* demonstrates the survival of a Mesozoic non-therian lineage into the late Cenozoic of South America. Proc Natl Acad Sci USA 109:20053–20058

Rowe T, Rich TH, Vickers-Rich P, Springer M, Woodburne MO (2008) The oldest platypus and its bearing on divergence timing of the platypus and echidna clades. Proc Natl Acad Sci USA 105:1238–1242

Schultz JA, Martin T (2014) Function of pretribosphenic and tribosphenic mammalian molars inferred from 3D animation. Naturwissenschaften 101:771–781

Schultz JA, Menz U, Winkler DE, Schulz-Kornas E, Engels S, Kalthoff DC, von Koenigswald W, Ruf I, Kaiser TM, Kullmer O, Südekum K-H, Martin T (2018) Modular wear facet nomenclature for mammalian post-canine dentitions. Hist Biol 30:30–41

Schultz JA, Bhullar B-AS, Luo Z-X (2019) Re-examination of the Jurassic mammaliaform *Docodon victor* by computed tomography and occlusal functional analysis. J Mammal Evol 26:9–38

Setoyama E, Kanungo S (2020) Mesozoic biochronostratigraphy and paleoenvironment of the South Atlantic: a revised framework based on 20 DSDP and ODP deep-water sites. J S Am Earth Sci 99:102511

Shaw G (1799) The naturalist's miscellany. Printed for Nodder & Co, London

Simpson GG (1928) Mesozoic Mammalia, XI; *Brancatheralum tendagurense* Dietrich. Am J Sci 5:303–308

Simpson GG (1929) The dentition of *Ornithorhynchus* as evidence of its affinities. Am Mus Novit 390:1–15

Simpson GG (1936) Studies of the earliest mammalian dentitions. Dental Cosmos 78:791–800, 940–953

Simpson GG (1961) Evolution of Mesozoic mammals. In: Vandebroek G (ed) International colloquium on the evolution of lower and non-specialized Mammals. Koninklijke Vlaamse Academie voor Wetenschappen, Letteren en Schone Kunsten van België, Part 1, Leuven, pp 57–95

Simpson GG (1971) Concluding remarks: Mesozoic mammals revisited. Zool J Linnean Soc 50:181–189

Wang H, Meng J, Wang Y (2019) Cretaceous fossil reveals a new pattern in mammalian middle ear evolution. Nature 576:102–105

Wood CB, Rougier GW (2005) Updating and recoding enamel microstructure in Mesozoic mammals: in search of discrete characters for phylogenetic reconstruction. J Mammal Evol 12:433–460

Woodburne MO (2003) Monotremes as pretribosphenic mammals. J Mammal Evol 10:195–248

Woodburne MO, Tedford RH (1975) The first Tertiary monotreme from Australia. Am Mus Novit 2588:1–11

Woodburne MO, MacFadden BJ, Case JA, Springer MS, Pledge NS, Power JD (1993) Land mammal biostratigraphy and magnetostratigraphy of the Etadunna Formation (late Oligocene) of South Australia. J Vertebr Paleontol 13:483–515

Woodburne MO, Rich TH, Springer MS (2003) The evolution of tribospheny and the antiquity of mammalian clades. Mol Phylogenet Evol 28:360–385

Zeller U (1989) Die Enwicklung und morphologie des Schädels von *Ornithorhynchus anatinus* (Mammalia: Prototheria: Monotremata). Abh Senckenb Naturforsch Ges 545:1–188

# Chapter 5
# "Triconodonts"

*Triconodonts not only had a long phylogenetic history, they also played a central role in the development of paleontological theories. Among the first fossil mammals to be discovered in Mesozoic rocks was a triconodont, and the group subsequently figured prominently in the writings of Richard Owen, O. C. Marsh, H. F. Osborn, W. K. Gregory, and others concerned with the origin and evolution of mammals.*
Farish A. Jenkins Jr and Alfred W. Crompton
*Triconodonta*, 1979
*In: Mesozoic Mammals: The First Two-Thirds of Mammalian History*

**Abstract** "Triconodonts" is used here for mammaliaforms with three main cusps aligned along the mesiodistal axis of the postcanines, or forming a very broad (obtuse) triangle. This is not a natural group, but some of the smaller clades are, for example, Eutriconodonta and Amphilestheria. "Triconodonts" were abundant in Laurasian landmasses during the Jurassic and to a lesser degree during the Cretaceous. In contrast, the South American fossil record is scarce and the two known taxa come from a single locality in the Early–Middle Jurassic of central Patagonia; even there they are rare members of the fauna and the materials rather poorly preserved. In this chapter, we summarize the known species from Argentina, which includes an amphilestherian and an eutriconodontan. Putative "triconodonts" from the Late Cretaceous of Argentina are regarded as premolars of meridiolestidans following recent re-interpretation (see Chap. 6).

**Keywords** Amphilestheria · Eutriconodonta · Alticonodontinae · Jurassic

## 5.1 Introduction

Triconodonta is a paraphyletic group of mammaliaforms originally characterized by having cheek teeth with three main cusps arranged in a line or forming a broad triangle, which are often flanked by mesial and distal accessory cusps totaling up to five major cusps (Osborn 1887, 1888; Simpson 1928, 1929; Kielan-Jaworowska et al. 2004). The primitive crown morphology of upper and lower cheek teeth is dominated



G. W. Rougier et al., *Mesozoic Mammals from South America and Their Forerunners*, Springer Earth System Sciences, https://doi.org/10.1007/978-3-030-63862-7_5

by a tall cusp A/a, mesially and distally flanked by cusps B/b and C/c, respectively, of subequal size, and by a mesial cusp E/e and distal cusp D/d, which are again subequal (Owen 1871; Simpson 1928, 1929). This architecture in cheek teeth, with some variation on cusp size and symmetry, is present in several non-mammaliaform cynodonts and basal mammaliaforms, including thrinaxodontids, probainognathids, brasilodontids, morganucodontids, triconodontids, "symmetrodonts", and "amphilestids" (e.g., Crompton 1974; Rougier et al. 2003, 2007a; Bonaparte et al. 2005; Abdala et al. 2013; Martinelli et al. 2017; Gao et al. 2010; Gaetano and Rougier 2012).

Recognizing "triconodonts" were a paraphyletic group; Kermack et al. (1973) divided the original Triconodonta into Morganucodonta, occupying a basal position among mammaliaforms, and Eutriconodonta, a group of mammals nested in the crown group (Kielan-Jaworowska et al. 2004) (Fig. 5.1). Eutriconodonta is a clade formed by the common ancestor of the Triconodontidae (sensu stricto) plus

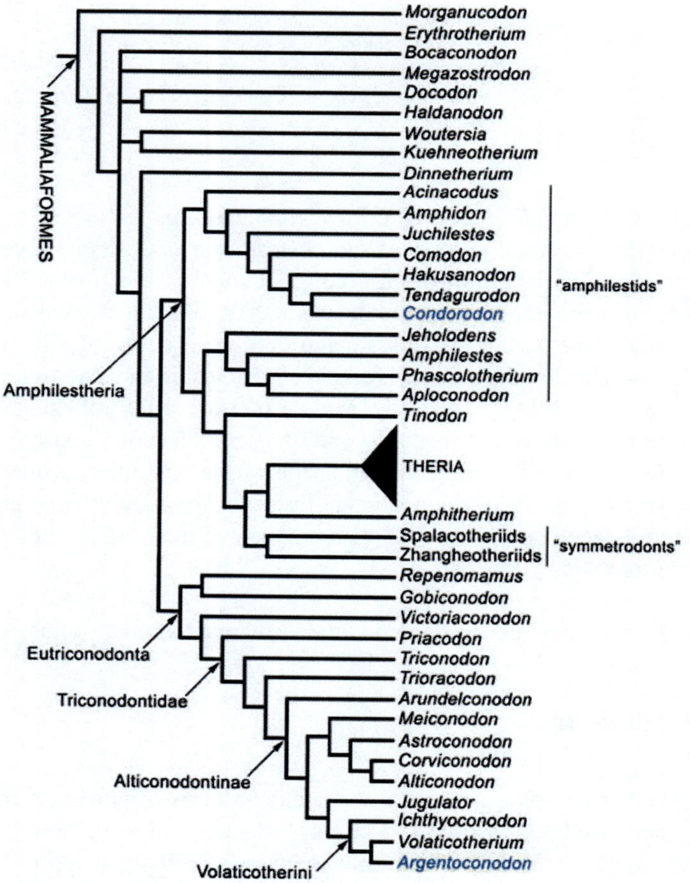


**Fig. 5.1** Phylogenetic tree of "triconodonts" with the SA taxa, *Condorodon spanios* and *Argentoconodon fariasorum* (based on Gaetano 2013; see also Gaetano and Rougier 2011, 2012)

any taxa more closely related to the Triconodontidae than to morganucodontans, spalacotheriids, tinodontids, and multituberculates (Kielan-Jaworowska et al. 2004). As such, eutriconodontans have a complex mastication, transitional middle ear/jaw suspensorium, and derived postcranium (e.g., Ji et al. 1999; Meng et al. 2006, 2011; Luo et al. 2007; Chen and Wilson 2015; Martin et al. 2015; Meng and Hou 2016; Chen et al. 2017). Triconodontids are monophyletic, a nested clade of eutriconodontans (Ji et al. 1999; Rougier et al. 2001, 2007b; Luo et al. 2002, 2007; Kielan-Jaworowska et al. 2004; Hu et al. 2005; Montellano et al. 2008; Kusuhashi et al. 2009; Gao et al. 2010; Gaetano and Rougier 2011, 2012; Meng et al. 2011). *Argentoconodon fariasorum* from the Early–Middle Jurassic (Toarcian–Bajocian?), Patagonia, Argentina is the only representative for the group in SA. Non-"amphilestid" eutriconodontans are also known in Gondwana by *Dyskritodon indicus* from the Lower Jurassic Kota Formation of India (Prasad and Manhas 2002), and *Dyskritodon amazighi* and *Ichthyoconodon jaworowskorum* from the Lower Cretaceous Séquence B des Couches Rouges (Synclinal d'Anoual) of Morocco (Sigogneau-Russell 1995).

A subset on non-triconodontid "triconodonts" is often regarded collectively as "Amphilestida", which depending on the resolution of the trees may or may not be a natural group; in one of the options, they have been recovered as successive stem taxa to cladotherians (Ji et al. 1999; Rougier et al. 2001, 2007a, b; Montellano et al. 2008; Kusuhashi et al. 2009; Gao et al. 2010; Gaetano and Rougier 2011; Meng et al. 2011; but see Luo et al. 2002, 2007; Kielan-Jaworowska et al. 2004; Hu et al. 2005; Martin et al. 2015 for alternative views). If amphilestids are stem therians, Amphilestheria is the most inclusive group and includes those taxa more closely related to Cladotheria than to *Triconodon mordax* (Gaetano and Rougier 2011) (Fig. 5.1). The radiation of the stem cladotherian "amphilestids" has been recorded mostly from the Middle Jurassic to the Late Cretaceous, from the Northern Hemisphere (Owen 1838, 1859, 1871; Simpson 1925a, b, 1928, 1929; Zhou et al. 1991; Engelmann and Callison 1998; Kretzoi and Kretzoi 2000; Rougier et al. 2007a; Gao et al. 2010; Lopatin et al. 2010; Hooker and Lawson 2011; Martin et al. 2015). Because of their complex occlusion and on occasion somewhat triangular arrangement of the main cusps, they have played an important role in building hypotheses regarding the origin of triangular molariforms culminating in tribosphenic molars (Osborn 1887, 1907; Crompton 1971; Osborn 1973; Crompton and Kielan-Jaworowska 1978; Davis 2011). In the Southern Hemisphere only, a few poorly represented "amphilestids" are known: *Tendagurodon janenschi* from the Late Jurassic (Kimmeridgian-Tithonian) of Tanzania (Heinrich 1998) and *Condorodon spanios* from the Early–Middle Jurassic (Toarcian–Bajocian?) of Patagonia, Argentina. Another putative member of this group is *Paikasigudodon yadagirii* from the Lower Jurassic Kota Formation of India (Prasad and Manhas 2002). This species is based on an upper molariform originally referred to as *Kotatherium haldanei* (Prasad and Manhas 1997). *Paikasigudodon* was considered an "amphilestid" but its phylogenetic position is certainly tentative and, in fact, it could belong to a basal mammaliaform (e.g., kuehneotheriid; Kielan-Jaworowska et al. 2004).

The fossil provenance and the phylogenetic relationships of Amphilestheria and Eutriconodonta suggest that their radiation took place in the Triassic or the earliest Jurassic in a continental configuration with only limited fragmentation and provincialism (Gaetano and Rougier 2011). Global dispersion by the Early Jurassic is supported by the putative affinities of the SA *Condorodon* and the African *Tendagurodon*, as well as the similarities between the Patagonian *Argentoconodon* and *Volaticotherium antiquus* from the likely Middle Jurassic Daohugou beds of China (Meng et al. 2006).

To date, *Condorodon spanios* and *Argentoconodon fariasorum* are the only SA "triconodonts" (i.e., "amphilestids" and triconodontids, respectively). Originally, Bonaparte (1986, 1992; see also Bonaparte and Migale 2010, 2015) reported "triconodonts" from the Upper Cretaceous Los Alamitos Formation. The two species, *Austrotriconodon mckennai* and *Austrotriconodon sepulvedai*, were, at the time, considered to be represented by isolated upper and lower cheek teeth. The identification of this material is controversial (Rougier et al. 2007b, 2011) and a recent revision (e.g., Gaetano et al. 2013; see also Chap. 6) concluded that the specimens from Los Alamitos correspond instead to mesial premolars of dryolestoids, perhaps taxa already described based on molars or otherwise yet to be formally recognized. Here, we agree with Gaetano et al. (2013) that there is no record of Cretaceous "triconodonts" in SA, and those specimens are discussed as dryolestoids (Chap. 6).

## 5.2 Systematics

Mammaliaformes Rowe 1988
Amphilestheria Gaetano and Rougier 2011
*Condorodon* Gaetano and Rougier 2012

Type species: *Condorodon spanios* Gaetano and Rougier 2012.

Included species: The type only.

*Condorodon spanios* Gaetano and Rougier 2012
(Fig. 5.2)

Holotype: MPEF-PV 2365, complete lower left molariform (Fig. 5.2).

Locality and horizon: Queso Rallado locality, about 3 miles northwest of the Cerro Cóndor village, Chubut Province, Patagonia, Argentina. Lower Member of the Cañadón Asfalto Formation, Toarcian–Bajocian? (Lower to Middle Jurassic).

Diagnosis (from Gaetano and Rougier 2012): "Amphilestids" with compressed lower molariforms, with five mesiodistally aligned cusps (a–e). Cusp a dominant, centered on the crown, and slightly recumbent. Cusp b smaller than c. Cusps b and c pointing mesially and distally, respectively. Cusps d and e relatively small. Cusp e not supported completely by the mesial root. Mesial interlocking structures absent. Enamel thicker lingually than labially. Unlike *Tendagurodon*, *Condorodon* has larger cusps b and c relative to cusp a, higher cusp c when compared to b, better-developed cusp d, broader and deeper notches between main cusps, more vertically

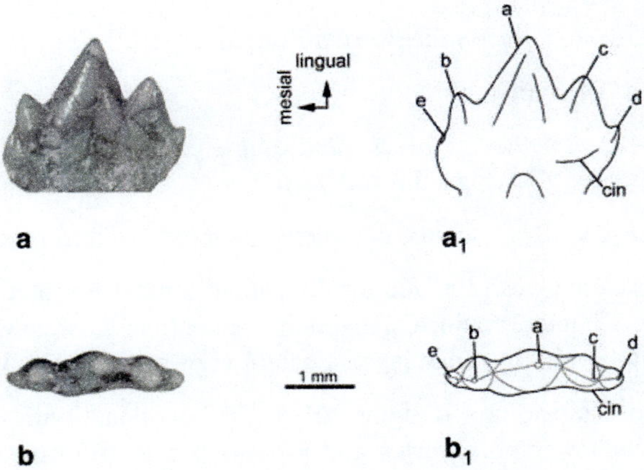


**Fig. 5.2** *Condorodon spanios* from Queso Rallado locality, Chubut, Argentina; Cañadón Asfalto Formation, Toarcian–Bajocian? (Lower to Middle Jurassic). Holotype, MPEF-PV 2365, isolated lower left molariform in labial (**a**) and occlusal (**b**) views and accompanying line drawings ($a_1$–$b_1$), from Gaetano and Rougier (2012). Abbreviations: a, b, c, d, e, g, triconodont cusp nomenclature; cin, cingulum

oriented roots that are slender with respect to the crown length, and lacking cingula below cusps b and e; the absence of cusp f (a feature also common to *Juchilestes*, *Aploconodon*, spalacotheriids, and zhangheotheriids) and the relative size of cusps b and c (shared with *Tendagurodon*) distinguishes *Condorodon* from other amphilestherians. *Condorodon* shares with *Argentoconodon*, *Hakusanodon*, *Tendagurodon*, *Volaticotherium*, spalacotheriids, and zhangheotheriids the lack of a mesial embayment indicative of an interlocking mechanism involving the fit of cusp d of the preceding tooth between mesial crests or cusps.

Comments: *Condorodon* represents the first and to date only known "amphilestid-triconodont" from SA. The lower molariform pattern closely resembles that of *Tendagurodon janenschi* from the Late Jurassic of Tanzania, the holotype of which has also been based on an isolated lower(?) molariform (Heinrich 1998). Both Jurassic Gondwanan taxa are recorded as sister taxa in the phylogenetic analyses (Gaetano and Rougier 2012; Gaetano 2013), forming a monophyletic clade together with Late Jurassic NA (*Comodon* and *Amphidon*) and Early Cretaceous Asian (*Acinacodus*, *Hakusanodon*, and *Juchilestes*) taxa (Fig. 5.1). Currently, the information on both Gondwanan taxa is very limited.

Eutriconodonta Kermack et al. 1973

Triconodontidae Marsh 1887

Alticonodontinae Fox 1976

Volaticotherini Meng et al. 2006

*Argentoconodon* Rougier et al. 2007b

Type species: *Argentoconodon fariasorum* Rougier et al. 2007b.

Included species: The type only.

*Argentoconodon fariasorum* Rougier et al. 2007b
(Figs. 5.3 and 5.4)

Holotype: MPEF-PV 1877, isolated complete upper left molariform (Fig. 5.3a–c).

Locality and horizon: Queso Rallado locality, about 3 miles northwest of the Cerro Cóndor village, Chubut Province, Patagonia, Argentina. Lower Member of the Cañadón Asfalto Formation, Toarcian–Bajocian? (Lower to Middle Jurassic).

Diagnosis (from Gaetano and Rougier 2011): "Triconodont" with imbricated (en echelon) upper and lower molariforms, and posterior premolariforms. Simple unicuspated incisors. Canines taller than other teeth, lower ones uniradiculated and conical, upper canines with a constricted root and more labiolingually compressed. Premolariforms without mesial accessory cusps and cingula, with one simple or constricted root or two roots. Molariforms extremely labiolingually compressed and lacking conspicuous cingula. Upper molariforms pentacuspidated with three main cusps (A, B, and C) and two accessory cusps (D and E). Cusp A subequal or taller than cusp C, both distally recumbent. Cusp B almost erect and lower than cusps A and C. Accessory cusps (D and E) project from the base of the crown. Lower molariforms with four cusps. Main cusps (a, b, and c) strongly recumbent. Distal accessory cusp (d) overhanging distally as a flange-like projection. Premolariforms of *Argentoconodon* differ from those of "amphilestids" and other tricondontids by lacking a mesial accessory cusp; lower molariforms of *Argentoconodon* differ from those of *Volaticotherium* and other non-volaticotherine alticonodontines by lacking well-developed cingula and having main cusps separated by wider valleys; further differences with non-Volaticotherini alticonodontines include more labiolingually compressed lower molariforms bearing four cusps (a–d) that are not subequal and are set apart by relatively wide valleys; differences with the lower molariforms of *Ichthyoconodon* also comprise cusp proportions and recumbence; upper molariforms of *Argentoconodon* are distinguished from those of *Volaticotherium* by the presence of well-developed cusps D and E and cusp B less recumbent than the other main cusps; and unlike *Corviconodon* and *Astroconodon*, in *Argentoconodon* cusps A–D are well-separated, cusp E is present, cusp D is proportionally lower than the main cusps and overhangs distally, and cingula are absent; *Argentoconodon* differs from *Priacodon*, *Trioracodon*, and *Triconodon* in having more labiolingually compressed molariforms with more recumbent main cusps and lacking well-developed cingula, and by the presence of cusp E in upper molariforms.

Comments: *Argentoconodon fariasorum* was originally known by an isolated upper molariform MPEF-PV 1877 (Rougier et al. 2007b); however, soon after, a partial skeleton, including maxillae, dentaries, teeth, and postcranium, was discovered at Queso Rallado locality, Cañadón Asfalto Formation, augmenting considerably the current knowledge on the species (Gaetano and Rougier 2011; Gaetano 2013). The

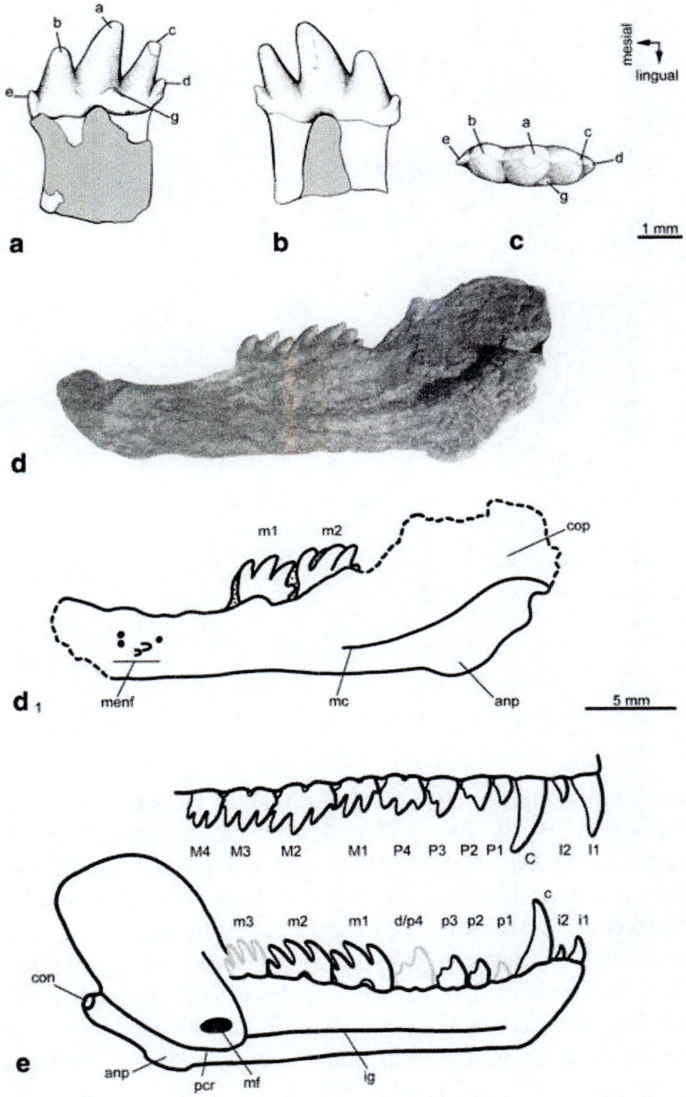


**Fig. 5.3** *Argentoconodon fariasorum* from Queso Rallado locality, Chubut, Argentina; Cañadón Asfalto Formation, Toarcian–Bajocian? (Lower to Middle Jurassic). Holotype, MPEF-PV 1877, isolated upper left molariform in lingual (**a**), labial (**b**), and occlusal (**c**) views, from Rougier et al. (2007b). MPEF-PV 2363, left dentary (**d**) and accompanying line drawing (**d**$_1$) in lateral view. Reconstruction of the dentary in medial view and upper and lower dentition of *Argentoconodon* (**e**), from Gaetano and Rougier (2011). Teeth in gray are not known. Abbreviations: a, b, c, d, e, g, triconodont cusp nomenclature; anp, angular process; C/c, upper and lower canines; con, condyle; cop, coronoid process; I/i, upper and lower incisors; ig, internal groove; M/m, upper and lower molariforms; mc, masseteric crest; menf, mental foramina; mf, mandibular foramen; P/p, upper and lower premolariform (d refers to deciduous premolar); pcr, pterygoid crest. Grey pattern indicates the matrix

**Fig. 5.4** *Argentoconodon fariasorum*, artistic reconstruction by Jorge L. Blanco

dental formula of *Argentoconodon* was restored on the basis of the available material as I2?/i2, C1/c1, P4/p4 M4/m3 (Fig. 5.3g). The phylogenetic results placed *Argentoconodon* as the sister taxon to *Volaticotherium* (Gaetano and Rougier 2011, 2012; Gaetano 2013). Both species together with *Ichthyoconodon jaworowskorum*, from the Cretaceous of Morocco, constitute a monophyletic group, and together with several Laurasian taxa (e.g., *Alticonodon*, *Arundelconodon*, *Austroconodon*, *Corviconodon*, and *Meiconodon*), can be considered derived members of the Alticonodontinae (Fig. 5.1) (Gaetano and Rougier 2011). Alticonodontines are moderately diverse and common in particular from the Early and Late Cretaceous of NA (Patterson 1951; Fox 1969, 1976; Winkler et al. 1990; Jacobs et al. 1991; Cifelli et al. 1997, 1998, 1999; Cifelli and Madsen 1998; Turnbull and Cifelli 1999). Alticonodontines have tall crowns, with the main cusps of similar height and forming a lamina in palisade, considered to reflect a piscivorous diet (Sigogneau-Russell 1995; Meng et al. 2011). Considering the rich Late Cretaceous mammalian record in SA, dominated by dryolestoids and to a lesser extent allotherians (see Chaps. 6 and 8), it seems likely that alticonodontines became extinct first in SA, then in Laurasia and the rest of Gondwana (Gaetano and Rougier 2011).

The femur of both *Volaticotherium* and *Argentoconodon* has a poorly differentiated head in line with the main shaft and confluent with the greater and lesser trochanters (Meng et al. 2006; Gaetano and Rougier 2011). These features, similar to those of living gliding mammals, coupled with the preservation of a likely patagium, led Meng et al. (2006) to regard *Volaticotherium* as a gliding mammal. The striking similarities of the femora and the sister group relationship of *Volaticotherium* and *Argentoconodon* suggest similar locomotory habits for *Argentoconodon* (Gaetano and Rougier 2011) (Fig. 5.4).

*Volaticotherium* and *Argentoconodon* are deeply nested within the "triconodont" phylogeny (Meng et al. 2006; Gaetano and Rougier 2012) and as such they create an extensive ghost lineage regarding other alticonodontines, which are Cretaceous in age; indeed, *Argentoconodon* is the oldest relatively well-preserved eutriconodontan. *Argentoconodon* and the henosferids from Cañadón Asfalto are currently the oldest mammals from SA. Therefore, the extensive ghost linage and the sizable eutriconodontan diversity recovered by most current phylogenetic trees must have an older origin than the current minimal age from the Cañadón Asfalto Formation, Toarcian, about 179 Ma (Cúneo et al. 2013; Figari et al. 2015). The fauna from Queso Rallado with its great antiquity and relatively derived taxa suggests extremely old and undocumented radiation for the major mammalian lineages.

## 5.3 Concluding Remarks

Both "triconodonts" from SA, *Condorodon spanios* and *Argentoconodon fariasorum*, come from the same layers in the Queso Rallado quarry (Rauhut et al. 2002; Rougier et al. 2007b), from the lower levels of the Cañadón Asfalto Formation of Toarcian–Bajocian? (Lower–Middle Jurassic) age (Cúneo et al. 2013). The quarry has been intensively worked on a yearly basis since the original discovery in 2002 and despite the protracted effort, there is a single, relatively well-preserved specimen assigned to *Condorodon*. On the other hand, *Argentoconodon* is known by a partial, disarticulated skeleton in a slab plus a handful of isolated teeth.

"Triconodonts" are obviously rare in the Queso Rallado fauna and in the Jurassic of Gondwana in general. Significantly, the closest relative of *Condorodon* is *Tendagurodon* from the Late Jurassic of Africa and only distantly related to *Argentoconodon* (Gaetano and Rougier 2012); both of these mammals are among the oldest members of their respective clades, Amphilestheria and Alticonodontidae, respectively. Intercalated among these deeply nested SA "triconodonts" are a variety of Jurassic and Cretaceous "triconodonts" from most major landmasses except Antarctica and Australia. Paleobiogeographical analysis of "triconodonts" by Gaetano and Rougier (2012) suggests a Pangeic paleogeography in place for the last common ancestor of the Patagonian "triconodonts" of a minimal earliest Jurassic, and more likely Late Triassic age. The geographical and age distribution of the more inclusive clades suggest the absence of dispersal barriers between eastern Asia and Gondwanan landmasses, the existence of extensive ghost lineages and/or wide distribution of ancestral forms. All of these alternatives are viable given the meager record of Gondwanan Jurassic mammals.

Jurassic mammals display a broad array of locomotory strategies, adapted to specialized environments (Luo 2007; Chen and Wilson 2015). *Argentoconodon*, the gliding *Volaticotherium* from the Jurassic of China (Meng et al. 2006), and *Ichthyoconodon* from the Early Cretaceous of Africa (Sigogneau-Russell 1995) are dentally very similar, clustering together in a seemingly natural group: Volaticotherini (Gaetano and Rougier 2011). They may (Gaetano and Rougier 2011) or may not (Martin et al. 2015), in turn, be related to alticonodontine "triconodonts", which show enhanced sectorial functions either related to carnivorous or piscivorous adaptations (Fox 1969, 1976; Slaughter 1969; Cifelli and Madsen 1998). *Argentoconodon* and relatives have tall, recurved cusps, but lack the more specialized hypertrophy of cusp d and similar interlocking that characterizes the derived alticonodontines (Cifelli and Madsen 1998; Gaetano and Rougier 2011; Meng et al. 2011). The tooth morphology of volaticotherines is compatible with a fish-based diet, and the dental similarities between *Volaticotherium* and *Argentoconodon* are very detailed, extending beyond general morphology. The jaws are also similar, and among the isolated elements found in the slab bearing the best specimen of *Argentoconodon,* there is a femur with a peculiar femoral head morphology matching closely that described for *Volaticotherium* (Meng et al. 2006: Fig. 5.3d) and used to argue for gliding in the Chinese mammal. The distinct possibility of a piscivorous clade of gliding "triconodonts" is tantalizing but hard to accommodate in a vicariance biogeographic model (Gaetano and Rougier 2012). The presence of such an old, highly specialized, and long-lived group in areas as remote from each other as they can be is surprising. The above suggests that much of the early mammalian diversification occurred substantially earlier than our oldest records and that much is yet to be discovered (in Gondwana, in particular), which should shorten the very long ghost lineages resulting from current phylogenetic hypotheses.

## References


Abdala NF, Jasinoski SC, Fernandez V (2013) Ontogeny of the Early Triassic cynodont *Thrinaxodon liorhinus* (Therapsida): Dental morphology and replacement. J Vertebr Paleontol 33:1408–1431

Bonaparte JF (1986) Sobre *Mesungulatum houssayi* y nuevos mamíferos cretácicos de Patagonia. 4° Congreso Argentino de Paleontología y Bioestratigrafía, Mendoza, Actas 2:48–61

Bonaparte JF (1992) Una nueva especie de Triconodonta (Mammalia), de la Formación Los Alamitos, Provincia de Río Negro y comentarios sobre su fauna de mamíferos. Ameghiniana 29:99–110

Bonaparte JF, Migale LA (2010) Protomamíferos y mamíferos Mesozoicos de América del Sur. Museo de Ciencias Naturales Carlos Ameghino de Mercedes, Buenos Aires. 1° Edition

Bonaparte JF, Migale LA (2015) Protomamíferos y mamíferos Mesozoicos de América del Sur. Fundación de Historia Natural Felix de Azara, Buenos Aires. 2° Edition

Bonaparte JF, Martinelli AG, Schultz CL (2005) New information on *Brasilodon* and *Brasilitherium* (Cynodontia, Probainognathia) from the Late Triassic of southern Brazil. Rev Bras Paleontol 8:25–46

Chen M, Wilson GP (2015) A multivariate approach to infer locomotor modes in Mesozoic mammals. Paleobiology 41:280–312

Chen M, Luo Z-X, Wilson GP (2017) The postcranial skeleton of *Yanoconodon allini* from the Early Cretaceous of Hebei, China, and its implications for locomotor adaptation in eutriconodontan mammals. J Vertebr Paleontol 37:e1315425

Cifelli RL, Madsen SK (1998) Triconodont mammals from the medial Cretaceous of Utah. J Vertebr Paleontol 18:403–411

Cifelli RL, Gardner JD, Nydam RL, Brinkman DL (1997) Additions to the vertebrate fauna of the Antlers Formation (Lower Cretaceous), southeastern Oklahoma. Oklahoma Geol Notes 57:124–131

Cifelli RL, Wible JR, Jenkins FA Jr (1998) Triconodont mammals from the Cloverly Formation (Lower Cretaceous), Montana and Wyoming. J Vertebr Paleontol 18:237–241

Cifelli RL, Lipka TR, Schaff CR, Rowe TB (1999) First early Cretaceous mammal from the eastern seaboard of the United States. J Vertebr Paleontol 19:199–203

Crompton AW (1971) The origin of the tribosphenic molar. In: Kermack DM, Kermack KA (eds) Early Mammals. Zool J Linn Soc 50, suppl 1:65–87

Crompton AW (1974) The dentitions and relationships of the southern African Triassic mammals, *Erythrotherium parringtoni* and *Megazostrodon rudnerae.* Bull Br Mus Nat Hist Geol 24:397–437

Crompton AW, Kielan-Jaworowska Z (1978) Molar structure and occlusion in Cretaceous therian mammals. In: Butler PM, Joysey KA (eds) Studies in the development, function and evolution of teeth. Academic Press, London, pp 249–287

Cúneo R, Ramezani J, Scasso R, Pol D, Escapa IH, Zavattieri AM, Bowring SA (2013) High-precision U-Pb geochronology and a new chronostratigraphy for the Cañadón Asfalto Basin, Chubut, central Patagonia: Implications for terrestrial faunal and floral evolution in Jurassic. Gondwana Res 24:1267–1275

Davis B (2011) Evolution of the tribosphenic molar pattern in early mammals, with comments on the "dual-origin" hypothesis. J Mammal Evol 18:224–227

Engelmann GF, Callison G (1998) Mammalian faunas of the Morrison Formation. Mod Geol 23:343–379

Figari EG, Scasso RA, Cúneo RN, Escapa IH (2015) Estratigrafía y evolución geológica de la Cuenca de Cañadón Asfalto, provincia del Chubut, Argentina. Lat Am J Sedimentol Basin Anal 22:135–169

Fox RC (1969) Studies of late Cretaceous vertebrates. III. A triconodont mammal from Alberta. Can J Zool 47:1253–1256

Fox RC (1976) Additions to the mammalian local fauna from the upper Milk River Formation (Upper Cretaceous), Alberta. Can J Earth Sci 13:1105–1118

Gaetano LC (2013) *Argentoconodon fariasorum*, un mamaliaforme del Jurásico de Patagonia, Argentina: Descripción, relaciones filogenéticas e implicancias evolutivas y paleobiogeográficas. Unpublished PhD thesis, University of Buenos Aires

Gaetano LC, Rougier GW (2011) New materials of *Argentoconodon fariasorum* (Mammaliaformes, Triconodontidae) from the Jurassic of Argentina and its bearing on triconodont phylogeny. J Vertebr Paleontol 31:829–843

Gaetano LC, Rougier GW (2012) First amphilestid from South America: a molariform from the Jurassic Cañadón Asfalto Formation, Patagonia, Argentina. J Mammal Evol 19:235–248

Gaetano LC, Marsicano CA, Rougier GW (2013) A revision of the putative Late Cretaceous triconodonts from South America. Cretac Res 46:90–100

Gao C-L, Wilson GP, Luo Z-X, Murat Maga A, Meng Q, Wang X (2010) A new mammal skull from the Early Cretaceous of China with implications for the evolution of obtuse-angled molars and "amphilestid" eutriconodonts. Proc R Soc Lond B 276:237–246

Heinrich W-D (1998) Late Jurassic mammals from Tendaguru, Tanzania, East Africa. J Mammal Evol 5:269–290

Hooker JJ, Lawson AG (2011) An "eutriconodontan" mammal from the UK Cenomanian (Late Cretaceous). Spec Pap Palaeontol 86:255–261

Hu Y, Meng J, Wang Y-Q, Li C (2005) Large Mesozoic mammals fed on young dinosaurs. Nature 433:149–152

Jacobs LL, Winkler DA, Murry PA (1991) On the age and correlation of the Trinity mammals, Early Cretaceous of Texas, USA. Newsl Stratigr 24:35–43
Ji Q, Luo Z-X, Ji S-A (1999) A Chinese triconodont mammal and mosaic evolution of mammalian skeleton. Nature 398:326–330
Kermack KA, Mussett F, Rigney HW (1973) The lower jaw of *Morganucodon*. Zool J Linn Soc 53:87–175
Kielan-Jaworowska Z, Cifelli RL, Luo Z-X (2004) Mammals from the age of dinosaurs. Origins, evolution, and structure. Columbia University Press, New York
Kretzoi M, Kretzoi M (2000) Fossilium Catalogus 1: Animalia Pars 137—Index Generum et Subgenerum Mammalium. Backhuys Publishers, Leiden
Kusuhashi N, Hu Y, Wang Y, Hirasawa S, Matsuoka H (2009) New triconodontids (Mammalia) from the lower Cretaceous Shahai and Fuxin formations, northeastern China. Geobios 42:765–781
Lopatin AV, Maschenko EN, Averianov AO (2010) A new genus of triconodont mammals from the Early Cretaceous of western Siberia. Dokl Biol Sci 433:282–285
Luo Z-X (2007) Transformation and diversification in early mammal evolution. Nature 450:1011–1019
Luo Z-X, Kielan-Jaworowska Z, Cifelli RL (2002) In quest for a phylogeny of Mesozoic mammals. Acta Palaeontol Pol 47:1–78
Luo Z-X, Chen P, Li G, Chen M (2007) A new eutriconodont mammal and evolutionary development in early mammals. Nature 446:288–293
Marsh OC (1887) American Jurassic mammals. Am J Sci 33:326–348
Martin T, Marugan-Lobon J, Vullo R, Martin-Abad H, Luo Z-X, Buscalioni AD (2015) A Cretaceous eutriconodont and integument evolution in early mammals. Nature 526:380–384
Martinelli AG, Soares MB, Oliveira TV, Rodrigues PG, Schultz CL (2017) The Triassic eucynodont *Candelariodon barberenai* revisited and the early diversity of stem prozostrodontians. Acta Palaeontol Pol 62:527–542
Meng J, Hou S (2016) Earliest known mammalian stapes from an Early Cretaceous eutriconodontan mammal and implications for transformation of mammalian middle ear. Palaeontol Pol 67:181–196
Meng J, Hu Y-M, Wang Y, Wang X, Li C (2006) A Mesozoic gliding mammal from northeastern China. Nature 444:889–893
Meng J, Wang Y, Li C (2011) Transitional mammalian middle ear from a new Cretaceous Jehol eutriconodont. Nature 472:181–185
Montellano M, Hopson JA, Clark JM (2008) Late Early Jurassic mammaliaforms from Huizachal Canyon, Tamaulipas, Mexico. J Vertebr Paleontol 28:1130–1143
Osborn HF (1887) On the structure and classification of the Mesozoic Mammalia. Proc Acad Nat Sci Phila 38:282–292
Osborn HF (1888) On the structure and classification of the Mesozoic Mammalia. J Acad Nat Sci Philadelphia 9:186–265
Osborn HF (1907) Evolution of mammalian molar teeth. MacMillan and Company, New York
Osborn JW (1973) The evolution of dentitions. Am Sci 61:548–559
Owen R (1838) On the jaws of the *Thylacotherium prevostii* (Valenciennes) from Stonesfield. Proc Geol Soc Lond 3:5–9
Owen R (1859) Palaeontology. Encyclopaedia Britannica 8th Edition, Vol 17. Adam and Black, Edinburgh, pp 91–176
Owen R (1871) Monograph of the fossil Mammalia of the Mesozoic formations. Monogr Palaeontol Soc 33:1–115
Patterson B (1951) Early Cretaceous mammals from northern Texas. Am J Sci 249:31–46
Prasad GVR, Manhas BK (1997) A new symmetrodont mammal from the Lower Jurassic Kota Formation, Pranhita Godavari Valley, India. Geobios 30:563–572
Prasad GVR, Manhas BK (2002) Triconodont mammals from the Jurassic Kota Formation of India. Geodiversitas 24:445–464

Rauhut OW, Martin T, Ortiz-Jaureguizar E, Puerta PF (2002) A Jurassic mammal from South America. Nature 416:165–168

Rougier GW, Novacek MJ, McKenna MC, Wible JR (2001) Gobiconodonts from the Early Cretaceous of Oshih (Ashile), Mongolia. Am Mus Novit 3348:1–30

Rougier GW, Spurlin BK, Kik PK (2003) A new specimen of *Eurylambda aequicrurius* and considerations on "symmetrodonts" dentition and relationships. Am Mus Novit 3398:1–15

Rougier GW, Isaji S, Manabe M (2007a) An Early Cretaceous mammal from the Kuwajima Formation (Tetori Group), Japan, and a reassessment of triconodont phylogeny. Ann Carnegie Mus 70:73–115

Rougier GW, Garrido A, Gaetano L, Puerta P, Corbitt C, Novacek MJ (2007b) First Jurassic triconodont from South America. Am Mus Novit 3580:1–17

Rougier GW, Gaetano L, Drury BR, Colella R, Gómez RO, Páez Arango N (2011) A review of the Mesozoic mammalian record of South America. In: Calvo J, Porfiri J, González Riga B, Dos Santos D (eds) Dinosaurios y paleontología desde América Latina. Editorial de la Universidad Nacional de Cuyo, Mendoza, pp 195–214

Rowe TB (1988) Definition, diagnosis, and origin of Mammalia. J Vertebr Paleontol 8:241–264

Sigogneau-Russell D (1995) Two possibly aquatic triconodont mammals from the Early Cretaceous of Morocco. Acta Palaeontol Pol 40:149–162

Simpson GG (1925a) Mesozoic Mammalia. I. American triconodonts: Part 1. Am J Sci 10:145–165

Simpson GG (1925b) Mesozoic Mammalia. I. American triconodonts: Part 2. Am J Sci 10:334–358

Simpson GG (1928) A catalogue of the Mesozoic Mammalia in the Geological Department of the British Museum. Trustees of the British Museum, London

Simpson GG (1929) American Mesozoic Mammalia. Mem Peabody Mus Yale Univ 3:1–235

Slaughter BH (1969) *Astroconodon*, the Cretaceous triconodont. J Mammal 50:102–107

Turnbull WD, Cifelli RL (1999) Triconodont mammals of the Aptian-Albian Trinity Group, Texas and Oklahoma. In: Mayhall JT, Heikkinen T (eds) Dental morphology'98. University of Oulu Press, Oulu, pp 252–272

Winkler DA, Murry PA, Jacobs LL (1990) Early Cretaceous (Comanchean) vertebrates of central Texas. J Vertebr Paleontol 10:95–116

Zhou M-Z, Cheng Z-W, Wang Y-Q (1991) A mammalian lower jaw from the Jurassic of Lingyuan, Liaoning. Vertebr Pal Asia 29:165–175

# Chapter 6
# Dryolestoids

*La clasificacion de los mamíferos se encuentra atravesando una etapa de profunda renovación como consecuencia de la acumulación de nuevos hechos y de reinterpretaciones tanto con respecto a las formas extinguidas como a las relaciones entre taxones vivientes y extinguidos.*
[The classification of mammals is in a stage of profound changes as consequence of the acumulation of new facts and re-interpretations involving extinct creatures as well as the phylogenetic relationships between living and extinct taxa.]
Osvaldo A. Reig
*Teoría del origen y desarrollo de la fauna de mamíferos de América del Sur, 1981*

**Abstract** Dryolestoids are iconic members of the Mesozoic mammalian associations in South America. They achieved a large taxonomic diversity in this region with disparate dental and cranial morphotypes ranging from the classical role of sharp-toothed insectivores to bunodont, complex dentitions reflecting omnivore/herbivore adaptations. The South American radiation of dryolestoids, the meridiolestidans, are among the most abundant Cretaceous mammals, surviving the K/Pg mass extinction and continuing until the Miocene as minor members of the South American biotas. New specimens have been recently discovered, some of them including associated upper and lower jaws, and exceptionally preserved skulls. These high-quality fossils provide crucial intraspecific dental variation, both along the tooth row and from upper to lower, allowing critical re-interpretation of some taxa originally named on the basis of isolated teeth or very incomplete material. The Cretaceous diversity of meridiolestidans has been grossly overestimated, with taxa based on different dental positions of what was later determinied to be a single taxon. One relatively poorly known Late Cretaceous taxon, *Groebertherium*, shares many features with the classical Holartic dryolestoids and may represent a Late Jurassic/Early Cretaceous foundational morphology expected for meridiolestidans.

**Keywords** Dryolestoidea · Dryolestidae · Meridiolestida · Mesungulatoidea

 
G. W. Rougier et al., *Mesozoic Mammals from South America and Their Forerunners*, Springer Earth System Sciences, https://doi.org/10.1007/978-3-030-63862-7_6

## 6.1 Introduction

Dryolestoids are an important and diverse group of mostly Jurassic and Cretaceous cladotherian mammals, representing a stem lineage to marsupials and placentals. The group is known from the Middle Jurassic to the Late Cretaceous in Europe, Asia, Africa, NA, and SA, surviving into the Cenozoic in SA and Antarctica (e.g., Marsh 1878, 1880; Simpson 1929, 1945; Krebs 1971; Freeman 1979; Prothero 1981; Bonaparte 1986a, 1990, 2002; Lillegraven and McKenna 1986; Ensom and Sigogneau-Russell 1998; Martin 1999; Rougier et al. 2011a, b; Martinelli et al. 2014; Averianov et al. 2016).

Like other cladotherians, dryolestoids have an upper and lower dentition formed by a series of reversed triangles, with upper molars wider than lower molars (Simpson 1929; Martin 1999; Luo et al. 2002; Kielan-Jaworowska et al. 2004) (Fig. 6.1) and complex mastication (Crompton 1971; Crompton et al. 1994; Davis 2011; Schultz and Martin 2011). The lower jaw is relatively advanced with a distinct angular process and sophisticated musculature (Prothero 1981; Schultz and Martin 2014), associated with an emphasis on a transverse masticatory component in the chewing cycle (Crompton 1971; Davis 2011; Schultz and Martin 2014). In Jurassic dryolestoids,

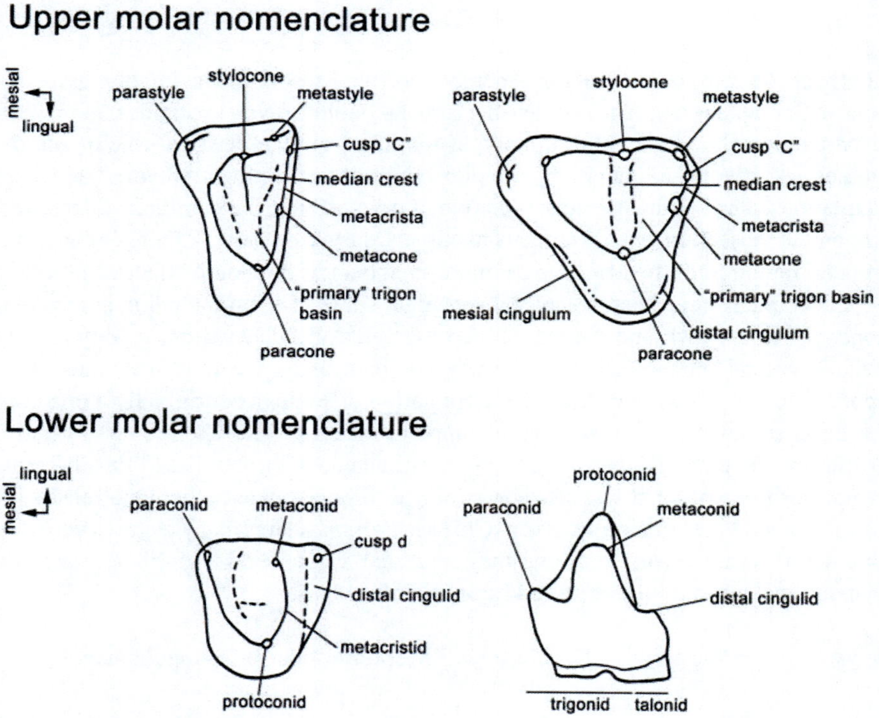


**Fig. 6.1** Upper and lower molar structure and terminology. Modified from Kielan-Jaworowska et al. (2004)

the dentary bears scars for the attachment of coronoid and splenial bones and a defined Meckelian groove for a patent Meckel's cartilage. Geologically younger taxa have a shallower Meckelian groove and likely lacked coronoid and splenial bones (Martin 1999; Kielan-Jaworowska et al. 2004). The dental formula varies between the groups, not only by the number of premolars, but also by the number of molars. Dryolestids have more than five molars (usually between seven and nine; e.g., Prothero 1981; Martin 1999), while paurodontids have a much smaller number, typically around four, although the number depends on interpretation of tooth loci and membership of the group. The meridiolestidans have few molars, three being present in basal forms and in the derived and highly nested taxa (e.g., Rougier et al. 2009b). Following the terminology of Martin (1999), the upper molars have one lingual cusp, the paracone, and two labial cusps: the parastyle and the metastyle, at the mesiolabial and distolabial corners of the tooth, respectively. Another labial cusp, the stylocone, is the primitive cusp forming the mesiolabial corner of the upper molar in "symmetrodonts" and basal cladotherians, and it is found in that position (or immediately distal to the parastyle) in most Laurasian dryolestoids (Butler 1939; Patterson 1956). However in some taxa, particularly in those where the stylocone is shifted distally and occupies a more central position, the stylocone is linked to the paracone via a median crest (mesocrista sensu Bonaparte 1990; medianer Grat sensu Martin 1999; median ridge sensu Kielan-Jaworowska et al. 2004) dividing the trigon basin into two valleys. This basin represents a "primary trigon" since it is associated to the paracone, and it is not homologous to the trigon in relation to the protocone of tribosphenic molars (Patterson 1956; Crompton 1971; Kielan-Jaworowska et al. 2004). The lower molars of dryolestoids have a trigonid, formed by the protoconid, paraconid, and metaconid, clearly differentiated from the talonid, with a single cusp (cusp d—homolog of the hypoconid or hypoconulid of tribosphenic teeth) (Crompton 1971; Kielan-Jaworowska et al. 2004; Davis 2011) (Fig. 6.1).

As with many other Mesozoic groups, the number and morphology of the premolars play an important role in the systematics of dryolestoids and related taxa. The concept of molar was originally developed by Owen (1845) and he specifically referenced replacement (or lack thereof) as the defining criterion for recognizing molar from premolars. However, Owen (1845) also considered the typical seven postcanine count of crown therians homologous; therefore, the anterior premolars of marsupials that are never replaced did not fully match his own criteria for premolars. Operationally, a more lax adaptation of Owen's definition of molars is currently used and any postcanine tooth distal to the last element showing replacement is considered a molar; while premolars would be the postcanines preceding the first molar. Bi et al. (2016) emended this definition as a molar can be defined as any tooth posterior to the last postcanine showing evidence of replacement, considering that in some groups the anterior postcanines are not replaced (e.g., P1 in eutherians), a likely derived condition. Recognition of the first molar in a dental series pertains directly to the taxon under study. It is possible, however, that a premolar, for instance, the last premolar, loses its replacement; technically, it becomes a molar. This developmental process is the likely path leading from a full replacement for all the postcanine positions present in non-mammalian cynodonts (e.g., Abdala et al. 2013) to a limited replacement affecting only the anterior dentition and thus determining molars and ante-molar dentition (including incisors, canines, and premolars) characteristic of therians. A more neutral term "molariform" (MF/mf) is often used to describe just the morphology present in a given element without a direct reference to its permanent or deciduous nature. It is the case in other Mesozoic mammals that the first molariform is in fact a premolar, as shown in *Maotherium* (Rougier et al. 2003a), *Zhangheotherium* (Luo and Ji 2005), and "amphilestids" (Rougier et al. 2001, 2007a, b). This appears also to be the case in several dryolestoids, particularly in meridiolestidans, in which the separation between premolars and molars is based on wear (a proxy for replacement, when the actual event is unrecorded in fossils). In the case of a "paurodont" from the Morrison Formation, as well as in *Cronopio, Necrolestes,* and *Peligrotherium,* the wear pattern suggests that the first molariform erupts after several of the posterior elements have been fully functional for an extended time. We have chosen to follow earlier proposals for meridiolestidans (Rougier et al. 2009a, b, 2011a, 2012) which regard this position as a premolar. Averianov et al. (2013) argued for calling these anterior molariforms molars because "in spite of its replacement in some taxa, is homologous to the first molar of cladotherians". This later view is not followed here, but it is in principle an alternative and acceptable interpretation of the nature of the transition premolar/molar and similar to the process by which the P3/M4 marsupial dental formula originates from the P5/M3 thought to be primitive for Theria and Eutheria (e.g., see review in O'Leary et al. 2013). This approach works best within a relatively narrow comparative framework, where tooth positions can be thought of as homologous regardless of the presence of replacement, an assumption defensible for therians. However, it becomes much harder when disparity plays a larger role, the groups to be compared are morphologically very different and the tooth count quite dissimilar. Each approach has its merits and problems; an absolute resolution seems difficult to defend until specimens showing early stages of replacement and more complete ontogenies of Mesozoic mammals become available. Such compelling evidence is lacking and not likely to be amassed in the near future; until then, we follow our earlier attempts to provide a consistent comparative framework for dryolestoid dentitions.

## 6.2 South American Dryolestoids

In SA, dryolestoids are a significant mammalian component of the Mesozoic terrestrial ecosystems. They are first registered in the early Late Cretaceous (Cenomanian; Rougier et al. 2011a) and are the dominant mammal group by the latest Cretaceous (Campanian–Maastrichtian; Bonaparte 1986a, b, 1990, 1994, 2002; Rougier et al. 2009a, b, 2011b). Younger and relictual taxa are found during the Cenozoic, extending into the late early Miocene (Burdigalian; Rougier et al. 2012; Wible and Rougier 2017).

Bonaparte (1994, 2002) formalized and elaborated the phylogenetic affinities between dryolestoids from the Cretaceous of SA and those from the Jurassic of

Laurasia, which he had advanced more tentatively in the original descriptions of the individual taxa (Bonaparte and Soria 1985; Bonaparte 1986a, 1990, 2002). In recent phylogenies (Rougier et al. 2011a, 2012; Averianov et al. 2013; O'Meara and Thompson 2014; Wible and Rougier 2017; but see Chimento et al. 2012; Fig. 6.2), the only taxon that appears to be deeply nested within the northern dryolestoids is *Groebertherium*. All other SA dryolestoids are recovered in a monophyletic group, the

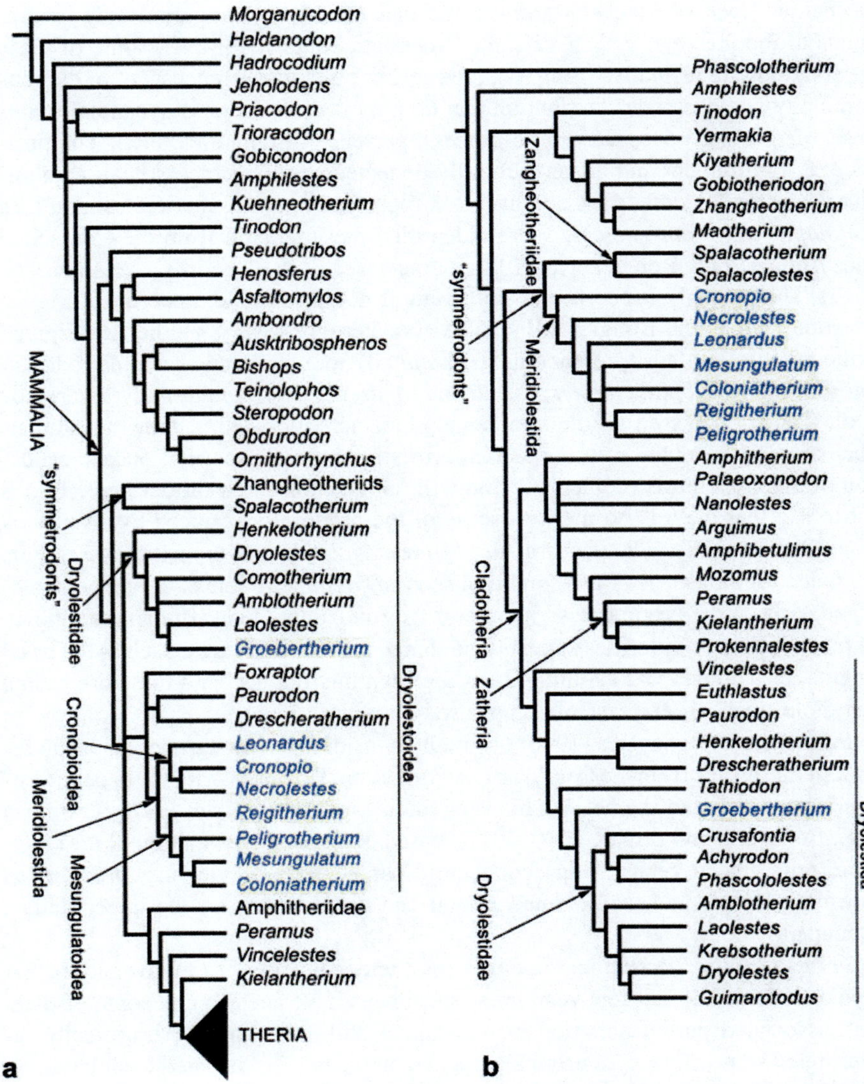


**Fig. 6.2** Phylogenetic tree with alternative hypotheses on the relationships of the Meridiolestida. Wible and Rougier (2017); see also Rougier et al. (2011a, b), with Meridiolestida as part of Dryolestoidea (**a**). Averianov et al. (2013), with Meridiolestida as "symmetrodonts" (**b**)

endemic Meridiolestida, sharing a more distant common ancestor with Laurasian groups (Rougier et al. 2011a; O'Meara and Thompson 2014; Wible and Rougier 2017; Fig. 6.2a). Averianov et al. (2013), however, have challenged the dryolestoid affinities of the meridiolestidans by suggesting that they are the sister taxon of "symmetrodonts" (Fig. 6.2b). This hypothesis was based on Meridiolestida having lower molars with mesiodistally compressed roots, prominent mesial and distal cingulids, absence of a distinct talonid, dentary with alleged absence of angular process (but see below), lack of Meckel's groove, and masseteric process purportedly homologous to the masseteric shelf of some "symmetrodonts". However, some of these observations are factually wrong (e.g., the absence of angular process) or of dubious homology and character distribution that do not optimize them as synapomorphies (e.g., Meckel's groove, masseteric process, presence of cingulids, etc.). The presence of a prominent and posteroventrally deflected angular process is an apomorphic feature of cladotherians and has been reported in several meridiolestidan taxa (*Cronopio*, *Peligrotherium*, and an unidentified mesungulatid from the Coniacian; Páez Arango 2008; Rougier et al. 2011a; Forasiepi et al. 2012; *contra* Averianov et al. 2013). This strongly supports a cladotherian, if not dryolestoid, ancestry of Meridiolestida (Wible and Rougier 2017). The masseteric process is neither widespread in the meridiolestidans (present only in *Cronopio*) nor in holarctic "symmetrodonts" (present solely in *Spalacolestes*; Cifelli and Madsen 1999). Additionally, Averianov et al. (2013) dismissed all the non-dental characters included in previous studies. This strategy, in addition to the factual errors in character scoring, makes evaluation of this hypothesis problematic and will not be considered further here. This is not to say that there is no merit to some of the proposals advanced by Averianov et al. (2013), as the radicular structure of mesungulatids and the extensive cingula are indeed reminiscent of the condition seen in "symmetrodonts", a fact acknowledged early in the original descriptions of the fundational taxa (Bonaparte 1986a). At present, those similarities appear to be shared primitive features, such as the basal cingula, or a convergent evolution between "symmetrodonts" and the more nested meridiolestidans, in the case of compressed roots.

Furthermore, Bonaparte (1990) originally considered symmetrodontan affinities for several species from the Los Alamitos Formation. However, with the exception of *Bondesius ferox* (but see below), his later studies reassigned them to Dryolestoidea (e.g., Bonaparte and Migale 2010, 2015), based on the relative position of the stylocone, pronounced central ridge connecting paracone and stylocone, pronounced parastylar hooks, and well-defined mesial and distal basins on the upper molars (Bonaparte 2002).

In this chapter, we summarize our current understanding of the diversity of SA dryolestoids. In the last few years, new specimens have been found, some of them with associated partial dentition in both upper and lower jaws. We critically reinterpreted some of the taxa originally named on the basis of isolated teeth based on the evidence provided by these more complete taxa. We summarize in Table 6.1 the list of valid genera and species of Dryolestoidea in SA, recognized after the review of this contribution.

**Table 6.1** Genera and species of South American dryolestoids recognized as potentially valid and their junior synonyms

| Genus and Species | Synonym | Clade | Type specimen | Mainly referred specimens | Horizon and age |
|---|---|---|---|---|---|
| *Groebertherium stipanicici* | *Groebertherium novasi*, *Brandonia intermedia*, *Alamitherium bishopi*, *Barberenia araujoae* | Dryolestidae | MACN-RN 13, left Mf | MACN-RN 19, left Mf (holotype *Groebertherium novasi*); MACN-RN 164, right mesial Mf (holotype *Brandonia intermedia*); MACN-RN 166, left penultimate deciduous P (holotype *Barberenia araujoae*); MACN-RN 170, fragmentary right dentary with possible two deciduous p; MACN-RN 1034, left penultimate P (holotype *Alamitherium bishopi*); MML-PV 14, right Mf | Los Alamitos and Allen Fms, Upper Cretaceous |
| *Groebertherium allenensis* | *Barberenia allenensis* | Dryolestidae | MML-PV 13, right penultimate P (formerly holotype *Barberenia allenensis*) | | Allen Fm, Upper Cretaceous |
| *Leonardus cuspidatus* | *Rougietherium tricuspes* | Meridiolestida | MACN-RN 172, fragment of right maxilla with last two P and the posterior alveolous of preceding P, M1–M2, alveoli of M3 | MACN-RN 162, left last M (holotype *Rougietherium tricuspes*), MACN-RN 246 + MACN-RN 1097, fragment right dentary with penultimate and last p and m1–m2 | Los Alamitos Fm., Upper Cretaceous |

(continued)

**Table 6.1** (continued)

| Genus and Species | Synonym | Clade | Type specimen | Mainly referred specimens | Horizon and age |
|---|---|---|---|---|---|
| *Cronopio dentiacutus* | | Meridiolestida | MPCA 454, incomplete skull with upper dentition | MPCA 450, partial left lower jaw with partial teeth; MPCA 453, incomplete skull with right lower jaw | Candeleros Fm., lower Upper Cretaceous |
| *Necrolestes patagonensis* | | Meridiolestida | MACN-A 5742–5753 (single specimen), both dentaries with dentition and partial postcranium | YPM PU 15065, YPM PU 15384, and YPM PU 15699, all including partial skull, jaws, and postcranium | Santa Cruz Fm., lower Miocene |
| *Necrolestes mirabilis* | | Meridiolestida | MLP 92-X-10-14, fragmentary left dentary with deciduous and erupting last p, and m1–m3 alveoli | MPEF-PV 4706, Mf; MPEF-PV 6030, mf | Sarmiento Fm., lower Miocene |
| *Mesungulatum houssayi* | *Austrotriconodon sepulvedai* | Meridiolestida, Mesungulatoidea | MACN-RN 1, left M2 | MACN-RN 01, left M2; MACN-RN 03, left M1; MACN-RN 05, left M3; MACN-RN 06, fragmentary left dentary with m1–m2; MACN-RN 181, right m3; MACN-RN 183, left P2; MACN-RN 239, mesial left P (holotype *Austrotriconodon sepulvedai*) | Los Alamitos Fm., Upper Cretaceous |
| *Mesungulatum lamarquensis* | | Meridiolestida, Mesungulatoidea | MML-PV 10, right M, probably M1 | | Allen Fm., Upper Cretaceous |

(continued)

**Table 6.1** (continued)

| Genus and Species | Synonym | Clade | Type specimen | Mainly referred specimens | Horizon and age |
|---|---|---|---|---|---|
| *Coloniatherium cilinskii* | | Meridiolestida, Mesungulatoidea | MPEF-PV 2087, fragmentary right dentary with alveoli for c and p1, roots of p2, complete p3, and roots of m1–m3 | MPEF-PV 2059, right p2; MPEF-PV 2064, left m1; MPEF-PV 2066, left P1; MPEF-PV 2073, left p3; MPEF-PV 2078, left M1; MPEF-PV 2081, right P3; MPEF-PV 2085, fragmentary right dentary with remnants of m1–m3, and alveolus of p3; MPEF-PV 2088, left P2; MPEF-PV 2092, left m2; MPEF-PV 2138, right m3; MPEF-PV 2163, right M3; MPEF-PV 2183, right M2; MPEF-PV 2230, right p1 | La Colonia Fm., Upper Cretaceous |
| *Reigitherium bunodontum* | | Meridiolestida, Mesungulatoidea | MACN-RN 173, fragmentary right m2 | MPEF-PV 2020, fragmentary left dentary with p4 and m1; MPEF-PV 2072, right P4; MPEF-PV 2237, right m2; MPEF-PV 2238, left M1; MPEF-PV 2317, right m1; MPEF-PV 2338, right p4; MPEF-PV 2339, right P3; MPEF-PV 2341, right M2; MPEF-PV 2369, fragmentary left M3; MPEF-PV 2376, right p3. | Los Alamitos and La Colonia Fms, Upper Cretaceous |
| *Peligrotherium tropicalis* | | Meridiolestida, Mesungulatoidea | UNPSJB-PV 914, fragment of right dentary with p2–m1 | MPEF-PV 2351, partial skull and jaws | Salamanca Fm., lower Paleocene |

(continued)

**Table 6.1** (continued)

| Genus and Species | Synonym | Clade | Type specimen | Mainly referred specimens | Horizon and age |
|---|---|---|---|---|---|
| *Casamiquelia rionegrina* | | Meridiolestida *incertae sedis* | MACN-RN 163, right MF | | Los Alamitos Fm., Upper Cretaceous |
| *Quirogatherium major* | | Meridiolestida *incertae sedis* | MACN-RN 171, penultimate right P | | Los Alamitos Fm., Upper Cretaceous |
| *Paraungulatum rectangularis* | | Meridiolestida *incertae sedis* | MACN-RN 180, left M | Tentative: MACN-RN 10, left m; MML-PV 1298, M; MML-PV 1303, M | Los Alamitos and Allen Fms, Upper Cretaceous |
| *Bondesius ferox* | | Meridiolestida *incertae sedis* | MACN-RN 161, right p | Tentative: MACN-RN 1142, fragmentary right edentulous dentary | Los Alamitos Fm., Upper Cretaceous |
| *Austrotriconodon mckennai* | | Meridiolestida *incertae sedis* | MACN-RN 21, left p | | Los Alamitos Fm., Upper Cretaceous |

## 6.3 Systematics

Mammalia Linnaeus 1758
Dryolestoidea Butler 1939
Dryolestidae Marsh 1879
*Groebertherium* Bonaparte 1986a

Type species: *Groebertherium stipanicici* Bonaparte 1986a.

Included species: The type species and *G. allenensis* (Rougier et al. 2009a).

*Groebertherium stipanicici* Bonaparte 1986a
(Fig. 6.3a–g)

Synonyms: *Groebertherium novasi* Bonaparte 1986a; *Brandonia intermedia* Bonaparte 1990; *Barberenia araujoae* Bonaparte 1990; *Alamitherium bishopi* Bonaparte 2002.

Holotype of *Groebertherium stipanicici*: MACN-RN 13, left upper molariform (Fig. 6.3a).

Holotype of *Groebertherium novasi*: MACN-RN 19, left upper molariform (Fig. 6.3b).

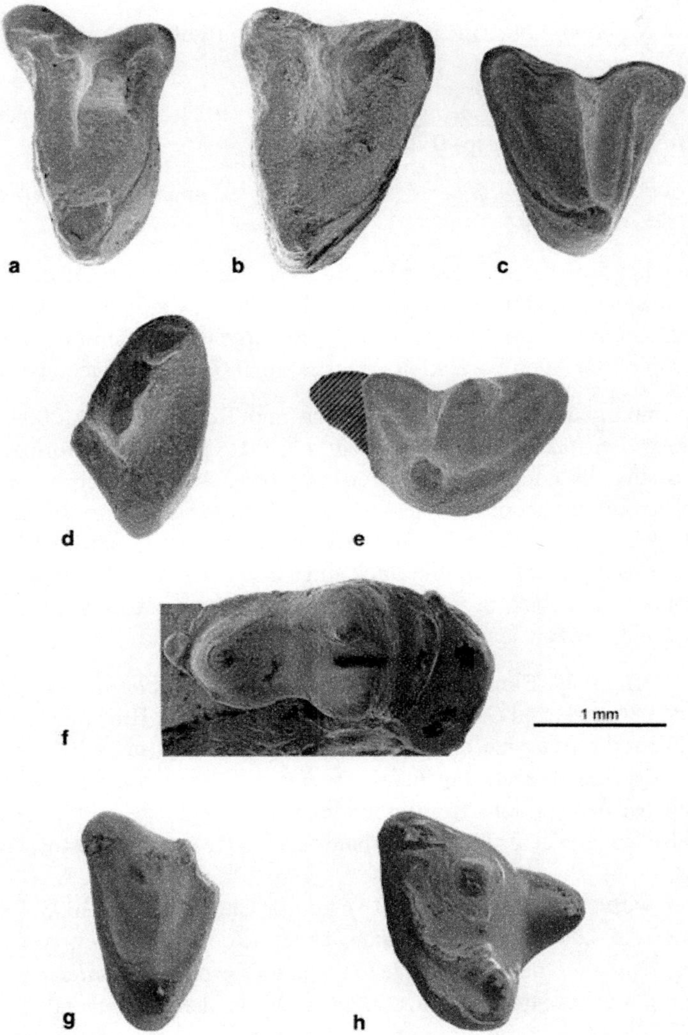


**Fig. 6.3** *Groebertherium stipanicici* from Ea. Los Alamitos and Cerro Tortuga, Río Negro, Argentina; Los Alamitos and Allen formations, Campanian–Maastrichtian, Late Cretaceous. Holotype of *Groebertherium stipanicici*, MACN-RN 13, left upper molariform (**a**); MACN-RN 19 (holotype of *Groebertherium novasi*), left upper molariform (**b**); MACN-RN 164 (holotype of *Brandonia intermedia*, upper right mesial molariform (**c**); MACN-RN 1034 (holotype of *Alamitherium bishopi*), left penultimate premolar, with broken mesial cusp (**d**); MACN-RN 166 (holotype of *Barberenia araujoae*), upper left deciduous penultimate premolar (**e**). MACN-RN 170, fragment of right dentary with a possible penultimate deciduous premolar (**f**). *Groebertherium stipanicici* from Cerro Tortuga, Río Negro, Argentina; Allen Formation, Maastrichtian, Upper Cretaceous; MML-PV 14, right upper molariform in occlusal view (**g**). *Groebertherium allenensis* from Cerro Tortuga, Río Negro, Argentina; Allen Formation, Campanian–Maastrichtian, Upper Cretaceous. Holotype, MML-PV 13, penultimate right upper premolar (**h**). Oblique lines indicate breakage

Holotype of *Brandonia intermedia*: MACN-RN 164, right upper mesial molariform (Fig. 6.3c).

Holotype of *Alamitherium bishopi*: MACN-RN 1034, left upper penultimate premolar (missing mesial cusp) (Fig. 6.3d).

Holotype of *Barberenia araujoae*: MACN-RN 166, left upper deciduous penultimate premolar (Fig. 6.3e).

Locality and horizon: Estancia Los Alamitos, west slope of Cerro Cuadrado, Arroyo Verde, Río Negro, Argentina. Los Alamitos Formation, Campanian–Maastrichtian (Upper Cretaceous). Cerro Tortuga, Río Negro, Argentina. Allen Formation, 38 m below the top of the Allen Formation, Maastrichtian (Upper Cretaceous).

Diagnosis (modified from Bonaparte 1986a and Rougier et al. 2009a): *Groebertherium stipanicici* is a small-sized dryolestid. Upper molariforms with tall crown, mesiodistally compressed, "trigon" approximately symmetrical in occlusal view, with partial cingula or completely absent. The parastyle forms a small hook. There is a stylocone separated from the paracrista and a long, narrow trigon basin. The mesocrista, represented by the thick lingual extension of the lingual slope of the stylocone, can contact, or not, the paracone. Lower molariforms mesiodistally compressed, with narrow anterior cingulum and small talonid.

Comments: Originally, Bonaparte named two species: *Groebertherium stipanicici* (Bonaparte 1986a: 51) and *G. novasi* (Bonaparte 1986a: 52). However, their distinction (subtle differences in size, relative robustness of molariforms, relative gracility of cusps, and sharpness of crests; Fig. 6.3a, b) seems to respond to variations in the tooth series and/or wear conditions from individuals of a single species (Chornogubsky 2003; see also Rougier et al. 2009a; Averianov et al. 2013), and so are now considered conspecific.

We agree with Averianov et al. (2013), *contra* Chornogubky (2003: 28–30, see also Rougier et al. 2009a), in that it is unnecessary to designate a new type for *Groebertherium stipanicici* (despite damage to the holotype of *G. stipanicici* MACN-RN 13, portions of it exist in the collection, and in addition, there are several photographs and plastic casts of the original material). As long as the discussion involves two separate specimens, they cannot be made into an objective synonym.

The species *Brandonia intermedia*, holotype MACN-RN 164 (Fig. 6.3c) and other referred specimens (e.g., Bonaparte 1990; Bonaparte and Migale 2010), was characterized by having upper molariforms slightly mesiodistally longer than in *Groebertherium stipanicici* (holotype), stylocone closer to the labial border of the tooth, deep ectoflexus, mesocrista connecting the stylocone and paracone, and anterior cingulum converging in the parastyle. Comparisons with other dryolestoids, particularly the more complete *Cronopio*, *Coloniatherium*, and *Peligrotherium* allow one to judge the validity of the characters used to diagnose *Brandonia*. We consider here that the morphology of *Brandonia* and the features used to diagnose it apply to a more mesial position in the tooth series of *Groebertherium stipanicici*, considering MACN-RN 13 and MACN-RN 19 more distal elements than MACN-RN 164.

*Alamitherium bishopi* was created by Bonaparte (2002) on the basis of a single specimen, its holotype MACN-RN 1034 (Fig. 6.3d). Originally, it was indicated as a complete crown, however, close restudy of the single specimen demonstrates that the mesial border of the tooth is broken and missing the mesial cusp typical of the penultimate premolars of other dryolestoids (see *Groebertherium allensis*, MML-PV 13; Fig. 6.3h and *Coloniatherium* P2). Diagnostic features originally indicated for *Alamitherium* (Bonaparte 2002) included the lack of the anterior basin of the trigon and short paracrista (features that are similar to the penultimate upper premolar of *Groebertherium allensis*, MML-PV13; Fig. 6.3h). *Alamitherium* MACN-RN 1034 has a vertical facet on the mesial face of the paracone, similar to other dryolestoid penultimate premolars (e.g., *Coloniatherium* P2). In addition, crown size and the stylocone larger than the paracone resembles *Groebertherium stipanicici* (holotype). In short, these arguments suggest that MACN-RN 1034 (Fig. 6.3d) corresponds to a left penultimate premolar, missing the mesial cusp by breakage, of *Groebertherium*, and thus we regard it as a junior synonym.

*Barberenia araujoae*, holotype MACN-RN 166 (Fig. 6.3e) and other referred specimens (Bonaparte 1990; Bonaparte and Migale 2010, 2015), has its crown much longer mediodistally than buccolingually, deep ectoflexus, paracone larger than stylocone and both cusps close to each other, stylocone positioned on the distal half of the crown, broad trigon basin divided into anterior and posterior parts by the mesocrista, well-developed, hook-like parastyle, broad anterior cingulum, and small metacone when present. Several authors have suggested that the particular morphology associated to the molars referred to *Barberenia* may actually correspond to deciduous premolars (Martin 1999; Bonaparte 2002; Averianov 2002; Chornogubsky 2003; Kielan-Jaworowska et al. 2004; Bonaparte and Migale 2010, 2015), possibly from *Brandonia* (Martin 1999; Chornogubsky 2003; Kielan-Jaworowska et al. 2004) or *Groebertherium* (Bonaparte and Migale 2010), an opinion with which we concur. Alternatively, Averianov et al. (2013) mentioned the hypothesis of considering *Barberenia* as the deciduous dentition of *Leonardus*. However, the size of MACN-RN 166 is considerably smaller than *Leonardus* and in the range of *Groebertherium*. In addition, the large stylocone closer to the labial side of the tooth, presence of an anterior cingulum, and trigon basin divided by the mesocrista recall the morphology of this taxon.

Among the referred material of *Groebertherium stipanicici*, there is an interesting specimen, MACN-RN 170 (Fig. 6.3f) from the Los Alamitos Formation, originally assigned to *Barberenia araujoae* (Bonaparte 1990). This specimen is a fragment of right dentary with one elongated tooth and another triangular tooth allegedly of the same individual (Bonaparte 1990). The elongated tooth was originally identified by Bonaparte as a fifth molariform—mf5—while the triangular tooth as a third molariform—mf3 (Bonaparte 1990: Fig. 8). After re-examination, Chornogubsky (2003) argued that the dentary has two contiguous molariforms, likely the last deciduous premolar (mf5 from Bonaparte 1990) and first molar (mf3 from Bonaparte 1990), in reverse order to what Bonaparte originally interpreted. By then considering *Barberenia* as a synonym of *Brandonia*, Chornogubsky (2003) interpreted MACN-RN 170 as part of *Brandonia*. Averianov et al. (2013) supported the mesial position of the elongated molariform as a premolar (p2), and the triangular tooth as a first molariform. They considered the possibility of a missing tooth between each preserved element (as originally suggested by Bonaparte 1990). However, they attribute MACN-RN 170 to *Leonardus*, which is untenable by direct inspection of the specimens including differences in size. We agree with Chornogubsky (2003), and we regard both positions in MACN-RN 170 as continuous elements, the elongated tooth as a deciduous—a possible penultimate deciduous premolar. However, unlike Chornogubsky (2003), we consider the triangular element (Bonaparte's mf3; Chornogubsky's m1) to also be deciduous—an ultimate deciduous premolar. This triangular tooth has an anterior cingulid that does not exist in *Groebertherium* nor in *Leonardus*, and a small mesial root, not well-developed and partially re-absorbed, representing a fully molarized last deciduous premolar. Based on the size, we refer this specimen to *Groebertherium* (= *Brandonia*, in line with Chornogubsky 2003) and not to *Leonardus* (such as Averianov et al. 2013).

*Groebertherium stipanicici* is known mostly from material from the Los Alamitos Formation. In addition, an isolated right upper molariform (MML-PV 14; Fig. 6.3g) referred to this species was recovered from the Cerro Tortuga locality, Allen Formation (Rougier et al. 2009a).

*Groebertherium allenensis* (Rougier et al. 2009a)
(Fig. 6.3h)

Synomys: *Barberenia allenensis* Rougier et al. 2009a.

Holotype: MML-PV 13, penultimate right upper premolar (Fig. 6.3h).

Locality and horizon: Cerro Tortuga, Río Negro, Argentina. Allen Formation, 38 m below the top of the Allen Formation, Maastrichtian (Upper Cretaceous).

Diagnosis (from Rougier et al. 2009a): Small-sized dryolestoid, slightly larger than *G. stipanicici* with more robust and bulging cusps. It has a skewed compressed basin, similar to *Quirogatherium major* and *Groebertherium stipanicici.* Differs from *Quirogatherium* in the more gracile cusp pattern, smaller size, and more skewed cusp arrangement. *Groebertherium allenensis* is similar to *G. stipanicici* MACN-RN 1034 in having sharp, tall cusps, but the Allen Formation species has a less mesiodistally elongated triangular outline with smaller parastylar area and parastylar hook, and less developed anterior cingulum.

Comments: *Groebertherium allenensis,* holotype MML-PV 13 (Fig. 6.3h), is regarded as a penultimate upper premolar (Rougier et al. 2009a). It follows that *Alamitherium bishopi* (holotype, MACN-RN 1034) is regarded as the penultimate premolar of *Groebertherium stipanicici. Groebertherium allenensis* is larger than the teeth of *G. stipanicici,* but they are within acceptable range. In addition, both specimens differ in the sharpness and bulging of the stylocone (more robust in MML-PV 13) and the presence of an accessory cusp on the labial margin of the tooth (in *G. stipanicic*i, MACN-RN 1034). Averianov et al. (2013) suggested this premolar morphology could represent the penultimate premolar of *Leonardus* instead. This is a viable and attractive possibility given the larger size and large stylocone of MML-PV13; however, at present, there is no *Leonardus*-type molariform in the Allen Formation, but *Groebertherium* is present. Until new specimens recording a *Leonardus* morphology are discovered, we prefer to keep MML-PV 13 as a species of *Groebertherium.*

Meridiolestida Rougier et al. 2011a
*Leonardus* Bonaparte 1990

Type species: *Leonardus cuspidatus* Bonaparte 1990.

Included species: The type only.

*Leonardus cuspidatus* Bonaparte 1990
(Fig. 6.4)

Synonyms: *Rougietherium tricuspes* Bonaparte 2002.

Holotype of *Leonardus cuspidatus*: MACN-RN 172, fragment of right maxilla with last two premolars and the posterior alveolus of the preceding premolar, first two molars (M1–M2), alveoli of the posterior molar (Fig. 6.4a).

Holotype of *Rougietherium tricuspes*: MACN-RN 162, complete left last upper molar (Fig. 6.4b).

Locality and horizon: Estancia Los Alamitos, west slope of Cerro Cuadrado, Arroyo Verde, Río Negro Province, Argentina. Los Alamitos Formation, Campanian–Maastrichtian (Upper Cretaceous).

Diagnosis (modified from Bonaparte 1990): Small-sized dryolestoid with upper molars mesiodistally narrow, lacking parastylar hook and with subequal parastylar and metastylar cusps, large stylocone of central position; shallow ectoflexus; molars widely separated from each other, paracone mesiodistally narrow; absence of a metacone, and further reduction of the posterior root of molars, absence of cingulae in both upper and lower molars.

Comments: The holotype of *L. cuspidatus* MACN-RN 172 was originally interpreted by Bonaparte (1990; see also Chornogubsky 2003, 2011; Averianov et al. 2013) as a left maxilla, preserving the penultimate molar and three preceding teeth for a total of at least five. However, there are several features that suggest that the specimen is a right maxilla (see also Páez Arango 2008). (1) On the lingual border of the maxilla, a small broken part of the base of the zygomatic arch is preserved and distally projected (Fig. 6.4a). The maxilla was collected by one of us (GWR) while doing picking; at the time of collection, the remnant of the zygomatic arch was larger and supported the interpretation of the maxilla as a right one; unfortunately, a portion of it was soon lost by handling during original study and before illustration (in Bonaparte 1990). (2) The major axis of the molars is anteromedially oriented. Consequently, the relatively shorter paracrista of the upper molar works against the shorter metacristid of the homologous lower molar and the strongly oblique metacrista of the uppers against the strongly oblique paracristid of the lowers (Fig. 6.4c). A similar oblique orientation

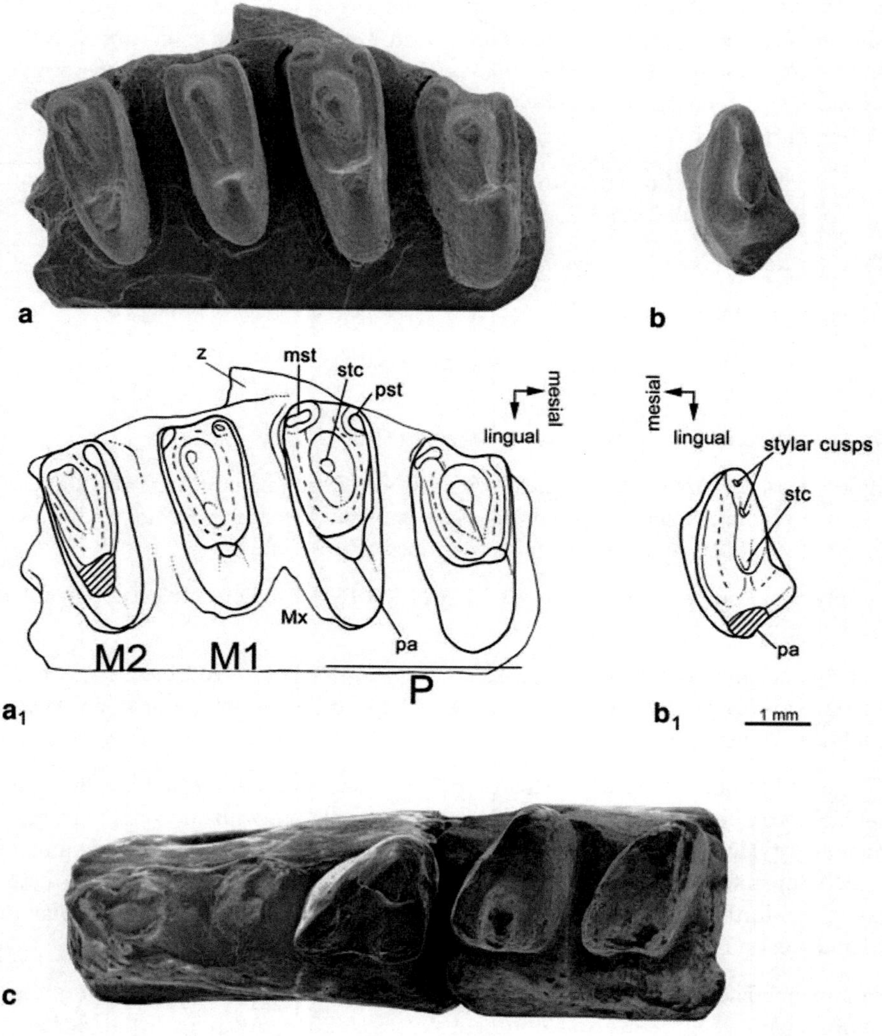


**Fig. 6.4** *Leonardus cuspidatus* from Ea. Los Alamitos, west slope of Cerro Cuadrado, Río Negro, Argentina; Los Alamitos Formation, Campanian–Maastrichtian, Upper Cretaceous. Holotype, MACN-RN 172, fragment of right maxilla with last two premolars and the posterior alveolus of preceding premolar, first two molars, and alveoli of the posterior molar (**a**) and accompanying line drawing (**a$_1$**); MACN-RN 162 (holotype of *Rougietherium tricuspes*), last upper left ultimate molar (**b**) and accompanying line drawing (**b$_1$**). Fragmentary right dentary with dentition (same specimen but catalogued separately; see Forasiepi et al. 2012): MACN-RN 246, fragment of right dentary with alveoli for lost element and one tooth (probably the alveoli for penultimate premolar and ultimate premolar, respectively); MACN-RN 1097, fragment of right dentary with two molariforms (probable m1 and m2) (**c**). Abbreviations: M1–M2, first and second upper molars; mst, metastyle; Mx, maxilla; pa, paracone; pst, parastyle; P, upper premolar; stc, stylocone; z, root of zygomatic arch. Oblique lines indicate breakage

of the upper molars is present in the dentitions of *Henkelotherium* (Krebs 1991), *Drescheratherium* (Krebs 1998), and several other species of dryolestoids from the Jurassic of England, USA, and Portugal (Martin 1999, 2000) (see discussion in Páez Arango 2008). Among meridiolestidans, *Cronopio* also has teeth with an oblique axis (Fig. 6.5b), especially those at the level of the zygomatic arch, and becoming distally more transverse. The specimen has been mounted on a pin so that the dorsal aspect of

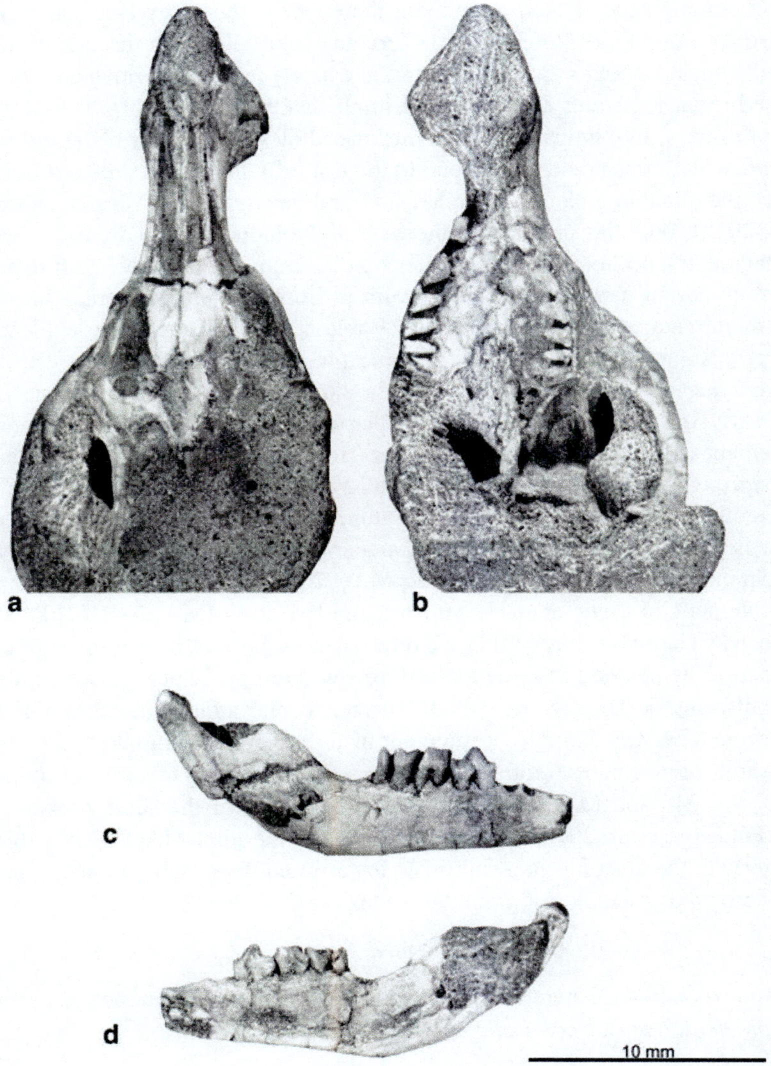


**Fig. 6.5** *Cronopio dentiacutus* from La Buitrera, Río Negro, Argentina; Candeleros Formation, Cenomanian, lower Upper Cretaceous. Holotype, MPCA 454, incomplete skull with upper dentition in dorsal (**a**) and ventral (**b**) views; MPCA 453, right dentary in labial (**c**) and lingual (**d**) views

the maxillary platform was not easily accessible. The curator of the MACN collection at the time, Dr. A. Kramarz, allowed the specimen to be unglued, which shows in dorsal view the path of the infraorbital canal. In mammals, the infraorbital canal starts in the orbit at the maxillary foramen medial to the tooth row and continues anteriorly to pierce the maxilla into the rostrum lateral to the tooth row. The orientation of the infraorbital canal in MACN-RN 172 supports our interpretation of the specimen as a right maxilla.

As a consequence of this orientation, the element shown by Bonaparte (1990; see also Chornogubsky 2003, 2011; Averianov et al. 2013) as the last preserved "molar" almost lacking wear could represent a newly replaced permanent premolar, followed by at least four elements. The immediately distal tooth is the largest of the tooth series; by comparing with other meridiolestidans (e.g., *Coloniatherium*, *Peligrotherium*), this could correspond to the last fully molarized premolariform. If so, then the remaining elements are M1–M2 and the alveoli for M3 (Rougier et al. 2011a, 2012). With this orientation, the wear of the molars would diminish distally.

Finally, the morphology of the palate and the arrangement of the teeth in arcade suggest an abrupt narrowing of the rostrum in front of the premolars, a condition seen among taxa with elongated snouts (Wible et al. 2004) and very evidently in *Cronopio* (Rougier et al. 2011a) and *Necrolestes*, among Meridiolestida.

*Rougietherium tricuspes* is based on the holotype and only known specimen, MACN-RN 162 (Fig. 6.4b, an isolated upper tooth; Bonaparte 2002). We interpret this element as the upper left ultimate molar (Bonaparte interpreted it as a right one) of *Leonardus cuspidatus* in agreement with its size. The reduction of the distal part of the tooth and consequent transverse alignment of paracone, stylocone, and stylar cusps suggest it as a last element. The paracone is slightly broader in MACN-RN 162 than in the holotype of *L. cuspidatus* MACN-RN 172.

A fragment of right dentary with two molariforms (MACN-RN 1097) was described by Chornogubsky (2011) and referred to *Leonardus*, an opinion with which we concur. The position of these molariforms was left open, being referred simply as molariforms in the text, but called first and second molar in a table. Later on, the specimen MACN-RN 246, a fragment of right dentary with one tooth and the two alveoli for an anterior tooth, was mentioned as belonging to *Leonardus*; in fact, MACN-RN 246 and MACN-RN 1097 are two pieces of the same broken specimen (Forasiepi et al. 2012; Fig. 6.4c). We regard the joint MACN-RN 246/1097 as preserving the alveoli of a penultimate lower premolar (double rooted, long and narrow crown predicted), last premolar, and m1–m2.

Cronopioidea nov.

Definition: Established here as a clade formed by the last common ancestor of *Cronopio dentiacutus*, *Necrolestes* spp., and all its descendants.

*Cronopio* Rougier et al. 2011a

Type species: *Cronopio dentiacutus* Rougier et al. 2011a.

Included species: The type only.

*Cronopio dentiacutus* Rougier et al. 2011a
(Figs. 6.5, 6.6 and 6.7)

Synomys: *Pronopio denticulatus* Bonaparte and Migale 2015; *lapsus calami.*

Holotype: MPCA 454, incomplete skull with upper dentition (Fig. 6.5a, b).

Locality and horizon: La Buitrera locality, Río Negro Province, Argentina. Candeleros Formation, Neuquén Group, Cenomanian (lower Upper Cretaceous).

Diagnosis (from Rougier et al. 2011a): Medium-sized dryolestoid, with high orbits, tall zygoma, small temporal area, extremely elongated rostrum, and extended edentulous portion of the premaxilla. *Cronopio* and *Leonardus* share the abrupt mesial narrowing of the rostrum and the incurving of the maxilla with the major axis of the molars directed mesiolingually. Dental formula I2/i?, C1/c1, P4/p3+ , M3/m3. Upper incisors peg-like. Canine very long and moderately compressed buccolingually. Upper and lower premolars with two roots and molars with a single labiolingually broad root. P1 separated from P2 by extensive diastema, P1–P2 lacking anterior accessory cusps but present in P3–P4. Molars strongly mesiodistally compressed; parastylar hook small and parastyle absent, except on M1. Molars of *Cronopio* differ from other SA dryolestoids like *Leonardus cuspidatus* in having contact between stylocone and paracrista, stylocone not as large, and presence of a distinct small parastylar hook. *Cronopio* and other SA dryolestoids share the presence of a posterior premolar (P3 in *Cronopio*) with a small anterior cusp supported by a circular root and a distinct mesiodistal basin supported by a transverse one. *Cronopio* and other SA dryolestoids share with all other dryolestoids a much taller labial crown height than lingual in the lower molars (the reverse in the uppers), a small talonid, a relatively transverse metacristid, and mesiodistally compressed trigonids. The jaw of *Cronopio* is unusual among dryolestoids, with a small coronoid process, small angular process, and a very elevated condyle. The angular process is long, hook-like, and ventromedially inflected (but without forming a medial shelf as seen in metatherians). The most distinctive feature is a large and prominent masseteric process that juts out laterally forming a broad platform and defining a deeply excavated masseteric fossa. A masseteric process is known among some "symmetrodonts" (Cifelli and Madsen 1999; Averianov et al. 2013) but it is much smaller and likely not homologous under standard topologies of phylogenetic trees of dryolestoids and "symmetrodonts".

Comments: The fossil material of *Cronopio* includes the most complete skull of a Mesozoic dryolestoid known from SA, including two partial skulls, dentary, upper and lower dentition, and many details of the basicranial anatomy. Its skull shows a combination of primitive and highly specialized mammalian features, including the presence of a septomaxilla, anterior lamina of the petrosal, lateral flange, and all the elements of cranial circulation thought to be primitive for therians, such as a fully developed stapedial system (Rougier et al. 2011a). On the contrary, the long and narrow snout, the palate with posterior premolars and molars forming a broad arch, resembling living elephant shrews, and the very long and slender upper canines are derived features, unique to this extinct form. The dryolestid *Drescheratherium*

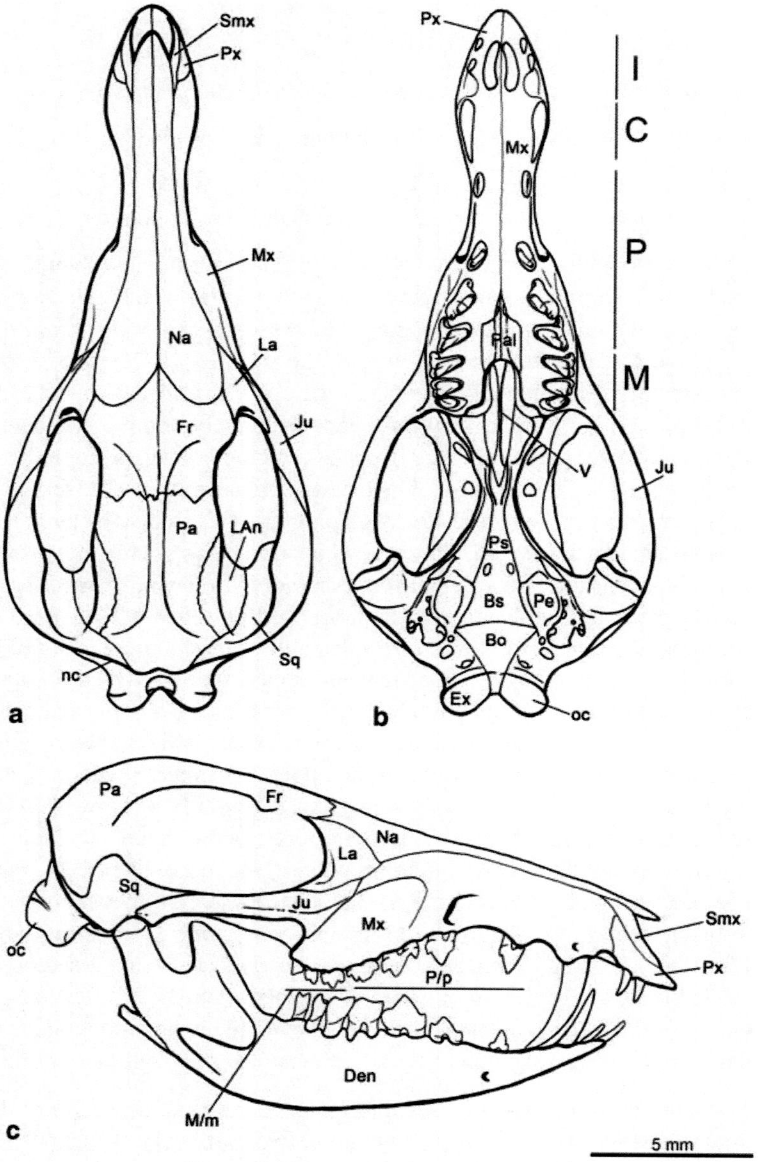


**Fig. 6.6** Skull reconstruction of *Cronopio dentiacutus*, in dorsal (**a**), ventral (**b**), and lateral (**c**) views, modified from Rougier et al. (2011a). Abbreviations: Bo, basioccipital; Bs, basisphenoid; C, upper canine; Den, dentary; Ex, exoccipital; Fr, frontal; I, upper incisors; Ju, jugal; La, lacrimal; LAn, lamina anterior of the petrosal; M/m, upper and lower molars; Mx, maxilla; Na, nasal; nc, nuchal crest; oc, occipital condyle; P/p, upper and lower premolars; Pa, parietal; Pal, palatine; Pe, petrosal; Ps, presphenoid; Px, premaxilla; Smx, septomaxilla; Sq, squamosal; V, vomer

**Fig. 6.7** *Cronopio dentiacutus*, artistic reconstruction by Jorge L. Blanco

(Krebs 1998) from the Jurassic of Portugal is somewhat similar to *Cronopio* by having very long canines and *Leonardus* may have a similarly shaped dental arcade. The finding of *Cronopio* revealed the existence of a large morphological disparity of the Mesozoic mammals in the southern continent with previously unrecognized morphotypes hinting to a great ecological diversity not easily gleamed from dentitions alone.

*Necrolestes* Ameghino 1891

Type species: *Necrolestes patagonensis* Ameghino 1891.

Included species: The type species and *N. mirabilis* Goin et al. 2007.

*Necrolestes patagonensis* Ameghino 1891
(Figs. 6.8, 6.9 and 6.10)

Holotype: MACN-A 5742–5753 (all numbers represent a single specimen), right and left dentaries with dentition and partial postcranium (Fig. 6.8a).

Locality and horizon: Several localities (10 miles South of Coy Inlet, 5 miles South of Coy Inlet, Killik Aike, Ea. La Costa, and Monte León), Santa Cruz, Argentina. Santa Cruz Formation, lower Miocene (Santacrucian SALMA, Burdigalian).

Diagnosis (from Rougier et al. 2012): Meridiolestidan that shares the following features with *Cronopio*: presence of long rostrum, globular braincase, dentary with

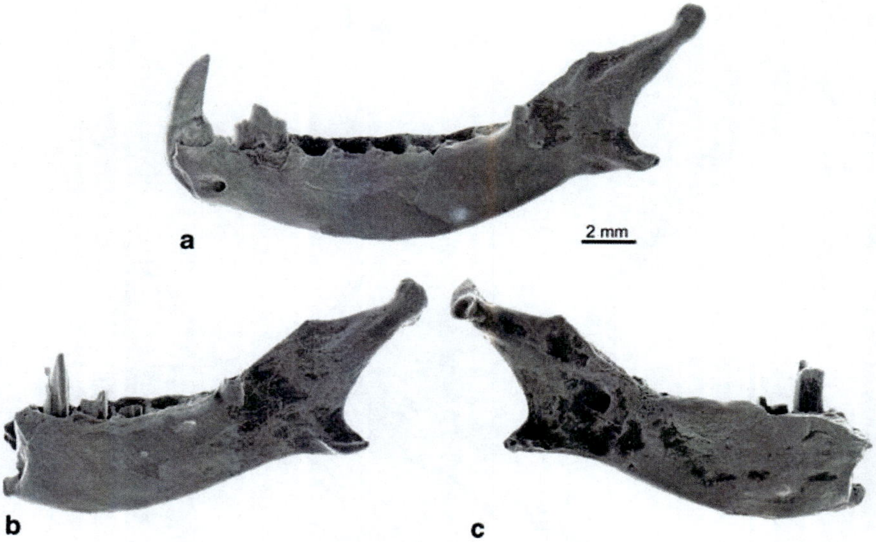


**Fig. 6.8** *Necrolestes patagonensis* from Santa Cruz, Argentina; Santa Cruz Formation, Burdigalian, lower Miocene. Holotype, MACN-A 5742, left dentary in labial view (**a**) MACN-A 10252, left dentary in labial (**b**) and lingual (**c**) views

long, relatively horizontal condylar process, angular process with some medial inflection (but not shelf-like, unlike the medially inflected angular process of metatherians), dentition with double-rooted anterior premolar, single-rooted molars, very tall molar crowns, no recognizable distinction between molar crowns and roots, and enamel extending deep into the molar alveoli (hypsodonty). It shares with *Leonardus* the presence of two molarized premolars and absence of talonids or other accessory cusps. *Necrolestes*, *Cronopio*, and *Leonardus* share a curved postcanine dental arcade. *Necrolestes* differs from other meridiolestidans by the presence of massive subtriangular canines, simple tricuspid (triangular) molar pattern, the presence of a characteristic upturned rostrum, and a prenasal process of the premaxilla. Dental formula I5/i4, C1/c1, P3/p3, M3/m3.

Comments: The Santa Cruz Formation in southern Patagonia, Argentina, where *Necrolestes patagonensis* comes from, is the richest high-quality fossiliferous unit of the Cenozoic of SA (Vizcaíno et al. 2012). However, *Necrolestes* material is scarce, representing one of the vertebrates with fewer specimens (Tauber et al. 2005). Even to date, the best preserved cranial material is that collected by J. B. Hatcher and O. A. Peterson during the 1896–1899 Princeton Expeditions to Patagonia (Hatcher 1903; Scott 1905; Fig. 6.9).

*Necrolestes* has several primitive features in the skull (i.e., presence of a septomaxilla, internarial bar of premaxilla, large ventral opening for cavum epiptericum, well-developed anterior lamina and lateral flange, and primitive vascular system with a dominant arteria diploetica magna supplying orbit and basicranium, including a

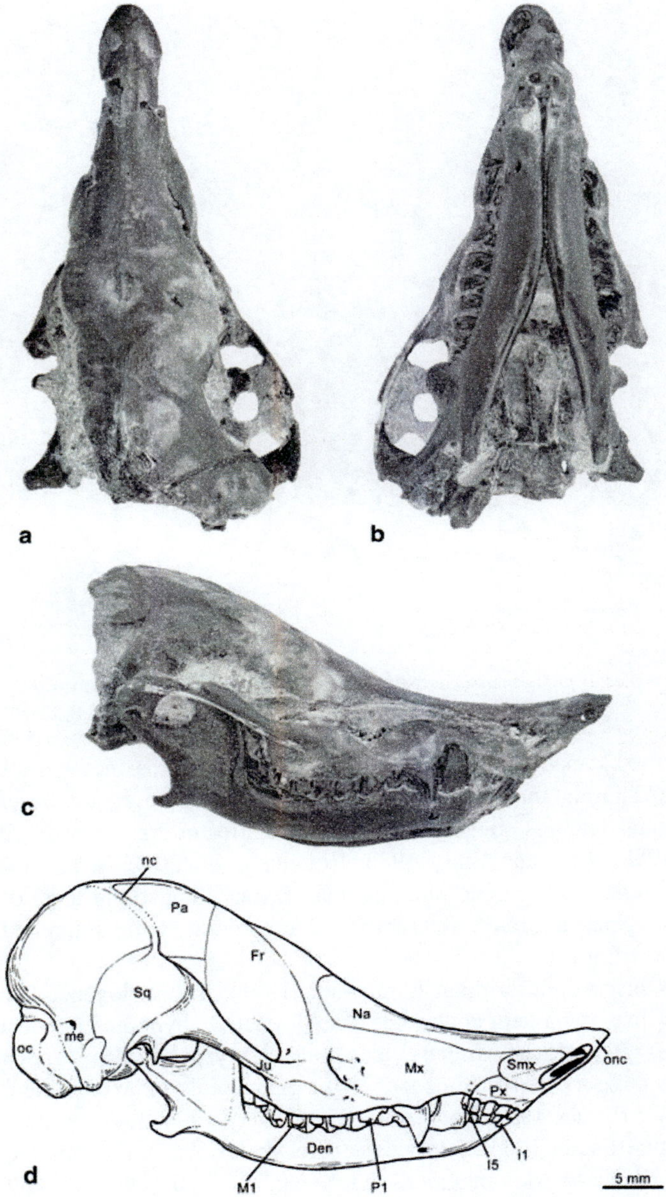


**Fig. 6.9** *Necrolestes patagonensis*, from Santa Cruz, Argentina; Santa Cruz Formation, Burdigalian, lower Miocene. YPM PU 15699, skull and dentary in dorsal (**a**), ventral (**b**), and right lateral (**c**) views. Skull reconstruction of *N. patagonensis* (**d**), based on Rougier and Wible (2017). Abbreviations: Den, dentary; Fr, frontal; i1, first lower incisor; I5, fifth upper incisor; Ju, jugal; M1, first upper molar; me, mastoid exposure of petrosal; Mx, maxilla; Na, nasal; nc, nuchal crest; oc, occipital condyle; onc, ossified nasal cartilage; P1, first upper premolar; Pa, parietal; Px, premaxilla; Smx, septomaxilla; Sq, squamosal

**Fig. 6.10** *Necrolestes patagonensis*, artistic reconstruction by Jorge L. Blanco

fully developed stapedial system), and derived traits (i.e., ossified nasal cartilages, long upturned nasals) that make its anatomical construction unique. Several cranial and postcranial features suggested subterranean life for *Necrolestes* (Scott 1905; Patterson 1958; Rose and Emry 1983; Rougier et al. 2012; Wible and Rougier 2017), presenting it as a mole-like mammal. In fact, this strong level of morphological convergence has interfered for more than a century in the interpretation of its phylogenetic affinities.

Since its original description (Ameghino 1891), the phylogenetic position of *Necrolestes* has remained controversial. Alternative hypotheses considered *N. patagonensis* related to the African golden moles, Chrysochloridae (Ameghino 1891; Scott 1905; Gregory 1910; Simpson 1945), marsupials or extinct stem relatives (Winge 1941; Romer 1945; Patterson 1958; Szalay 1994; Ladevèze et al. 2008), palaeanodonts (Saban 1954), xenarthans (McDowell 1958), therians (Asher et al. 2007), or even non-therian mammals (McKenna and Bell 1997; Asher et al. 2007; Goin et al. 2007), including gondwanatherians (Van Valen 1988). However, after further preparation of the original cranial material described by Scott (1905) and ultimately the finding of *Cronopio* (Rougier et al. 2011a), the morphology of *Necrolestes* was re-analyzed and re-interpreted under a new framework of comparisons. These more recent studies, in turn, suggested that *N. patagonensis* is a late surviving member of the dryolestoid Meridiolestida (Rougier et al. 2012; see also Chimento et al. 2012;

Averianov et al. 2013, though these authors include Meridiolestida among "symmetrodonts"; O'Meara and Thompson 2014; Wible and Rougier 2017) and in most of these results, it is grouped together with *Cronopio* and *Leonardus* (Rougier et al. 2012; O'Meara and Thompson 2014; Wible and Rougier 2017). Clear imprints of meridolestidan affinities are in the ear region, sidewall of the braincase, snout, and dentition (Wible and Rougier 2017). *Necrolestes* is the youngest surviving remnant of emblematic Mesozoic archaic lineages surviving up to the early Miocene in likely marginal environments otherwise dominated by endemic eutherians and metatherians in an island continent (Rougier et al. 2012; see also Chap. 10).

*Necrolestes mirabilis* Goin et al. 2007
(Fig. 6.11)

Holotype: MLP 92-X-10-14, fragment of left dentary with deciduous (dp3) and erupting last premolar (p3) in place, and the alveoli of m1–m3 (Fig. 6.11a).

Locality and horizon: Puesto Almendra, Gran Barranca (= Barranca sur del Lago Colhué Huapi), Chubut, Argentina. Sarmiento Formation, Colhué Huapi Member; lower Miocene (Colhuehuapian SALMA, Burdigalian).

Diagnosis (from Goin et al. 2007): Slightly larger than *N. patagonensis*, dentary with ventral border of the horizontal ramus straighter, alveoli of lower molars with labial border slightly more curved, larger size difference between last premolar (p3) and molars, lower postcanines with protoconids more separated from each other.

Comments: The dentition in *Necrolestes patagonensis* includes six postcanines: one double-rooted tooth and five single-rooted molariforms. Discussions have addressed the homologies of those teeth (Patterson 1958; Goin et al. 2007). However, the presence in the third postcanine locus of a deciduous element and its permanent tooth in replacement in the holotype of *N. mirabilis* was strong evidence to conclude

**Fig. 6.11** *Necrolestes mirabilis* Puesto Almendra, Gran Barranca, Chubut, Argentina; Sarmiento Formation, Colhué-Huapi, Burdigalian, lower Miocene. Holotype, MLP 92-X-10-14, fragment of left mandible with deciduous premolar (dp3) in replacement by the permanent premolar (p3) and the alveoli of m1–m3 (**a**); MPEF-PV 4706, upper molariform in occlusal view (**b**); MPEF-PV 6030, lower molariform in occlusal view (**c**) Modified from Goin et al. (2007)

that the postcanines correspond to three premolars (one non-molariform premolar and two fully molarized premolars) and three molars (Goin et al. 2007). The hypothesis proposed that a likely mesial premolar was missing on the basis of a large diastema between canine and first premolar (Goin et al. 2007), but no further evidence suggests a fourth missing premolar in the dental formula of *Necrolestes*. In any case, the holotype of *N. mirabilis*, MLP 92-X-10-14, in an exceptional ontogenetic stage, provided evidence to Goin et al. (2007) suggesting that *Necrolestes* had no features that unequivocally support metatherian affinities. The replacement pattern can then either be regarded as showing similarities with eutherians or with extinct lineages not referable to Metatheria or Eutheria, which eventually led Goin et al. (2007) to place Necrolestidae as Mammalia *incertae sedis*. The meridiolestidan affinities resolve these inconsistencies and render *Necrolestes* as a close relative of the equally bizarre *Cronopio* (Rougier et al. 2012; Wible and Rougier 2017).

Mesungulatoidea Rougier et al. 2011a
*Mesungulatum* Bonaparte and Soria 1985

Type species: *Mesungulatum houssayi* Bonaparte and Soria 1985.

Included species: The type species and *M. lamarquensis* Rougier et al. 2009a.

*Mesungulatum houssayi* Bonaparte and Soria 1985
(Figs. 6.12 and 6.13)

Synonyms: *Austrotriconodon sepulvedai* Bonaparte 1992.

Holotype of *Mesungulatum houssayi*: MACN-RN 1, isolated left M2 (Fig. 6.12a).

Holotype of *Austrotriconodon sepulvedai*: MACN-RN 239, mesial left upper premolar (Fig. 6.12d).

Locality and horizon: Estancia Los Alamitos, west slope of Cerro Cuadrado, Arroyo Verde, Río Negro Province, Argentina. Los Alamitos Formation, Campanian–Maastrichtian (Upper Cretaceous).

Diagnosis (modified from Bonaparte and Soria 1985; Chornogubsky 2003; Bonaparte and Migale 2010): Large meridiolestid, only slightly smaller than *Coloniatherium cilinskii*. Upper and lower molars quadrangular in shape with broad cingula/cingulids on the mesial and distal faces. Upper molars with large stylocone on the labial side of the teeth and connected by crests to paracone, parastyle, and metastyle cusps. Upper molars lacking metacone, like most meridiolestidans. Lower molars with metaconid transversely aligned with protoconid.

Comments: Originally *Mesungulatum houssayi* was described as an ungulate-like "condylarth" (Bonaparte and Soria 1985); however, this view was quickly revised and the taxon was considered as Dryolestoidea (Bonaparte 1986a). *Mesungulatum houssayi* and in fact MACN-RN 1 was the first definitive Mesozoic mammal recovered for SA (see Chap. 2), in clear association with hadrosaur, titanosaur, and theropod dinosaurs, and other Mesozoic vertebrates. The finding of *Mesungulatum* opened a

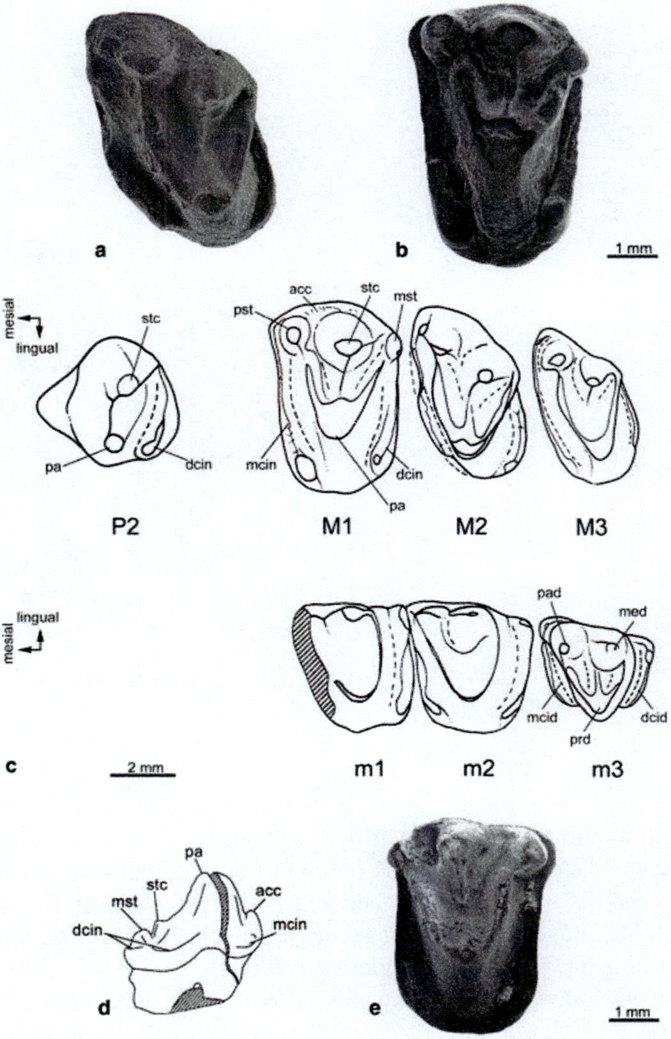


**Fig. 6.12** *Mesungulatum houssayi* from Ea. Los Alamitos, west slope of Cerro Cuadrado, Río Negro, Argentina; Los Alamitos Formation, Campanian–Maastrichtian, Upper Cretaceous. Holotype, MACN-RN 1, left M2 (**a**) and MACN-RN 03, left M1 (**b**). Reconstruction of the tooth sequence of *M. houssayi* (**c**), line drawing of MACN-RN 183, left P2; MACN-RN 03, left M1; MACN-RN 01, left M2; MACN-RN 05, left M3; MACN-RN 06, left dentary with m1 and m2; MACN-RN 181, right m3 (inverted). MACN-RN 239 (holotype of *Austrotriconodon sepulvedai*), mesial left upper premolar (**d**) (line drawing from Gaetano et al. 2013). *Mesungulatum lamarquensis* from Cerro Tortuga, Río Negro, Argentina; Allen Formation, Maastrichtian, Upper Cretaceous. Holotype, MML-PV 10, right upper molar, probably M1 (**e**). Abbreviations: acc, accessory cusp; dcid, distal cingulid; dcin, distal cingulum; mcid, mesial cingulid; mcin, mesial cingulum; med, metaconid; mst, metastyle; pa, paracone; pad, paraconid; prd, protoconid; pst, parastyle; stc, stylocone. Oblique lines indicate breakage

**Fig. 6.13** *Mesungulatum houssayi*, artistic reconstruction by Jorge L. Blanco

new research line in SA consisting of new challenges on its systematics, biogeographical history, and paleoecological interpretations. Shared similarities in tooth morphology suggest close affinities between *Mesungulatum* and *Paraungulatum* from the Los Alamitos Formation (Bonaparte and Migale 2010), and *Coloniatherium* from the La Colonia Formation (Rougier et al. 2009b).

*Austrotriconodon sepulvedai* holotype MACN-RN 239 (Fig. 6.12d) was considered a "triconodont" lower left molariform (Bonaparte 1992). However, the finding of taxa with associated, or referable, premolar and molar series (e.g., the meridiolestidans *Coloniatherium*, *Peligrotherium*, and *Cronopio*; Rougier et al. 2009b, 2011a; the eutriconodont *Argentoconodon fariasorum* Gaetano and Rougier 2011), suggested that MACN-RN 239 corresponds to a meridiolestidan P1 tooth (Rougier et al. 2011a; Gaetano et al. 2013). The tooth of *A. sepulvedai* is labiolingually compressed with a mesial paracone, small metastyle similar in size to the mesial accessory cusp, small stylocone on the crest that connects paracone and metastyle, cingula labial and lingual to the metastyle, and mesial root smaller than the distal one (Gaetano et al. 2013). Considering the taxa from the Los Alamitos Formation, the size and general morphology of *A. sepulvedai*, robust aspect of the tooth, and well-developed cingula led to the suggestion of conspecificity with *Mesungulatum houssayi* (Gaetano et al. 2013), which is here accepted.

The left upper premolar MACN-RN 183 (Bonaparte and Migale 2010: 336, 2015: 356), interpreted as indeterminate dryolestoid, is an extraordinarily robust tooth with thick enamel and peculiar morphology that matches the tooth interpreted as P2 in both *Coloniatherium* and *Peligrotherium* (Figs. 6.15b and 6.20, respectively) and it is here assigned to *Mesungulatum* (Fig. 6.12, P2). This tooth shows a robust distal cingulum and narrow basin stretching between stylocone and paracone and a small very highly placed anterior cusp. This is a characteristic morphology of mesungulatids.

*Mesungulatum lamarquensis* Rougier et al. 2009a
(Fig. 6.12e)

Holotype: MML-PV 10, right upper molar, probably M1 (Fig. 6.12e).

Locality and horizon: Cerro Tortuga, Río Negro Province, Argentina. Allen Formation (Malargüe Group), Campanian–Maastrichtian (Upper Cretaceous).

Diagnosis (from Rougier et al. 2009a): *Mesungulatum lamarquensis* is similar in size to *M. houssayi*, differing from the latter in the features that follow (based on M1). The stylocone is more labially located; consequently, its labial wall is more vertical and its apex is almost on the labial margin of the tooth. The cingula are broader in *M. lamarquensis* than in *M. houssayi*, determining a rectangular outline of the crown. The mesial and distal cingula culminate lingually in robust and distinct cuspules in *M. lamarquensis*, but they are not as developed in *M. houssayi*. The parastyle is fully independent with a separate base and a labial surface distinct from the stylocone in relatively unworn specimens of *M. houssayi* (MACN-RN 03); on the contrary, in *M. lamarquensis* the parastyle is lower, more rounded, and not so sharply separated from the stylocone by a broad groove. There are two cusps in the metastylar area of *M. lamarquensis*, a metastyle and a distal cingular cusp; the metastyle is prominent and the distal cingular cusp is well-differentiated. In contrast, both cusps are poorly developed in *M. houssayi,* and the metastyle is missing in most specimens. The metastyle is positioned more labially and closely appressed in *M. houssayi* than in *M. lamarquensis*, where they are almost in line. The mesocrista on the labial slope of the paracone is poorly developed in *M. lamarquensis* but is sharp and distinct in unworn specimens of *M. houssayi*.

Comments: *Mesungulatum lamarquensis* and *M. houssayi* from the Los Alamitos Formation are very similar in morphology, the same as observed with the association in general. These differences are changes in proportions and relationships of relatively minor morphological features suggesting that both faunas are close systematically and temporally (Rougier et al. 2009a).

*Coloniatherium* Rougier et al. 2009b

Type species: *Coloniatherium cilinskii* Rougier et al. 2009b.

Included species: The type only.

*Coloniatherium cilinskii* Rougier et al. 2009b
(Figs. 6.14, 6.15 and 6.16)

Holotype: MPEF-PV 2087, fragmentary right lower jaw with alveoli for double-rooted canine and p1, root fragments of p2, complete p3, and root fragments of m1–m3 (Fig. 6.14a, b).

Locality and horizon: El Uruguayo quarry, Mirasol Chico Canyon, Chubut Province, Argentina. La Colonia Formation, Campanian–Maastrichtian (Upper Cretaceous).

Diagnosis (modified from Rougier et al. 2009b): Large mesungulatid with dental formula I?/i?, C1/c1, P3/p3, M3/m3. Length of the postcanine dental series 230 mm (average of three specimens); double-rooted canine; P1 and p1 small, transversely narrow, and implanted obliquely without diastema behind the canine; P2 and p2 subtriangular with acute mesial angle bearing a cingular cusp and with a broad posterior cingulid; p3 large, bulbous, fully molarized, with supernumerary roots, and larger than m1; m1 with inflated crown, with small mesial and distal cingulids and variable number of small roots between the two principal ones; M3 mesiodistally compressed, with the stylocone distally located, directly connected to the metacrista. Some upper molars with small labial rugosity or cingulum. Differs from *Mesungulatum* spp. by its larger size, more bunoid appearance, proportionately broader molars, and a greater posterior reduction of the lower molars (assuming MACN-RN 6 represents m1–m2). The cingula of the upper and lower molars are broader and more elevated into the crown; cingular cusps are not as well defined as in *Mesungulatum.* The lingual crest of the paracone that would contribute to the mesocrista, very distinct in *Mesungulatum,* is less conspicuous in *Coloniatherium.* The mesial cingulum in *Coloniatherium* extends to the base of the parastyle, while it lies mesial to the parastyle in *Mesungulatum.* The parastyle and metastyle are similar in size in *Coloniatherium,* while the parastyle is proportionately higher in *Mesungulatum*; the stylocone is relatively higher in *Coloniatherium.*

Comments: The original description of *Coloniatherium* was based on isolated teeth, lower jaws with partial dentition, and five isolated petrosals referred to this taxon by general size and abundance (Rougier et al. 2009b). Finding of other meridiolestidans with associated premolars and molars such as *Peligrotherium, Cronopio,* and *Necrolestes* (Páez Arango 2008; Rougier et al. 2011a, 2012; Wible and Rougier 2017) permitted reconstruction of the dental morphology of *Coloniatherium* in some detail, at least for the upper and lower postcanine series (Figs. 6.15 and 6.16). The dentition is clearly mesungulatid-like (Fig. 6.15) with bunodont cusps, supernumerary roots in the penultimate and ultimate premolars, largely molarized last upper and lower premolars, mesiodistally compressed roots in lower molars, large pre- and post-cingula/cingulids, lack of talonid, robust and almost bifid metaconid, and thick prismatic enamel with a small AP zone (Rougier et al. 2009b, 2011a, 2012).

The morphology of the petrosals assigned to *Coloniatherium* was surprisingly derived, suggesting a phylogenetic position similar to *Vincelestes,* but sharing some

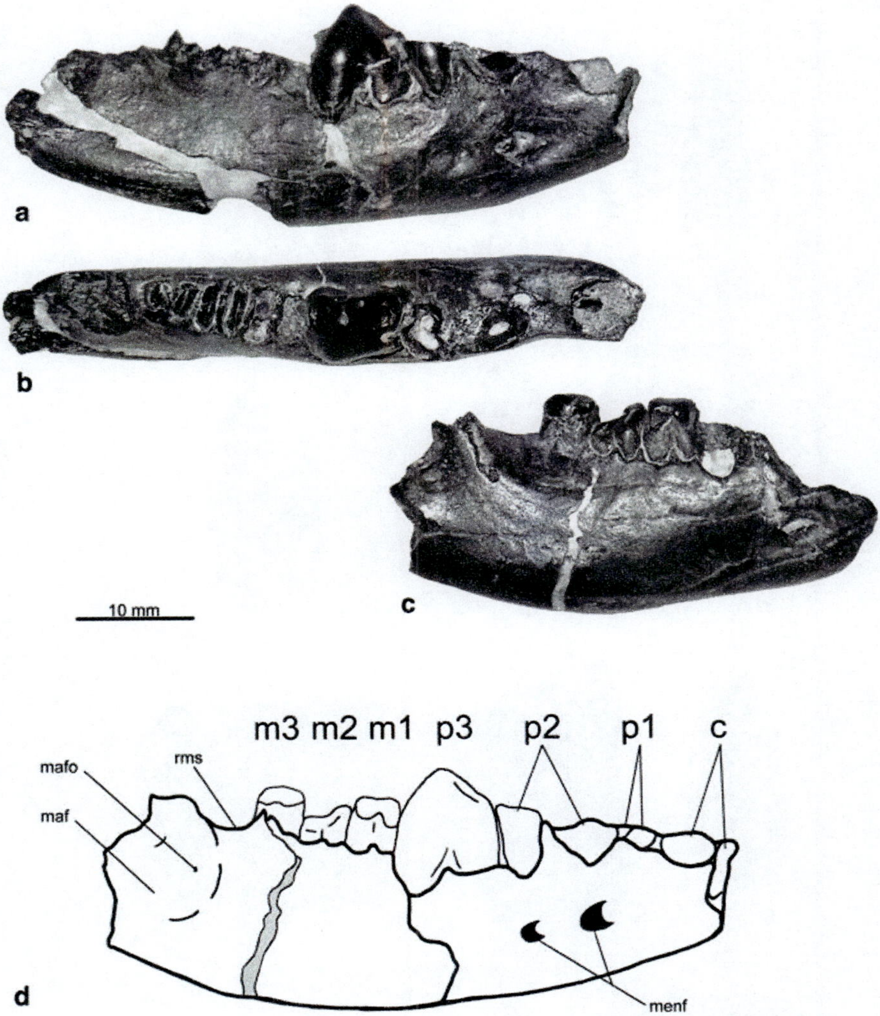


**Fig. 6.14** *Coloniatherium cilinskii* from Mirasol Chico Canyon, Chubut, Argentina; La Colonia Formation, Maastrichtian, Upper Cretaceous. Holotype, MPEF-PV 2087, right dentary with complete p3, root fragments of p2, m1–m3, and alveoli for p1 and double rooted canine in labial (**a**) and occlusal (**b**) views. MPEF-PV 2085, fragmentary right dentary with very worn remnants of m1–m3 in situ, and alveolus for p3 (**c**). Reconstruction of the dentary based on the two previous specimens. The fracture line posterior to the p3 indicates the junction of the two different elements (modified from Rougier et al. 2009b) (**d**). Abbreviations: c, lower canine; m1–m3, lower molars; maf, masseteric fossa; mafo, masseteric foramen; menf, mental foramina; p1–p3, lower premolars; rms, retromolar space

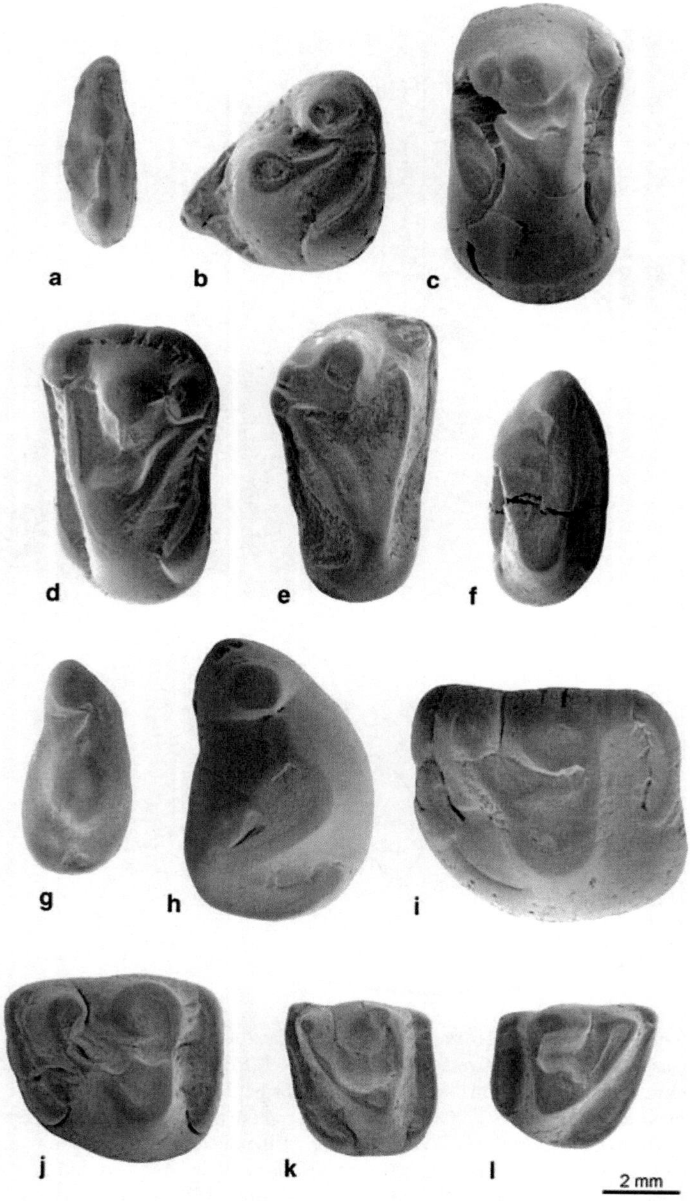


**Fig. 6.15** *Coloniatherium cilinskii* from Mirasol Chico Canyon, Chubut, Argentina; La Colonia Formation, Maastrichtian, Upper Cretaceous. Upper and lower dentition: MPEF-PV 2066, left P1 (**a**); MPEF-PV 2088, left P2 (**b**); MPEF-PV 2081, right P3 (**c**); MPEF-PV 2078, left M1 (**d**); MPEF-PV 2183, right M2 (**e**); MPEF-PV 2163, right M3 (**f**); MPEF-PV 2230, right p1 (**g**); MPEF-PV 2059, right p2 (**h**); MPEF-PV 2073, left p3 (**i**); MPEF-PV 2064, left m1(**j**); MPEF-PV 2092, left m2 (**k**); MPEF-PV 2138, right m3 (**l**)

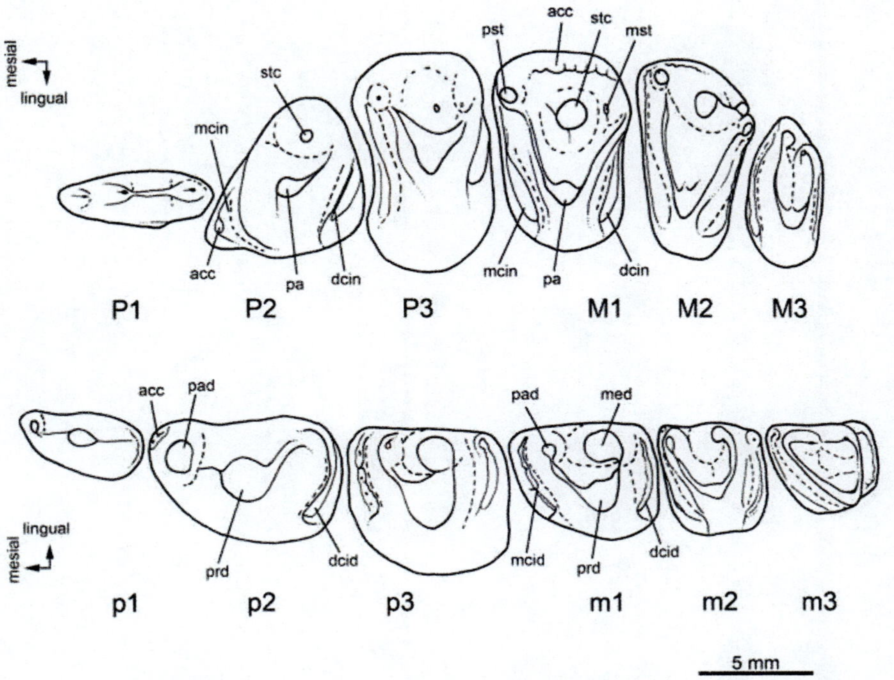


**Fig. 6.16** *Coloniatherium cilinskii*. Reconstruction of the left upper and lower tooth sequence on the basis of specimens from Fig. 6.13 (in this drawing, MPEF-PV 2081, P3; MPEF-PV 2183, M2; MPEF-PV 2163, M3; MPEF-PV 2230, p1; MPEF-PV 2059, p2; and MPEF-PV 2138, m3 are inverted). Abbreviations: acc, accessory cusp/cuspid; dcid, distal cingulid; dcin, distal cingulum; mcid, mesial cingulid; mcin, mesial cingulum; med, metaconid; mst, metastyle; pa, paracone; pad, paraconid; prd, protoconid; pst, parastyle; stc, stylocone

derived features, possibly convergent, with therians (Rougier et al. 2009b), including a cochlear endocast with at least one full turn.

*Reigitherium* Bonaparte 1990

Type species: *Reigitherium bunodontum* Bonaparte 1990.

Included species: The type only.

*Reigitherium bunodontum* Bonaparte 1990
(Figs. 6.17 and 6.18)

Holotype: MACN-RN 173, fragmentary right m2 (Fig. 6.17a).

Locality and horizon: Estancia Los Alamitos, west slope of Cerro Cuadrado, Arroyo Verde, Río Negro Province, Argentina. Los Alamitos Formation, Campanian–Maastrichtian (Upper Cretaceous). El Uruguayo quarry, Mirasol Chico Canyon, and Anfiteatro 1 locality, southeastern slope of the Sierra de la Colonia, Chubut Province,

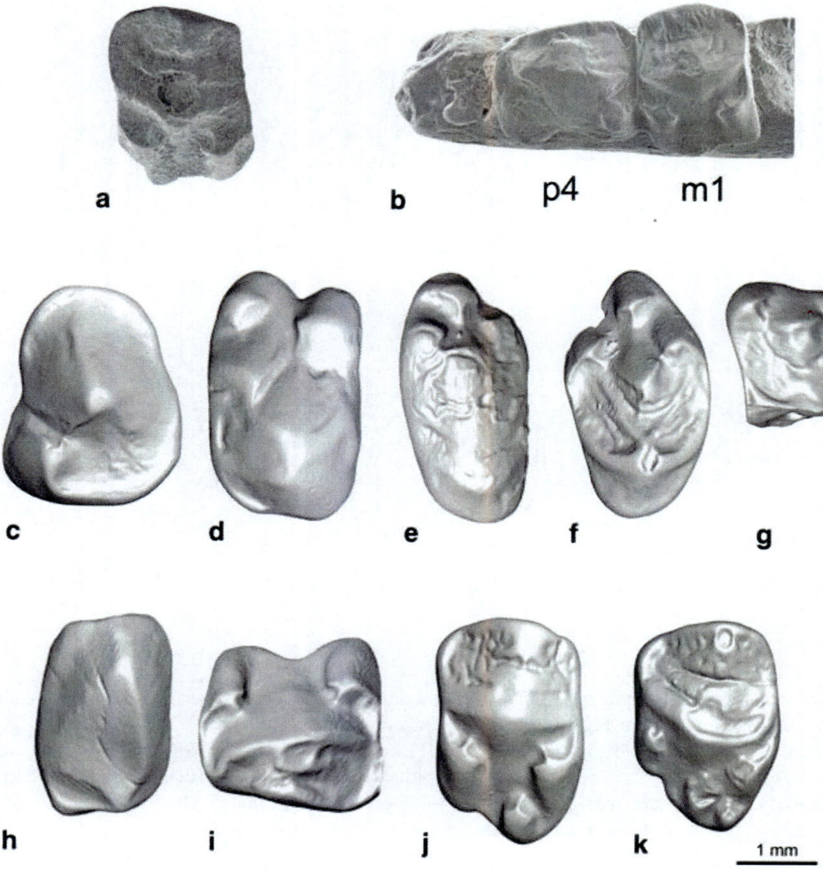


**Fig. 6.17** *Reigitherium bunodontum* from Ea. Los Alamitos, Río Negro and Mirasol Chico Canyon and Anfiteatro 1, Chubut, Argentina; Los Alamitos and La Colonia formations, Campanian–Maastrichtian, Upper Cretaceous. Holotype, MACN-RN 173, fragment of right m2 (**a**); MPEF-PV 2020, fragment of left dentary with last premolar (p4) and m1 (**b**); MPEF-PV 2339, right P3 (**c**); MPEF-PV 2072, right P4 (**d**); MPEF-PV 2238, left M1 (**e**); MPEF-PV 2341, right M2 (**f**); MPEF-PV 2369, fragmentary left M3 (**g**); MPEF-PV 2376, right p3 (**h**); MPEF-PV 2338, right p4 (**i**); MPEF-PV 2317, right m1 (**j**); and MPEF-PV 2237, right m2 (**k**)

Argentina. Indeterminate locality, La Colonia Formation, Campanian–Maastrichtian (Upper Cretaceous).

Diagnosis (from Harper et al. 2019): Very small mesungulatoid with dental formula I?/i?, C1/c1, P4/p4, M3/m3. Simple premolars increasing in size distally to an enlarged molariform P4/p4, and molars with complex and buccolingually extended crowns, decreasing in size distally. Compared to the better-known mesungulatids, *Coloniatherium* (Rougier et al. 2009b) and *Peligrotherium* (Páez Arango 2008), *Reigitherium* is much smaller and has the following derived features: (1) interradicular

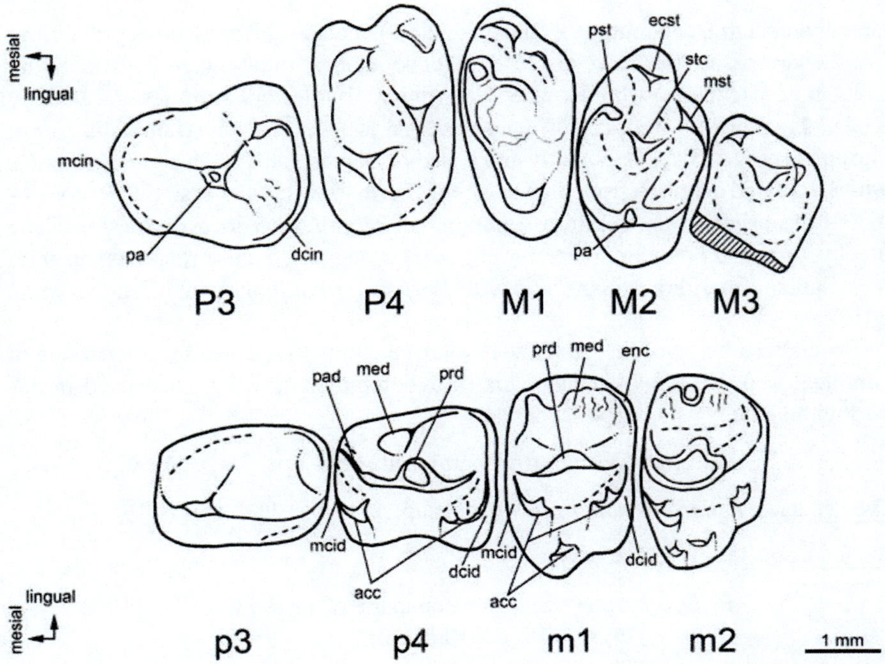


**Fig. 6.18** *Reigitherium bunodontum.* Reconstruction of the left upper and lower tooth sequence on the basis of specimens from Fig. 6.17 (in this drawing, MPEF-PV 2339, P3; MPEF-PV 2072, P4; MPEF-PV 2341, M2; MPEF-PV 2376, p3; MPEF-PV 2338, p4; MPEF-PV 2317, m1; and MPEF-PV 2237, m2 are inverted). Abbreviations: acc, accessory cusp/cuspid; dcid, distal cingulid; dcin, distal cingulum; ecst, ectostyle; enc, enceinte; mcid, mesial cingulid; mcin, mesial cingulum; med, metaconid; mst, metastyle; pa, paracone; pad, paraconid; prd, protoconid; pst, parastyle; stc, stylocone. Oblique lines indicate breakage

crests connecting the roots of upper and lower canine and postcanine elements (understood as additional roots); (2) highly crenulated trigonids and primary trigons, with an enclosing enceinte structure in the lower molars, and (3) neomorphic ectostyles on the first and second upper molars, and neomorphic accessory cuspulids (also seen in *Peligrotherium*) distributed within the labial portion of the lower molariforms.

Comments: *Reigitherium* has a very complex molariform morphology, and consequently, its holotype from the Los Alamitos Formation was originally misidentified. The MACN-RN 173 is a fragmentary right lower molar missing part of the lingual edge of the crown; however, originally it was described as a left upper molariform (Bonaparte 1990). The report of a fragmentary left dentary with three sequential lower postcanines (p3, p4, and m1 based on Harper et al. 2019) from the La Colonia Formation made clear the identity of the holotype as a lower molar (Pascual et al. 2000).

The interpretation of the phylogenetic affinities of *Reigitherium* has been controversial. Originally, this taxon was considered a Dryolestoidea (Bonaparte 1990), but later changed to Docodonta (Pascual et al. 2000), a clade currently unrecorded from SA (however, see Martin et al. 2013). Docodontan affinities were questioned by Rougier et al. (2003b) who defended the original dryolestoid affinities and later by Kielan-Jaworowska et al. (2004) who regarded *Reigitherium* as Mammalia (sensu lato) *incertae sedis*. More recently and based on new isolated teeth and jaw remains with associated dentition from the La Colonia Formation, Harper et al. (2019) corroborated the original taxonomic assignment of *Reigitherium* as a dryolestoid. The brachyodont and bunodont posterior cheek teeth suggested close relationships with mesungulatid meridiolestidans, and with *Peligrotherium* in particular (Harper et al. 2019).

Specialized herbivorous habits were interpreted for *Reigitherium*, comparable in complexity to some fossil marsupials (Polydolopimorphia) that succeeded in SA during the Paleogene (Harper et al. 2019).

*Peligrotherium* Bonaparte et al. 1993

Type species: *Peligrotherium tropicalis* Bonaparte et al. 1993.

Included species: The type only.

*Peligrotherium tropicalis* Bonaparte et al. 1993
(Figs. 6.19, 6.20 and 6.21)

Holotype: UNPSJB-PV 914, fragment of right dentary with p2–m1 (Fig. 6.19).

Locality and horizon: Punta Peligro locality, Chubut Province, Argentina. Salamanca Formation, Hansen Member, BNI (Banco Negro Inferior), lower Paleocene (Peligran SALMA, Danian).

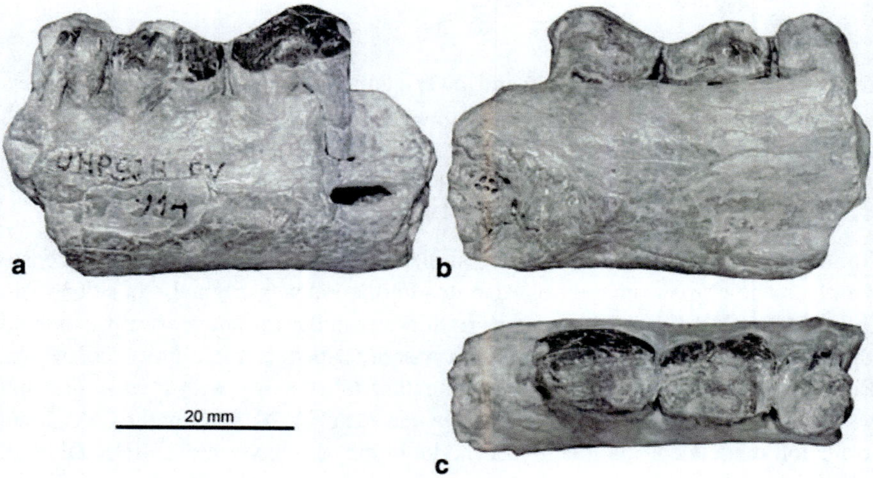


**Fig. 6.19** *Peligrotherium tropicalis* from Punta Peligro, Chubut, Argentina; Salamanca Formation, Danian, lower Paleocene. Holotype, UNPSJB-PV 914, fragment of right dentary with p2–m1 in labial (**a**), lingual (**b**), and occlusal (**c**) views

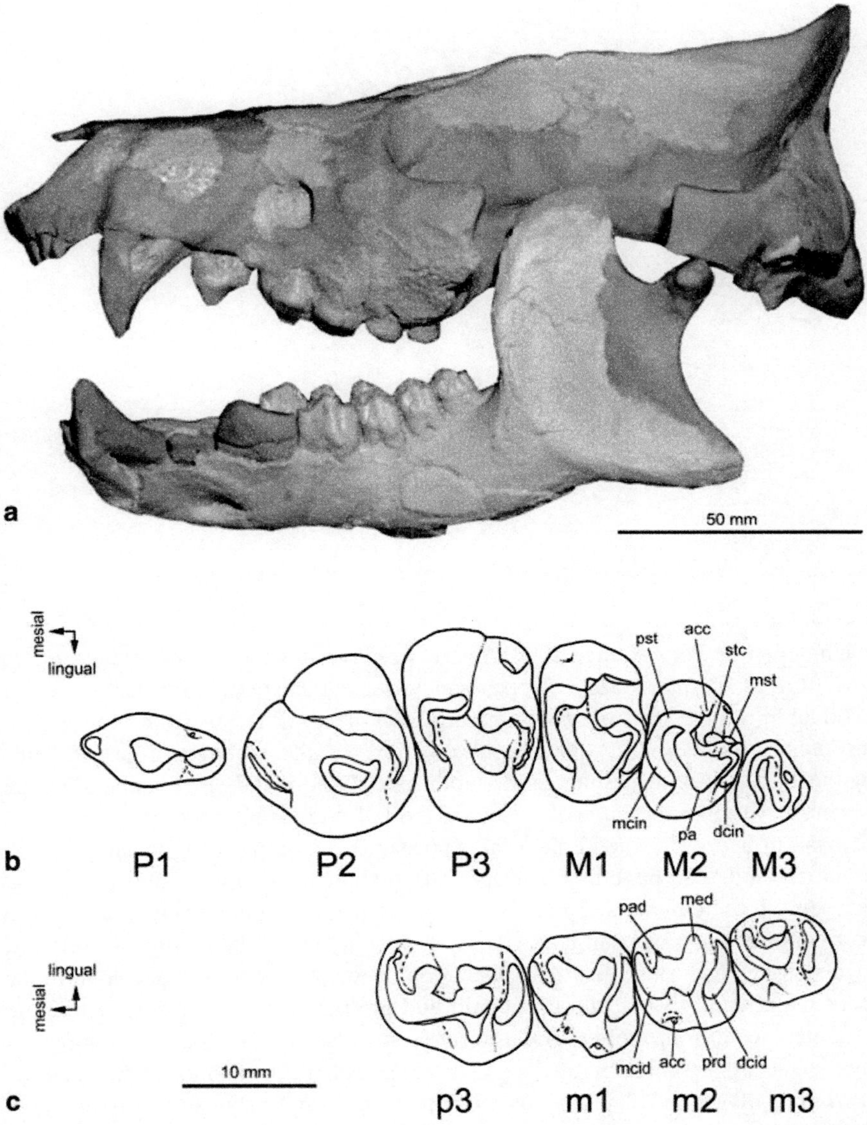


**Fig. 6.20** *Peligrotherium tropicalis* from Punta Peligro, Chubut, Argentina; Salamanca Formation, Danian, lower Paleocene. Skull reconstruction of *P. tropicalis* in lateral view, based on specimen MPEF-PV 2351 (**a**); left upper dentition (right M3 inverted) (**b**) and lower dentition (**c**) in occlusal view. Abbreviations: acc, accessory cusp/cuspid; dcid, distal cingulid; dcin, distal cingulum; mcid, mesial cingulid; mcin, mesial cingulum; med, metaconid; mst, metastyle; pa, paracone; pad, paraconid; prd, protoconid; pst, parastyle; stc, stylocone

**Fig. 6.21** *Peligrotherium tropicalis*, artistic reconstruction by Jorge L. Blanco

Diagnosis (from Páez Arango 2008): Large-sized mesungulatid with dental formula I4?/i?, C1/c1, P3/p3, M3/m3. Upper and lower canines extremely large, robust, and double rooted. Last premolars greatly molarized, differing from more posterior molariforms by less quadrangular outline of the crown and less defined trigon/trigonid. As in mesungulatids, both mesial and distal cingula/cingulids are present, but unlike the latter the cingula are tall and rapidly become part of the occlusal surface, even when little wear is present. Main cusps of trigon, and to lesser extent trigonid, low, bunodont, and less differentiated, forming a continuous wear surface with the cingula and main crests of the trigon/trigonid. Young adults lose most cusps of the crown; in later stages of wear, a central, labiolingually oriented ridge survives and represents the connection of the stylocone and paracone in the upper molars and that of the metaconid and protoconid in the lowers. P3/p3 are the largest teeth in the series. As in *Reigitherium*, but unlike other mesungulatids, there are accessory cusps on the labial surface of both upper and lower teeth. Molar crowns labially extended, projecting far from the base of the root and with very thick enamel. Petrosal with a full turn of the cochlea and extremely robust, with a spacious postpromontorial tympanic sinus, lateral flange and anterior lamina, and shallow but expansive subarcuate fossa. Differing from therians, *Peligrotherium* has a fenestra semilunaris and a relatively broad promontorium with vascular impressions and a persisting large prootic canal; the posttemporal canal opens into the middle ear cavity via an extremely broad canal that becomes confluent with a narrow facial sulcus.

Comments: *Peligrotherium tropicalis* is the largest known dryolestoid, with massive skull and jaws, and bunodont teeth. Originally, Bonaparte et al. (1993) assigned it to the eutherian "condylarth" Periptychida, based on what they interpreted as tribosphenic lower molars. However, further studies and material demonstrated dryolestoid affinities for the taxon (Rougier et al. 2000; Gelfo and Pascual 2001). The basicranium of *Peligrotherium* retains some primitive features shared with other stem Theria, such as having a fenestra semilunaris connecting the cavum suracochleare and cavum epiptericum, broad posttemporal canal opening ventrally into the middle ear cavity and confluent with a narrow facial sulcus; broad promontorium with vascular impressions; large prootic canal; presence of an anterior lamina, and a thick lateral flange (Páez Arango 2008). However, the coiling of the cochlea is 360° or more, implying an independent acquisition of this feature between at least the derived mesungulatids and therians, while the vascular pattern shows a reduced or modified stapedial system with loss of the stapedial artery (Páez Arango 2008), possibly related to the relatively large size of *Peligrotherium* (Diamond 1989). The bunodont postcanines and large-sized distal premolars suggest a diet with an important herbivorous component, perhaps associated with hard elements, such as seeds and pods (Páez Arango 2008). *Peligrotherium* is an example of convergent evolution of an archaic lineage and ungulate-like "condylarths" or some recent medium-sized suids (pigs) and tayasuids (peccaries); represents the start of the large herbivorous mammalian body architecture that will be dominant, subsequently, during the Cenozoic (Páez Arango 2008).

In SA, the biotic response to the Cretaceous/Paleogene extinction event is much milder compared to other parts of the world (Rougier et al. 2000; Pascual et al. 2001; Pascual and Ortiz-Jaureguizar 2007). The mammalian composition of the lower Paleocene Salamanca Formation is represented by the survival of dryolestoids, such as *Peligrotherium*, accompanied by the monotreme, *Monotrenatum sudamericanum* (Pascual et al. 1992; Forasiepi and Martinelli 2003; see also Chap. 4) and the gondwanatherian *Sudamerica ameghinoi* (Scillato-Yané and Pascual 1985; Krause and Bonaparte 1993; Pascual et al. 1999; Gurovich 2005; Gurovich and Beck 2009; see also Chap. 8), inhabiting together with metatherians and eutherians (Bond et al. 1995; Bonaparte and Morales 1997; Goin et al. 2002, 2004; Gelfo 2007; Gelfo et al. 2007; Forasiepi and Rougier 2009; see also Chap. 10). In other parts of the world, archaic lineages of mammals become extinct together with dinosaurs and other Mesozoic elements at the end of the Era, with the exception of multituberculates (e.g., Prothero 1981; Miao 1988; Alroy 1996).

Meridiolestida *incertae sedis*
*Casamiquelia* Bonaparte 1990

Type species: *Casamiquelia rionegrina* Bonaparte 1990.

Included species: The type only.

*Casamiquelia rionegrina* Bonaparte 1990
(Fig. 6.22a)

Holotype: MACN-RN 163, right upper molariform (Fig. 6.22a).

**Fig. 6.22** Meridiolestida indet. from Ea. Los Alamitos, west slope of Cerro Cuadrado, Río Negro, Argentina; Los Alamitos Formation, Campanian–Maastrichtian, Upper Cretaceous. *Casamiquelia rionegrina*. Holotype, MACN-RN 163, right upper molariform (**a**). *Quirogatherium major*. Holotype, MACN-RN 171, penultimate right upper premolar (**b**). *Paraungulatum rectangularis*. Holotype, MACN-RN 180, left upper molariform (**c**)

Locality and horizon: Estancia Los Alamitos, west slope of Cerro Cuadrado, Arroyo Verde, Río Negro Province, Argentina. Los Alamitos Formation, Campanian–Maastrichtian (Upper Cretaceous).

Diagnosis (from Bonaparte 1990, 2002): Small dryolestoid. Upper molariform with acute-angled upper molariforms, mesiodistally longer than labiolingually wide, and with reduction of the mesial basin of the trigon. Pronounced ectoflexus defining two subequal lobes. Parastyle absent or smaller than metastyle. Stylocone larger than paracone located in the middle of the crown, connected with paracone by a short mesocrista, and separated from both metacrista and paracrista. Paracrista and metacrista well defined, both without cusps.

Comments: *Casamiquelia rionegrina* was erected based on MACN-RN 163, an isolated right upper molariform with a very peculiar morphology (Bonaparte 1990; Fig. 6.22a). Subsequently, specimens MACN-RN 1032 and 1033 were included in the hypodygm (Bonaparte 2002; though soon after the later specimen was moved to *Alamitherium bishopi* by Chornogubsky 2003); we tentatively consider both belonging to *Groebertherium*.

Bonaparte (2002) erected the family Casamiquelidae to include *Casamiquelia rionegrina*, *Rougietherium tricuspes* (however see *Leonardus*), and *Alamitherium bishopi* (however, see *Groebertherium*), characterized by a reduction of the mesial basin of the trigon from the labial toward the lingual side, dorsal displacement of the labial area of the paracrista, and lack of parastyle. More recently, Averianov et al. (2013) considered that *Casamiquelia* holotype MACN-RN 163 represents an ultimate left molar of *Leonardus*. Re-examination of the specimen, however, suggests this hypothesis is unlikely. The MACN-RN 163 is minute, considerably smaller than *Leonardus* and in the size range of *Groebertherium*. The molariform has two subequal roots, mesiodistally compressed; this feature resembles mesungulatids more than *Groebertherium*. The crown in *Casamiquelia* is very low and the enamel very thin, which could also imply a deciduous tooth. Since alternative homologies for

*Casamiquelia* are controversial, and no likely synonymy can be established with known taxa, we prefer to maintain the name of this taxon (see also Bonaparte and Migale 2010, 2015; *contra* Averianov et al. 2013). Perhaps more material would provide decisive evidence on the taxonomic status of *Casamiquelia*, which shows at least superficial similarities with *Microdersodon* and *Atlasodon* from the Cretaceous of Africa (Sigogneau-Russell 1991; Bonaparte 2002; Rougier et al. 2011a; Chimento et al. 2016).

*Quirogatherium* Bonaparte 1990

Type species: *Quirogatherium major* Bonaparte 1990.

Included species: The type only.

*Quirogatherium major* Bonaparte 1990
(Fig. 6.22b)

Holotype: MACN-RN 171, penultimate right upper premolar (Fig. 6.22b).

Locality and horizon: Estancia Los Alamitos, west slope of Cerro Cuadrado, Arroyo Verde, Río Negro Province, Argentina. Los Alamitos Formation, Campanian–Maastrichtian (Upper Cretaceous).

Diagnosis (from Bonaparte 1990): Paracone and stylocone near one to each other and connected by a sharp, tall crest. Narrow distal basin defined between the line formed by the stylocone and the two stylar cusps and the sharp and high metacrista. Large distal cingulum.

Comments: *Quirogatherium major* is known only by its holotype, MACN-RN 171, an isolated upper tooth from a medium-sized Meridiolestida. The morphology of *Quirogatherium* is characterized by a mesial lobe with a low cusp, modified parastyle, small, subcircular mesial root, and transversely, well-developed distal root. The distal portion of the tooth is dominated by two cusps: paracone and stylocone, plus an extra distal stylar cusp on the distolingual corner of the crown that determines the boundaries of a narrow and oblique basin.

The taxonomic validity of *Quirogatherium major* has been repeatedly questioned. Described originally as a member of the "symmetrodont" family Barbereniidae (Bonaparte 1990), *Quirogatherium* was alternatively considered to possibly represent a posterior molarized permanent upper premolar of *Mesungulatum houssayi* (Bonaparte 2002). This hypothesis was followed by Chornogubsky (2003), who established the synonym with *M. houssayi* and referred the holotype of *Quirogatherium* MACN-RN 171 to an upper molariform of this species. In turn, Averianov (Averianov 2002; see also Martin 1999) suggested that the tooth in question corresponds to an upper deciduous element. Accordingly, Kielan-Jaworowska et al. (2004) tentatively referred MACN-RN 171 to a deciduous DP3? of *Mesungulatum houssayi*. This interpretation was followed by Averianov et al. (2013) and alternatively accepted by Bonaparte and Migale (2010, 2015); although, these authors emphasized that a synonym is still a premature decision, an opinion with which we concur. The general morphology of *Quirogatherium major* is similar to the deciduous premolars of Late Jurassic dryolestoids from Europe and North America (Martin 1999; see also Rougier et al. 2011a).

Rougier et al. (2011a) compared *Quirogatherium* with the penultimate upper premolar of other Meridiolestida (P3 in *Cronopio*; P2 in *Coloniatherium* and *Peligrotherium*). The sharp constriction of the rostrum of *Cronopio* correlates with a much lower mesial accessory cusp departed from the moderately basined distal half of the tooth. This morphology suggests that MACN-RN 171 could correspond to a penultimate upper right premolar and likely related to a proportionally slender snout (Rougier et al. 2011a). However, the presently available material of either *Cronopio* or *Peligrotherium* does not allow determination of whether the tooth in question is a retained deciduous or a second-generation tooth (Rougier et al. 2011a).

Comparing with taxa from the Los Alamitos Formation known by molars, size is smaller and tooth MACN-RN 171is more mesiodistally elongated than in *Mesungulatum* (Fig. 6.12); the latter also shows low bunodont cusps covered by thick enamel, which are lacking in *Quirogatherium.* Finally, a suitable P2 is known for *Mesungulatum* (see above, MACN-RN 183), and it does not seem to be a deciduous tooth; therefore, it appears unlikely that *Quirogatherium* could be referred to *Mesungulatum.* By size and elongation, *Quirogatherium major* recalls *Paraungulatum rectangularis* (Fig. 6.22c); however, we prefer to avoid any taxonomic conclusion until new material becomes available.

*Paraungulatum* Bonaparte 2002

Type species: *Paraungulatum rectangularis* Bonaparte 2002.

Included species: The type only.

*Paraungulatum rectangularis* Bonaparte 2002
(Fig. 6.22c)

Holotype: MACN-RN 180, left upper molar (Fig. 6.22c).

Locality and horizon: Estancia Los Alamitos, westslope of Cerro Cuadrado, Arroyo Verde, Río Negro Province, Argentina. Los Alamitos Formation, Campanian–Maastrichtian (Upper Cretaceous).

Diagnosis (from Bonaparte 2002): Upper molar with rectangular crown in occlusal view and with the main axis mesiodistally long. There are five cusps in line on the labial side, in addition to the buccal "trigon" cusps, the mesial and distal cingula are also terminated in distinct cusps. The stylocone is of moderate size, not very different from the remaining cusps; the well-developed mesial and distal cingula are very distinct.

Comments: The only available material from *Paraungulatum rectangularis* is the holotype, MACN-RN 180 (Bonaparte 2002; see also Bonaparte 1994). The morphology of *Paraungulatum* is distinctive indeed and easily differentiated from other dental remains collected from the Los Alamitos Formation. *Paraungulatum* is either a valid taxon or an unknown molariform position of a previously known species; in this case, it could be either a permanent or a deciduous element. To date, the best-known examples of *bona fide* deciduous premolars in Dryolestoidea are those from the Jurassic of Portugal (Martin 1997, 1999). These elements are narrower bucollingually and longer mesiodistally than the permanent molars.

In *Paraungulatum,* the broad cingula, large centrally positioned stylocone, and broad basin with a prominent mesocrista are features recalling mesungulatids. Excluding *Paraungulatum,* currently only one mesungulatid, *Mesungulatum houssayi,* has been described from the Los Alamitos fauna. However, *Paraungulatum* is much smaller than the premolars or mesial molars of *Mesungulatum* (Fig. 6.12). This size difference appears to rule out the possibility of *Paraungulatum* being a permanent P3 of *Mesungulatum,* and we have another dental morphology we believe is the P2 (see Fig. 6.12); therefore, in our view, it seems unlikely that *Paraungulatum* is a permanent premolar of *Mesungulatum.* A deciduous ultimate or penultimate premolar, likely DP2/DP3 of *Mesungulatum* is, in principle, still possible.

In the collection from Los Alamitos, there is a variation in size among the teeth recognized as *Mesungulatum houssayi* and some of the smallest specimens (e.g., MACN-RN 10, lower left molariform) would make for viable lower molar candidates to be *Paraungulatum.* Based on *Cronopio* (Rougier et al. 2011a), *Necrolestes* (Wible and Rougier 2017), *Peligrotherium* (Páez Arango 2008), and *Coloniatherium* (Rougier et al. 2009b), we know that in meridiolestidans the molars diminish in size posteriorly. It is conceivable that these smaller teeth would represent posterior molar positions of a smaller mesungulatid, potentially *Paraungulatum.* If this were to be the case, *Paraungulatum* could be a posterior upper premolar of such a taxon.

In the smaller collection of dryolestoids from the Allen Formation, there are two upper molars (MML-PV 1298 and MML-PV 1303) almost identical in size and shape to MACN-RN 180, which for all intents and purposes can be considered *Paraungulatum* sp. Just as in the Los Alamitos collection, it seems that there is too large a size gap between these *Paraungulatum*-like teeth and *Mesungulatum lamarquensis* (MML-PV 10), so far the only recognized mesungulatid from the Allen Formation. A *Paraungulatum* morphotype is also present among isolated elements from La Colonia, where smaller mesungulatids than those referred to *Coloniatherium cilinskii* make a rare appearance suggesting a greater diversity of smaller-sized taxa.

The intrinsic morphology of *Paraungulatum* offers little help to determine unambiguously if its holotype, MACN-RN 180, and eventually MML-PV 1298 and 1303, are permanent or deciduous teeth. Based on enamel thickness and erosion of the roots on both specimens and the La Colonia molariform, it is possible that the *Paraungulatum* morph is indeed a deciduous element. Furthermore, as commented above, the specimen MACN-RN 170 here assigned to *Groebertherium stipanicici* is with some confidence interpreted as deciduous. This tooth, although lower instead of upper as MACN-RN 180, shows that meridiolestidans share the generalized mammalian condition of having more molarized and mesiodistally elongated deciduous premolars (see Martin 1997, 1999), features that are present in *Paraungulatum.*

In sum, given the uncertainties on the serial homology of the molariform MACN-RN 180, considering the fact that likely the diversity of mesungulatid is underestimated, and that we lack any definitive argument to subsume *Paraungulatum rectangularis* under any other genus with priority, we opt to tentatively retain the name as a valid taxon.

*Bondesius* Bonaparte 1990

Type species: *Bondesius ferox* Bonaparte 1990.

Included species: The type only.

*Bondesius ferox* Bonaparte 1990
(Fig. 6.23a–c)

Holotype: MACN-RN 161, right lower premolar (Fig. 6.23a–c).

Locality and horizon: Estancia Los Alamitos, westslope of Cerro Cuadrado, Arroyo Verde, Río Negro Province, Argentina. Los Alamitos Formation, Campanian–Maastrichtian (Upper Cretaceous).

Diagnosis (modified from Bonaparte 1990): Obtuse angled crown, with an asymmetrical distribution of cusps. Metaconid larger than paraconid and nearer to the protoconid. Small distal cingulid and accessory cusp that do not form a talonid.

Comments: *Bondesius ferox* is known only by its holotype, MACN-RN 161, an isolated right lower tooth. At first (Bonaparte 1990), the holotype of *Bondesius* was accompanied by an upper molar MACN-RN 162; however, this was later designated as the holotype of *Rougietherium tricuspes* (Fig. 6.4b; see *Leonardus*).

Originally Bonaparte (1990, 1992, 1994; see also Chornogubsky 2003; Bonaparte and Migale 2010, 2015) referred *Bondesius* to "Symmetrodonta". The dominant structure of the crown in MACN-RN 161 (Fig. 6.23a–c) is a robust protoconid, slightly curved posteriorly. This is associated with the paraconid and a larger metaconid appressed to the protoconid. There is a distal cingulum, lower in the crown, with an accessory distal cuspid (cusp d), but lingual cingulids are absent. This general morphology recalls some "symmetrodonts", such as *Kuehneotherium*, however, several apomorphic features raised doubts regarding the original referral (Kielan-Jaworowska et al. 2004). Averianov (2002) suggested that the holotype of *Bondesius ferox* is a deciduous premolar of a Dryolestoidea, considering the two slender roots, circular in section. Alternatively, this tooth could belong to a mesial permanent premolar of another Dryolestoidea from the Los Alamitos fauna.

Comparisons with material from Los Alamitos at MACN suggested a match between the roots of *Bondesius ferox*, MACN-RN 161, and the edentulous dentary MACN-RN 1142 (Bonaparte and Migale 2010:262, 2015:316—more complete jaw in those figures, which were formerly referred to "triconodonts"), with the mesial and distal roots of MACN-RN 161, occupying the second pair of the preserved alveoli (i.e., with the roots housed in the alveoli directly anterior and posterior to the mental foramen, respectively). Comparing MACN-RN 161 and the dentary referred to *Leonardus cuspidatus* (Fig. 6.4c; MACN-RN 246/1097; Chornogubsky 2011;

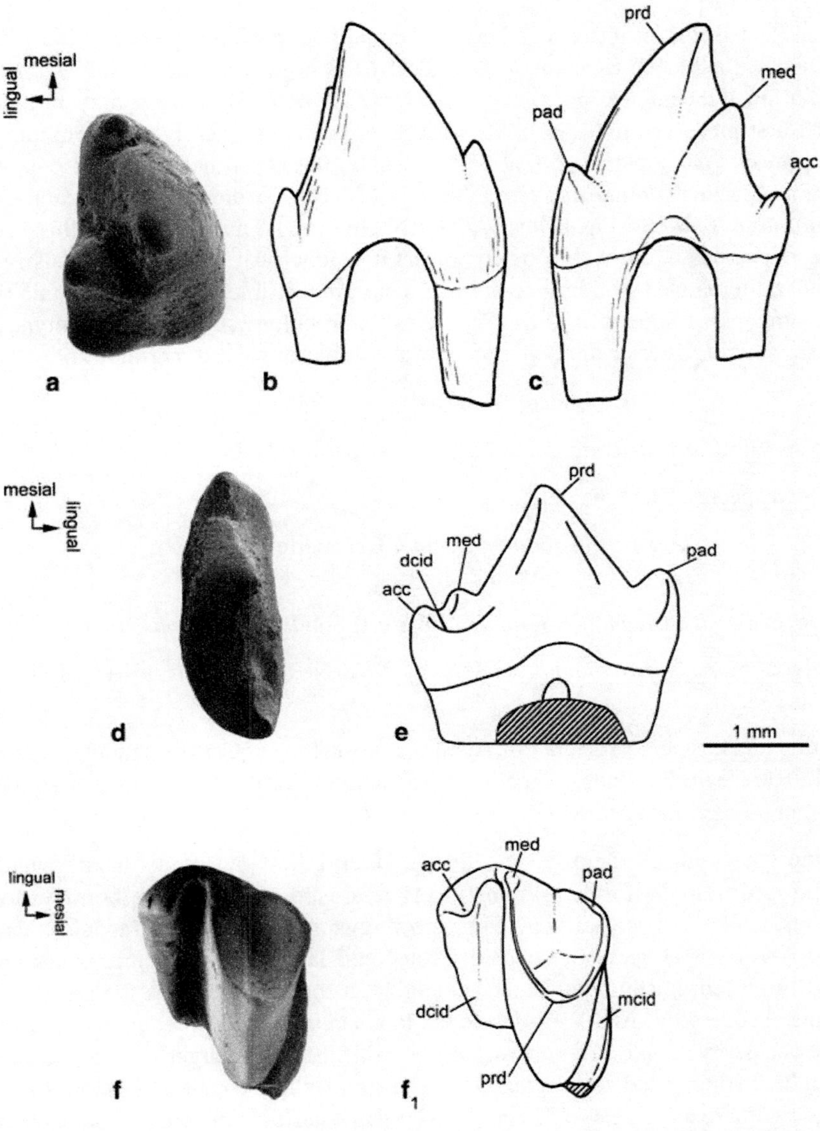


**Fig. 6.23** Dryolestoidea indet. from Ea. Los Alamitos, west slope of Cerro Cuadrado, Río Negro, Argentina; Los Alamitos Formation, Campanian–Maastrichtian, Upper Cretaceous. MACN-RN 161 (holotype of *Bondesius ferox*), ultimate right lower premolar in occlusal view (**a**); MACN-RN 21 (holotype of *Austrotriconodon mckennai*) left lower premolar in occlusal (**b**) and lingual (**c**) views (line drawing from Gaetano et al. 2013). Scale corresponds to **a**–**c**. Dryolestoidea indet from Seymour Locality IAA 90/1, Seymour Island, Antarctica Peninsula, La Meseta Formation, Ypresian–Lutetian, lower–middle Eocene. MLP 91-II-4-3, lower right molariform (**d**) and accompanying line drawings ($D_1$). Not to scale. Abbreviations: acc, accessory cuspid; dcid, distal cingulid; mcid, mesial cingulid; med, metaconid; pad, paraconid; prd, protoconid. Oblique lines indicate breakage or covered surfaces

Forasiepi et al. 2012), and establishing the homologous position by the position of the opening of the mental foramen, MACN-RN 161 is slightly smaller than *Leonardus*. In addition, if the inferred position is correct, MACN-RN 161would occupy the locus of the first preserved tooth in MACN-RN 246, and consequently, the morphology of *Bondesius* is simpler than that of *Leonardus*. Based on these comparisons, we are not able to fit *Bondesius ferox*, MACN-RN 161, in the attributed dentary of *Leonardus cuspidatus* (Fig. 6.4c; MACN-RN 246/1097). The possibility of this tooth being deciduous is still viable, but at present it is not clearly supported nor falsified. Because alternative taxonomic re-arrangements are possible, we opt to maintain the taxonomic name *Bondesius ferox*. The overall morphology suggests the holotype of *Bondesius* is a posterior lower premolar of a non-mesungulatid meridiolestid.

*Austrotriconodon* Bonaparte 1986a

Type species: *Austrotriconodon mckennai* Bonaparte 1986a.

Included species: The type only.

*Austrotriconodon mckennai* Bonaparte 1986a
(Fig. 6.23d, e)

Synonyms: *Austrotriconodon ferox* Bonaparte 1986b *lapsus calami*.

Holotype of *Austrotriconodon mckennai*: MACN-RN 21, left lower premolar (Fig. 6.23d, e).

Locality and horizon: Estancia Los Alamitos, west slope of Cerro Cuadrado, Arroyo Verde, Río Negro Province, Argentina. Los Alamitos Formation, Campanian–Maastrichtian (Upper Cretaceous).

Diagnosis (from Gaetano et al. 2013): Meridiolestidan with labiolingually compressed premolariforms, asymmetric in lateral aspect. Lingual face more concave than labial face in occlusal view. Four crown cusps: protoconid, paraconid, metaconid, and a distal accessory cuspid. Protoconid larger than the other cusps and mesially placed. Distinct paraconid, emerging at the same level as the metaconid and distal accessory cusp. Well-developed lingual cingulid between protoconid and distal accessory cuspid. Two roots present, with the distal larger than the mesial. It can be distinguished from *Cronopio dentiacutus*, the most similar taxon, by the presence of a relatively large paraconid, a larger posterior than anterior root, and the bases of the paraconid, metaconid, and distal accessory cuspid at the same level.

Comments: *Austrotriconodon mckennai* was originally described only by its holotype MACN-RN 21, an isolated left lower tooth; however, later other specimens were included in the hypodigm of the species (Bonaparte 1992; Bonaparte and Migale 2010, 2015). *Austrotriconodon mckennai* together with *A. sepulvedai* (Fig. 6.12d; but see *Mesungulatum*) were originally described as "triconodont" mammals and the unique members of the Family Austrotriconodontidae. However, later comparisons with some premolars assigned to *Coloniatherium cilinskii* and *Cronopio dentiacutus* raised doubts about their alleged "triconodont" nature (Rougier et al. 2011a). A more recent analysis (Gaetano et al. 2013) contends that *Austrotriconodon* specimens are likely premolars of Meridiolestida. Several meridiolestidans known from the similar aged Los Alamitos, Allen, and La Colonia formations are based on isolated teeth or incomplete dentitions, mostly molariform loci, and eventually some of them could represent different tooth positions of the same taxon (e.g., Rougier et al. 2011a). The holotype of *Austrotriconodon mckennai* was not possible to unambiguously assign to any of these taxa on the basis of the available evidence (Gaetano et al. 2013). Consequently, following Bonaparte (1986a, 1992), *Austrotriconodon mckennai* is tentatively recognized as a valid taxon but in the clade Meridiolestida (Gaetano et al. 2013).

## 6.4 Unnamed SA Dryolestoids and Corollaries on Distribution and Age

### *6.4.1 Dryolestoids from Bolivia (Maastrichtian)*

The locality of Pajcha Pata in south-central Bolivia has provided a very rich assortment of plants, invertebrates, and vertebrates including microvertebrates from the lower member of the El Molino Formation (middle Maastrichtian) (Gayet et al. 1991, 2001; see also Chap. 2). Among them, Gayet et al. (2001: Fig. 16) briefly described two complete and one partial mammalian upper teeth (specimens MHNC 8588, 8589, and 8590). The authors suggested non-therian affinities for MHNC 8588 and 8590, and likely dryolestoid nature for MHNC 8588 (Gayet et al. 2001), an opinion we follow here. However, therian affinities were suggested for MHNC 8589. The crown is dominated by a massive cusp in central position and a distal smaller cusp adjoined to it. This tooth can be interpreted as having a broad paracone (Gayet et al. 2001: Fig. 16c, d); if this is the case, it resembles instead SA Late Cretaceous dryolestoids (Rougier et al. 2011a), in particular, the more robust mesungulatids. The specimen is ultimately very poor, and though tantalizing it may be to interpret it as a mesungulatid, other alternatives are possible. Given that meridiolestidans are part of the SA fauna from at least early in the Late Cretaceous (Rougier et al. 2011a), it is to be expected they were part of other Cretaceous extra-Patagonian communities (e.g., Brazil, Bolivia, and Peru) where Cretaceous fossils are known.

### *6.4.2 Dryolestoids from Antarctica (Eocene)*

Among the mammals (gondwanatherians, marsupials, pilosans, litopterns, and astrapotherians) found from the lower–middle Eocene La Meseta Formation, Seymour Island, Antarctica Peninsula, there is a small isolated tooth (MLP 91-II-4-3), from Seymour Locality IAA 90/1, with peculiar "zalambdodont morphology" (Fig. 6.23f), with zalambdocone, mesiostyle (parastyle), stylocone, distostyle (metastyle), and pre- and post-cingulum (MacPhee et al. 2008). This tooth was originally interpreted as a possible upper molar of a "bat or insectivore" (Goin and Reguero 1993), although after subsequent comparisons it was assigned to Mammalia *incertae sedis* with possible placental affinities (i.e., "Insectivore"; MacPhee et al. 2008; Reguero et al. 2013). Sadly, the only known specimen is lost (Reguero pers comm 2014). Consequently, the only information available about this peculiar tooth is the descriptions and SEM pictures (MacPhee et al. 2008; Reguero et al. 2013). After the first publication of its photographs, Martinelli et al. (2014) presented an alternative interpretation, in which the specimen represents a molar of a non-therian Dryolestoidea, perhaps related to, or member of, the clade Meridiolestida. It was interpreted as a right lower molar (Fig. 6.23f). The protoconid is flanked mesially by the paracristid and distally by the metacristid, which reaches the paraconid and metaconid, respectively (Martinelli et al. 2014). Both crests form an acute angle, and there is no distinctive notch at mid-way. The labial wall of the protoconid is convex while the lingual face is slightly concave. The paraconid is worn out and lower than the metaconid. The flexid is conspicuous and forms a V-shaped notch between paraconid and metaconid. The protoconid and metaconid are similar in height. The metaconid connects by means of a crest with the talonid. The talonid has a hook-like distolingual projection with a large cusp. The mesial cingulum starts as a faint line below the paraconid and gets wider ventrolabially (Martinelli et al. 2014). Due to preservation, it is unknown if the cingulum continues on the labial slope of the protoconid. Information on root morphology is partial; there is a portion of root preserved (MacPhee et al. 2008) below the paraconid-protoconid, which seems to be transversely wide.

An alternative interpretation is to consider MLP 91-II-4-3 a left upper molar, in which the "talonid" in fact represents the parastyle. If so, the crown lacks a centrally located stylocone and the medial crest characteristic of derived meridiolestidans, recalling instead the molars of *Cronopio* (Rougier et al. 2011a). Although, in *Cronopio* and *Necrolestes* (Rougier et al. 2012; Wible and Rougier 2017) the parastyle is smaller, and the crown is mesiodistally compressed and much labiolingually broader than in MLP 91-II-4-3. *Cronopio* and *Necrolestes* have roughly similar upper and lower tall molars with a relatively simple triangular morphology, that if isolated, their recognition as part of the upper or lower dentition can be difficult to determine. On this basis, the mesial and distal cingula and the lack of stylocone/medial crest in MLP 91-II-4-3 are features more reminiscent of more bunodont meridiolestidans (e.g., *Mesungulatum*). We are confident on the non-therian nature of the molariform and its similarities with dryolestoids, but less certain regarding its position in the dental arcades.

If the interpretation of MLP 91-II-4-3 as a meridiolestidan is correct, it constitutes the second Mesozoic lineage surviving until the Eocene in Antarctica, further increasing the already taxonomically diverse mammalian assemblage of the La Meseta Formation (Martinelli et al. 2014).

### 6.4.3 *Putative Oldest Mesungulatid from Lago Los Barreales (Coniacian), Patagonia*

The Lago Los Barreales locality has provided a mammalian edentulous left lower jaw (MCF-PVPH 412; Fig. 6.24) from the Los Bastos Formation, lower to middle Coniacian age (Garrido 2010, 2011; Forasiepi et al. 2012). The specimen is small-sized. The dentary has a flat angular process and no evidence of Meckelian groove. Only the roots of the canine and the roots of the third postcanine have been preserved. The count of the alveoli suggests the presence of three incisors, one biradiculate canine, and six biradiculate postcanines (Forasiepi et al. 2012), likely three premolars and three molars as in *Coloniatherium* (Rougier et al. 2009b) and *Peligrotherium* (Páez Arango 2008).

The sizes and shapes of the postcanine alveoli suggest the first four loci would house a small p1 and a large p2 (the largest tooth of the series), and that the last four loci would house the penultimate and the ultimate double-rooted molariforms, which clearly diminish in size posteriorly (Forasiepi et al. 2012). Similar characteristics to the dental pattern occur in *Coloniatherium* and *Peligrotherium* (Páez Arango 2008; Rougier et al. 2009b).

The genus and species of the specimen from Lago Los Barreales were left in open nomenclature but the authors suggested affinities with mesungulatid meridiolestidans (Forasiepi et al. 2012). If so, this finding represents the geologically oldest mesungulatid known to date.

### 6.4.4 *Mesungulatid indet. from Cerro Yeso (Maastrichtian), Patagonia*

A fragment of an edentulous Mesungulatidae lower jaw (MLP 89-XI-1-1) was collected from the vicinities of Cerro Yeso, the Coli Toro Formation (Maastrichtian) (Casamiquela 1978). The specimen has been only informally described in the context of a scientific meeting (Chornogubsky and Gelfo 2011). It has preserved the mesiodistally compressed alveoli of the last three molariforms, from which a reduction in size of the last molar was interpreted (Chornogubsky and Gelfo 2011). The size of the specimen is intermediate between *Mesungulatum* and *Peligrotherium*, and according to Chornogubsky and Gelfo (2011), it may correspond to a new taxon.

### 6.4.5 *Meridiolestida indet. from Paso Córdoba (Campanian), Patagonia*

Goin et al. (1986) described an incomplete edentulous right lower jaw (MC RN 11342) as a likely marsupial, collected from the Paso Córdoba locality, Anacleto

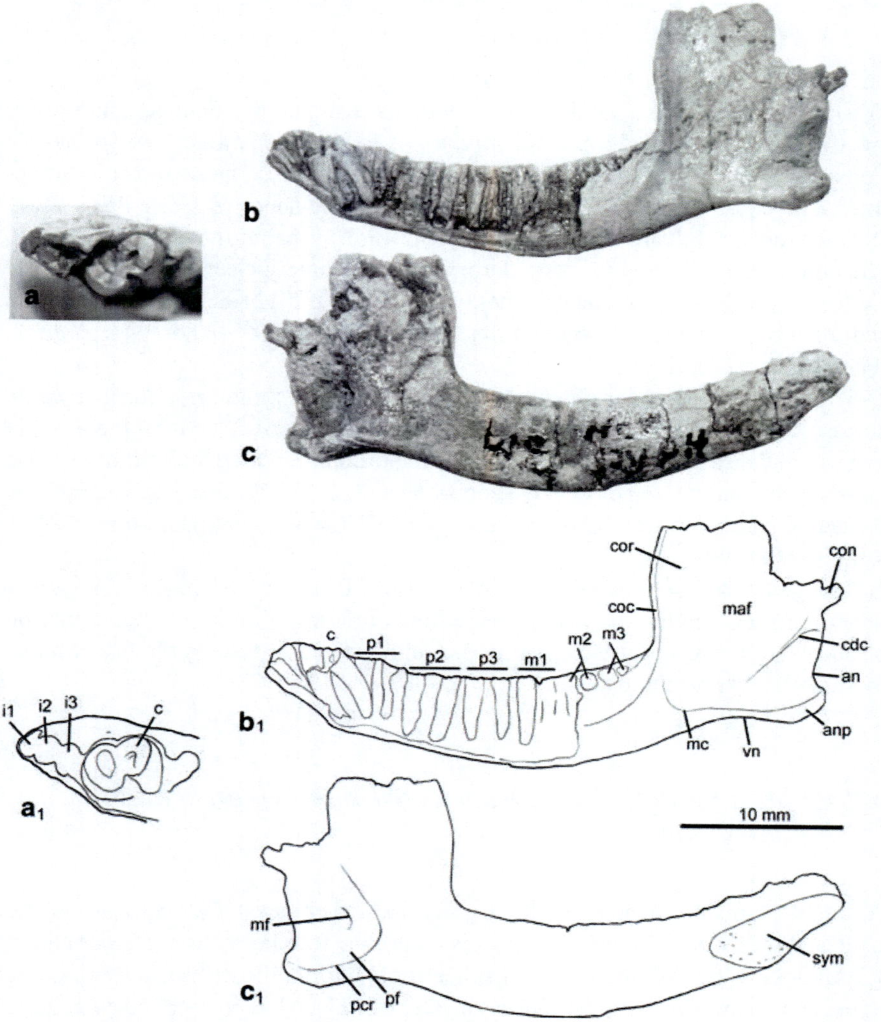


**Fig. 6.24** Dryolestoidea gen. et sp. indet. from Los Barreales Lake, Neuquén, Argentina; Los Bastos Formation, Coniacian, Upper Cretaceous. MCF-PVPH 412, edentulous left dentary. Close-up of the end of the dentary with detail of the incisor alveoli (**a**), dentary in labial (**b**) and lingual (**c**) views, and accompanying line drawings (**$a_1$**, **$b_1$**, **$c_1$**). Abbreviations: an, angular notch; anp, angular process; c, lower canine; cdc, condyloid crest; coc, coronoid crest; con, condylar process; cor, coronoid process; i1–i3, lower incisors; m1–m3, lower molars; maf, masseteric fossa; mc, masseteric crest; mf, mandibular foramen; p1–p3, lower premolars; pcr, pterygoid crest; pf, pterygoid fossa; sym, mandibular symphysis; vn, ventral notch. Modified from Forasiepi et al. (2012)

Formation (Campanian). However, Martinelli and Forasiepi (2004) believed the jaw probably belongs to a meridiolestidan. The specimen is tiny and has preserved four alveoli (Goin et al. 1986). The roots are compressed in the penultimate molar but less so in the ultimate one (Goin et al. 1986), similar to mesungulatids, but unlike sharper-toothed meridiolestidans, like *Leonardus*. In fact, the preserved portion of the jaw is reminiscent of the Lagos Los Barreales dentary (Forasiepi et al. 2012) and those of the recently described dentaries of *Reigitherium* which also show subcircular posterior molar roots and labial/buccal dentary margins strongly unequal in height (Harper et al. 2019: Figs. 4 and 5).

## 6.5 Concluding Remarks

To date, meridiolestidans are the most diverse group of Mesozoic mammals in SA. The fossil record, incomplete as it may be, verifies a long history for the group in the continent comprising more than 90 Ma, with the oldest representatives unearthed from early Late Cretaceous (Rougier et al. 2011a) and the youngest from the early Miocene, both from Patagonia. Considering that in Laurasia dryolestoids flourished during the Jurassic and became extinct in the Cretaceous (Simpson 1929, 1945; Martin 1999; Kielan-Jaworowska et al. 2004), the temporal distribution in SA is remarkable. SA dryolestoids are unknown in the Jurassic, but they survived the K/Pg mass extinction (Pascual et al. 2001; Páez Arango 2008), living in therian-dominated faunas as relictual representatives until the early Miocene (Burdigalian; Rougier et al. 2012; Wible and Rougier 2017).

In SA, the dominance of the group is recorded by the Late Cretaceous (Campanian–Maastrichtian), when dryolestoids flourished as major components of the mammalian associations (Bonaparte 1986a, b, 1994, 2002). The classic fossil locality of Los Alamitos yielded 35 years ago an isolated mammalian molar, later described as *Mesungulatum houssayi* (Bonaparte and Soria 1985), the first for a SA Mesozoic mammal. Befittingly, the first fossil is the patronym of the ubiquitous group of bunodont SA meridiolestidans, the mesungulatids. The relatively large mammalian collection from the Los Alamitos Formation is a challenging one; it consists of isolated teeth and a few maxilla/dentary fragments, occasionally bearing more than one tooth. Consequently, the taxonomical diversity was originally overestimated by recognizing twelve species of dryolestoids: *Alamitherium bishopi, Barberenia araujoae, Brandonia intermedia, Casamiquelia rionegrina, Groebertherium stipanicici, G. novasi, Leonardus cuspidatus, Mesungulatum houssayi, Paraungulatum rectangularis, Quirogatherium major*, *Reigitherium bunodontum,* and *Rougietherium tricuspes,* one "symmetrodont": *Bondesius ferox,* and two "triconodonts": *Austrotriconodon sepulvedai* and *A. mckennai,* plus three Multituberculata/Gondwanatheria: *Gondwanatherium patagonicus, Ferugliotherium windhauseni,* and *Vucetichia gracilis* (Bonaparte and Soria 1985; Bonaparte 1986a, b, 1990, 1994, Bonaparte 2002; Bonaparte and Migale 2010, 2015; Table 6.1; see also Chap. 8). Our revision of the Los Alamitos collection informed by more complete material from other fossil sites (e.g., *Coloniatherium cilinskii* from the La Colonia Formation, Rougier et al. 2009b; *Peligrotherium tropicalis* from the Salamanca Formation, Gelfo and Pascual 2001; Páez Arango 2008; *Cronopio dentiacutus* from the Candeleros Formation; Rougier et al. 2011a) lead us to recognize a smaller number of valid taxa (Table 6.1). We regard four species from the Los Alamitos Formation as valid taxa: *Groebertherium stipanicici*, *Leonardus cuspidatus*, *Mesungulatum houssayi*, and *Reigitherium bunodontum*, while we are uncertain about five other species (i.e., *Austrotriconodon mckennai*, *Bondesius ferox*, *Casamiquelia rionegrina*, *Quirogatherium major*, *Paraungulatum rectangularis*) that may indeed represent valid taxa or deciduous/permanent premolars of a taxon already described for the unit. Regardless of the ultimate affinities of these five species, they are all dryolestoids of one kind or another. There are no "symmetrodonts" or "triconodonts" among the mammals from the Los Alamitos Formation. Species originally allocated in these groups were based on simpler premolars of dryolestoids (Rougier et al. 2011a; Gaetano et al. 2013).

With the exception of *Groebertherium* spp., which clusters with Laurasian dryolestids, other SA dryolestoids were part of an endemic radiation: the Meridiolestida (Rougier et al. 2011a, b, 2012; Wible and Rougier 2017). Dryolestoids in SA display great disparity in body size, ranging from the tiny shrew-sized dryolestids (like *Groebertherium* spp.) to the large dog-sized meridiolestidan mesungulatids culminating in *Peligrotherium tropicalis*. The group also reveals very disparate dental and cranial morphotypes, expanding from the plesiomorphic sharp-toothed insectivores (*Cronopio*, *Groebertherium* spp., *Leonardus cuspidatus*) to bunodont, complex dentitions (e.g., *Reigitherium bunodontum*, *Peligrotherium tropicalis*, *Coloniatherium cilinskii*, *Mesungulatum* spp.), reflecting a range of omnivore/herbivore adaptations. Certainly, the morphotypes of SA dryolestoids were much more disparate than those in Laurasia, achieving complex tooth-on-tooth occlusion similar in complexity to that of Cenozoic therians (Harper et al. 2019). Much is likely yet remaining to be discovered about these strange and highly autapomorphic dryolestoids that prospered in isolation in SA during much of the Cretaceous. The recently discovered *Cronopio dentiacutus* (Rougier et al. 2011a) illustrates this case. The high-quality material preserves a distinctive cranial morphotype unique among mammals: short and broad braincase, very long and narrow shrew-like snout, with long, recurving pointy canines, a jaw with a partially inflected angle and a peculiar masseteric process; such a creature would not be predicted from isolated teeth alone. Bizarre mammals like this are more easily found in movies (as *Scrat* in the "Ice Age" film series) or fiction literature than in the desolate slopes of a hill, or the collection drawer of a museum (*Cronopios* are fictional creatures central to the wrintings of Julio Cortázar, *Historia de Cronopios y de Famas*, 1962).

To date, reliable Mesozoic fossil dryolestoids are restricted to Patagonia, Argentina; however, the combination of a potential find in the tropics (Gayet et al. 2001), in conjunction with the lack of geological evidence for a stark separation of Patagonia from the rest of SA during the Cretaceous, suggests that dryolestids were likely more widely distributed. The Cretaceous outcrops of Bolivia, Peru,

Ecuador, Brasil, and Chile ought to have dryolestoids as members of their Cretaceous ecosystems. If *Groebertherium* is indeed a surviving member of the Jurassic radiation of Holarctic dryolestoids and the relatively generalized molar morphology of *Cronopio* and *Leonardus* optimizes as primitive for meridiolestidans, a biogeographic inference emerges. The last common ancestor of Dryolestidae and Meridiolestidae must have a minimal Middle Jurassic age and be of small size, with simple, sharp, triangular molariforms devoid of prominent cingula, a small unbasined talonid and unequal tooth roots. The origin of meridiolestidans is likely pre-Gondwanic, and most certainly predate the full disintegration of Gondwana in its current major blocks and the full opening of the Southern Atlantic (Granot and Dyment 2015; Foulger 2018). The presence of a dryolestoid in Antarctica (Martinelli et al. 2014) supports the idea that dryolestoids reached southern latitudes, and perhaps meridiolestidans are not necessarily a SA product. The broader distribution of other archaic groups such as gondwanatherians and monotremes in the daughter masses from Gondwana supports the idea of fluid contacts among faunas during the Mesozoic. Madagascar, because of its early separation, may have been more isolated from the rest contributing to the insular character of its Mesozoic fauna, including mammals (Krause et al. 2014, 2020). The role of Africa in this story is unclear; similarities between the SA dryolestoids and *Donodon* from the Early Cretaceous of Africa (Sigogneau-Russell et al. 1990, Sigogneau-Russell 1991) and potentially *Thereuodon* (Sigogneau-Russell and Ensom 1998) from the Purbeck (England) have been pointed out by Bonaparte (1994, 2002; Martin 1999; Chornogubky 2011). These specimens are a few isolated teeth, variously regarded as molars, deciduous premolars of dryolestoids and/or "symmetrodonts". We have not been able to personally study the specimens, but the similarities of *Donodon* with meridiolestidans (Bonaparte 1994; Chornogusky 2011) seem to warrant further attention. *Thereuodon* on the other hand seems to be a deciduous premolar (Martin 1999). The material evidence is too fragmentary at this point to conclusively establish affinities between meridiolestidans and other non-tribosphenic cladotherians outside SA.

Africa is centrally located among the fragments of Gondwana but isolated early from the rest of the fragments of the supercontinent except from SA (Gheerbrant and Rage 2006; Torsvik and Cocks 2013). The prediction is that the stem lineages of Meridiolestida, if not meridiolestidans themselves, must have been present in Africa. Other vertebrate groups support close relationships between Africa and SA during the Cretaceous (e.g., pipoid anurans, mesoeucrocodylians, abelisaur and carcharodontosaurid theropods, diplodocoid and titanosaur sauropods; e.g., Bonaparte 1986b; Sereno et al. 2004; Trueb et al. 2004; Sereno and Brusatte 2008; Sereno and Larsson 2009; Fanti et al. 2013; Gorscak and O'Connor'2016). It is hard to imagine the biogeographical history of African and SA Mesozoic mammals not being deeply intertwined; however, the bulk of the SA record comes from its southern tip, Patagonia, which today and during most of the Cenozoic can be hardly considered to be representative of SA as a whole. The patchy and ever poorer African record does not help to paint a more optimistic picture. Filling geographical and temporal gaps is a necessity to obtain a more realistic view of dryolestoid evolution in the southern continents. Implementation of different techniques specifically targeting microvertebrates (Chap. 1) has the potential to dramatically increase what we know about dental diversity and temporal distribution of this iconic group of SA Mesozoic mammals. Experience says that if you are to find a single isolated tooth in the Cretaceous of SA, chances are it will be a meridiolestidan or a gondwanatherian (Chap. 8).

## References


Abdala NF, Jasinoski SC, Fernández V (2013) Ontogeny of the Early Triassic cynodont *Thrinaxodon liorhinus* (Therapsida): Dental morphology and replacement. J Vertebr Paleont 33:1408–1431

Alroy J (1996) Constant extinction, constrained diversification, and uncoordinated stasis in North American mammals. Palaeogeogr Palaeoclimatol Palaeoecol 127:285–311

Ameghino F (1891) Nuevos restos de mamíferos fósiles descubiertos por Carlos Ameghino en el Eoceno inferior de la Patagonia austral. Especies nuevas adiciones y correcciones. Rev Argent Hist Nat 1:289–328

Asher RJ, Horovitz I, Martin T, Sánchez-Villagra MR (2007) Neither a rodent nor a platypus: a reexamination of *Necrolestes patagonensis* Ameghino. Am Mus Novit 3546:1–40

Averianov AO (2002) Early Cretaceous "symmetrodont" mammal *Gobiotheriodon* from Mongolia and the classification of "Symmetrodonta". Acta Palaeontol Pol 47:705–716

Averianov AO, Martin T, Lopatin AV (2013) A new phylogeny for basal Trechnotheria and Cladotheria and affinities of South American endemic Late Cretaceous mammals. Naturwissenschaften 100:311–326

Averianov AO, Martin T, Skutschas PP, Danilov IG, Schultz JA, Schellhorn R, Obraztsova E, Lopatin AV, Sytchevskaya EK, Kuzmin IT, Krasnolutskii SA, Ivantsov SV (2016) Middle Jurassic vertebrate assemblage of Berezovsk coal mine in Western Siberia (Russia). Glob Geol 19:187–204

Bi S, Zheng X, Meng J, Wang X, Robinson N, Davis B (2016) A new symmetrodont mammal (Trechnotheria: Zhangheotheriidae) from the Early Cretaceous of China and trechnotherian character evolution. Sci Rep 6:26668

Bonaparte JF (1986a) Sobre *Mesungulatum houssayi* y nuevos mamíferos cretácicos de Patagonia. 4° Congreso Argentino de Paleontología y Bioestratigrafía, Mendoza, Actas 2:48–61

Bonaparte JF (1986b) History of the terrestrial Cretaceous vertebrates of Gondwana. 4°Congreso Argentino de Paleontología y Bioestratigrafía, Mendoza, Actas 2:63–95

Bonaparte JF (1990) New Late Cretaceous mammals from the Los Alamitos Formation, northern Patagonia. Natl Geogr Res 6:63–93

Bonaparte JF (1992) Una nueva especie de Triconodonta (Mammalia) de la Formación Los Alamitos, Provincia de Río Negro y comentarios sobre su fauna de mamíferos. Ameghiniana 29:99–110

Bonaparte JF (1994) Approach to the significance of the Late Cretaceous mammals of South America. Berlin Geowiss Abt E 13:31–44

Bonaparte JF (2002) New Dryolestida (Theria) from the Late Cretaceous of Los Alamitos, Argentina, and paleogeographical comments. Neues Jahrb Geol Paläontol 224:339–371

Bonaparte JF, Migale LA (2010) Protomamíferos y mamíferos Mesozoicos de América del Sur. Museo de Ciencias Naturales Carlos Ameghino de Mercedes, Buenos Aires. 1° Edition

Bonaparte JF, Migale LA (2015) Protomamíferos y mamíferos Mesozoicos de América del Sur. Fundación de Historia Natural Felix de Azara, Buenos Aires. 2° Edition

Bonaparte JF, Morales J (1997) Un primitivo Notonychopidae (Litopterna) del Paleoceno Inferior de Punta Peligro, Chubut, Argentina. Est Geol 53:183–286

Bonaparte JF, Soria MF (1985) Nota sobre el primer mamífero del Cretácico Argentino, Campaniano-Maastrichtiano (Condylarthra). Ameghiniana 21:177–183

Bonaparte JF, Van Valen LM, Kramarz A (1993) La fauna local de Punta Peligro, Paleoceno Inferior, de la Provincia de Chubut, Patagonia, Argentina. Evol Monogr 14:1–61

Bond M, Carlini AA, Goin FJ, Legarreta L, Ortiz-Jaureguizar E, Pascual R, Uliana MA (1995) Episodes in South American land mammal evolution and sedimentation: Testing their apparent concurrence in a Paleocene succession from Central Patagonia. 6° Congreso Argentino de Paleontología y Bioestratigrafía, Trelew, Actas, 47–58

Butler PM (1939) The teeth of the Jurassic mammals. Proc Zool Soc Lond B 109:329–356

Casamiquela RM (1978) La zona litoral de la transgresión Maástrichtense en el norte de Patagonia. Aspectos ecológicos. Ameghiniana 15:137–148

Chimento NR, Agnolín FL, Novas FE (2012) The Patagonian fossil mammal *Necrolestes*: a Neogene survivor of Dryolestoidea. Rev Mus Argent Cienc Nat "B Rivadavia" 14:261–306

Chimento N, Agnolín FL, Martinelli AG (2016) Mesozoic mammals from South America: Implications for understanding early mammalian faunas from Gondwana. Mus Argent Cienc Nat "B Rivadavia". Contribuciones 6:199–209

Chornogubsky L (2003) Revisión preliminar de los mamíferos de la Formación Los Alamitos (Campaniano–Maastrichtiano, provincia de Río Negro, Argentina). Bachelor thesis, University of Buenos Aires

Chornogubsky L (2011) New remains of the dryolestoid mammal *Leonardus cuspidatus* from the Los Alamitos Formation (Late Cretaceous, Argentina). Paläontol Z 85:343–350

Chornogubsky L, Gelfo JJ (2011) Un nuevo mesungulátido (Mammalia, Dryolestoidea) del Cretácico Tardío de Argentina. Ameghiniana 48(Suppl):9R

Cifelli RL, Madsen SK (1999) Spalacotheriid symmetrodonts (Mammalia) from the medial Cretaceous (upper Albian or lower Cenomanian) Mussentuchit local fauna, Cedar Mountain Formation, Utah, USA. Geodiversitas 21:167–214

Crompton AW (1971) The origin of the tribosphenic molar. In: Kermack DM Kermack KA (eds) Early mammals. Zool J Linn Soc 50(Suppl 1):65–87

Crompton AW, Wood CB, Stern DN (1994) Differential wear of enamel: a mechanism for maintaining sharp cutting edges. Adv Comp Environ Physiol 18:321–346

Davis B (2011) Evolution of the tribosphenic molar pattern in early mammals, with comments on the "dual-origin" hypothesis. J Mammal Evol 18:224–227

Diamond MN (1989) Coarctation of stapedial artery: an unusual adaptative respose to competing functional demands in the middle ear of some eutherians. J Morphol 200:71–86

Ensom PC, Sigogneau-Russell D (1998) New dryolestoid mammals from the basal Cretaceous Purbeck Limestone group of southern England. Palaeontology 41:35–55

Fanti F, Cau A, Hassine M, Contessi M (2013) A new sauropod dinosaur from the early cretaceous of tunisia with extreme avian-like pneumatization. Nat Comm 4:2080

Forasiepi AM, Martinelli AG (2003) Femur of a monotreme (Mammalia, Monotremata) of the early Paleocene Salamanca Formation of Patagonia, Argentina. Ameghiniana 40:625–630

Forasiepi AM, Rougier GW (2009) Additional data on Early Paleocene metatherians (Mammalia) from Punta Peligro (Salamanca Formation, Argentina): comments based on the morphology of the petrosal bone. J Zool Syst Evol Res 47:391–398

Forasiepi AM, Coria RA, Hurum J, Currie PJ (2012) First dryolestoid (Mammalia, Dryolestida) dentary from the Coniacian of Patagonia, Argentina. Ameghiniana 49:497–504

Foulger GR (2018) Origin of the South Atlantic igneous province. J Volcanol Geotherm Res 355:2–20

Freeman EF (1979) A Middle Jurassic mammal bed from Oxfordshire. Palaeontology 22:135–166

Gaetano LC, Rougier GW (2011) New materials of *Argentoconodon fariasorum* (Mammaliaformes, Triconodontidae) from the Jurassic of Argentina and its bearing on triconodont phylogeny. J Vertebr Paleontol 31:829–843

Gaetano LC, Marsicano CA, Rougier GW (2013) A revision of the putative Late Cretaceous triconodonts from South America. Cretac Res 46:90–100

Garrido AC (2010) Estratigrafía del Grupo Neuquén, Cretácico Superior de la Cuenca Neuquina (República Argentina): Nueva propuesta de ordenamiento litoestratigráfico. Rev Mus Argent Cienc Nat "B Rivadavia", ns 12:121–177

Garrido AC (2011) Grupo Neuquén (Cenomaniano–Campaniano). In: Leanza HA, Arregui C, Carbone O, Danieli JC, Vallés JM (eds) Geología y recursos naturales de la Provincia del Neuquén, 18° Congreso Geológico Argentino, Neuquén, Relatorio, pp 231–244

Gayet M, Marshall LG, Sempere T (1991) The Mesozoic and Paleocene vertebrates of Bolivia and their stratigraphic context: a review. In: Suarez-Soruco R (ed) Fósiles y facies de Bolivia. Vertebrados. Rev Tec YPFB 12:393–433

Gayet M, Marshall LG, Sempere T, Meunier FJ, Cappetta H, Rage J-C (2001) Middle Maastrichtian vertebrates (fishes, amphibians, dinosaurs and other reptiles, mammals) from Pajcha Pata (Bolivia). Biostratigraphic, palaeoecologic and palaeobiogeographic implications. Palaeogeogr Palaeoclimatol Palaeoecol 169:39–68

Gelfo JN (2007) The condylarth *Raulvaccia peligrensis* (Mammalia: Didolodontidae) from the Paleocene of Patagonia, Argentina. J Vertebr Paleontol 27:651–660

Gelfo JN, Pascual R (2001) *Peligrotherium tropicalis* (Mammalia, Dryolestida) from the Early Paleocene of Patagonia, a survival from a Mesozoic Gondwanan radiation. Geodiversitas 23:369–379

Gelfo JN, Ortiz-Jaureguizar E, Rougier GW (2007) New remains and species of the "condylarth" genus *Escribania* (Mammalia: Didolodontidae) from the Early Paleocene of Patagonia, Argentina. Earth Env Sci T R So 98:127–138

Gheerbrant E, Rage J-C (2006) Paleobiogeography of Africa: How distinct from Gondwana and Laurasia? Palaeogeogr Palaeoclimat Palaeoecol 241:224–246

Goin FJ, Reguero MA (1993) Un "enigmático insectívoro" del Eoceno de Antártida. Ameghiniana 30(Suppl):108R

Goin FJ, Carlini AA, Pascual R (1986) Un probable marsupial del Cretácico Tardío del Norte de Patagonia, Argentina. 4° Congreso Argentino de Paleontología y Bioestratigrafía. Mendoza, Actas 2:43–47

Goin FJ, Forasiepi AM, Candela AM, Ortiz Jaureguizar E, Pascual R, Archer M, Godthelp H, Muirhead J, Augee M, Hand S, Wroe S (2002). Earliest Paleocene marsupials from Patagonia. 1° International Palaeontological Congress, Sydney, Abstracts, 68

Goin FJ, Ortiz Jaureguizar E, Pascual R, Rougier GW, Cladera G (2004) El primer registro de un Bonapartheriidae (Marsupialia, Polydolopimorphia) en el Paleoceno Inferior (Formación Salamanca, "Edad-Mamífero" Peligrense) de América del Sur. Ameghiniana 41(Suppl):49R

Goin FJ, Abello A, Bellosi E, Kay R, Madden R, Carlini AA (2007) Los Metatheria sudamericanos de comienzos del Neógeno (Mioceno Temprano, Edad-Mamífero Colhuehuapense). Parte 1: Introducción. Didelphimorphia y Sparassodonta. Ameghiniana 44:29–71

Gorscak E, O'Connor' PM (2016) Time-calibrated models support congruency between Cretaceous continental rifting and titanosaurian evolutionary history. Biol Lett 12:20151047

Granot R, Dyment J (2015) The cretaceous opening of the South Atlantic Ocean. Earth Planet Sc Lett 414:156–163

Gregory WK (1910) The orders of mammals. Bull Am Mus Nat Hist 27:1–524

Gurovich Y (2005) Bio-evolutionary aspects of Mesozoic mammals: description, phylogenetic relationships and evolution of the Gondwanatheria (Late Cretaceous and Paleocene of Gondwana). Unpublished Ph.D.as thesis, University of Buenos Aires

Gurovich Y, Beck RMD (2009) The phylogenetic affinities of the enigmatic mammalian clade Gondwanatheria. J Mammal Evol 16:25–49

Harper T, Parras A, Rougier GW (2019) *Reigitherium* (Meridiolestida, Mesungulatoidea) an enigmatic Late Cretaceous mammal from Patagonia, Argentina: morphology, affinities, and dental evolution. J Mammal Evol 26:447–478

Hatcher JB (1903) Narrative of the expeditions: eography of southern Patagonia. Reports of the Princeton University Expedition to Patagonia, 1896–1899, vol I. Princeton University Press, Princeton

Kielan-Jaworowska Z, Cifelli RL, Luo Z-X (2004) Mammals from the age of dinosaurs. Origins, evolution, and structure. Columbia University Press, New York

Krause DW, Bonaparte JF (1993) Superfamily Gondwanatherioidea: a previously unrecognized radiation of multituberculate mammals in South America. Proc Natl Acad Sci USA 90:9379–9383

Krause DW, Hoffmann S, Wible JR, Kirk EC, Schultz JA, von Koenigswald W, Groenke JR, Rossie JB, O'Connor PM, Seiffert ER, Dumont ER, Holloway WL, Rogers RR, Rahantarisoa RJ, Kemp AD, Andriamialison H (2014) First cranial remains of a gondwanatherian mammal reveal remarkable mosaicism. Nature 515:512–517

Krause DW, Hoffmann S, Hu Y, Wible JR, Rougier GW, Kirk EC, Groenke JR, Rogers RR, Rossie JB, Schultz JA, Evans AR, von Koenigswald W, Rahantarisoa LJ (2020) Skeleton of Cretaceous mammal from Madagascar reflects long-term insularity. Nature 581:421–427

Krebs B (1971) Evolution of the mandible and lower dentition in dryolestoids (Pantotheria, Mammalia). In: Kermack DM, Kermack KA (eds) Early mammals. Zool J Linnean Soc 50(Suppl 1):89–102

Krebs B (1991) Das Skelett von *Henkelotherium guimarotae* gen. et sp. nov., (Eupantotheria, Mammalia) aus dem Oberen Jura von Portugal. Berl Geowiss Abh A 133:1–121

Krebs B (1998) *Drescheratherium acutum* gen. et sp. nov., ein neuer Eupantotherier (Mammalia) aus dem Oberen Jura von Portugal. Berl Geowiss Abh E 28:91–111

Ladevèze S, Asher RJ, Sánchez-Villagra MR (2008) Petrosal anatomy in the fossil mammal *Necrolestes*: evidence for metatherian affinities and comparisons with the extant marsupial mole. J Anat 213:686–697

Lillegraven JA, McKenna MC (1986) Fossil mammals from the "Mesaverde" Formation (Late Cretaceous, Judithian) of the Bighorn and Wind River basins, Wyoming, with definitions of Late Cretaceous North American land mammal "ages" Am Mus Novit 2840:1–68

Linnaeus C (1758) Systema naturae per regna tria naturae, secundum classes, ordines, genera, species, cum characteribus, differentiis, synonymis, locis. Vol. 1: Regnum animale. Editio decima, reformata. Laurentii Salvii, Stockholm

Luo Z-X, Ji Q (2005) New study on dental and skeletal features of the Cretaceous "symmetrodontan" mammal *Zhangheotherium*. J Mammal Evol 12:337–357

Luo Z-X, Kielan-Jaworowska Z, Cifelli RL (2002) In quest for a phylogeny of Mesozoic Mammals. Acta Palaeontol Pol 47:1–78

MacPhee R, Reguero MA, Strgnac P, Nishida M, Jacobs P (2008) Out of Antarctica: Paleontological reconnaissance of Livingston island (South Shetlands) and Seymour island (James Ross Group). J Vertebr Paleont 28(Suppl):110A

Marsh OC (1878) Fossil mammal from the Jurassic of the Rocky Mountains. Am J Sci 15:459

Marsh OC (1879) Notice of new Jurassic mammals. Am J Sci 20:396–398

Marsh OC (1880) Notice on Jurassic mammals representing two new orders. Am J Sci 20:235–239

Martin T (1997) Tooth replacement in late Jurassic Dryolestidae (Eupantotheria, Mammalia). J Mammal Evol 4:1–18

Martin T (1999) Dryolestidae (Dryolestoidea, Mammalia) aus dem Oberen Jura von Portugal. Abh Senckenberg Naturforsch Ges 550:1–119

Martin T (2000) The dryolestids and the small "peramurid" from the Guimarota coal mine. In: Martin T, Krebs B (eds) Guimarota. Verlag, München, A Jurassic ecosystem, pp 109–120

Martin T, Goin FJ, Chornogubsky L, Gelfo JN, Schultz J (2013) Early Late Cretaceous (Cenomanian) mammals and other vertebrates from the Mata Amarilla Formation of Southern Patagonia (Argentina). J Vertebr Paleontol 33(Suppl):172A

Martinelli AG, Forasiepi AM (2004) Late Cretaceous vertebrates from Bajo de Santa Rosa (Allen Formation), Río Negro Province, Argentina, with the description of a new sauropod dinosaur (Titanosauridae). Rev Mus Argent Cienc Nat "B Rivadavia", ns 6:257–305

Martinelli AG, Chornogubsky L, Abello A, Goin FJ, Reguero M (2014) The first non-therian dryolestoid from Antarctica. In: 2014 SCAR open science conference, Auckland, Abstracts, p 432

McDowell SB Jr (1958) The Greater Antillean insectivores. Bull Am Mus Nat Hist 115:113–214

Mckenna MC, Bell SK (1997) Classification of mammals above the species level. Columbia University Press, New York

Miao D (1988) Skull morphology of *Lambdopsalis bulla* (Mammalia, Multituberculata) and its implications to mammalian evolution. Contrib Geol. Univ Wyoming Spec Pap 4:1–104

O'Leary' MA, Bloch JI, Flynn JJ, Gaudin TJ, Giallombardo A, Giannini NP, Goldberg SL, Kraatz BP, Luo ZX, Meng J, Ni X, Novacek MJ, Perini FA, Randall ZS, Rougier GW, Sargis EJ, Silcox MT, Simmons NB, Spaulding M, Velazco PM, Weksler M, Wible JR, Cirranello AL (2013) The placental mammal ancestor and the post-K-Pg radiation of placentals. Science 339:662–667

O'Meara RN, Thompson RS (2014) Were there Miocene meridiolestidans? Assessing the phylogenetic placement of *Necrolestes patagonensis* and the presence of a 40 million year merdiolestidan ghost lineage. J Mammal Evol 21:271–284

Owen R (1845) Odontography; or a treatise on the comparative anatomy of the teeth; their physiological relations, mode of development, and microscopic structure, in the vertebrate animals. Hippolyte Ballière, London

Páez Arango N (2008) Dental and craniomandibular anatomy of *Peligrotherium tropicalis*: The evolutionary radiation of South American dryolestoid mammals. Unpublished Ms thesis, University of Louisville

Pascual R, Ortiz-Jaureguizar E (2007) The Gondwanan and South American episodes: two major moments in the history of South American mammals. J Mammal Evol 14:75–137

Pascual R, Archer M, Ortiz-Jaureguizar E, Prado JL, Godthelp H, Hand SH (1992) First discovery of monotremes in South America. Nature 356:704–705

Pascual R, Goin FJ, Krause DW, Ortiz-Jaureguizar E, Carlini AA (1999) The first gnathic remains of *Sudamerica*: implications for gondwanathere relationships. J Vertebr Paleontol 19:373–382

Pascual R, González P, Ardolino A, Puerta PF (2000) A highly derived docodont from the Patagonian Late Cretaceous: evolutionary implications for Gondwanan mammals. Geodiversitas 22:395–414

Pascual R, Balarino ML, Udrizar Sauthier DE (2001) The K/T and Tertiary-Pleistocene South American mammalian turnovers. Similar phenomena? 7° International Symposium on Mesozoic Terrestrial Ecosystems. Asociación Paleontológica Argentina, Buenos Aires, Publicación Especial 7:151–156

Patterson B (1956) Early Cretaceous mammals and the evolution of mammalian molar teeth. Fieldiana Geol 13:1–105

Patterson B (1958) Affinities of the Patagonian fossil mammal *Necrolestes*. Breviora 94:1–14

Prothero DR (1981) New Jurasic mammals from Como Bluff, Wyoming, and the interrelationships of non-tribosphenic Theria. Bull Am Mus Nat Hist 167:277–325

Reguero MA, Goin FJ, Hospitaleche CA, Dutra T, Marenssi SA (2013) Late Cretaceous/Paleogene west Antarctica terrestrial biota and its intercontinental affinities. Springer, Dordrecht

Romer AS (1945) Vertebrate paleontology. University of Chicago Press, Chicago

Rose KD, Emry RJ (1983) Extraordinary fossorial adaptations in the Oligocene palaeanodonts *Epoicotherium* and *Xenocranium* (Mammalia). J Morphol 175:33–56

Rougier GW, Novacek MJ, Pascual R, Gelfo JN, Cladera G (2000) New Late Cretaceous mammals from Argentina and the survival of Mesozoic lineages in the Patagonian Early Tertiary. J Vertebr Paleontol 20:65A

Rougier GW, Novacek MJ, McKenna MC, Wible JR (2001) Gobiconodonts from the early Cretaceous of Oshih (Ashile), Mongolia. Am Mus Novit 3348:1–30

Rougier GW, Qiang J, Novacek MJ (2003a) A new symmetrodont mammal with fur impressions from the Mesozoic of China. Acta Geol Sin 77:7–14

Rougier GW, Novacek MJ, Ortiz-Jaureguizar E, Pol D, Puerta P (2003b) Reinterpretation of *Reigitherium bunodontum* as a Reigitheriidae dryolestoid and the interrelationships of South American dryolestoids. J Vertebr Paleont 23(Suppl):90A

Rougier GW, Martinelli AG, Forasiepi AM, Novacek MJ (2007a) New Jurassic mammals from Patagonia, Argentina: A reappraisal of australosphenidan morphology and interrelationships. Am Mus Novit 3566:1–54

Rougier GW, Garrido A, Gaetano L, Puerta P, Corbitt C, Novacek MJ (2007b) First Jurassic triconodont from South America. Am Mus Novit 3580:1–17

Rougier GW, Chornogubsky L, Casadio S, Paéz Arango N, Giallombardo A (2009a) Mammals from the Allen Formation, Late Cretaceous, Argentina. Cretac Res 30:223–238

Rougier GW, Forasiepi AM, Hill RV, Novacek MJ (2009b) New mammalian remains from the Late Cretaceous La Colonia Formation, Patagonia, Argentina. Acta Palaeontol Pol 54:195–212

Rougier GW, Apesteguía S, Gaetano LC (2011a) Highly specialized mammalian skulls from the Late Cretaceous of South America. Nature 479:98–102

Rougier GW, Gaetano L, Drury BR, Colella R, Gómez RO, Páez Arango N (2011b) A review of the Mesozoic mammalian record of South America. In: Calvo J, Porfiri J, González Riga B, Dos Santos D (eds) Dinosaurios y paleontología desde América Latina. Editorial de la Universidad Nacional de Cuyo, Mendoza, pp 195–214

Rougier GW, Wible JR, Beck RMD, Apesteguía S (2012) The Miocene mammal *Necrolestes* demonstrates the survival of a Mesozoic nontherian lineage into the late Cenozoic of South America. Proc Natl Acad Sci USA 109:20053–20058

Saban R (1954) Phylogénie des insectivores. Bull Mus Natl Hist Nat, serie 2(26):419–432

Schultz JA, Martin T (2011) Wear pattern and functional morphology of dryolestoid molars (Mammalia, Cladotheria). Paläontol Z 85:269–285

Schultz JA, Martin T (2014) Function of pretribosphenic and tribosphenic mammalian molars inferred from 3D animation. Naturwissenschaften 101:771–781

Scillato-Yané GJ, Pascual R (1985) Un peculiar Xenarthra del Paleoceno medio de Patagonia (Argentina). Su importancia en la sistemática de los Paratheria. Ameghiniana 21:316–318

Scott WB (1905). Mammalia of the Santa Cruz beds. Part II. Insectivora. In: Scott WB (ed) Reports of the Princeton University Expedition to Patagonia, 1896–1899. Vol. 5, Paleontology II. Princeton University Press, Princeton, pp 365–383

Sereno PC, Brusatte SL (2008) Basal abelisaurid and carcharodontosaurid theropods from the Lower Cretaceous Elrhaz Formation of Niger. Acta Palaeontol Pol 53:15–46

Sereno PC, Larsson HCE (2009) Cretaceous Crocodyliformes from the Sahara. ZooKeys 28:1–143

Sereno PC, Wilson JA, Conrad JL (2004) New dinosaurs link southern landmasses in the Mid-Cretaceous. Proc R Soc Lond B 271:1325–1330

Sigogneau-Russell D (1991) Nouveaux Mammifères theriens du Crétacé Inférieur du Maroc. C R Acad Sci 313:279–285

Sigogneau-Russell D, Ensom P (1998) *Thereuodon* (Theria, Symmetrodonta) from the Lower Cretaceous of North Africa and Europe, and a brief review of symmetrodonts. Cretac Res 19:445–470

Sigogneau-Russell D, Monbaron M, de Kaenel E (1990) Nouvelles données sur le gisement à mammifères Mésozoïque du Haut-Atlas Marocain. Geobios 23:461–483

Simpson GG (1929) American Mesozoic Mammalia. Peabody Mus Nat Hist, Memoirs 3:1–235

Simpson GG (1945) The principles of classification and a classification of mammals. Bull Am Mus Nat Hist 85:1–350

Szalay FS (1994) Evolutionary history of the marsupials and an analysis of osteological characters. Cambridge University Press, Cambridge

Tauber AA, Luna CA, Palacios ME (2005) El registro de *Necrolestes patagonensis* Ameghino, 1891 (Mammalia) en La Formación Santa Cruz (Mioceno), Patagonia Austral. In: Cabaleri N, Cingolani CA, Linares E, López de Luchi MG, Ostera HA, Panarello HO (eds) 16° Congreso Geológico Argentino, La Plata, Actas, vol 5, pp 157–164

Torsvik TH, Cocks RM (2013) Gondwana from top to base in space and time. Gondwana Res 24:999–1030

Trueb L, Ross CF, Smith RMH (2004) A new pipoid anuran from the Late Cretaceous of South Africa. J Vertebr Paleont 25:533–547

Van Valen LM (1988) Faunas of a southern world. Nature 333:113

Vizcaíno SF, Kay RF, Bargo MS (eds) (2012) Early Miocene paleobiology in Patagonia. Cambridge University Press, Cambridge

Wible JR, Rougier GW (2017) Craniomandibular anatomy of the subterranean meridolestidan *Necrolestes patagonensis* Ameghino, 1891 (Mammalia, Cladotheria) from the Early Miocene of Patagonia. Ann Carnegie Mus 84:183–251

Wible JR, Novacek MJ, Rougier GW (2004) New data on the skull and dentition in the Mongolian Late Cretaceous eutherian mammal *Zalambdalestes*. Bull Am Mus Nat Hist 281:1–144

Winge H (1941) The interrelationships of the mammalian genera. Volume I. Monotremata, Marsupialia, Insectivora, Chiroptera, Edentata. C.A. Reitzels Forlag, København

# Chapter 7
# Stem Therians

*From a biogeographic point of view, the presence of Vincelestes in Early Cretaceous beds of Patagonia, and the absence of Prototribosphenida into the Late Cretaceous ones, is enigmatic.*
Rosendo Pascual and Eduardo Ortiz-Jaureguizar
*The Gondwanan and South American Episodes: Two major and unrelated moments in the history of the South American mammals*, 2007

**Abstract** The cladothere *Vincelestes neuquenianus*, from the La Amarga Formation (Barremian–lower Aptian, Lower Cretaceous) of Patagonia, Argentina, is known by several nearly complete skulls, lower jaws, and abundant postcranial elements. This exceptional material is crucial in shaping our knowledge of early stem therians and high-level mammalian phylogeny. *Vincelestes* retains a relatively primitive braincase, similar to what is expected for the last common ancestor of all mammals, and a derived but unequivocally non-therian ear region and dentition.

**Keywords** Cladotheria · Skull · Skeleton · *Vincelestes neuquenianus*

## 7.1 Introduction

Most SA Mesozoic mammalian lineages are represented by fragmentary jaws and isolated teeth; only a handful of taxa are known by relatively well-preserved craniomandibular remains. Most of those relatively complete specimens are part of the bizarre radiation of SA dryolestoids (e.g., *Cronopio dentiacutus, Necrolestes patagonensis, Peligrotherium tropicalis*; see Chap. 6). The only exception, a non-dryolestoid mammal, is *Vincelestes neuquenianus*, a taxon from the Barremian–lower Aptian (Lower Cretaceous) known by six nearly complete skulls, 17 lower jaws, and several postcranial elements, all from a single quarry from the La Amarga Formation, Patagonia, Argentina (Bonaparte 1986; Bonaparte and Rougier 1987; Rougier 1993).

The first specimen of *Vincelestes neuquenianus* (MACN-N 01, which would later become the holotype; Bonaparte 1986), a jaw and a few postcranial bones, was found in 1985, the last day in the field of the 9° Paleontological Expedition to



G. W. Rougier et al., *Mesozoic Mammals from South America and Their Forerunners*, Springer Earth System Sciences, https://doi.org/10.1007/978-3-030-63862-7_7

Patagonia organized by José F. Bonaparte. The main goal of the expedition was to complete the excavation and collection of the skeleton of the dinosaur *Amargasaurus cazaui*, discovered by one of us (GWR) and partially extracted the year before. The mammalian remains were found by Mr. Martín Vince (PVL, CONICET), Bonaparte's long-term preparator and technician, who continued to play a pivotal role in his field program even after Bonaparte's move from Tucumán to Buenos Aires in the late seventies. The genus—*Vincelestes*—honors Mr. Vince's talent and sustained effort (Bonaparte 1986). The following year a quarry was opened in the place of the original find; however, for a day or two no fossils came to light. Funding for the field season was provided in part by Dr. Malcolm C. McKenna, who also participated during the initial opening of the quarry and fieldwork, but left before any more specimens were found. Soon afterwards we found a cluster of bones in the quarry (see vignette below), which would eventually become at least nine specimens including the first skulls and associated postcranial material. Given the completeness and exceptional quality of preservation, *Vincelestes* has been studied in some detail. The external anatomy of the skull and skeleton was described (Bonaparte and Rougier 1987; Rougier 1993), as well as the sidewall of the braincase (Rougier et al. 1992; Hopson and Rougier 1993; Rougier and Wible 2006; Crompton et al. 2018), the petrosal (Rougier et al. 1992; Wible and Hopson 1993), the pattern of cranial vasculature (Rougier et al. 1992), the internal cranial morphology and cavities (Macrini et al. 2007), and its tooth enamel microstructure (Wood and Rougier 2005). Consequently, *Vincelestes* has been included in several comprehensive studies of mammalian evolution, being critical to polarize basal therian characters (e.g., Rowe 1993; Rougier et al. 1996, 1998, 2004, 2011, 2012; Novacek et al. 1997; Horovitz 2000, 2003; Wible et al. 2001; Horovitz and Sánchez-Villagra 2003; Luo et al. 2011; Krause et al. 2014; Wible and Rougier 2017; Huttenlocker et al. 2018; King and Beck 2020).

*Vincelestes* has been considered a stem therian by most authors (Bonaparte 1986; Butler 1990; Rougier et al. 1992, 1998, 2011, 2012; Rowe 1993; Luo et al. 2002; Kielan-Jaworowska et al. 2004; Beck and Lee 2014; O'Meara and Thompson 2014; Chimento et al. 2016; Wible and Rougier 2017; Huttenlocker et al. 2018) (Fig. 7.1). However, Bonaparte (2008) has supported australosphenidan affinities on the basis of selected characters from the lower jaw and lower dentition of *Vincelestes* and some other Laurasian and Gondwanan taxa. On the other hand, Averianov et al. (2013) in their reassessment of the Cretaceous taxa from SA recovered *Vincelestes* as a basal dryolestidan among other striking results. Our most recent results (Wible and Rougier 2017) and the majority of recent papers still support *Vincelestes* as a stem therian (O'Leary et al. 2013; Huttenlocker et al. 2018; King and Beck 2020), and we follow this interpretation here.

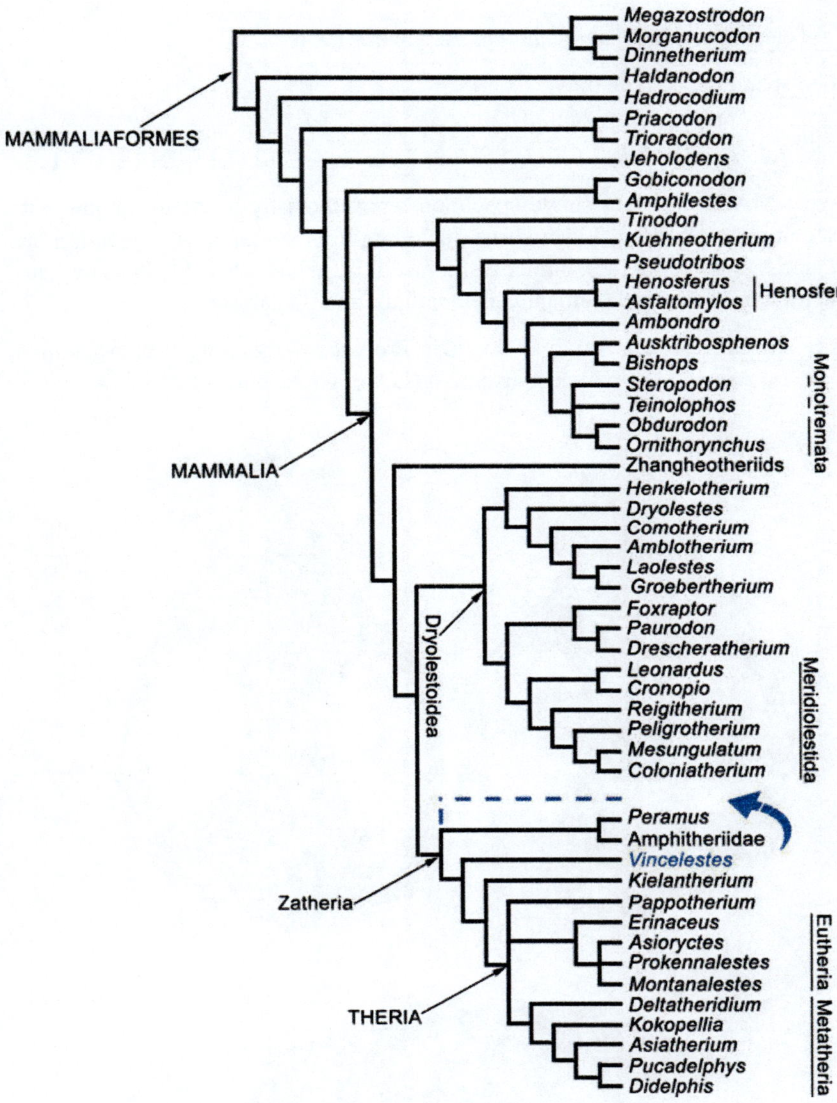


**Fig. 7.1** Phylogenetic tree of Mammalia showing alternative phylogenetic positions for *Vincelestes* as stem Theria (Modified from Rougier et al. 2011)

## 7.2 Systematics

Mammalia Linnaeus 1758
Cladotheria McKenna 1975
*Vincelestes* Bonaparte 1986

Type species: *Vincelestes neuquenianus* Bonaparte 1986.

Included species: The type only.

*Vincelestes neuquenianus* Bonaparte 1986
(Figs. 7.2, 7.3, 7.4, 7.5, 7.6, 7.7, 7.8, 7.9, 7.10, 7.11, 7.12, 7.13, 7.14 and 7.15)

Holotype: MACN-N 01, a single specimen represented by a left lower jaw with almost complete dentition, 30 vertebrae (including fragment of atlas, dorsal, lumbar, sacral, and caudal vertebrae), humerus, radius, ulna, pelvis, tibia, fibula, astragalus, calcaneus, carpal or tarsal elements, metapodials, and phalanges.

Locality and horizon: La Amarga, Neuquén Province, Argentina. Puesto Antigual Member, La Amarga Formation, Barremian (Lower Cretaceous).

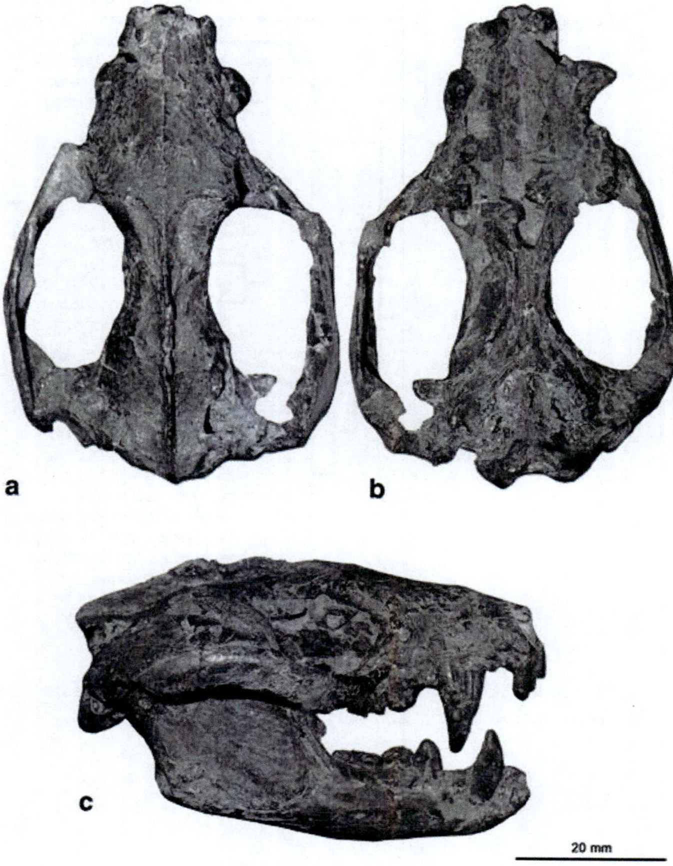


**Fig. 7.2** *Vincelestes neuquenianus* from La Amarga, Neuquén, Argentina; La Amarga Formation, Barremian–lower Aptian, Lower Cretaceous. MACN-N 05, skull in dorsal (**a**), ventral (**b**), and lateral (**c**) views

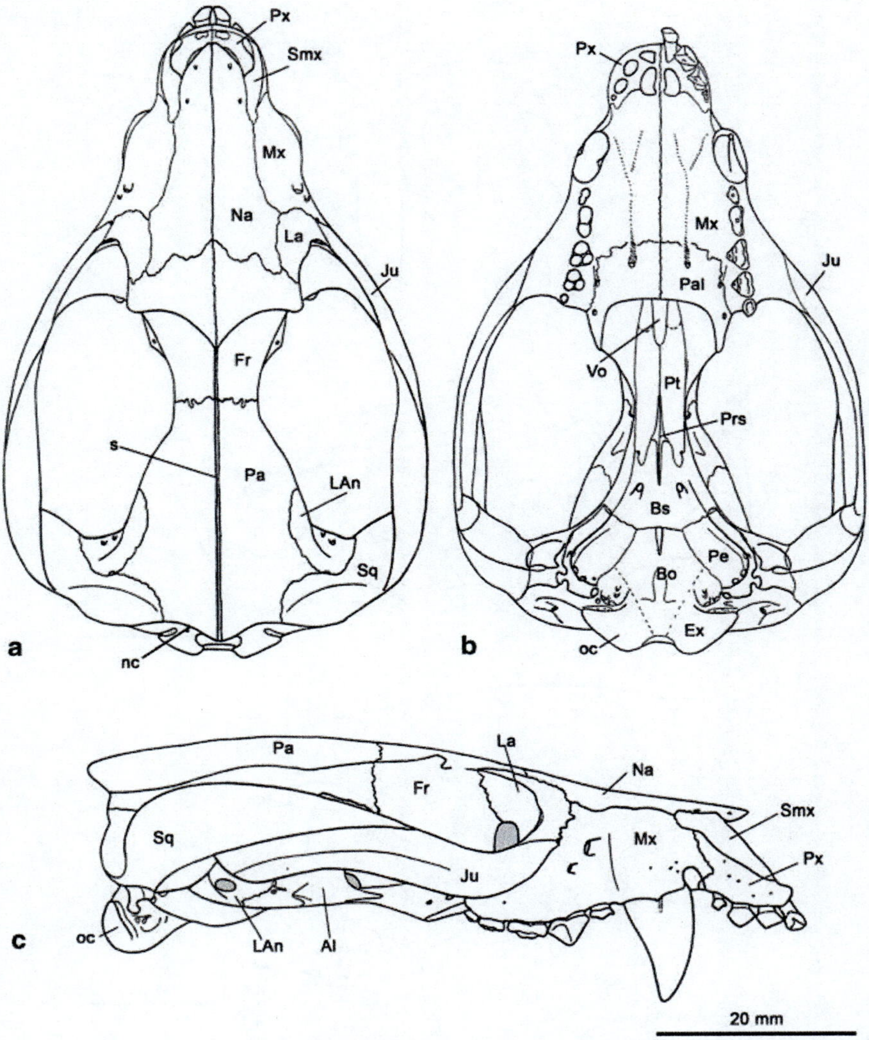


**Fig. 7.3** Skull reconstruction of *Vincelestes neuquenianus* in dorsal (**a**), ventral (**b**), and lateral (**c**) views, based on Rougier (1993). Abbreviations: Al, alisphenoid; Bo, basioccipital; Bs, basisphenoid; Ex, exoccipital; Fr, frontal; Ju, jugal; La, lacrimal; LAn, lamina anterior of the petrosal; Mx, maxilla; Na, nasal; nc, nuchal crest; oc, occipital condyle; Pa, parietal; Pal, palatine; Pe, petrosal; Prs, parasphenoid; Pt, pterygoid; Px, premaxilla; s, sagittal crest; Smx, septomaxilla; Sq, squamosal; Vo, vomer

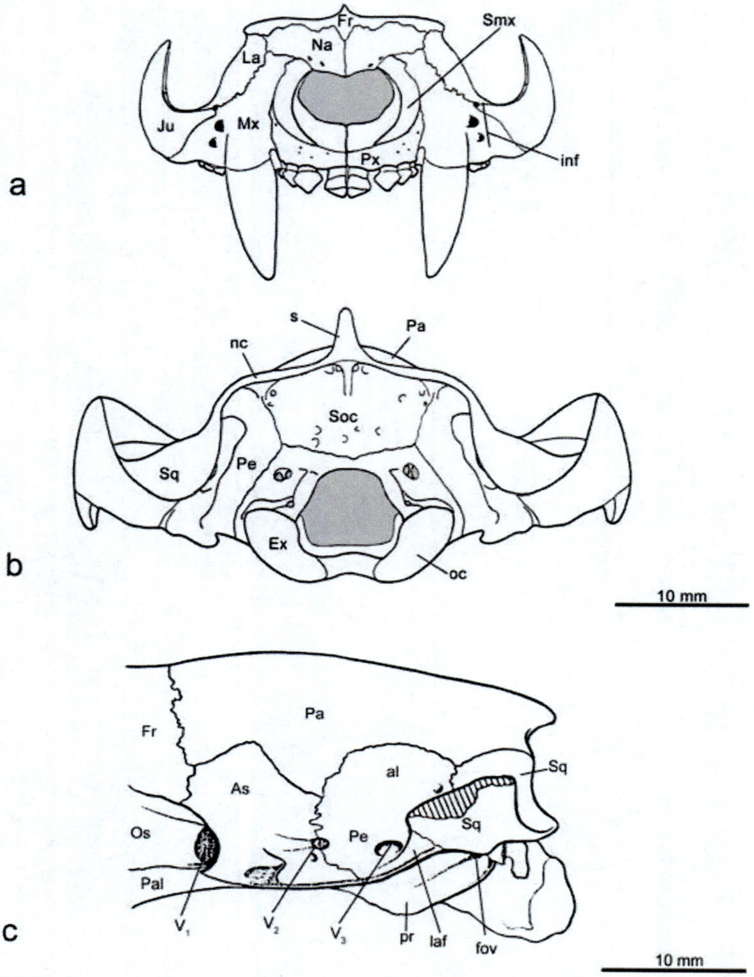


**Fig. 7.4** Skull reconstruction of *Vincelestes neuquenianus* in rostral (**a**), occipital (**b**) views, detail of the lateral wall of the braincase in lateral view (**c**), based on Rougier (1993). Abbreviations: al, anterior lamina of petrosal; As, alisphenoid; Ex, exoccipital; fov, fenestra ovalis; Fr, frontal; inf, infraorbital foramina; Ju, jugal; La, lacrimal; laf, lateral flange of the petrosal; Mx, maxilla; Na, nasal; nc, nuchal crest; oc, occipital condyle; Os, orbitosphenoid; Pa, parietal; Pal, palatine; Pe, petrosal; pr, promontorium; Px, premaxilla; s, sagittal crest; Smx, septomaxilla; Soc, supraoccipital; Sq, squamosal; $V_1$, $V_2$, $V_3$, foramen for ophthalmic, maxillary, and mandibular branches of trigeminal nerve, respectively

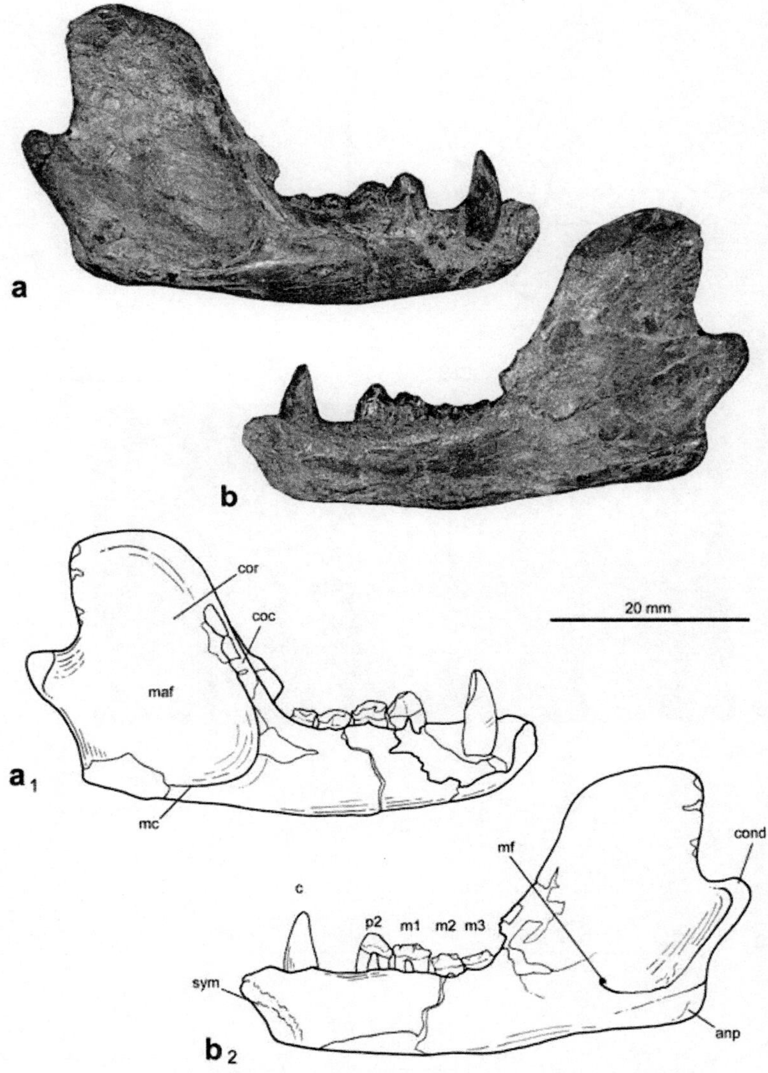


**Fig. 7.5** *Vincelestes neuquenianus*. MACN-N 05, right dentary in labial (**a**) and lingual (**b**) views, with accompanying line drawings (**a**$_1$, **b**$_1$). Abbreviations: anp, angular process; c, lower canine; coc, coronoid crest; cond, condylar process; cor, coronoid process; m1–3, lower molars; maf, masseteric fossa; mc, masseteric crest; mf, mandibular foramen; p2, second lower premolars; sym, mandibular symphysis

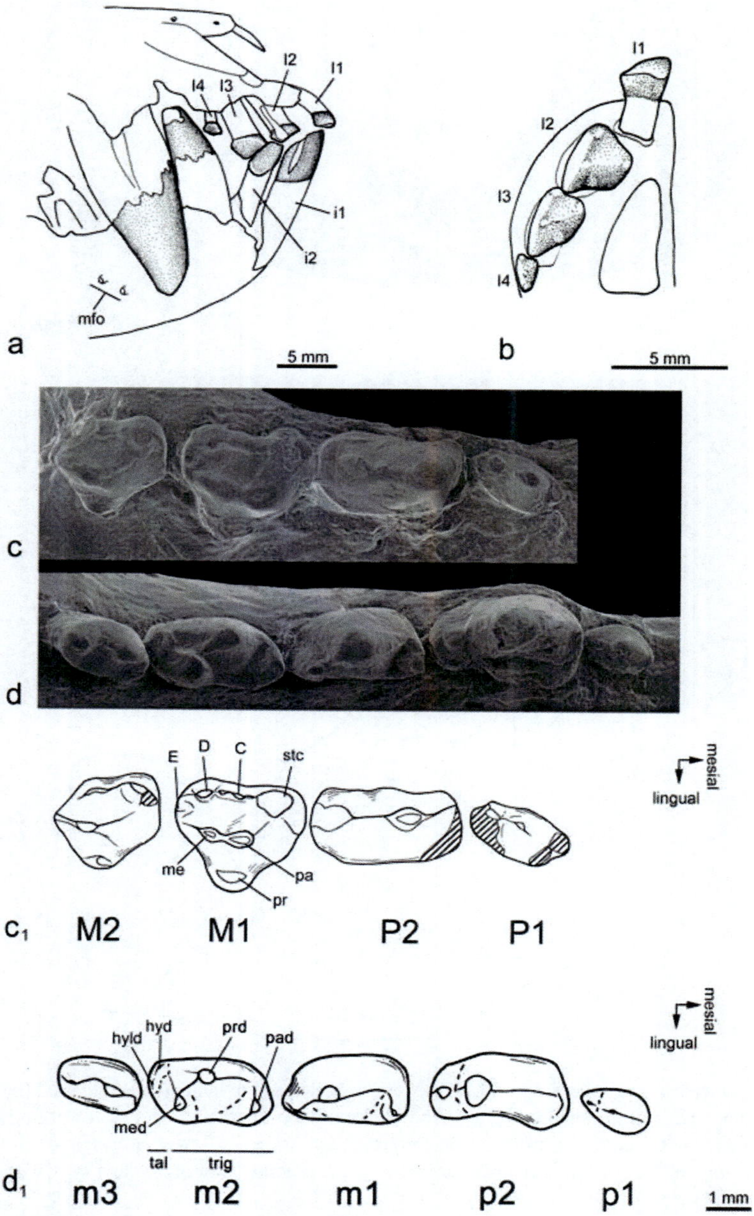


**Fig. 7.6** *Vincelestes neuquenianus.* MACN-N 09, anterior portion of skull and jaw with dentition in occlusion, in lateral view (**a**) and reconstruction of right upper incisors, based on different specimens, in occlusal view (**b**), based on Rougier (1993). MACN-N 06, right upper dentition (**c**) and holotype, MACN-N 01, left lower dentition (**d**), with accompanying line drawings (**c**$_1$, **d**$_1$). Abbreviations: C, stylar cusp C; D, stylar cusp D; E, stylar cusp E; hyd, hypoconid; hyld, hypoconulid; I/i, upper/lower incisor; M/m, upper/lower molar; me, metacone; med, metaconid; mfo, mental foramen; P/p, upper/lower premolar; pa, paracone; pad, paraconid; pr, protocone; prd, protoconid; stc, stylocone; tal, talonid; trig, trigonid. Oblique lines indicate breakage

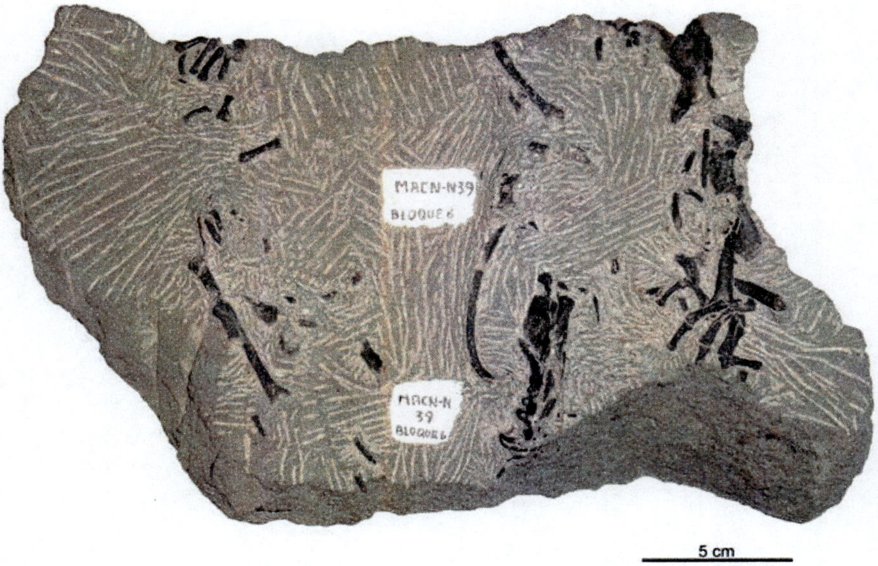


**Fig. 7.7** *Vincelestes neuquenianus.* Block 6 with several bones of specimen MACN-N 39. Note the partial lower jaw (bottom middle) and left hand (bottom right)

Diagnosis: *Vincelestes* is the sole member of the monotypic Family Vincelestidae (Bonaparte 1986). A cladothere with a dental formula of I4/i3, C1/c1, P2/p2, M3/m3. *Vincelestes* differs from other cladotheres by its complex incisors: pyramidal in nature, triangular in occlusal outline, with distinct cuspules at each vertex. An additional basal cuspule is on the distolingual face of the distal incisors. Very robust double-rooted canine. The two upper premolars are patterned similarly to the incisors, but with better-developed cusps. The last premolar is low, thick, and robust, being the longest tooth of the series. The upper molars have aligned paracone-metacone, with the paracone slightly more robust and taller in the fresh molar. There is a small protocone at the base of the crown that lacks conules and does not form a full basin, but it is supported by a lingual root. Parastyle and parastylar cusp are missing, while the stylocone is large and robust. Only two functional molars, the M3 is peg-like. Lower incisors are procumbent and the i1 strongly concave and spatulated in the unworn specimens. The i2 has a median ridge and a basal cusp, i3 is minute but relatively complex. The p1 is very small and single rooted. The lower molars are low with a larger paraconid than metaconid flanking a dominant and postclivus protoconid. The paracristid is divided into two by a deep notch, with the portion attached to the protoconid (preprotocristid) parallel to the jaw and the mesial portion of it arranged almost at a right angle ending in the paraconid. A small, incomplete crest descends from the distal slope of the protoconid directed toward the buccal portion of the talonid. The metaconid is conical, blunt, and with a prominent distal metacristid that is confluent with a low cristid obliqua. The talonid is basined with only two cuspids, poorly separated from each other: the hypoconid, that anchors the cristid

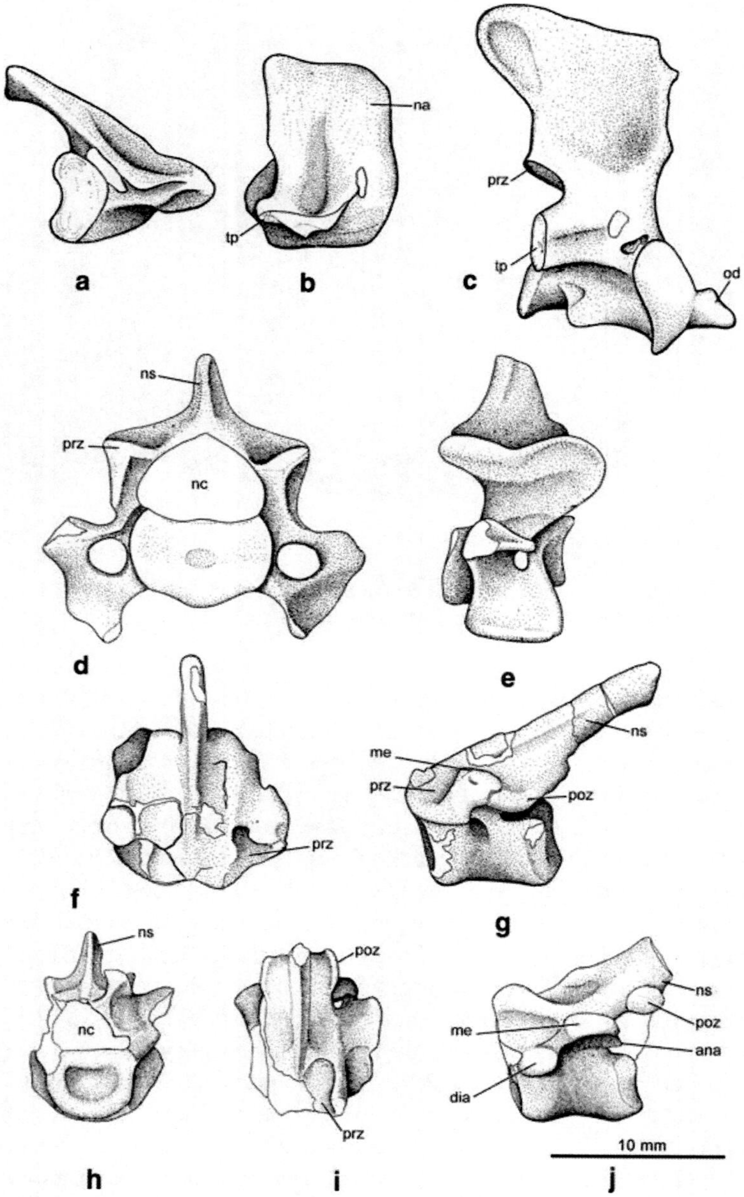


**Fig. 7.8** *Vincelestes neuquenianus*. MACN-N 01, fragmentary right arch of atlas in posterior (**a**) and lateral (**b**) views. MACN-N 39, axis in lateral view (**c**). MACN-N 39, sixth cervical vertebra in anterior (**d**) and lateral (**e**) views. MACN-N 01, anterior dorsal vertebra (prediaphragmatic) in dorsal (**f**) and lateral (**g**) views. MACN-N 01, posterior dorsal vertebra (postdiaphragmatic) in anterior (**h**), dorsal (**i**) and lateral (**j**) views. Abbreviations: ana, anapophysis; dia, diapophysis; me, metapophysis; na, neural arch; nc, neural canal; ns, neural spine; od, odontoid process; poz, postzygapophysis; prz, prezygapophysis; tp, transverse process

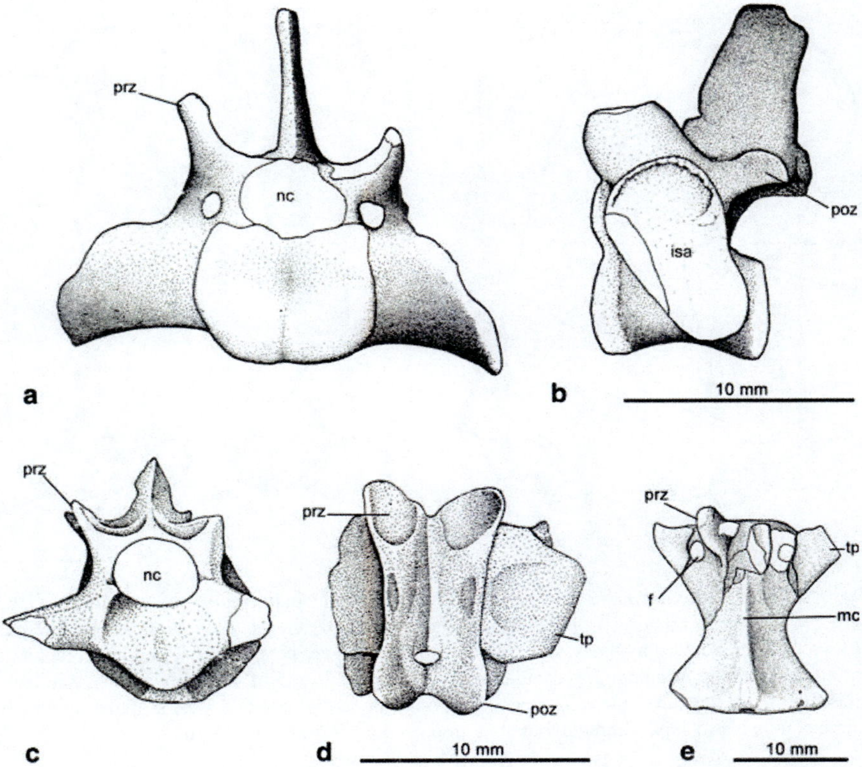


**Fig. 7.9** *Vincelestes neuquenianus*. MACN-N 01, first sacral vertebra in anterior (**a**) and lateral (**b**) views. MACN-N 01, second(?) caudal vertebra in anterior (**c**) and dorsal (**d**) views. MACN-N 39, middle caudal vertebra in dorsal view (**e**). Abbreviations: f, foramen; isa, iliosacral articulation; mc, medial crest; nc, neural canal; poz, postzygapophysis; prz, prezygapophysis; tp, transverse process

obliqua, and the poorly differentiated hypoconulid. A low crest bounds the distal and lingual aspect of the talonid, with no thickening that can be recognized as an entoconid. No wear in the talonid basin. Deep jaw with no scars for the attachment of sizable postdentary elements and distinct but poorly projected angular process. Petrosal retains primitive circulatory pattern with prootic canal, stapedial artery, and cochlea coiled only about 270°. Lateral wall of the braincase formed by extensive prootic and alisphenoid.

Comments: *Vincelestes* is a relatively large mammal, in the range of medium size opossums (e.g., *Lutreolina crassicaudata*). Its body mass was estimated ranging between 619 and 1228 gr (Rougier 1993; Macrini et al. 2007), depending on the specimen used and the formula employed, which rely on skull length or limb-bone circumference (Alexander et al. 1979; Luo et al. 2001). Considering the body mass

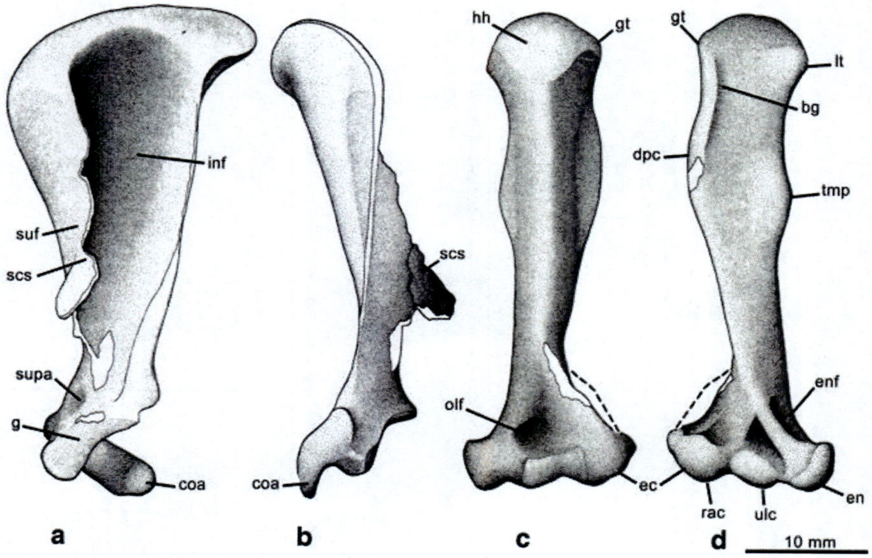


**Fig. 7.10** *Vincelestes neuquenianus.* MACN-N 39, left scapula in lateral (**a**) and anterior (**b**) views. MACN-N 39, right humerus in posterior (**c**) and anterior (**d**) views. Abbreviations: bg, bicipital groove; coa, coracoid apophysis; dpc, deltopectoral crest; ec, ectepicondyle; en, entepicondyle; enf, entepicondylar foramen; g, glenoid cavity; hh, humeral head; inf, infraspinous fossa; lt, lesser tubercle; gt, greater tubercule; olf, olecranon fossa; rac, radial condyle; scs, scapular spine; suf, supraspinous fossa; supa, supraglenoid area; tmp, teres major process; ulc, ulnar condyle

and the endocranial reconstruction, Macrini et al. (2007) calculated the EQ for *Vincelestes* (0.37). The value overlaps with the lower range of EQs from basal eutherians (0.36–0.80; Novacek 1982, 1986; Kielan-Jaworowska 1984, 1986), and is in the range of the EQs calculated for the basal metatherian *Pucadelphys* (0.32) and *Didelphis virginiana* (0.34) (see Macrini et al. 2007). These values also suggest that larger EQ values evolved several times independently across Mammalia (e.g., Monotremata, Placentalia, Marsupialia) (Macrini et al. 2007).

*Vincelestes* has a robust, short-faced skull (Figs. 7.2, 7.3 and 7.4), built with the same bones present in mammaliaforms and we believe conserved in the mammalian last common ancestor. *Vincelestes* retains a well-developed septomaxilla bordering the nares, an extensive anterior lamina forming a substantial portion of the lateral wall of the braincase, and a broad lateral trough and distinct lateral flange bounding laterally the ear region (Rougier et al. 1992; Hopson and Rougier 1993) (Fig. 7.4c). A primitive circulatory pattern was present with a large vertical prootic canal and ramus superior of the stapedial artery that is a smaller subsidiary of the arteria diploetica magna (Rougier et al. 1992). The inner ear shows a cochlear canal with less than one full turn and the inner cranial vault retains at least a partially ossified primary wall. All the features listed above are predicted not to be present in the last common ancestor of marsupials and placentals, but they are either present in monotremes,

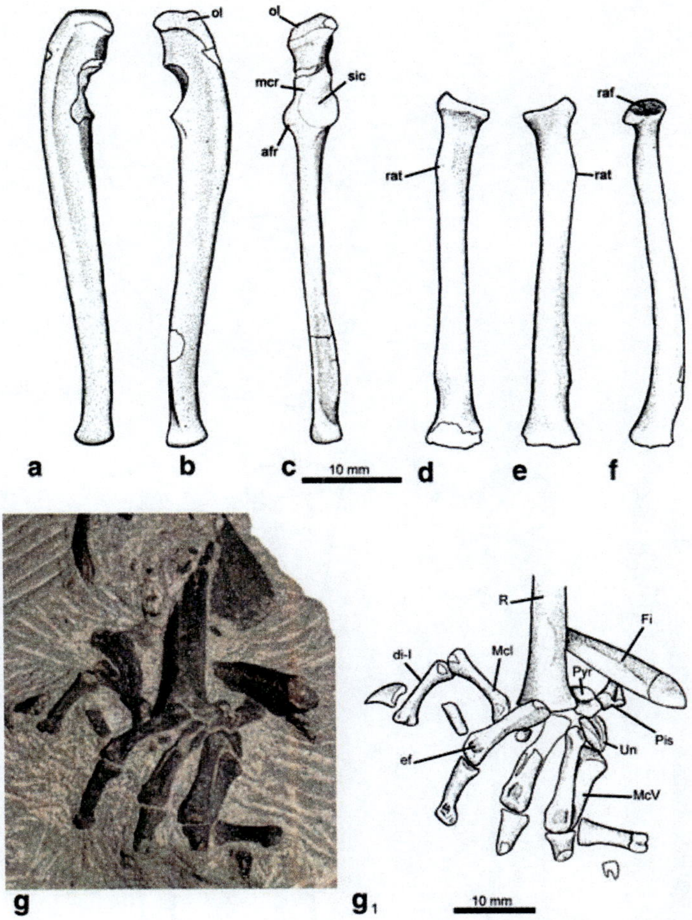


**Fig. 7.11** *Vincelestes neuquenianus*. MACN-N 01, right ulna lateral (**a**), medial (**b**), and anterior (**c**) views. MACN-N 01, right radius in posterior (**d**), anterior (**e**), and external (**f**) views. MACN-N 39, partially articulated left hand in dorsal view (**g**) and accompanying line drawing ($\mathbf{g_1}$). Abbreviations: afr, articular facet for radius; di-I, digit I; ef, extensor fossa; Fi, fibula; mcr, medial crest; McI, metacarpal I; McV, metacarpal V; ol, olecranon; Pyr, Pyramidal; Pis, Pisiform; R, radius; raf, radial fovea; rat, radial tubercle; sic, sigmoid cavity; Un, unciform

expected in the last common ancestor of mammals, or part of a morphocline of character transformations between the hypothetical mammalian ancestor and therians. The dentition is highly autapomorphic, particularly because of the low number of postcanine teeth (Fig. 7.6). The molar morphology is viable as a starting point for the development of the tribosphenic molar that characterizes therians and their closest relatives; the basic geometry and major cusps are there. *Vincelestes* on its own right suggests a scenario where the cranial morphology is very conservative retaining a basic plesiomorphic morphology until the origin of the tribosphenic dentition or

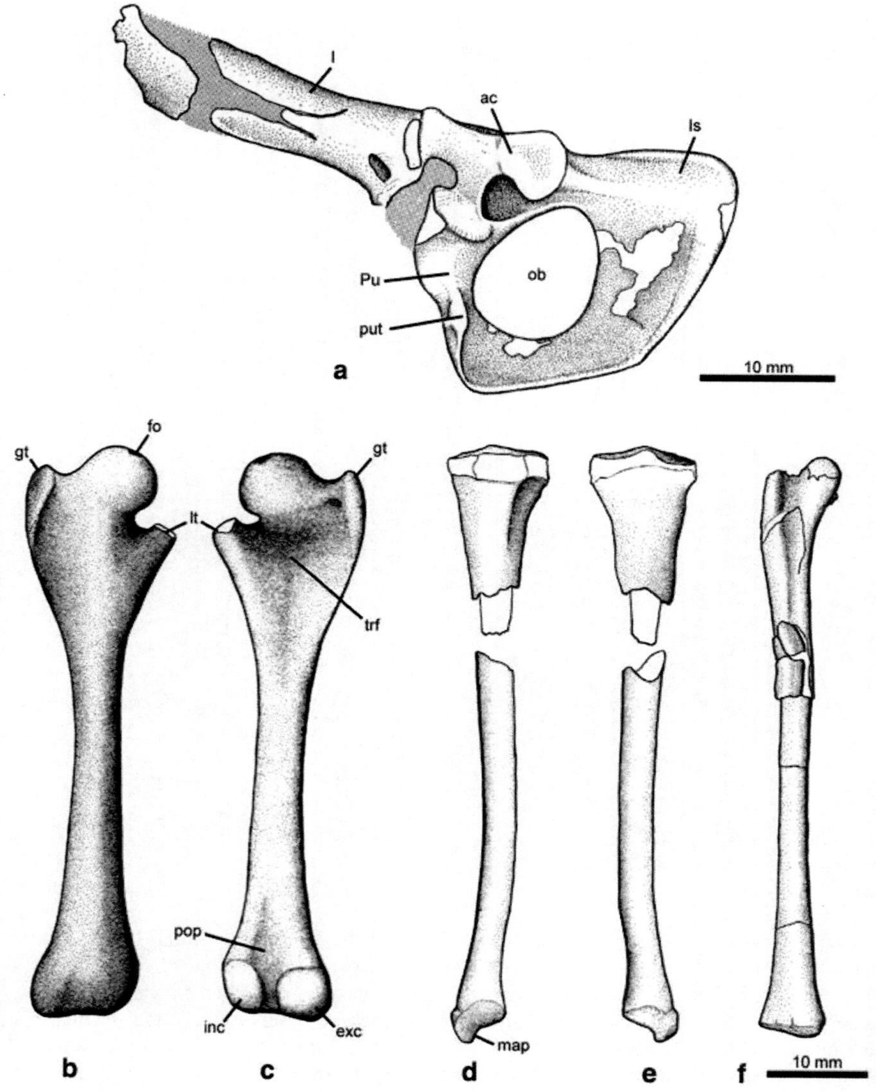


**Fig. 7.12** *Vincelestes neuquenianus*. MACN-N 39, left pelvis in slightly ventrolateral view (**a**). MACN-N 38, right femur in anterior/dorsal (**b**) and posterior/ventral (**c**) views. MACN-N 01, left tibia in posterior (**d**) and anterior (**e**) views. MACN-N 38, left fibula in anterolateral view (**f**). Abbreviations: ac, acetabulum; exc, external tuberosity; fo, fovea capitis; I, ilium; Is, ischium; inc, internal condyle; lt, lesser trochanter; map, maleolar process; gt, greater trochanter; ob, obturador foramen; pop, popliteal fossa; Pu, pubis; put, pubic tubercle; trf, trochanteric fossa

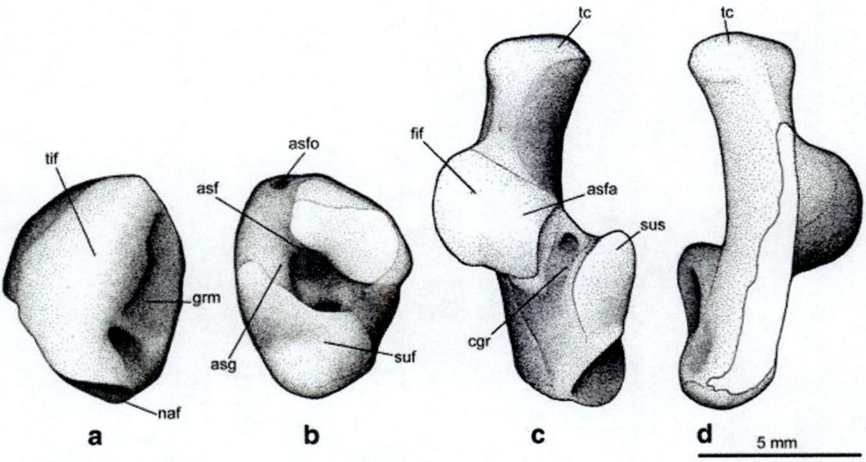


**Fig. 7.13** *Vincelestes neuquenianus*. MACN-N 01, right astragalus in dorsal (**a**) and ventral (**b**) views. MACN-N 01, right calcaneous in dorsal (**c**) and ventral (**d**) views. Abbreviations: asf, astragalar fossa; asfa, astragalar facet; asfo, astragalar foramen; asg, astragalar groove; cgr, calcanear groove; fif, fibular facet; grm, groove for maleolar process; naf, navicular facet; susf, sustentacular facet; sus, sustentaculum tali; tc, tuber calcanear; tif, tibial facet

quite close to it. If that were the case, it would be logical to associate the acquisition of the iconic therian tribosphenic molar with the radical transformation of cranial morphology that distinguished *Vincelestes* from therians. The alternative, so far rejected by current phylogenetic trees, is that the dental features of *Vincelestes* are independent acquisitions from those in the therian stem lineage. To this scenario should be added the presence of full-fledged tribosphenic lower molars and primitive jaw morphology in the australosphenidans from SA (see Chap. 4). There is no simple tree topology congruent with all the character systems mentioned here. Among skull structure, dentition, and jaw morphology, one, or more than one, of those character clusters is convergent, complicating overall assessment of the origin of these quintessential mammalian traits.

The enamel structure in *Vincelestes* is derived and can be described as prismatic (Wood and Rougier 2005). A unique feature is that the enamel prisms are erratically spaced, but intimately associated with convoluted and discontinuous incremental growth lines that are otherwise seen roughly parallel to the dentine-enamel junction (DEJ) and the outer enamel surface (OES) in other mammals (Wood and Rougier 2005).

The postcranial skeleton of *Vincelestes* was described in detail and extensively compared in the PhD thesis of one of us (Rougier 1993). Bonaparte and Migale (2010, 2015) highlighted the relevance of the unpublished work and significance of the available postcranial material and reproduced several of Rougier's (1993) original figures in their books. Also, the postcranium of *Vincelestes* was scored in the data matrix of several studies dealing with high-level mammalian relationships (e.g., Horovitz and Sánchez-Villagra 2003; Luo et al. 2011; O'Leary et al. 2013;

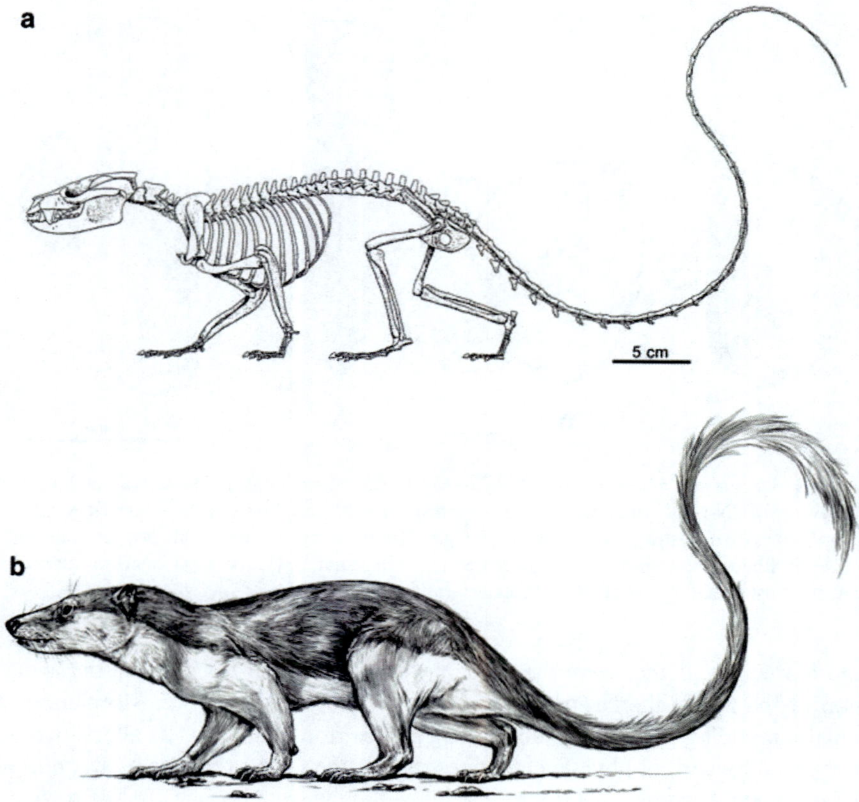


**Fig. 7.14** *Vincelestes neuquenianus*, line drawing of the skeleton (**a**) from Rougier (1993) and artistic reconstruction by Jorge L. Blanco (**b**)

Krause et al. 2014, 2020; Wible and Rougier 2017; Huttenlocker et al. 2018). The postcranium was partially removed from the rock for study, but some elements still remain in partially prepared rock blocks (Fig. 7.7).

The axial skeleton is relatively well known, including elements of the entire vertebral column (a composite from different specimens). The atlas is based on two hemi-atlases of different individuals (MACN-N 01 and MACN-N 38). The best preserved (Fig. 7.8a, b) has the transverse process that extends posteriorly to the level of the posterior facets for the axis. The occipital articular facets are deeply concave, covering the condyles dorsally. The neural arch is well developed, lacking a neural spine with a small contact area for a small, yet unidentified intercentrum. There is an atlantal foramen but there is no transverse foramen. The axis is available only in specimen MACN-N 39, being almost complete (Fig. 7.8c). The vertebral body is short and wide, with a prominent odontoid process, with articular facets (prezygapophysis) that extend ventrally to the dens. The ventral surface of the body has

**Fig. 7.15** *Vincelestes neuquenianus* in its nest by Jorge L. Blanco

three longitudinal, prominent crests, with the hypapophysis in the sagittal plane. The neural arch is tall and robust, with a tall neural spine that extends beyond the level of the postzygapophysis and a robust transverse process with a large oval facet for a free rib. Other cervical vertebrae known are the fifth (?), sixth (Fig. 7.8d,e), and seventh, plus other poorly preserved fragments, which may correspond to the third and fourth vertebrae. *Vincelestes* likely had seven cervical vertebrae, as in most mammals, which are robust, with a low neural spine, and prominent fused post-axial cervical ribs. The bodies are short and their articular surface is slightly larger than the size of the neural canal. The neural arch is high, with robust pre- and postzygapophyses.

There is a sequence of eight articulated dorsal vertebrae with partial ribs (MACN-N 36) and isolated vertebrae; consequently, the exact number of dorsal elements is unknown. In Fig. 7.8f–j, an anterior (prediaphragmatic) and a posterior (postdiaphragmatic) dorsal vertebra are illustrated. Two fused sacral vertebrae are present in the holotype MACN-N 01 of *Vincelestes*. In addition, a partial second sacral vertebra is also known for another specimen. The sacral vertebrae are robust elements. The wing of the sacrum is thick and bears an ovoid surface for contact with the iliac blade (Fig. 7.9a,b). The neural arch is low but with a tall and laminar neural spine.

The number of caudal vertebrae is unknown but it definitively had a long and robust tail, as indicated by semi-articulated mid-distal tail sections. Several distinct caudal morphologies are known (Fig. 7.9c–e), including some haemal arches. The bones of the sternal apparatus have also been found, including two manubria and more posterior sternebrae.

The forelimbs are relatively well documented for *Vincelestes*. There are four partial scapular girdles that of specimen MACN-N 39 being fairly complete (Fig. 7.10a, b). The coracoid process is hook-like. The scapular blade is divided by the spine, delimiting a smaller supraspinous and a larger infraspinous fossae. The major axis of the glenoid forms a strong angle with the lamina of the scapula so that the bone has a general twisted aspect. The process for the teres major muscle is prominent in the disto-vertebral corner of the scapulocoracoid. The humerus is mostly complete in MACN-N 39 (Fig. 7.10c, d), with the ectepicondylar crest broken, but represented by other fragmentary material. It is a slender bone, with a large and spherical humeral head. The greater and lesser tubercles are small. The deltopectoral crest is not particularly prominent, extending ventrally to the mid-length of the shaft. The bicipital groove is narrow, and its deepest point is close to the humeral head. There is a prominent process for the attachment of the teres major at the mid-length of the shaft. The distal epiphysis of the humerus is transversely wide, with spherical radial and ulnar condyles, separated by a deep and narrow intercondylar notch. The ulnar condyle is smaller than the radial condyle, with its major axis slightly oblique. There is a shallow olecranon fossa. The entepicondyle is larger than the, bearing a large entepicondylar foramen (Fig. 7.10c, d).

The ulna and radius of *Vincelestes* is known by several specimens; in particular, those of the holotype MACN-N 01 are well preserved (Fig. 7.11a–f). The ulna is long and slender, laterally compressed, with a sigmoid shape. The sigmoid cavity is deep, concave, and faces anteriorly, and the olecranon process is well developed and square in section (Fig. 7.11a–c). The radius is shorter and stouter than the ulna, also with sigmoid shape. The radial fovea is roughly circular and concave. The radial tubercle is prominent. The distal end is incomplete in most specimens (Fig. 7.11d–f).

Three partially complete and articulated hands are known for *Vincelestes*. The partial left hand of MACN-N 39 is shown in Fig. 7.11g. Several proximal and distal carpals (i.e., pyramidal, semilunar, pisiform, ?central proximal/distal, ?trapezoid, ?trapezium, ?scaphoid, unciform) are also known. The hand has five digits, with similar robustness, the smaller being digits I and V, which are about 30 and 15% smaller in length, respectively, than digit III (the largest). The ungual phalanges are pointy claws, similar in shape for each finger, laterally compressed, and dorsoventrally tall.

The pelvis is known by six specimens, with MACN-N 39 fairly complete (Fig. 7.12a). The pelvis has a typical mammalian configuration, with an elongated iliac blade and reduced pubis and ischium, delimiting a large obturator foramen. The tuberosity for the muscle rectus femoris does not particularly protrude. The acetabulum is circular, concave, and walled, with a deep ilioischiatic notch. In the pubis, there is a subtle facet, which can be interpreted for the attachment of the prepubis (this bone is known by at least two elements). The obturator foramen is larger in size than the acetabulum.

The femur is complete in MACN-N 39 (Fig. 7.12b, c). It is slender with a narrow shaft and slightly expanded proximal and distal epiphyses. The femoral head is spherical, anteromedially oriented, and bears a fovea capitis. The greater and lesser trochanters are separated from the head by conspicuous notches, with the greater trochanter placed in a higher position than the lesser trochanter. The greater trochanter is lower than the head. The trochanteric fossa is shallow, between both trochanters, below the head base. The distal epiphysis is narrower than the proximal epiphysis, with the lateral condyle slightly larger and more distally protruding than the medial one. The intercondylar notch is shallow.

Complete tibiae are unknown but proximal and distal portions are available for several specimens (Fig. 7.12d, e). The available material, and the fibula, suggests that the tibia was not shorter than the femur. It is a long and slender bone, almost straight with the distal end slightly anterolaterally oblique. In the proximal view, the proximal epiphysis is larger mediolaterally than anteroposteriorly; this bears the surface for articulation with the femoral condyles. A facet for the fibula is not distinct, possibly due to preservation artifact. In the distal epiphysis, the medial maleolar process is well developed. The distal articulation of the tibia is spiral; functionally, this bone rotates around its own proximo-distal axis similar to the condition described for some multituberculates (Jenkins and Krause 1983) and not too different from the morphology present in generalized non-therians (Jenkins and Parrington 1976; Szalay 1994; Horovitz and Sánchez-Villagra 2003). The fibula is badly preserved in most specimens. It is long and slender (Fig. 7.12f). The proximal part of the shaft is flat and flares forming the flavelliform process, which may articulate with the tibia and femur.

The astralagus and calcaneus are well preserved in the holotype MACN-N 01 (Fig. 7.13). The astragalus has a shape similar to basal mammaliforms (e.g., *Morganucodon*), being oval and semispherical, with the dorsal surface being strongly convex (Fig. 7.13a, b), indicating the articular surface for the tibia (there is no angle between medial and lateral facets for the tibia). There is a depression for the maleolar process of the tibia and the facet for the navicular bone. The ventral surface has two main facets, separated by the astragalar groove. The facets are subequal in size. The medial portion of the bone bears the sustentacular facet. There is not a well-defined astragalar neck.

The calcaneus shows a mixture of primitive and derived morphology (Fig. 7.13c, d); it resembles basal mammaliaforms in the lack of full overlap, but the contact with the astragalus is slanted, beyond the condition seen in monotremes and more basal taxa (Lewis 1983; Szalay 1994, 2006; Hurum et al. 2006). The calcaneus is elongated and dorsoventrally flat, with a prominent calcaneal tuberosity, which is neither very long nor excessively curved. The calcaneal sustentacular facet develops over the sustentaculum, which is placed at the anterior end of the bone. The calcaneal facet for the fibula is present, facing dorsally and continuous with the astragalar facet; an extensive peroneal process was present, but the full extent is not known because of damage to the specimen.

Rougier (1993) noted the strong resemblances of the skeleton of *Vincelestes* with that of *Henkelotherium*, a Late Jurassic dryolestid from Portugal (Krebs 1991; Vázquez-Molinero et al. 2001; Jäger et al. 2020), which exhibited primitive features (e.g., morphology of astragalus and calcaneus) in relation to therian mammals. However, *Vincelestes* has a long and likely prehensile tail (Figs. 7.14 and 7.15), the opposability of digit I in the hand and foot, and the possible reversion of the foot, suggesting the ability to navigate uneven substrate or even arboreality. These features are also present in some scansorial mammals (Jenkins 1974; Jenkins and Krause 1983); consequently, a more likely scansorial-arboreal behavior can be proposed for *Vincelestes*. The taphonomy, however, does not favor a purely arboreal way for life; the fossils represent a family group with a diversity of ages and sexual dimorphism. It is unlikely such a number of individuals could be buried together if they were mostly arboreal. The pelvis has a typical mammalian configuration present also in early mammaliaforms. The pelvic girdle has undergone fewer modifications than the scapular girdle, which shows more derived features (e.g., supraspinous fossa reaching the glenoid cavity, but still smaller than the infraspinous fossa; acromial process located over the glenoid cavity; ventral orientation of the glenoid cavity; procoracoid absent) in relation to basal mammaliaforms.

## 7.3 Concluding Remarks

*Vincelestes neuquenianus* is an important taxon in the history of SA Mesozoic mammals; it followed quickly (Bonaparte 1986) the announcement of the discovery of the first-ever skeletal remains of mammals (Bonaparte and Soria 1985). It was originally recognized as a "eupantothere" based on the type lower jaw and dentition, that is, a non-tribosphenic stem therian in modern parlance. This idea of relatively close affinities with the last common ancestor of marsupials and placentals received further support following the preliminary description of cranial material (Bonaparte and Rougier 1987) and further advanced through detailed description, in particular of the braincase and ear region (Rougier et al. 1992; Hopson and Rougier 1993). The phylogenetic position of this taxon has been relatively stable, being recovered around peramurans, on occasion closer to Theria than *Peramus* and allies (e.g., Wible and Rougier 2017), for others with *Peramus* closer to therians than *Vincelestes* (e.g., Luo et al. 2002; Huttenlocker et al. 2018). The phylogenetic variations are relatively minor, but reflect the difficulty in fully integrating the dentition of *Vincelestes* in the overall story of therian dental evolution. Rougier (1993) defended previous identification of the lingualmost cusp of the upper molars as a protocone, supported by its own root, but of relatively small size, quickly worn down, and not occluding into the talonid basin of the lower molar but labial to the cristid obliqua. The size and relationships are atypical for the protocone of tribosphenic mammals but a reasonable primitive morphology for elaboration of a full tribosphenic dentition. The number of teeth, size, and overall morphology of *Vincelestes* dentition are highly derived, with basically just three functional postcanines, the last premolar and the first two molars. The very low postcanine count fits poorly with the predicted primitive number expected in the last common ancestor of therians, usually assumed to be eight (McKenna 1975; O'Leary et al. 2013). No matter the scenario, *Vincelestes* is a highly specialized representative of a clade that must include other taxa dentally more similar to what is predicted for the therian ancestor. Its braincase and sidewall, on the other hand, have no obvious striking autapomorphies and have been since its publication the sole representative of stem therian cranial morphology. The lateral wall is remarkably primitive regarding its composition, in particular for the large size of the anterior lamina, extensive alisphenoid, number and position of the trigeminal exits (Hopson and Rougier 1993; Rougier and Wible 2006), and the absence of an exposed optic foramen (Crompton et al. 2018). The ear region also preserves a plesiomorphic pattern, including a well-developed lateral flange, a ventral opening of the cavum epiptericum, incomplete separation of the space for cranial ganglia V and VII (fenestra semilunaris), and others (Rougier et al. 1992). However, there is a candy cane-shaped cochlea, a fully developed processus recessus, impression of a transpromontorial artery, among other features. If *Vincelestes* is truly very close to the last common ancestor of therians, its morphology suggests that the development of the dentition was precocial with regard to the braincase, and by and large, the ear region. The acquisition of a therian braincase with the lack of lateral flange, anterior lamina, and lateral trough and the concomitant expansion of the squamosal/forward shift of the glenoid would occur after a protocone was present in the upper molar and a talonid developed in the lowers. It is tempting to speculate about the possible causes of this sequence; however, we know next to nothing regarding the cranial anatomy of any other immediate therian outgroups, such as peramurids and "tribotheres" (see Kielan-Jaworowska et al. 2004, for a review of tribosphenic mammals of uncertain affinities) and alternatives can always be advanced. A few "symmetrodont" skulls and ear regions have been partially described (e.g., Rougier et al. 2003; Bi et al. 2016; Harper and Rougier 2019; Mao et al. 2020); however, not much detail has been furnished by these specimens yet, as they are badly flattened, incomplete, or partially described. Nevertheless, they seem to be constructed along the same general plan as *Vincelestes* with large anterior laminae on the side of the braincase and small squamosals. The less bulbous promontorium, lack of distinct transpromontorial artery, and cochlear shape seem all to be plesiomorphic regarding *Vincelestes* and therians.

Because of the completeness of the specimen and at times painfully detailed description (Rougier 1993), *Vincelestes* has served to illustrate what we consider to be the pre-therian condition. Based on our discussion above, we believe at present that *Vincelestes* is still the best taxon to illustrate the early stages of the rise of the tribosphenic molar and the sequence of radical cranial changes leading to therians.

Besides some sporadic odd results regarding the position of *Vincelestes* in studies addressing other issues (for instance Averianov et al. 2013; Krause et al. 2020), the stem therian position of *Vincelestes* has remained mostly unchallenged. However, Bonaparte (2008) revised his original position and upon revision of the materials of *Vincelestes* concluded that it could represent a basal australosphenidan; in fact in his analysis, *Vincelestes* is the basalmost member of this clade that is recovered as the sister group of Theria and *Peramus*. Other outcomes of the analysis are also very odd; dryolestoids, "symmetrodonts", "triconodonts", and gobiconodontids are all outside Mammalia. Bonaparte's proposal is a priori not without merit. As pointed out in the opening quote of this chapter, the presence of a relatively derived stem therian in the Early Cretaceous of Patagonia, which appears to have lax affinities with Laurasic taxa, is intriguing. The lack of any mammal with a similar dentition in the ever-expanding collections of SA Mesozoic mammals (e.g., Bonaparte and Kielan-Jaworowska 1987; Pascual and Ortiz-Jaureguizar 2007; Goin et al. 2012; Defler 2019), particularly those from the Late Cretaceous, implies either the extinction of the pretribosphenic lineages before then or a missing record. Both are possible, as the total number of specimens is relatively small and their stratigraphic distribution is patchy. On the other hand, the extinction of a lineage does not require a justification. So far, it appears to be a defensible hypothesis that no tribosphenic mammal is recorded in the Late Cretaceous of Patagonia and likely in SA. It should be noted that this does not apply necessarily to the Jurassic: *Asfaltomylos* and *Henosferus* have lower molars compatible with a full-fledged tribosphenic dentition (see Chap. 4). For some obscure reason tribosphenic, or near-tribosphenic, mammalian lineages present in earlier strata of Patagonia are not abundant, not present in the areas sampled, or extinct in the Late Cretaceous faunas studied at this time. Non-tribosphenic therians, mostly dryolestoid lineages, diversified dentally and in size, covering a broad spectrum of niches ranging from traditional insectivores to complex herbivores (see Chap. 6).

The problem at the root of Bonaparte's (2008) proposal is a very unorthodox treatment of the dental and mandibular characters of these groups as formulated in the matrix by Luo et al. (2002), leading to results that have not been recovered in any of the many subsequent broad-scale phylogenetic analyses of early mammals. The characters and character changes argued by Bonaparte are included in those studies; however, many of the modifications to scores advanced by him are not accepted. Among others, the trees centered in SA Mesozoic mammals, meridiolestidans in particular (e.g., Rougier et al. 2011, 2012; Chimento et al. 2012), recovered *Vincelestes* in a traditional position. Not so Averianov et al. (2013), who placed *Vincelestes* as a sister group of northern dryolestoids, but still far removed from australosphenidans and monotremes. The relatively derived postcranial skeleton of *Vincelestes* (Rougier 1993; Figs. 7.7, 7.8, 7.9, 7.10, 7.11, 7.12 and 7.13) makes monotremes and australosphenidan affinities even more unlikely.

## 7.4 Vignette by GWR: The Pact of La Amarga, or How Careers Are Determined in Science!

The type specimen of *Vincelestes* was found by Martín Vince, long-term preparator and collaborator of J. F. Bonaparte from his time in Tucumán, in February of 1985. Bonaparte (1986) named the specimen after him as a tribute: "eficaz colaborador desde 1960, autor del hallazgo del material que describimos". Martín, like Bonaparte, was an exceedingly difficult person, and could easily and frequently be rough and unpleasant, particularly with young unseasoned students/preparators. We (GWR and others in the team) were regularly accused of bringing more cameras than excavation equipment! His complaints may have had a kernel of truth. Despite the tensions and difficulties that accompanied Martín, he was extremely experienced and talented; no good collector can truly be special without a lucky touch—he certainly had it.

Bonaparte and his team had been finishing excavating the skeleton of what would become the type of the peculiar sauropod *Amargasaurus cazaui*. I had found it the previous year, but at the time of the discovery, it was just the two of us, Bonaparte and me; after a few days of hard work, we managed to collect about half of the skeleton. We were driving a decrepit 1961 Jeep Wagoneer (see Chap. 2, Fig. 2.14a) and could simply carry no more. Bonaparte decided to leave the sacrum and extremities in place and return the next year. In 1985, I could not join the returning team, as I had been compulsorily conscripted to serve for a year in the Argentine Military. The draft lottery decided I was to be an Air Force recruit instead of an aspiring paleontologist.

The sauropod excavation completed, Bonaparte and his team dedicated what remained of the very last day to explore around. Martín found a group of small bones with a few fragments rolling down the steep cliff at La Amarga, quickly realizing it was something interesting. The specimen was jacketed and taken to Buenos Aires, with the suspicion it could be a mammal but it appeared to be larger than expected. Shortly after preparation began, the jaw of the type material was uncovered, leaving no doubts. I was shown the specimen during one of my leaves from service; Bonaparte was very happy about it and I was curious, but dinosaurs were my childhood dream. At the time, the mammalian jaw was just one more of the many surprising things that were being found on what I consider the golden years at the National Museum in Buenos Aires.

In 1986, Bonaparte had secured additional funding from the McKenna Foundation to return to the La Amarga Formation, with the idea of trying to recover more mammals. Malcolm C. McKenna, a world-famous mammalian specialist from the AMNH in New York, and his wife Priscilla came down to La Amarga and shared camp with us. Unknown at the time, almost ten years later I was to be a postdoc working with Malcolm and his colleagues at the AMNH.

Done with my service in the Air Force, I rejoined Bonaparte's team for the 1986 field season. A quarry was opened in the spot where the first specimen was found but no more fossils were recovered for several days. At the end of that short while, Malcolm and Priscilla McKenna were taken by Bonaparte to the airport to return home, a round trip that took most of the day. That day Luis Chiappe, now a renowned paleontologist at the Los Angeles County Museum, Pablo Puerta an expert preparator from MPEF, myself, and others kept working in the quarry with flagging hope. Late in the evening, in a deep corner of the excavation, we saw tell-tale black sections of small bones. We were all exultant, but Bonaparte came back too late that day to go check them. Over the next few days, we extracted several blocks from the quarry that clearly contained several skulls and a multitude of postcranial elements.

One night after dinner, Bonaparte told Luis Chiappe and me: "I have two students and two specimens that are ideal for Ph.D. theses: the skeleton of a Cretaceous bird

(*Patagopteryx*) and these mammals with skulls, jaws, and postcrania. Talk it over between yourselves and let me know who is going to do what". Luis and I were a little shocked; he was interested in crocodiles, having already published on them, and I was interested, naturally, in dinosaurs, having a lousy paper submitted at the time. It was clear, however, that our interests had to change. We both walked out and went to the nearby corral where sheep were kept. Luis said of the two, I prefer the bird, and I said of the two, I prefer the mammals. That was it! A couple of minutes later we were back inside, told Bonaparte, and we never looked back. Luis wrote his dissertation on *Patagopteryx,* and I got to work on *Vincelestes* for mine; I know we are both happy with our choices. Bonaparte's generosity in that moment, passing along materials that were clearly very important, is in my memory a rare bright spot, perhaps the event that casts him in the best light.

## References

Alexander RM, Jayes AS, Maloiy GMO, Wathuta EM (1979) Allometry of the limb bones of mammals from shrews (*Sorex*) to elephant (*Loxodonta*). J Zool 189:305–314

Averianov AO, Martin T, Lopatin AV (2013) A new phylogeny for basal Trechnotheria and Cladotheria and affinities of South American endemic Late Cretaceous mammals. Naturwissenschaften 100:311–326

Beck RMD, Lee MSY (2014) Ancient dates or accelerated rates? Morphological clocks and the antiquity of placental mammals. Proc R Soc Lond B 281:20141278

Bi S, Zheng X, Meng J, Wang X, Robinson N, Davis B (2016) A new symmetrodont mammal (Trechnotheria: Zhangheotheriidae) from the Early Cretaceous of China and trechnotherian character evolution. Sci Rep 6:26668

Bonaparte JF (1986) Sobre *Mesungulatum houssayi* y nuevos mamíferos cretácicos de Patagonia. 4° Congreso Argentino de Paleontología y Bioestratigrafía. Mendoza, Actas 2:48–61

Bonaparte JF (2008) On the phylogenetic relationships of *Vincelestes neuquenianus*. Hist Biol 20:81–86

Bonaparte JF, Kielan-Jaworowska Z (1987) Late Cretaceous dinosaur and mammal faunas of Laurasia and Gondwana. In: Currie PM, Koster EH (eds) 4° Symposium on Mesozoic Terrestrial Ecosystems, Short Papers, Occasional Papers of the Tyrrell Museum of Palaeontology 3:24–29

Bonaparte JF, Migale LA (2010) Protomamíferos y mamíferos Mesozoicos de América del Sur. Museo de Ciencias Naturales Carlos Ameghino de Mercedes, Buenos Aires. 1° Edition

Bonaparte JF, Migale LA (2015) Protomamíferos y mamíferos Mesozoicos de América del Sur. Fundación de Historia Natural Felix de Azara, Buenos Aires. 2° Edition

Bonaparte JF, Rougier GW (1987) Mamíferos del Cretácico Inferior de Patagonia. 4° Congreso Latinoamericano de Paleontología. Santa Cruz De La Sierra 1:343–359

Bonaparte JF, Soria MF (1985) Nota sobre el primer mamífero del Cretácico Argentino, Campaniano-Maastrichtiano (Condylarthra). Ameghiniana 21:177–183

Butler PM (1990) Early trends in the evolution of tribosphenic molars. Biol Rev 65:529–552

Chimento NR, Agnolín FL, Novas FE (2012) The Patagonian fossil mammal *Necrolestes*: a Neogene survivor of Dryolestoidea. Rev Mus Argent Cienc Nat "B Rivadavia", ns 14:261–306

Chimento NR, Agnolín FL, Martinelli AG (2016) Mesozoic mammals from South America: implications for understanding early mammalian faunas from Gondwana. Mus Argent Cienc Nat "B Rivadavia." Contribuciones 6:199–209

Crompton AW, Musinsky C, Rougier GW, Bhullar BAS, Miyamae JA (2018) Origin of the lateral wall of the mammalian skull: fossils, monotremes and therians revisited. J Mammal Evol 25:301–313

Defler T (2019) History of terrestrial mammals in South America. How South American mammalian fauna changed from the Mesozoic to Recent times. Springer, Dordrecht

Goin FJ, Gelfo JN, Chornogubsky L, Woodburne MO, Martin T (2012) Origins, radiations, and distribution of South American mammals. From Greenhouse to Icehouse worlds. In: Patterson BD, Costa LP (eds) Bones, clones, and biomes: an 80-million year history of Recent Neotropical mammals. The University of Chicago Press, Chicago, pp 20–50

Harper T, Rougier G (2019) Petrosal morphology and cochlear function in Mesozoic stem therians. PLoS One 14(8):e0209457

Hopson JA, Rougier GW (1993) Braincase structures in the oldest known skull of a therian mammal: implications for mammalian systematics and cranial evolution. Am J Sci 293A:268–299

Horovitz I (2000) The tarsus of *Ukhaatherium nessovi* (Eutheria, Mammalia) from the Late Cretaceous of Mongolia: an appraisal of the evolution of the ankle in basal therians. J Vertebr Paleontol 20:547–560

Horovitz I (2003) Postcranial skeleton of *Ukhaatherium nessovi* (Eutheria, Mammalia) from the Late Cretaceous of Mongolia. J Vertebr Paleontol 23:857–868

Horovitz I, Sánchez-Villagra MR (2003) A morphological analysis of marsupial mammal higher-level phylogenetic relationships. Cladistics 19:181–212

Hurum JH, Luo Z-X, Kielan-Jaworowska Z (2006) Were mammals originally venomous? Acta Palaeontol Pol 51:1–11

Huttenlocker AK, Grossnickle DM, Kirkland JI, Schultz JA, Luo Z-X (2018) Late-surviving stem mammal links the lowermost Cretaceous of North America and Gondwana. Nature 558:108–112

Jäger KRK, Luo Z-X, Martin T (2020) Postcranial skeleton of *Henkelotherium guimarotae* (Cladotheria. J Mammal Evol, Mammalia) and locomotor adaptation. J Mamm Evol 27:349–372

Jenkins FA Jr (1974) Tree shrew locomotion and arborealism. In: Jenkins FA (ed) Primate locomotion. Academic Press, New York, pp 85–115

Jenkins FA Jr, Krause DW (1983) Adaptations for climbing in North American multituberculates (Mammalia). Science 220:712–715

Jenkins FA Jr, Parrington FRS (1976) The postcranial skeleton of the Triassic mammals *Eozostrodon, Megazostrodon* and *Erytrotherium.* Philos T R Soc B 273:387–431

Kielan-Jaworowska Z (1984) Evolution of the therian mammals in the Late Cretaceous of Asia. VI. Endocranial casts of eutherian mammals. Palaeontol Pol 46:157–171

Kielan-Jaworowska Z (1986) Brain evolution in Mesozoic mammals. In: Flanagan KM, Lillegraven JA (eds) Vertebrates, phylogeny, and philosophy. Contributions to Geology, Special Paper 3. University of Wyoming, Laramie, pp 21–34

Kielan-Jaworowska Z, Cifelli RL, Luo Z-X (2004) Mammals from the age of dinosaurs. Origins, evolution, and structure. Columbia University Press, New York

King B, Beck RMD (2020) Tip dating supports novel resolutions of controversial relationships among early mammals. Proc R Soc Lond B 287:20200943

Krause DW, Hoffmann S, Wible JR, Kirk EC, Schultz JA, von Koenigswald W, Groenke JR, Rossie JB, O'Connor PM, Seiffert ER, Dumont ER, Holloway WL, Rogers RR, Rahantarisoa RJ, Kemp AD, Andriamialison H (2014) First cranial remains of a gondwanatherian mammal reveal remarkable mosaicism. Nature 515:512–517

Krause DW, Hoffmann S, Hu Y, Wible JR, Rougier GW, Kirk EC, Groenke JR, Rogers RR, Rossie JB, Schultz JA, Evans AR, von Koenigswald W, Rahantarisoa LJ (2020) Skeleton of Cretaceous mammal from Madagascar reflects long-term insularity. Nature 581:421–427

Krebs B (1991) Das Skelett von *Henkelotherium guimarotae* gen. et sp. nov. (Eupantotheria, Mammalia) aus dem Oberen Jura von Portugal. Berliner Geowiss Abh, a 133:1–110

Lewis OJ (1983) The evolutionary emergence and refinement of the mammalian pattern of foot architecture. J Anat 137:21–45

Linnaeus C (1758) Systema naturae per regna tria naturae, secundum classes, ordines, genera, species, cum characteribus, differentiis, synonymis, locis. Vol. 1: Regnum animale. Editio decima, reformata. Laurentii Salvii, Stockholm

Luo Z-X, Crompton AW, Sun A-L (2001) A new mammaliaform from the Early Jurassic and evolution of mammalian characteristics. Science 292:1535–1540

Luo Z-X, Kielan-Jaworowska Z, Cifelli RL (2002) In quest for a phylogeny of Mesozoic Mammals. Acta Palaeontol Pol 47:1–78

Luo Z-X, Yuan C-X, Meng Q-J, Ji Q (2011) A Jurassic eutherian mammal and divergence of marsupials and placentals. Nature 476:442–445

Macrini TE, Rougier GW, Rowe T (2007) Description of a cranial endocast from the fossil mammal *Vincelestes neuquenianus* (Theriiformes) and its relevance to the evolution of endocranial characters in therians. Anat Rec 290:875–892

Mao F, Hu Y, Li C, Wang Y, Chase MH, Smith AK, Meng J (2020) Integrated hearing and chewing modules decoupled in a Cretaceous stem therian mammal. Science 367:305–308

McKenna MC (1975) Toward a phylogenetic classification of the Mammalia. In: Luckett WP, Szalay FS (eds) Phylogeny of the Primates. Plenum Publishing Corporation, New York, pp 21–46

Novacek MJ (1982) The brain of *Leptictis dakotensis*, an Oligocene leptictid (Eutheria: Mammalia) from North America. J Paleontol 56:1177–1186

Novacek MJ (1986) The skull of leptictid insectivorans and the higher-level classification of Eutherian mammals. Bull Am Mus Nat Hist 183:1–112

Novacek MJ, Rougier GW, Wible JR, McKenna MC, Dashzeveg D, Horovitz I (1997) Epipubic bones in eutherian mammals from the Late Cretaceous of Mongolia. Nature 389:483–486

O'Leary MA, Bloch JI, Flynn JJ, Gaudin TJ, Giallombardo A, Giannini NP, Goldberg SL, Kraatz BP, Luo ZX, Meng J, Ni X, Novacek MJ, Perini FA, Randall ZS, Rougier GW, Sargis EJ, Silcox MT, Simmons NB, Spaulding M, Velazco PM, Weksler M, Wible JR, Cirranello AL (2013) The placental mammal ancestor and the post-K-Pg radiation of placentals. Science 339:662–667

O'Meara RN, Thompson RS (2014) Were there Miocene meridiolestidans? Assessing the phylogenetic placement of *Necrolestes patagonensis* and the presence of a 40 million year meridiolestidan ghost lineage. J Mammal Evol 21:271–284

Pascual R, Ortiz-Jaureguizar E (2007) The Gondwanan and South American episodes: two major moments in the history of South American mammals. J Mammal Evol 14:75–137

Rougier GW (1993) *Vincelestes neuquenianus* Bonaparte (Mammalia, Theria), un primitivo mamífero del Cretácico Inferior de la Cuenca Neuquina. Unpublished PhD thesis, University of Buenos Aires

Rougier GW, Wible JR (2006) Major changes in the ear region and basicranium of early mammals. In: Carrano MT, Gaudin TJ, Blob RW, Wible JR (eds) Amniote paleobiology: Perspectives on the evolution of mammals, birds, and reptiles. University of Chicago Press, Chicago, pp 269–311

Rougier GW, Wible JR, Hopson JA (1992) Reconstruction of the cranial vessels in the Early Cretaceous mammal *Vincelestes neuquenianus*: implications for the evolution of the mammalian cranial vascular system. J Vertebr Paleontol 12:188–216

Rougier GW, Wible JR, Hopson JA (1996) Basicranial anatomy of *Priacodon fruitaensis* (Triconodontidae, Mammalia) from the Late Jurassic of Colorado, and a reappraisal of mammaliaform interrelationships. Am Mus Novit 3183:1–38

Rougier GW, Wible JR, Novacek MJ (1998) Implications of *Deltatheridium* specimens for early marsupial history. Nature 396:459–463

Rougier GW, Qiang J, Novacek MJ (2003) A new symmetrodont mammal with fur impressions from the Mesozoic of China. Acta Geol Sin 77:7–14

Rougier GW, Wible JR, Novacek MJ (2004) New specimens of *Deltatheroides cretacicus* (Metatheria, Deltatheroida) from the Late Cretaceous of Mongolia. In: Dawson MR, Lillegraven JA (eds) Fanfare for an uncommon paleontologist: papers in honor of Malcolm C. McKenna. Bull Carnegie Mus Nat Hist 36:245–266

Rougier GW, Apesteguía S, Gaetano LC (2011) Highly specialized mammalian skulls from the Late Cretaceous of South America. Nature 479:98–102

Rougier GW, Wible JR, Beck RMD, Apesteguía S (2012) The Miocene mammal *Necrolestes* demonstrates the survival of a Mesozoic nontherian lineage into the late Cenozoic of South America. Proc Natl Acad Sci USA 109:20053–20058

Rowe TB (1993) Phylogenetic systematics and the early history of mammals. In: Szalay FS, Novacek MJ, McKenna MC (eds) Mammal phylogeny: Mesozoic differentiation, multituberculates, monotremes, early therians, and marsupials. Springer-Verlag, New York, pp 129–145

Szalay FS (1994) Evolutionary history of the marsupials and an analysis of osteological characters, 1st edn. Cambridge University Press, Cambridge.

Szalay FS (2006) Evolutionary history of the marsupials and an analysis of osteological characters, 2nd edn. Cambridge University Press, Cambridge.

Vázquez-Molinero R, Martin T, Fischer MS, Frey R (2001) Comparative anatomical investigations of the postcranial skeleton of *Henkelotherium guimarotae* Krebs, 1991 (Eupantotheria, Mammalia) and their implications for its locomotion. Zool Reihe 77:207–216

Wible JR, Hopson JA (1993) Basicranial evidence for early mammal phylogeny. In: Szalay FS, Novacek MJ, McKenna MC (eds) Mammalian phylogeny, vol 1. Mesozoic differentiation, multituberculates, monotremes, early therians, and marsupials. Springer, New York, pp 45–62

Wible JR, Rougier GW (2017) Craniomandibular anatomy of the subterranean meridiolestidan *Necrolestes patagonensis* Ameghino, 1891 (Mammalia, Cladotheria) from the Early Miocene of Patagonia. Ann Carnegie Mus 84:183–251

Wible JR, Rougier GW, Novacek MJ, McKenna MC (2001) Earliest eutherian ear region: a petrosal referred to *Prokennalestes* from the Early Cretaceous of Mongolia. Am Mus Novit 3322:1–44

Wood CB, Rougier GW (2005) Updating and recoding enamel microstructure in Mesozoic mammals: in search of discrete characters for phylogenetic reconstruction. J Mammal Evol 12:433–460

# Chapter 8
# Allotheria: Gondwanatherians and Multituberculates

*As one would expect following a quarter-century of research, the alpha-level diversity of known kinds (especially at the generic level) of Mesozoic mammals has skyrocketed. There have also been some big surprises, such as the discovery of Cretaceous mammals in South America having dentitions that look like they belonged in the Miocene...*
Jason A. Lillegraven and William A. Clemens
*Prologue*, 2004
*In: Mammals from the age of dinosaurs. Origins, evolution, and structure*

**Abstract** The enigmatic Gondwanatheria includes mammals with a mosaic of plesiomorphic and apomorphic cranial and dental features challenging our attempts to reconstruct their phylogenetic affiliation. They are generally perceived as sharing a closer ancestor with multituberculates than with therians in a variably conceived Allotheria. Two major groups are classically recognized among the South American Gondwanatheria: the brachyodont-toothed Ferugliotheriidae and the hypsodont-toothed Sudamericidae, although not all taxa fall easily in these categories. The affinities of the Ferugliotheriidae are, however, unsettled, with some authors favoring the hypothesis that they are indeed a derived branch of multituberculates. The original foundational Patagonian finds of gondwanatherians have recently been much improved by spectacular Late Cretaceous Malagasy materials, which increase dental diversity of the group, provide detailed cranial/postcranial morphology, and support allotherian affinities for the group.

**Keywords** Allotheria · Gondwanatheria · Multituberculata · Gondwana

## 8.1 Introduction

Gondwanatherians are one of the most bizarre and enigmatic groups of the SA Mesozoic, and in fact, one of the most surprising mammalian groups overall. They display distinctive dental features, including hypsodont incisors, molariforms with complex occlusal patterns, and deep rodentiform lower jaw. Some gondwanatherians have tall hypsodont open-rooted molariforms, others have tall crowns but distinct



G. W. Rougier et al., *Mesozoic Mammals from South America and Their Forerunners*, Springer Earth System Sciences, https://doi.org/10.1007/978-3-030-63862-7_8

roots are still recognizable, while others are brachyodont. The enamel can vary from extremely heavy and thick in *Gondwanatherium* to supposedly absent, as in *Galulatherium* (Scillato-Yané and Pascual 1985; Mones 1987; Bonaparte 1986a, b, c, 1987; Krause et al. 1992; Bonaparte et al. 1993; Krause and Bonaparte 1993; Gurovich 2005, 2008; Gurovich and Beck 2009; Rougier et al. 2009a; Goin et al. 2004, 2006, 2012; O'Connor et al. 2019). Although most gondwanatherians share a rodent-like appearance, they include a range of sizes with the Cretaceous Malagasy *Vintana* and *Adalatherium* weighting an estimated 9 kg and 3 kg, respectively (Krause et al. 2014a, 2020); on the other end of the spectrum, the Patagonian *Ferugliotherium* with molars under 1 mm in size likely weighed less than 70 g.

Scillato-Yané and Pascual (1985, see also Scillato-Yané and Pascual 1984) described *Sudamerica ameghinoi* as a peculiar xenarthran from the Paleocene of Patagonia, erecting the family Sudamericidae. Soon thereafter, Bonaparte (1986c) published *Gondwanatherium patagonicum* based on isolated teeth from the Late Cretaceous of Patagonia and erected the family Gondwanatheriidae. Both taxa were first thought to be related to placental xenarthrans (sloths, anteaters, and cingulates) or Paratheria (i.e., a group no longer supported, that in addition to xenarthrans included pangolins, aardvark, taeniodonts, among others), due to the gross similarity with the molariform teeth of the more recent xenarthran glyptodonts (Scillato-Yané and Pascual 1985; Bonaparte 1986a, b, c, 1987, 1988; Bonaparte and Pascual 1987; see also Bonaparte 2017). The distinctiveness of *Sudamerica* and *Gondwanatherium* from other mammals and their shared similarities led Mones (1987), to erect the xenarthran Order Gondwanatheria, after an unrelated visit to Buenos Aires collections and Bonaparte's candid discussion of the idea.

Bonaparte published *Gondwanatherium* in 1986 (Bonaparte 1986c); earlier the same year, he described in a separate paper *Ferugliotherium windhauseni* (Bonaparte 1986a) and erected the family Ferugliotheriidae, based on an isolated brachyodont tooth from the Late Cretaceous of Los Alamitos, Patagonia. He assigned *Ferugliotherium* to Allotheria, a clade in which members have molars with cusps in line and a rudimentary to fully gliriform morphology. Allotheria was coined by Marsh (1880), originally containing only two multituberculate species. Much later, Haramiyida (Hahn et al. 1989), and even more recently an assorted variety of forms from China, collectively called Euharamiyida/Eleutherodontida have also been included (e.g., Butler 2000; Krause et al. 2014a; Bi et al. 2014; Luo et al. 2015, 2017; Han et al. 2017; Huttenlocker et al. 2018; see however Sereno 2006 for a definition of Allotheria). At the time Bonaparte studied *Ferugliotherium*, affinities with multituberculates were the most reasonable option and with caveats he regarded *Ferugliotherium* as related to this group. By the late eighties, we had a bizarre lineage of Cretaceous/Paleocene xenarthrans (*Gondwanatherium*/*Sudamerica*), and a distinct group of likely Cretaceous multituberculates (*Ferugliotherium*). In a few years, these taxa were reinterpreted and a radically different scenario unfolded.

The non-therian affinities of gondwanatherians were first proposed by Krause and Bonaparte (1990, 1993; see also Sigogneau-Russell et al. 1991; Krause et al. 1992). They placed gondwanatherians (i.e., their Superfamily Gondwanatherioidea) within Multituberculata, and included the hypsodont Sudamericidae (Sudamericidae had priority over Gondwanatheriidae), *Sudamerica*, *Gondwanatherium*, and the brachyodont Ferugliotheriidae, *Ferugliotherium*. The main feature uniting these seemingly disparate groups is the similar gross anatomy of the occlusal surface of molariforms and inferred direction of jaw movement (i.e., palinal: distal movement of the jaw during the power stroke of the grinding cycle). The overall affinities of gondwanatherians as allotherians were supported by subsequent contributions on SA mammals, such as Gurovich (2005, 2008), Gurovich and Beck (2009), and Rougier et al. (2009a), although doubts were raised regarding the inclusion of *Ferugliotherium* among gondwanatherians (e.g., Gurovich and Beck 2009; Rougier et al. 2009a). Later on, gondwanatherians were considered close to Multituberculata, within allotherians (Krause et al. 1997, 2014a, 2019; Pascual and Ortiz-Jaureguizar 2007; Goin et al. 2012), related to Euharamiyida/Eleutherodontidae (Huttenlocker et al. 2018), or as Mammalia *incertae sedis* (Koenigswald et al. 1999; Pascual et al. 1999; Kielan-Jaworowska et al. 2004). More recently, Gondwanatheria was defined as the clade that includes the common ancestor of *Ferugliotherium*, *Sudamerica*, and all of its descendants (Chimento et al. 2015) (Fig. 8.1), or as the most inclusive clade including *Gondwanatherium* but not *Taeniolabis*, *Cifelliodon*, or *Shenshou* (Hoffmann et al. 2020).

Since the initial description of sudamericids and ferugliotheriids from the Late Cretaceous and Paleocene of Patagonia in the eighties, their record grew considerably

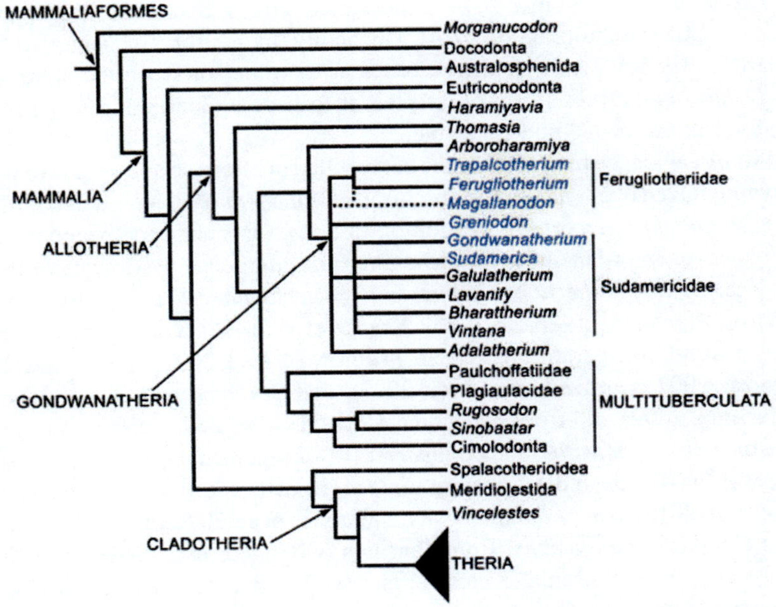


**Fig. 8.1** Phylogenetic tree highlighting the relationships of sudamericids and ferugliotheriids, based on Krause et al. (2014a)

by further findings in Patagonia and in the Late Cretaceous of Chile, India, Madagascar, and the Cenozoic of Argentina, Antarctica, and possibly Peru. Among these newer discoveries, one of the most remarkable is the exquisitely preserved skull of *Vintana sertichi* from Madagascar, which is thought to be related to the SA forms (e.g., Krause et al. 2014a, b, c, 2020). The specimen consists of a large-sized complete skull, lacking lower jaws, of a groundhog-like critter with a puzzling mosaic of primitive and derived characters. Considering that gondwanatherians (e.g., *Gondwanatherium* and *Sudamerica*) are known by isolated teeth and fragmentary lower jaws (unknown in *Vintana*), there is little overlap in comparable features and their close phylogenetic affinities rest solely in the "rodentiform" overall morphology and the presence of peculiar enamel inlets and enamel folds. Relationships between *Vintana* and gondwanatherians are attractive and have been recently supported and reinforced by the discovery of *Adalatherium hui* (Krause et al. 2020), from the Late Cretaceous of Madagascar. *Adalatherium* is represented by a stunningly complete skeleton with a very peculiar dentition and a poorly preserved skull sharing specialized traits with *Vintana* (Krause et al. 2020). *Adalatherium* is recovered in the phylogenetic tree as basal to sudamericids, and *Ferugliotherium* as sister taxon to that group or alternatively clustering with multituberculates. If *Vintana* and *Adalatherium* are members of Gondwanatheria or part of the stem leading to Gondwanatheria, the available skulls, dentition, and postcranial skeleton of *Vintana sertichi* and *Adalatherium hui* open a new window to the gross anatomy of sudamericid gondwanatherians (Hoffmann et al. 2014; Kirk et al. 2014; Krause 2014; Krause et al. 2014a, b, c, 2020; Koenigswald and Krause 2014; Schultz et al. 2014). Uncannily, by a stroke of genius or luck, Bonaparte's (1986c) Gondwanatheria, based at the time solely on the patronymic genus *Gondwanatherium*, is shown at present to have a Gondwanic fossil record during the late Mesozoic–mid-Cenozoic.

The sum of evidence from Gondwana has resulted in fifteen species, some of which were synonymized (e.g., *Vucetichia gracilis*, *Dakshina jederi*) or have dubious affinities (i.e., *Patagonia peregrina*), in addition to a few yet unnamed specimens. The Late Cretaceous record includes *Gondwanatherium patagonicum* (Bonaparte 1986c, 1988, 1990), *Ferugliotherium windhauseni* (=*Vucetichia gracilis*; =*Argentodites coloniensis*; Bonaparte 1986a, b, 1990; Krause et al. 1992; Krause 1993; Kielan-Jaworowska and Bonaparte 1996; Kielan-Jaworowska et al. 2007; Pascual and Ortiz-Jaureguizar 2007; Gurovich and Beck 2009), and *Trapalcotherium matuastensis* (Rougier et al. 2009a), all from Argentina, *Magallanodon baikashkenke* (Goin et al. 2020) from Chile, *Bharattherium bonapartei* (=*Dakshina jederi*; Prasad et al. 2007; Wilson et al. 2007) from India, *Lavanify miolaka* (Krause et al. 1997), *Vintana sertichi* (Krause et al. 2014a), and *Adalatherium hui* (Krause et al. 2020), from Madagascar, perhaps *Galulatherium jenkinsi* from Tanzania (O'Connor et al. 2019), plus indeterminate taxa from Tanzania (Krause et al. 2003), India (Verma et al. 2012), and Madagascar (Krause 2013). The Cenozoic record includes *Sudamerica ameghinoi* (Scillato-Yané and Pascual 1984, 1985; Koenigswald et al. 1999; Pascual et al. 1999; Bonaparte et al. 1993; Gurovich 2008) and perhaps *Greniodon sylvaticus* (Goin et al. 2012), and partial material from the Antarctic Peninsula (Goin et al. 2006). Other taxa, such as *Patagonia peregrina* (Pascual and Carlini 1987; Chimento et al. 2015), provides inconclusive phylogenetic hypotheses with the material at hand (see below), as happens with some records from Peru (Campbell et al. 2004; Goin et al. 2004; Antoine et al. 2012).

Other rodent-like mammals, such as *Groeberia minoprioi* and *G. pattersoni* (Metatheria) from the middle Eocene Divisadero Largo Formation (Mendoza, Argentina; e.g., Pascual et al. 1994 and references herein) were recently assigned to sudamericid gondwanatherians by Chimento et al. (2015). Their observations expanded on the bizarre dental morphology of *Groeberia* and vague rodentiform craniomandibular shape that led to the questioning of the classic alleged metatherian affinities (McKenna 1980; Reig 1981). However, Zimicz and Goin (2020; see also Beck 2017), recently reviewed in detail the known fossil material including not only the evidenced morphology, but also the interpretation of jaw movements; they provide conclusive support in favor of therian affinities and among them metatherian allegiance. Zimicz and Goin (2020) unambiguously interpreted for *Groeberia* a tribosphenic molar pattern in *Groeberia*, with a dental formula of I2/i1, C1/c1, P3/p1, M4/m4, which is compatible with the usual metatherian set of three premolars and four molars in the upper dentition, and with the common "pseudodiprotodont" lower postcanine formula of one premolar (p3) and four molars (Zimicz and Goin 2020). In our opinion the rejection of *Groeberia* as a gondwanatherian is conclusive.

Chimento et al. (2015) also recognized the enigmatic *Patagonia peregrina* (Pascual and Carlini 1987), from the Miocene of Patagonia as a gondwanatherian. *Patagonia* was originally referred to as a marsupial of its own monotypic family Patagoniidae and supra familiar taxon Patagonioidea. Other authors (Goin and Abello 2013; Goin et al. 2016, and references therein), have regarded *Patagonia* as related to argyrolagids. The known specimens of *Patagonia* are poor, consisting of a fragmentary lower jaw with dentition and isolated molariform teeth (Pascual and Carlini 1987). We regard the status of *Patagonia* as uncertain and it is included in this chapter and discussed below under Mammalia *incertae sedis* because it has been proposed as a gondwanatherian. In our view, the proposal of Chimento et al. (2015), holds no advantage over the traditional view of *Patagonia* as a metatherian, but it is almost logically impossible to disprove their conclusion with the available material. This chapter provides the best framework to revise their observations and conclusions on that taxon.

Despite the dramatic increase of information on likely gondwanatherians from Madagascar (Hoffmann et al. 2014; Krause et al. 2014a, b, 2019, 2020; Koenigswald and Krause 2014; Schultz et al. 2014), the unifying motif remains the rodentiform adaptations of large mesial incisors, reduced number of premolars, deep jaws, and coronal occlusal features related to palinal/propalinal movement, such as cusps in rows and transverse crests. There is enough cranial overlap in *Vintana* and *Adalatherium* to ensure they are part of a closely related group, but the remaining gondwanatherians fare much worse, relying only on dental features. *Ferugliotherium* and relatives are the most unstable members of Gondwanatheria, if they belong to the group at all, the alternative being they are a distinct group of multituberculates. Important questions are still pending: Are gondwanatherians related to multituberculates? How are ferugliotheriids related to other gondwanatherians? What are the relationships of Allotheria? Gondwanatherians are likely to remain an iconic and enigmatic group of Mesozoic SA mammals for years to come.

## 8.2 Systematics

Mammalia Linnaeus 1758
Gondwanatheria Mones 1987
Sudamericidae Scillato-Yané and Pascual 1985
*Gondwanatherium* Bonaparte 1986c

Type species: *Gondwanatherium patagonicum* Bonaparte 1986c.

Included species: The type only.

*Gondwanatherium patagonicum* Bonaparte 1986c
(Fig. 8.2)

Holotype: MACN-RN 22, complete upper molariform (Fig. 8.2a, b).

Locality and horizon: Estancia Los Alamitos, west slope of Cerro Cuadrado, Arroyo Verde, Río Negro, Argentina. Middle levels of the Los Alamitos Formation, Campanian–Maastrichtian (Upper Cretaceous).

Diagnosis (modified from Bonaparte 1986c and Gurovich 2008): Gondwanatherian mammals of relatively large size (~0.8 to 1.3 kg) with high crowned molariforms, prismatically shaped and open roots; molar crowns consisting of three transverse lophs that with wear turn into dentine islets surrounded by enamel, with an almost flat occlusal surface; buccal side of the crown with three lobes and lingual side with two lobes, separated by long furrows.

Comments: *Gondwanatherium patagonicum* is known by a few referred incisor fragments, several isolated molariforms, and a single incomplete anterior portion of the dentary (Bonaparte 1986c, 1990; Gurovich 2005), most of which are deposited in the MACN-PV collection. For several years, *Gondwanatherium* represented the oldest hypsodont mammal and the oldest Mesozoic gondwanatherian. However, older, but unrelated, hypsodont teeth have been reported for the Jurassic mammaliaform *Fruitafossor windscheffeli* (Luo and Wible 2005), and even for the Late Triassic traversodontid cynodont *Menadon besairiei* (Melo et al. 2019). In addition, hypsodonty was reported for the tall enamelless molariforms of *Galulatherium*, a putative gondwanatherian from the Turonian–Campanian of Tanzania (O'Connor et al. 2019).

The dental formula of *Gondwanatherium* is unknown. However, inferences based on the different tooth morphs present in the collection combined with the referred lower jaw, the lower jaw of *Sudamerica* (Pascual et al. 1999), and the Malagasy gondwanatherians *Vintana* and *Adalatherium* (Krause 2014; Krause et al. 2014a, c, 2020) suggest i1, c0, mf4 as a likely formula. Nevertheless, alternative scorings of

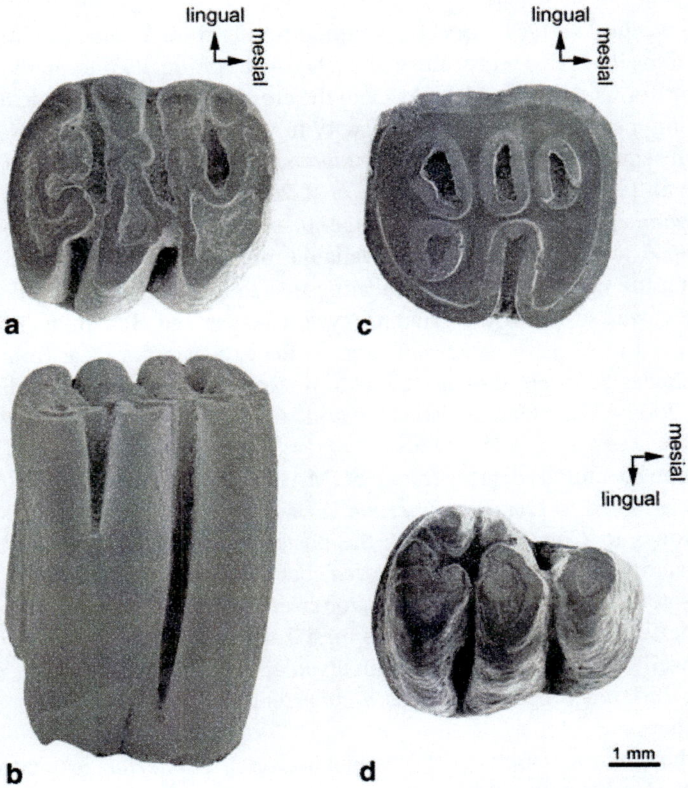


**Fig. 8.2** *Gondwanatherium patagonicum* from Ea. Los Alamitos, west slope of Cerro Cuadrado, Río Negro, Argentina; Los Alamitos Formation, Campanian–Maastrichtian, Upper Cretaceous. Holotype, MACN-RN 22, upper left second molariform–MF2 (based on Gurovich 2008), in occlusal (**a**) and labial (**b**) views; RN 1029, upper left molariform in occlusal view (**c**); MACN-RN 23, almost unworn, tentatively lower left first molariform–mf1, in occlusal view (**d**)

the lower dental formula for gondwanatherians, as for example, i1, c0, pmf2, mf2, have shown to yield not dramatically different phylogenetic results (Krause et al. 2020). The available isolated incisors of *Gondwanatherium* are hypsodont (open rooted) and bow-shaped (following Koenigswald et al. 1999). They are flat lingually and convex labially, with a thick band of enamel on the labial face and restricted to the mesial third of the lingual surface. Apparently, upper incisors have stronger curvatures than the lower ones, with a less-acute angle formed between the apical wear facet and the labial margin of the crown (Krause and Bonaparte 1993).

The molariforms are hypsodont, with high and prismatic crowns (Fig. 8.1). They have two to three transversely oriented lophs of enamel that define three/two labial lobes and two or one lingual ones, respectively, separated by long furrows. During wear, the pattern of the occlusal surface changes considerably, starting in small and subcircular figures that become complex and V-shaped containing distinctive dentine islets. The occlusal surface lacks cementum, differing from *Sudamerica* in which it fills most of the lobes (Koenigswald et al. 1999). As traditionally interpreted, wear on lower molariforms results in a labial side of the crown more elevated than the lingual one, defining a subtle labiolingual concavity in the occlusal surface. Based on the molariform sequence of *Sudamerica, Vintana,* and to a lesser degree *Adalatherium* (Pascual et al. 1999; Krause 2014; Krause et al. 2014a, 2020), the mesial molariform in *Gondwanatherium* is mesiodistally longer than wide, whereas they become more square-shaped toward the rear. Most available molariforms have distinctive wear striations on the occlusal surface, indicating that the dentary has moved palinally during the power stroke of the grinding cycle (Krause and Bonaparte 1993). The lophs and crest of *Gondwanatherium* form by the expansion of wear from isolated conical cuspules that in little-worn teeth allow the recognition of longitudinal rows (Gurovich 2005:423), a feature also present in *Magallanodon* (Goin et al. 2020, see below).

There is an edentulous dentary fragment (MACN-RN 228) from the Los Alamitos Formation assigned to *Gondwanatherium* (Bonaparte 1990; Gurovich 2005). If it indeed belongs to *Gondwanatherium,* the posterior molariforms would diminish in size posteriorly, as seen in *Sudamerica,* but the jaw would lack the rodentiform excavated diastema present in other gondwanatherians. The edentulous portion between the fragmentary alveolus of the mesial incisor and the first molariform is preserved well enough, forming a straight alveolar margin. This would be a sharply contrasting difference between *Gondwanatherium* and other known dentaries of gondwanatherians.

As in other gondwanatherians, the enamel in *Gondwanatherium* is thick and prismatic, with circular prisms of small size (<5 μm in diameter) and anastomosed interprismatic matrix surrounding the prisms, without interrow sheets (Krause and Bonaparte 1993; Sigogneau-Russell et al. 1991; Koenigswald and Krause 2014). The dentine has numerous parallel tubules, some of which continue into the enamel layer (Sigogneau-Russell et al. 1991).

Gurovich (2008) estimated the body mass of *Gondwanatherium* based on the occlusal surface of the first lower molariform, indicating ~ 1 to 1.3 kg using a wide sample of mammals and ~ 0.8 to 1.1 kg using only rodents as models.

*Gondwanatherium* is a historically important taxon, however, its systematic import is yet to be fully realized. There is no single specimen of *Gondwanatherium* with a tooth in situ, other than isolated teeth, and the interpretation of all the dental relationships upper/lower, buccal/lingual/mesial/distal are tentative at best.

*Sudamerica* Scillato-Yané and Pascual 1984 (*nomen nudum*) Scillato-Yané and Pascual 1985

Type species: *Sudamerica ameghinoi* Scillato-Yané and Pascual 1985.

Included species: The type only.

*Sudamerica ameghinoi* Scillato-Yané and Pascual 1984 (*nomen nudum*) Scillato-Yané and Pascual 1985
(Figs. 8.3 and 8.4)

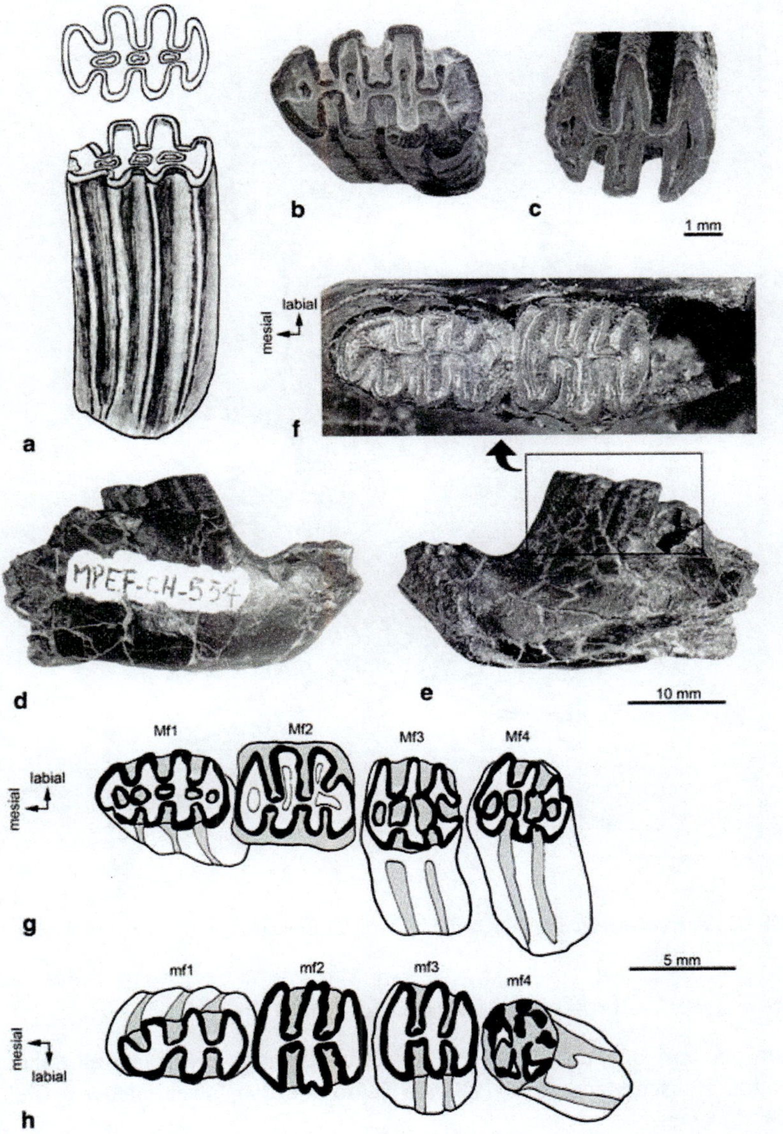


**Fig. 8.3** *Sudamerica ameghinoi* from Punta Peligro, Chubut Argentina; Salamanca Formation, Danian, lower Paleocene. Holotype, UNP 79-II-18-1, probably lower right molariform in occlusal and labial views (**a**), from Scillato-Yané and Pascual (1985); MACN-CH 1983, isolated left upper first molariform–Mf1, in occlusal view (**b**); MACN-CH 1943, isolated right lower third molariform–mf3, in occlusal view (**c**); MPEF-PV 534, right dentary with two mesial molariforms, two distal molariform alveoli, and an alveolus for a continuous growing incisor in labial (**d**), lingual (**e**) and detail of the dentition in occlusal view (**f**). Inferred reconstructions of the upper (**g**) and lower (**h**) molariform tooth row in occlusal view, based on isolated teeth, modified from Koenigswald et al. (1999). The root can be observed in some positions. Grey indicates cementum and the wide black line indicates enamel

**Fig. 8.4** *Sudamerica ameghinoi*, illustration by Jorge L. Blanco

Holotype: UNP 79-II-18-1, lower right molariform (Fig. 8.3a).

Locality and horizon: Punta Peligro locality, Chubut Province, Argentina. Salamanca Formation, Hansen Member, BNI (Banco Negro Inferior), lower Paleocene (Peligran SALMA, Danian).

Diagnosis (modified from Scillato-Yané and Pascual 1985; Pascual et al. 1999; Gurovich 2008): Gondwanatherian mammals of relatively large size (~0.9 to 1.4 kg), with hypsodont lower incisor; absence of canine; four lower and upper hypsodont molariform teeth with thick cementum in the occlusal surface; upper molariforms with three to four transversal lophs and lowers with two to four, separated by long furrows; lophs connected by a central crest; variation of enamel islets in the crown due to different stage of wear; presence of large diastema between lower incisor and first molariform.

Comments: *Sudamerica ameghinoi* was the first gondwanatherian described (Scillato-Yané and Pascual 1985). This species comes from the Paleocene of Patagonia and is a member of a group that survived the K/Pg extinction. Most known specimens are isolated teeth, which are quite common, in the classical locality of Punta Peligro, Salamanca Formation (Bonaparte et al. 1993; Koenigswald et al. 1999; Pascual et al. 1999; Gurovich 2008). The discovery of the right mandible MPEF-CH 534 (Fig. 8.3d, e), bearing a broken enlarged incisor, first and second molariforms, and the alveoli for two subequal molariforms (third and fourth) provided the first clues to decipher the morphological changes of the upper/lower dentition and the variation along the tooth row (Pascual et al. 1999; Koenigswald et al. 1999) (Fig. 8.3d–h). *Sudamerica* provided, thus, the first general framework for interpreting isolated gondwanatherian teeth, which to date constitute the bulk of the evidence.

The lower incisor in *Sudamerica* is large and hypsodont, laterally compressed with a thick layer of enamel restricted to the "ventro-labial" side (Pascual et al. 1999). The interalveolar portion of the incisor extends distally to the level of the last molariform. Incisor and molariforms are separated by a large diastema and molariform tooth row is positioned labial to the incisor. The upper and lower molariforms are hypsodont, with two to four transverse lophs defined by a thick layer of enamel (Fig. 8.3g, h). Each loph defines labial and lingual lobes, usually asymmetrical, separated by long furrows, filled with cementum. Mesial upper and lower molariforms have four lophs, triangular in shape, and more mesiodistally elongated than distal teeth. The Mf1/mf1 are the most asymmetrical teeth because the lophs define up to four lobes in the lingual side and up to three in the labial one. The Mf2 has four lophs, whereas the Mf3 and Mf4 have three (Koenigswald et al. 1999; Gurovich 2008). The mf2 and mf3 has three lophs and the mf4 has two (Koenigswald et al. 1999; Pascual et al. 1999; Gurovich 2008) (Fig. 8.3h). The external thick layer of enamel bordering the lophs is connected along the crown, and it is contacted medially by an enamel ridge. This condition differs from *Gondwanatherium* in which the enamel of each loph is separated. The shape of the islets on the occlusal surface changes considerably during tooth wear. The islets increase with wear in number and size (from circular to elongate); throughout all wear stages, the lophs are connected by enamel ridges (Fig. 8.3).

The microstructure of the enamel reveals a thick layer of prismatic enamel, with small-sized circular prisms surrounded by anastomosed interprismatic matrix, without interrow sheets (Sigogneau-Russell et al. 1991; Krause and Bonaparte 1993; Koenigswald and Krause 2014). The incisors have a thick layer of external prism-less enamel, the outer aprismatic enamel (=OES; Koenigswald and Krause 2014).

Gurovich (2008) estimated the body mass of *Sudamerica* using linear measurements of the occlusal surface of the mf1 and Legendre (1989)'s method. This resulted in ~1.2 to 1.4 kg using a wide sample of mammals and about 0.9–1.3 kg using only rodents as models. However, the body mass for *Sudamerica* based on Mf1 and Mf2, with McDermott et al. (2002)'s method, resulted in a much smaller critter of ~0.4 kg (Gurovich 2008).

The paleobiology of *Sudamerica ameghinoi* is difficult to assess (Fig. 8.4). Considering the hypsodont incisors and molariforms, Gurovich (2008) inferred folivorous diet, browser of abrasive vegetation; in analogy with beavers (*Castor canadensis*) or coypus (*Myocastor coypus*); perhaps *S. ameghinoi* was semiaquatic, or in analogy with tuco-tucos (*Ctenomys* spp.) perhaps a burrowing mammal.

Gondwanatheria Mones 1987 or Multituberculata Cope 1884
Ferugliotheriidae Bonaparte 1986a
*Ferugliotherium* Bonaparte 1986a

Type species: *Ferugliotherium windhauseni* Bonaparte 1986a.

Included species: The type only.

*Ferugliotherium windhauseni* Bonaparte 1986a
(Figs. 8.5a–e and 8.6)

Synonyms: *Vucetichia gracilis* Bonaparte 1990, *Argentodites coloniensis* Kielan-Jaworowska et al. 2007.

Holotype: MACN-RN 20, upper left second molariform (Fig. 8.5a), described originally as right m2.

Holotype of *Vucetichia gracilis*: MACN-RN 174, lower right first molariform (Fig. 8.5b), originally described as an upper molar.

Holotype of *Argentodites coloniensis*: MPEF 604, isolated left plagiaulacoid premolariform (pmf4) (Fig. 8.5d, e).

Locality and horizon: The holotypes of *Ferugliotherium windhauseni* and *Vucetichia gracilis* come from Estancia Los Alamitos, west slope of Cerro Cuadrado, Arroyo Verde, Río Negro, Argentina. Los Alamitos Formation, Campanian–Maastrichtian (Upper Cretaceous). The holotype of *Argentodites coloniensis* and specimens referred to *F. windhauseni* come from Mirasol Chico Canyon, Chubut Province, Argentina. La Colonia Formation, Campanian–Maastrichtian (Upper Cretaceous).

Diagnosis (modified from Krause et al. 1992; Kielan-Jaworowska and Bonaparte 1996; Gurovich 2008): Small-sized mammal (~65 to 70 gr) with large hypsodont lower incisor, laterally compressed, and restricted band of enamel; plagiaulacoid (blade-like) pmf4; brachyodont molariforms with prominent transverse ridges connecting cusps of adjacent rows and prominent transverse furrows extending nearly or completely across crowns between cusps of adjacent rows; crest forms a mesial triangle directed lingually in the upper molariforms and labially in the lowers. *Ferugliotherium* incisors have prismatic enamel with small circular prisms.

Comments: *Ferugliotherium windhauseni* is a poorly known taxon, first discovered from the Los Alamitos Formation (Fig. 8.6). This species was also mentioned for the La Colonia Formation by Pascual and Ortiz-Jaureguizar (2007), but the material has never been formally described.

*Ferugliotherium* was first allocated to its own family Ferugliotheriidae and tentatively referred to Multituberculata (Bonaparte 1986a). The holotype MACN-RN 20 is a brachyodont molariform with a quadrangular crown bearing lingual and labial

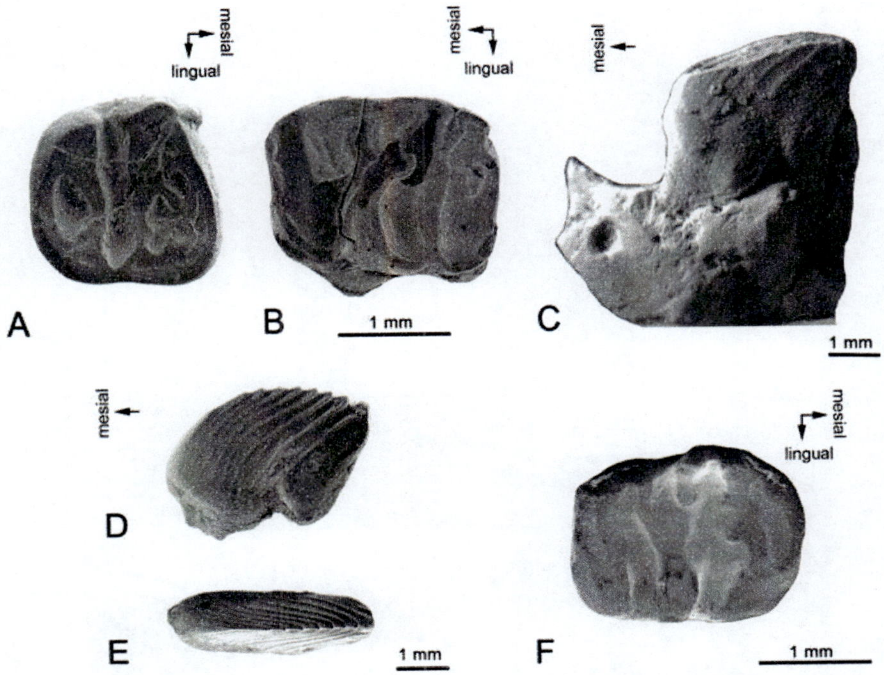


**Fig. 8.5** *Ferugliotherium windhauseni* from Ea. Los Alamitos and possible Mirasol Chico Canyon, Río Negro and Chubut, Argentina; Los Alamitos and La Colonia formations, Campanian–Maastrichtian, Upper Cretaceous. Holotype, MACN-RN 20, described as lower right second molariform–mf2, in occlusal view, but here considered as a likely right second upper molariform Mf2 (**a**); MACN-RN 174 (holotype of *Vucetichia gracilis*), lower right first molariform–mf1, in occlusal view (**b**); MACN-RN 975, fragment of left dentary with last premolariform–pmf4, in labial view (**c**). MPEF-PV 604 (holotype of *Argentodites coloniensis*, from Mirasol Chico Canyon, Chubut, Argentina; La Colonia Formation, Campanian–Maastrichtian, Upper Cretaceous), left lower last premolariform–pmf4, in labial (**d**) and occlusal (**e**) views, from Kielan-Jaworowska et al. (2007). *Trapalcotherium matuastensis* from Cerro Tortuga, Río Negro, Argentina; Allen Formation, Campanian–Maastrichtian, Upper Cretaceous. Holotype, MML-PV 16, lower left first molariform–mf1, in occlusal view (**f**)

worn cusps, and mesial and distal enamel islets separated by a transversal sulcus (Bonaparte 1986a) (Fig. 8.5a). This tooth was regarded as the last lower molar; however, it is likely an upper last molariform.

Later, Bonaparte (1990), named *Vucetichia gracilis*, based on brachyodont teeth originally considered as upper molariforms (holotype MACN-RN 174 and hypodigm MACN-RN 175) (Fig. 8.5b), and related this taxon to *Gondwanatherium.* The similarity in the occlusal figure of the enamel is real, in particular, the presence of a rather unusual triangular figure on the mesial portion of the tooth followed by transverse connections between lingual and labial cusp rows. The brachyodont molariforms of *Ferugliotherium* are much smaller than the highly hypsodont dentition characterizing *Gondwanatherium* and *Sudamerica* (Bonaparte 1990).

**Fig. 8.6** *Ferugliotherium windhauseni*, illustration by Jorge L. Blanco

Subsequently, Krause et al. (1992) published additional specimens from Los Alamitos including incisors, premolariforms, and molariforms which they referred to *Ferugliotherium widhauseni*. The new material provided a more complete view of positional variation and led the authors to suggest *Vucetichia gracilis* was likely a junior synonym of *Ferugliotherium*. This taxonomic rearrangement was later formally proposed (Krause 1993), considering that the holotype and hypodigm of *V. gracilis* (worn "upper molars") were instead mesial lower molariforms of *F. widhauseni* (see also Krause et al. 1992). The new material of *Ferugliotherium*

(=*Vucetichia*) allowed painting of a relatively complete picture of the species: it can be characterized as having hypsodont incisors with a thick layer of enamel mostly restricted to its labial side, brachyodont molariforms with four roots and at least two longitudinal rows of multiple cusps. There are at least two mesial upper premolars, they have conical or pyramidal cusps, some associated with faint radiating ridges. The jaw moved in a palinal (distal) direction of motion during the power stroke of the grinding cycle (Krause et al. 1992; Krause and Bonaparte 1993), a rather peculiar feature shared with multituberculates.

Subsequently, Krause and Bonaparte (1993) allied *Ferugliotherium*, *Gondwanatherium*, and *Sudamerica* in a single group in sharing a similar occlusal enamel pattern (see also Bonaparte 1994). The authors maintained *Gondwanatherium* and *Sudamerica* in Sudamericidae and *Ferugliotherium* in Ferugliotheriidae, both as Gondwanatherioidea, and the whole group related to plagiaulacoid multituberculates.

Later, Kielan-Jaworowska and Bonaparte (1996), described a fragmentary left dentary with partial alveolus of a large incisor, short diastema, and complete but worn plagiaulacoid tooth, interpreted as pmf4 (MACN-RN 975) (Fig. 8.5c). Because it was found in the same locality and stratigraphic unit as *Ferugliotherium* and the size was compatible with their known molariforms, the specimen was tentatively referred to *Ferugliotherium.* As result, the authors considered *Ferugliotherium* to have a rodentiform lower jaw with one large incisor, a long diastema, a plagiaulacoid pmf4, and at least two molariforms (deduced from isolated teeth). This morphology is congruent with that of cimolodont multituberculates, reinforcing presumed links between gondwanatherians and multituberculates. However, the referral of this fragmentary jaw to *Ferugliotherium* was not readily accepted by the community. Pascual et al. (1999; see also Kielan-Jaworowska et al. 2004; Pascual and Ortiz-Jaureguizar 2007), based on the lower jaw of *Sudamerica*, suggested that gondwanatherians (including *Ferugliotherium*) should have no blade-like premolars but only four molariforms. As such, the plagiaulacoid pmf4 of specimen MACN-RN 975 was inconsistent with the known lower jaw and dental morphology of *Sudamerica*, which in turn, could be made to match the known teeth of *Gondwanatherium* and the fragmentary dentary referred to it. In addition, isolated premolariforms earlier described by Krause et al. (1992), and referred to *Ferugliotherium* were discredited. Instead, those "premolars" were likely indicative of multituberculates in the Patagonian faunas, likely the same taxon represented by the plagiaulacoid lower premolar, whose molars were yet to be recovered.

Kielan-Jaworowska et al. (2007) described *Argentodites coloniensis*, based on an isolated plagiaulacoid premolar (Fig. 8.5d, e). They reported the presence of a normal-sized prismatic enamel (average prism diameter of 3.8 μm) as in generalized cimolodontan multituberculates and *Ferugliotherium.* The premolar lacks a distinct distal accessory cusp and compares favorably with multituberculates regarding overall morphology and number of serrations and crests, etc. The authors supported earlier conclusions by Pascual et al. (1999): the molars of *Feruglitherium* from Los Alamitos were of those of a brachyodont gondwanatherian, the dentary and worn out pmf4 assigned by Kielan-Jaworowska and Bonaparte (1996) to *Ferugliotherium* belonged to an as yet unnamed multituberculate, as well as those mesial premolars described by Krause et al. (1992). Hence, *Argentodites* was a multituberculate for the nearly coeval La Colonia Formation. Under these interpretations, multituberculate premolars were known in Los Alamitos and La Colonia, but not molariforms, while ferugliotheriid molariforms were known in the Los Alamitos, Allen, and La Colonia formations (see below). These arguments were not accepted by Gurovich and Beck (2009) or Rougier et al. (2009a), who supported the original proposals of Krause et al. (1992; see also Krause and Bonaparte 1993) and Kielan-Jaworowska and Bonaparte (1996). At its core, the argument revolved around the presence, or absence, of a plagiaulacoid premolar in ferugliotheriids.

There are very minor size differences in the plagiaulacoid premolars from the Los Alamitos and La Colonia formations; if meaningful, the difference would justify a species difference at most (Fig. 8.5c–e). Certainly, it is logically possible to have *Argentodites* (a multituberculate) in Los Alamitos and *Ferugliotherium* (a gondwanatherian) in the same locality. The collection of isolated teeth from Los Alamitos is relatively numerous (over one hundred and fifty) and it seems strange not to have any record of the multituberculate molars expected to go along with the dentary described by Kielan-Jaworowska and Bonaparte (1996), but instead, *Ferugliotherium* teeth that match the expected size. Rougier et al. (2009b) reported on a new collection of mammals from the La Colonia Formation coming from a rich new locality (El Uruguayo) which yielded about 350 mammalian specimens. Most of these new specimens are relatively large mesungulatids (see Chap. 6), but ferugliotheriids are present, as they are in the even more recently discovered La Colonia locality of Amphitheater 1 (Harper et al. 2019). Amphitheater 1 is very rich in small mammals, among which a species of *Ferugliotherium* is very abundant. Neither El Uruguayo nor Amphitheater 1, both of which have yielded plagiaulacoid premolars of *Argentodites*, have any non-ferugliotheriid multituberculate molar among the combined hundreds of specimens collected. We consider the association of the plagiaulacoid premolars from La Colonia and Los Alamitos with ferugliotheriid molariforms, an almost inescapable conclusion; as such, we believe subsuming *Argentodites* as a junior synonym of *Ferugliotherium* is the most fruitful path. Additionally, Pascual and Ortiz-Jaureguizar (2007), mentioned a possible molariform of *Ferugliotherium windhauseni* from La Colonia but the material was never formally described. This specimen must correspond to the *Ferugliotherium* material we recovered or are recovering from El Uruguayo and Amphitheater 1 localities.

There is a chance that more complete specimens from the Los Alamitos and La Colonia formations may reveal subtle differences presently masked by the multitude of isolated teeth in different states of preservation and wear recovered in both localities. The plagiaulacoid premolar from Los Alamitos is slightly larger than the type of *Argentodites*, a difference which we currently consider as individual variation. But it is possible that upon further study, those differences, plus potentially others, may justify the recognition in the future of two species of *Ferugliotherium.* Perhaps as a reaction to the rampant creation of species and taxonomic categories for single isolated teeth that characterized earlier taxonomic works, we choose to be conservative regarding diversity recognition.

The knowledge of ferugliotheriids is still limited; the association of isolated, but distinctive, dental types (incisors, blade-like premolariforms, brachyodont premolariforms, and molariforms) beyond any doubts to a single taxon will require new and better specimens. The recently reported Amphitheater 1 locality is still productive and occasionally surprises us with relatively well-preserved jaws with several teeth in place (Harper et al. 2019); this is currently the best hope to deliver the evidence needed to settle these vexing issues.

*Trapalcotherium* Rougier et al. 2009a

Type species: *Trapalcotherium matuastensis* Rougier et al. 2009a.

Included species: The type only.

*Trapalcotherium matuastensis* Rougier et al. 2009a
(Fig. 8.5f)

Holotype: MML-PV 16, lower left first molariform (Fig. 8.5f).

Locality and horizon: Cerro Tortuga, Río Negro, Argentina. Allen Formation (38 m below the top of the Allen Formation), Campanian–Maastrichtian (Upper Cretaceous).

Diagnosis (taken from Rougier et al. 2009a): Ferugliotheriid similar in size to *Ferugliotherium*, differs from the species from the Los Alamitos Formation in the occlusal figure and cusp proportions. The mesial triangle is more compressed, with a curved mesial crest in *Trapalcotherium*, while it is more open and equilateral in *Ferugliotherium*. The cusp bases are better-developed in *Ferugliotherium* than in *Trapalcotherium*, where they become part of the crests, and therefore, less differentiated as individual structures. There is a larger degree of height difference between the lingual and buccal cusp rows in *Trapalcotherium* than in *Ferugliotherium*.

Comments: Only one tooth is known for this species. Although it can be differentiated from *Ferugliotherium*, they represent similar taxa very close to each other in their morphology and coming from two stratigraphic units (Allen and Los Alamitos formations) geographically close and similarly aged. *Ferugliotherium* and *Trapalcotherium* are both included in Ferugliotheriidae; however, in recent phylogenetic analyses (Krause et al. 2014a, 2020), these two taxa do not always form a monophyletic clade; instead, *Ferugliotherium* and *Trapalcotherium* (see below) were recovered as successive sister taxa of sudamericids.

?Ferugliotheriidae Bonaparte 1986a
*Magallanodon* Goin et al. 2020

Type species: *Magallanodon baikashkenke* Goin et al. 2020.

Included species: The type only.

*Magallanodon baikashkenke* Goin et al. 2020
(Fig. 8.7)

Holotype: CPAP 5916, first lower molariform (Fig. 8.7), interpreted as of the left side by Goin et al. (2020).

Locality and horizon: Río de Las Chinas valley, Estancia Cerro Guido, about 100 km north from Puerto Natales, Última Esperanza Province, Patagonia, Chile. Dorotea Formation, upper Campanian–lower Maastrichtian (Upper Cretaceous).

Diagnosis (from Goin et al. 2020): Large-sized gondwanatherian (mesiodistal length of mf1: 8.34 mm) with the following association of features: brachyodont molariforms with four to five roots; molariforms with distinct transverse lophs, which in unworn condition are formed by multiple rounded adjacent cuspules (as in *Gondwanatherium*); mesial lower molariform crown with four main lobes, the mesial one D-shaped, the second formed by a longitudinal ridge, and the third and fourth

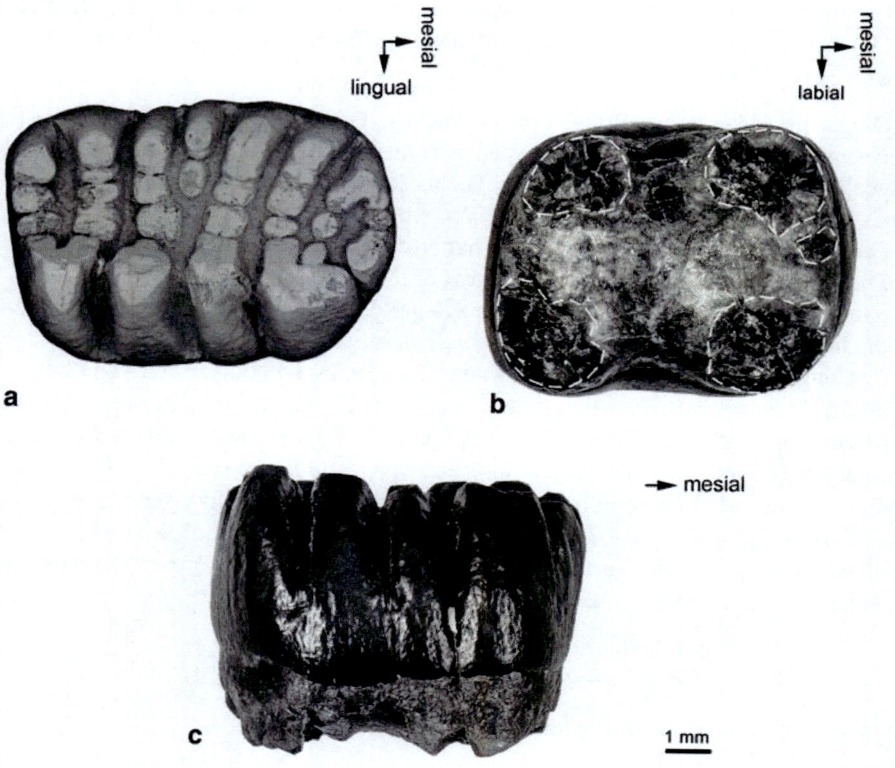


**Fig. 8.7** *Magallanodon baikashkenke* from Río de Las Chinas valley, Última Esperanza Province, Chile; Dorotea Formation, upper Campanian–lower Maastrichtian, Upper Cretaceous. Holotype CPAP 5916, first lower left molariform (mf1) in occlusal (**a**), radicular (**b**) and lingual (**c**) views. The tooth orientation follows Goin et al. (2020) that is based on interpretations made by Krause (2014) and Krause et al. (2014a) on *Vintana*

V/U-shaped in occlusal view, all lophs separated by continuous labiolingual grooves that open on the lingual wall of the crown as deep furrows; the lobes define seven cuspules on the labial margin of the crown; distalmost upper molariforms with two lobes divided by a transverse groove that starts in the labial edge and does not reach the lingual side; presence of enamel islets on the worn occlusal surface of the lobes; distalmost upper molariforms with crown distal width ¾ smaller than the mesial width.

Comments: *Magallanodon* is based on the holotype, a putative lower molariform, in addition to other referred specimens which include a partial upper incisor and two distal upper molariforms. This taxon represents the first Mesozoic mammal for Chile and the southernmost record of a gondwanatherian for this Era (Goin et al. 2020). *Magallanodon* has hypertrophied, rodent-like first upper incisor and molariforms with transverse lophs and long furrows. Based on the dental pattern described for *Vintana sertichi*, Goin et al. (2020), regarded the side of the crown with the deepest furrows in *Magallanodon* as labial in the upper molariforms; for the lower molariforms (not known in *Vintana*) the deepest furrows were considered to be the lingual. Following this criterion, the holotype was interpreted as a left lower molariform. *Vintana* and *Sudamerica* are the only two gondwanatherians with informative molariforms in situ, the bizarre *Adalatherium* is not very helpful in this regard. Goin et al. (2020) chose to use *Vintana* as a model because, as for *Magallanodon*, its molariforms are low-crowned and with distinct plesiomorphic root structure. Comparison with the available lower jaw of *Sudamerica*, which has two hypsodont molariforms, would indicate an opposite orientation for CPAP 5916: the deep furrows would be labial and the shallower ones the lingual; this would make the holotype of *Magallanodon* a right lower molariform. However, the morphology of both *Magallanodon* and *Sudamerica* are considerably different regarding overall dental morphology and occlusal pattern. Ultimately, the interpretations regarding the orientation and position (upper/lower, locus) of the isolated molariforms of ferugliotheriids and *Magallanodon* carry a measure of uncertainty due to the lack of postcanines in situ and the unstable phylogenetic relationships of these taxa within gondwanatherians. It is still possible that the holotype of *Magallanodon* is, in fact, an upper tooth; a distinct possibility we follow here for the holotype of *Ferugliotherium*. Determination of the mesiodistal orientation of CPAP 5916 was decided by the presence of a distinct interdental facet, recognized only on what is regarded as the distal face. The mesial side is more convex, lacking clear evidence of contact with another tooth, and was, therefore, interpreted as the mesialmost molariform (Goin et al. 2020). Additionally, differential wear is expected to be more pronounced on the mesial than in the distal portions of the crown; *Magallanodon* shows differential wear in the pole regarded as mesial, providing further confirmation of its orientation. The masticatory movements were inferred as palinal, based on striations left by wear in the cusps, which are roughly aligned with the tooth row. *Magallanodon* was tentatively referred to as Ferugliotheriidae due to the molariform occlusal pattern that is similar to that of *Ferugliotherium*, in conjunction with brachyodont crowns supported by transverse pairs of roots. *Magallanodon* is much larger than *Ferugliotherium*, with mesiodistal molariform length about five times larger in the Chilean taxon. At the same time, the occlusal pattern of *Magallanodon* closely resembles that of the hypsodont-toothed *Gondwanatherium.* This similarity was interpreted by Goin et al. (2020), as a result of closer phylogenetic affinities between ferugliotheriids and sudamericids (i.e., *Gondwanatherium* in particular), as originally proposed by Krause and Bonaparte (1993; see also Pascual et al. 1999; Pascual and Ortiz-Jaureguizar 2007; *contra* the diphyletic hypothesis of ferugliotheriids closer to multituberculates than to sudamericids discussed above).

More and better preserved specimens hopefully with dentition in situ of *Magallanodon*, as well as of the ferugliotheriids, will help to better understand the available material which at present consists mostly of isolated teeth for SA gondwanatherians. The relationships between the brachyodont and hypsodont gondwanatherians/multituberculates from the Late Cretaceous of Gondwana, if any, rest on the future discovery of such specimens. Chimento et al. (2020) referred an isolated tooth, discovered in the Upper Cretaceous Chorrillo Formation at the La Anita farm, Santa Cruz Province, Argentina, to *M. baikashkenke*. The specimen is tentatively identified as first upper molariform.

## *8.2.1 Gondwanatheria* Indet.

*Greniodon* Goin et al. 2012

Type species: *Greniodon sylvaticus* Goin et al. 2012.

Included species: The type only.

*Greniodon sylvaticus* Goin et al. 2012
(Fig. 8.8)

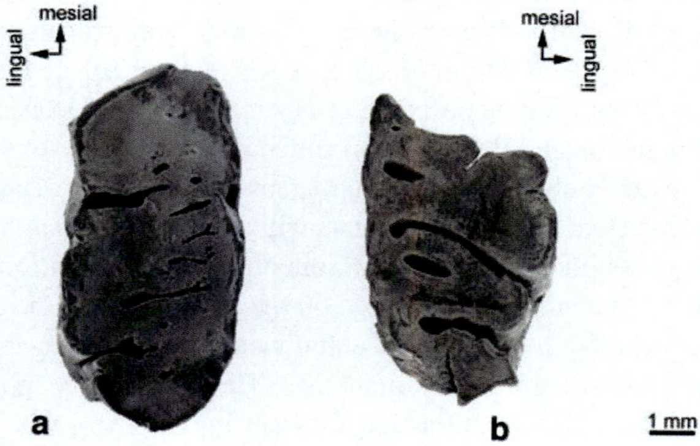


**Fig. 8.8** *Greniodon sylvaticus* from La Barda, Chubut, Argentina; Volcanic–Pyroclastic Complex, interbedded tuffs of the Andesitas Huancache Formation, earliest middle Eocene. Holotype LIEB-PV 2001, upper left molariform (?Mf3) (**a**) and LIEB-PV 2000, fragmentary upper right molariform (?Mf2) (**b**), in occlusal views, from Goin et al. (2012)

Holotype: LIEB-PV 2001, upper left third molariform (Fig. 8.8).

Locality and horizon: La Barda fossil site, Chubut Province, Argentina. Middle Chubut River Volcanic-Pyroclastic Complex, interbedded tuffs of the Andesitas Huancache Formation, earliest middle Eocene (Paso del Sapo Fauna, Ypresian–Lutetian).

Diagnosis (modified from Goin et al. 2012): Differs from Ferugliotheriidae by its larger size, mesodont to protohypsodont molariforms with better defined transverse lobes separated by deep, narrow grooves; wear in *Greniodon* is more pronounced in the central portion of the molariforms and all along its entire length. It differs from all Sudamericidae in having a much thinner enamel layer in molariforms. It differs from SA Sudamericidae in its larger size and specifically from *Sudamerica ameghinoi* in the marked asymmetry of its upper molariform grooves (exclusively lingual in *Greniodon*, both labial and lingual in *Sudamerica*). It differs from the Sudamericidae from Madagascar and India, by having a thinner enamel layer, more complex occlusal design of its molariforms, absence of perikymata (i.e., incremental growth lines on the surface of tooth enamel), and much deeper lingual grooves that divide the crown into several distinct lobes.

Comments: In addition to the holotype LIEB-PV 200, a possible Mf3, specimen LIEB-PV 2000 was referred to this taxon, corresponding to a possibly upper right molariform (?Mf2), lacking the mesialmost lobe. The incompleteness of the known specimens of *Greniodon* precludes a solid phylogenetic assignation. Goin et al. (2012) performed detailed comparisons with therians (e.g., polydolopid and abderitid marsupials, basal caviomorphs) and non-therian mammals (e.g., monotremes), they identified the strongest morphological similarities with gondwanatherians. In some regards, the hypsodonty of the Eocene *Greniodon* is similar to that of *Gondwanatherium*. However, the occlusal surface of *Greniodon* exhibits a deep longitudinal concavity (Fig. 8.8), that differs from *Gondwanatherium* and *Sudamerica* (Goin et al. 2012). According to Goin et al. (2012), the number and orientation of grooves and islets of enamels in *Greniodon* resemble the worn occlusal surface of specimen MACN-RN 248 of *Ferugliotherium*, a taxon with brachyodont molars. The resemblance of *Greniodon* with the hypsodont sudamericids on one hand, and the putative similar occlusal surface with worn out teeth of *Ferugliotherium* on the other, precludes the assignment to any of the two known groups (Goin et al. 2012). The phylogenetic analysis of Krause et al. (2014a), placed *Greniodon* within an unresolved sudamericid clade, a result mostly repeated in the recent study of the Malagasy gondwanatherian *Adalatherium* (Krause et al. 2020; Hoffmann et al. 2020). We regard this taxon as enigmatic, representing a distinctive linage, likely a gondwanatherian, poorly represented in the fossil record, and possibly relictual and exploiting a marginal environment, as seems to be the case with all the archaic lineages surviving the K/Pg extinction event.

#### 8.2.1.1 Marambio, Antarctica

Goin et al. (2006; see also Reguero et al. 2002) reported an anterior part of a left dentary with a partial rodent-like incisor (MLP 95-I-10-5) and an isolated ?left upper incisor (MLP 96-I-5-47) as belonging to cf. *Sudamerica ameghinoi*. The material comes from the locality IAA 1/90 (also known as "Ungulate site"; Marenssi et al. 1994) from the Cucullaea I Member of the La Meseta Formation, of middle Eocene age (Reguero et al. 2002), exposed at Marambio (Seymour) Island, Antarctic Peninsula. The specimens were compared with several mammalian groups (e.g., polydolopines), the authors finally concluded that the strongest similarities were shared with sudamericids, *Sudamerica ameghinoi* in particular (Goin et al. 2006). Shared features include dorsally-placed mental foramen, alveolar plane much higher than the diastema, large incisor medially flat, laterally convex, angled ventromedially, and rounded dorsally (Goin et al. 2006). These specimens constitute the southernmost record of gondwanatherians and one of the few for the Eocene. Notably, these putative gondwanatherians come from the same unit where an isolated tooth regarded as a dryolestoid was found (Martinelli et al. 2014; see Chap. 6). The presence of Eocene gondwanatherians and dryolestoids in Antarctica, and meridiolestidans surviving into the Miocene in continental SA, reinforces our view of a less-dramatic K/Pg transition in Gondwana than in Laurasia with archaic lineages that persisted during most of the Cenozoic as relictual members likely occupying highly specific and perhaps marginal environments.

### 8.2.2 *Mammalia* Incertae Sedis

*Patagonia* Pascual and Carlini 1987

Type species: *Patagonia peregrina* Pascual and Carlini 1987.

Included species: The type only.

*Patagonia peregrina* Pascual and Carlini 1987
(Fig. 8.9)

Holotype: MACN-CH 869, fragment of right mandibular ramus with three molariforms, alveolus of the ?canine/incisor, and intra-alveolar portion of the incisor (Fig. 8.9).

Locality and horizon: south portion of the Chubut River valley, near Gaiman, Chubut Province, Argentina. Sarmiento Formation, Trelew Member; lower Miocene (Colhuehuapian SALMA, Burdigalian).

Diagnosis (modified from Pascual and Carlini 1987): Small-sized mammal with hypsodont lower incisor, oval in cross section, strongly curved; incisor extends

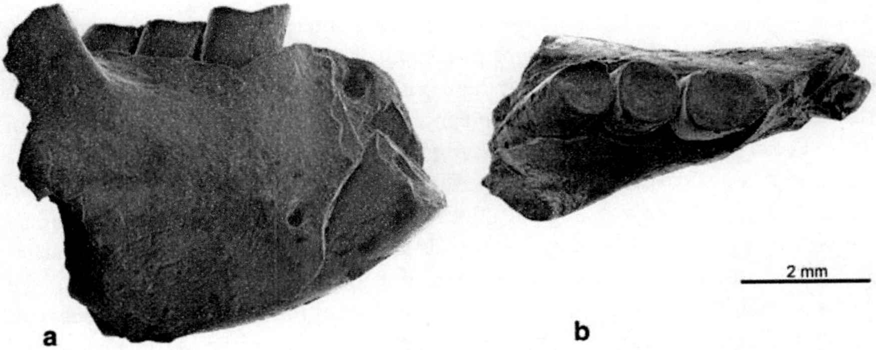


**Fig. 8.9** *Patagonia peregrina* from the Chubut River valley, near Gaiman, Chubut, Argentina; Sarmiento Formation, lower Miocene (Colhuehuapian SALMA). Holotype MACN-CH 869, partial right mandibular ramus with three molariforms, alveolus of caniniform, and intra-alveolar portion of the incisor in labial (**a**) and occlusal (**b**) views

lingually along the ventral border of the mandible to the level of the last molariform, where it forms a prominence similar to that of hystricognath rodents, but ventrally; small, lower caniniform, apparently incisor-like, closed-rooted, separated from cheek teeth by a short and crested diastema; lower molariforms oval in cross section, decreasing in size posteriorly, hypsodont, rootless, wholly surrounded by enamel and slightly curved, with the concavity forward. Short and deep horizontal ramus of dentary, increasing in height distally, particularly beneath the masseteric fossa, where the mandibular body becomes strongly convex and inflected; deep pterygoid fossa, limited ventrally by a flange-like process; strong, salient coronoid process; deep but reduced masseteric fossa, dorsally positioned with a prominent masseteric crest; unfused subvertical symphysis, with nearly smooth symphyseal surface.

Comments: In addition to the lower jaw of the holotype, nine more specimens are known including partial lower jaws and isolated teeth, three of which were tentatively referred to as upper molariforms (MACN-CH 879).

The dentary and dentition of *Patagonia* are rodentiform with highly derived features and simplified coronal pattern. The data at hand makes it challenging to decide between alternative phylogenetic positions, so it is of little surprise that its position is controversial. The dentary is extremely short and deep, the masseteric fossa is confined to a dorsal position, the coronoid process roots below the second molariform forming a salient lamina, the lingually inflected ventral border of the dentary suggests that the mandibular angle was also probably inflected. The mesial incisors are hypertrophied and rootless; this is followed by a shallow and conical alveolus for a vestigial rooted tooth, possibly a caniniform or a simple premolar. There is a short, crested diastema separating the mesial teeth from the molariforms. It lacks the height difference with the alveolar margin of the molariforms distinctive of most rodentiform jaws, including glires, multituberculates, euharamiyidans, and gondwanatherians. Molariforms are columnar and open-rooted, surrounded by enamel on all sides, and lacking lobes that could be interpreted as relicts of trigonid, talonid, or any sign of original cusps of a tribosphenic molar. The dentine forms a shallow basin surrounded by enamel; there is a deep mesiodistal groove that suggests principal propalinal jaw movements and shorter ectental stroke (Pascual and Carlini 1987). These features suggest that *Patagonia* was rodentiform in appearance.

However, what are the phylogenetic affinities of *Patagonia*? *Patagonia peregrina* was first thought to be a marsupial allocated within its own group, Patagoniidae, by monotypy, possibly related to groeberiids and argyrolagiids (Pascual and Carlini 1987). The authors emphasized the evidence of the inflected ventral border of the preserved jaw that then suggested the presence of an inflected angular process of dentary, typical of metatherians, but also reported to some degree in some Mesozoic therians and living placentals (e.g., Sánchez-Villagra and Smith 1997). Actually, the type of *Patagonia* does not preserve evidence of an inflected angle, the specimen being broken too far forward for that; it does, however, show an inflected position of the pterygoid crest, which in some taxa, like most marsupials, culminates in the angular process. Multituberculates on the other hand can have a robust pterygoid crest but lack an angular process (Kielan-Jaworowska 1971; Kielan-Jaworowska and Hurum 1997; Wible and Rougier 2000). Originally, Pascual and Carlini (1987), considered the presence of i1, c1, pm0, m3, a reduced formula that recalled groeberiids. In turn, the reduced caniniform of *Patagonia* recalled that of Epidolopidae and Prepidolopidae metatherians (Pascual and Carlini 1987). However, the homology of this tooth was not totally accepted and it was also considered as a possible second incisor (Sánchez-Villagra et al. 2000). In later studies, *Patagonia* was regarded as belonging to Argyrolagoidea or closely related to them (Koenigswald and Pascual 1990; Flynn and Wyss 1999; Goin and Abello 2013; Goin et al. 2016; Beck 2017). All these studies presuppose that the lack of evidence of the tribosphenic molar structure is obscured by wear or extensive phylogenetic derivation. However, more recently, Chimento et al. (2015), restudied the same specimens of *Patagonia* and concluded that most of its mandibular and dental features are reminiscent of non-tribosphenic gondwanatherian mammals. These similarities appear far-fetched but at present cannot be totally discounted.

Despite renewed interest, *Patagonia* remains an enigmatic mammal. Its highly apomorphic morphology prevents strongly supporting, or rejecting, any of the hypotheses advanced. As is often the case with interesting but obfuscating incomplete fossils, a better material basis is needed to move us forward from the current impasse.

#### 8.2.2.1 Santa Rosa, Peru

Goin et al. (2004) described an isolated brachyodont molariform (LACM 149371) as indeterminate Mammalia, but indicated that several details in cusps and crests resemble the pattern of the Late Cretaceous *Ferugliotherium windhauseni* (Feruglio-theriidae). LACM 149371 comes from the Santa Rosa fossil locality, Atalaya Province, Ucayali Department, Peru, from levels of the Yahuarango Formation of

?upper Eocene/lower Oligocene age. The tooth was interpreted as an upper molar, and as claimed by the authors, it is really enigmatic regarding its taxonomic assignment. In our view, the resemblances with ferugliotheriid are not as strong as purported. Although the cusp arrangement is distinct from that of other mammals, the chemical degradation of its surface (Goin et al. 2004), can preclude comparisons and obscure its morphology. Homology of the main cusps is problematic; however, a tribosphenic pattern, with a triangular occlusal outline, formed by a protocone (=E), paracone (=F), metacone (=D), and a large stylar area is viable, which would disqualify this fossil as a gondwanatherian. We regard the affinities of this tooth as uncertain.

#### 8.2.2.2 Contamana, Peru

Antoine et al. (2012: Fig. 1 and Suppl. Info) mentioned a probable sudamericid gondwanatherian, based on three fragmentary teeth (not figured), from CTA-27 locality, Loreto, near Contamana city, middle Eocene to lower Oligocene, Yahuarango Formation. Later on, Antoine et al. (2016), mentioned that one of these teeth would more likely be an upper premolar referable to a polydolopimorphian marsupial. The information on gondwanatherians remains unverified until a full description of the material is provided.

## 8.3 Concluding Remarks

Although recent findings of exquisitely preserved skeletal remains of gondwanatherians in the Late Cretaceous of Madagascar (Krause et al. 2014a, 2020), have drastically increased our knowledge of the gross anatomy of the group, the high-level phylogenetic relationships of gondwanatherians are far from settled. Different analyses yield alternative arrangements of gonwanatherians, euharamiyidans, and multituberculates (Krause et al. 2020; Hoffmann et al. 2020); within Gondwanatheria the position of ferugliotheriids, often clustering with multituberculates instead of certain gondwanatherians, is the most unstable and significant alternative. The resolution of the position of the brachyodont taxa, the presumably primitive morphology corroborated by character polarity in most trees, is likely to be crucial to determine gondwanatherian relationships among allotherians and other basal mammalian groups. The exceptional Late Cretaceous Malagasy material (Krause et al. 2014a, 2020), highlights the strangeness of gondwanatherians, making much of the anatomy more of a conundrum than a reliable guide to solve the evolutionary history of the group. The skulls of *Vintana* and *Adalatherium* show an enormous septomaxilla contacting the lacrimal, retaining features absent in mammaliaforms (such as a septomaxillary shelf and canal), a large premaxilla that, in the case of *Vintana*, contacts the palatine excluding the maxilla from the palate, retention of a fully developed primary wall of the braincase, and a very short but curved and anomalous cochlea. The dentition, the classical compass of mammalian phylogeny, is of little help in navigating gondwanatherian relationships within the broader framework of mammaliaform evolution. *Vintana* is similar to other gondwanatherians with thick enamel and fossae and pits in the occlusal surface. *Adalatherium* on the other hand is strikingly different from *Vintana*, and in fact, any other mammal described to the present (Krause et al. 2020). *Adalatherium*, originally placed in its own family Adalatheriidae (Krause et al. 2020), and later nested within a poorly resolved sudamericid clade (Hoffmann et al. 2020), shows a single rodentiform lower incisor and four cheek teeth. Krause et al. (2020) chose not to advance any homology for the four "postcanine" elements but tested the impact of different assumptions for the two most mesial molariforms. Assumptions ranged from both of them being premolar to both of them being molars. The impact on the trees was not dramatic (Krause et al. 2020). We regard here the first lower cheek tooth of *Adalatherium* as a premolar; the morphology is much simpler, dominated by a single cusp and a small talonid with three small cuspules. The mesial margin of this premolar is moderately convex and more trenchant than any other tooth. It follows that the central cusp of this premolar is the homolog of the primitive main cusp of lower molariforms, that is to say, the protoconid of triangular postcanines (Osborn 1907; Patterson 1956; Crompton 1971; Davis 2011). This mesial premolar resembles that of euharamiyidans (Bi et al. 2014; Luo et al. 2017; Mao and Meng 2019; Meng et al. 2019), in the trenchant nature of the mesial half and the talonid-like basin developed distally.

Despite the strange morphology of the more distal molariforms in *Adalatherium*, it appears certain that the mesial cusp in Krause et al. (2020), is homologous with the protoconid of the mesialmost premolar. The three cusps of the talonid basin in the mesial premolar are also arranged in a similar pattern in the more distal premolariforms and can be considered homologous. The peculiar mesiolabial basin is a secondary structure developed at cingular height that derives from the primitive diamond-shaped crown of the tooth. A dominant mesiolingual cusp in the lower molariforms is present in haramiyidans, like *Haramiyavia* (Jenkins et al. 1997; Luo et al. 2015) and euharamiyidans, like *Vilevolodon* (Luo et al. 2017), *Qishou* (Mao and Meng 2019), among others (e.g., Bi et al. 2014). The remaining molariforms are quite different and it is hard to attempt to further speculate regarding homologies. The same logic can be used to reach similar conclusions regarding the upper molariform pattern. Where known gondwanatherians have four lower molariforms (except, possibly, for *Ferugliotherium*; but see also Pascual et al. 1999 and Pascual and Ortiz-Jaureguizar 2007), and this is not the primitive number for multituberculates, in which basal taxa may have six lower postcanines. Euharamiyidans on the other hand seems to have a single premolariform and two molariforms (Krause et al. 2020), for a total number of three postcanines. Any of these cheek teeth counts are problematic for gondwanatherians.

We think it is reasonable to assume the first lower postcanine of *Adalatherium* as a premolar and the two most distal postcanines as molars; the second postcanine is transitional in morphology and the case is not clear cut. Therefore, there are two viable lower postcanine dental formula for *Adalatherium*, p2, m2 or p1, m3. If *Adalatherium* is the sister group to Sudamericidae or a basal Sudamericidae (Krause et al. 2020;

Hoffmann et al. 2020), the most mesial of the lower molariforms in *Sudamerica* and supposedly in *Gondwanatherium* (e.g., Bonaparte 1986c; Krause and Bonaparte 1993; Koenigswald et al. 1999; Pascual et al. 1999; Gurovich 2008; Gurovich and Beck 2009), must derive from a simple premolariform and almost certainly should be a premolar. Further, complicating this issue is the strong advocacy for gondwanatherian affinities of *Ferugliotherium* and allies (Krause and Bonaparte 1990; Krause et al. 1992, 2014a, but see Krause et al. 2020 for mixed results and Goin et al. 2020). We are confident that the distinct plagiaulacoid premolars recognized in the Los Alamitos and La Colonia formations (Kielan-Jaworowska and Bonaparte 1996; Kielan-Jaworowska et al. 2007), belong to a ferugliotheriid that likely has only two lower molariforms (see *Ferugliotherium* section above; but see also Pascual et al. 1999; Pascual and Ortiz-Jaureguizar 2007; Goin et al. 2020). It seems almost certain that the most mesial molariform of sudamericids should derive from a premolar and if ferugliotheriids (i.e., with plagiaulacoid tooth) are related to gondwanatherians, likely the two most mesial elements. *Ferugliotherium* as a sister group of gondwanatherians implies that it has lost a premolar mesial to the plagiaulacoid tooth, unless the four postcanines of gondwanatherians are explained by a de novo acquisition of a mesial premolar, an unlikely situation known to occur in other mammalian groups, but not among allotherians (Rougier et al. 2016; Csiki-Sava et al. 2018; Wible et al. 2019). The presence of a plagiaulacoid tooth in the antepenultimate position in *Ferugliotherium* implies it is homologous with the antepenultimate tooth of gondwanatherians; given that plagiaulacoid teeth are universally regarded as premolars, it follows that the two most mesial teeth in gondwanatherians should be premolar, or premolariform, in origin. Originating the second molariform of gondwanatherians from a plagiaulacoid tooth similar to that of *Ferugliotherium* and multituberculates is a major and unprecedented transformation for which we have at present no evidence, weakening the putative links between ferugliotheriids and gondwanatherians. Presence of more than two molariforms in *Ferugliotherium* would weaken this argument. This is a problem likely to be resolved by newly collected material in La Colonia Formation.

If *Ferugliotherium* is removed from the equation and *Adalatherium* is placed as a basal Sudamericidae (Hoffmann et al. 2020), the transitional nature of the second postcanine in *Adalatherium* argues in favor of p2, m2 as basal for gondwanatherians. As recognized by Pascual et al. (1999), this is a prospect involving radical morphological changes, a corollary which still rings true, regardless of the tree topology recovered.

Among gondwanatherians, *Sudamerica* and *Gondwanatherium* have a relatively compressed first lower molariform with regard to more posterior molariforms; in the case of *Gondwanatherium* this tooth is also relatively simple (Bonaparte 1986c: Fig. 1b; Gurovich 2008). Specimen MACN-RN 23 (Fig. 8.2d) of *Gondwanatherium* (Bonaparte 1986c) shows a low mesial cusp with minor wear, a main, larger central cusp, followed distally by another smaller cusp less worn than the central one and a ringlet of lower and smaller conical cusps connecting the central cusp and the more distal cusp (Fig. 8.2d). Serial homology suggests that the three main cusps of the MACN-RN 23 are homologous with the three lophs of distal molariforms.

*Sudamerica* (MPEF-CH 534; Pascual et al. 1999) demonstrates that the triangular occlusal figure is mesial in the first molariform, with a single buccal cusp (or vertex), and two lingual cusps or vertices. Accepting the attributions, positions, and the recent phylogenetic framework, a few tentative conclusions can be drawn from these specimens. Basal gondwanatherians have relatively simple mesial premolariforms dominated by a central main cusp and a relatively broad multicuspated talonid; more derived taxa like *Gondwanatherium* and *Sudamerica* further elaborate that central cusp and talonid into the middle and distal lophs. The mesial, triangular loph has a single cusp of buccal position; this mesial portion of the tooth is absent in *Adalatherium*, present as a single cusp in *Gondwanatherium*, and as a full-fledged triangle in *Sudamerica*. The main row of cusps is lingual in the lower dentition of *Adalatherium*, and by inference in the left mesial molariform MACN-RN 23 of *Gondwanatherium* and the *Sudamerica* jaw MPEF-CH 534. The neomorphic portion of the molariforms is buccal in gondwanatherians; the mesiobuccal basin of *Adalatherium* would be autapomorphic. The distolingual cusp (or vertex) of the mesial occlusal triangle of the first postcanine of *Sudamerica* is homologous to the single mesial cusp of *Gondwanatherium* MACN-RN 23 and it is absent in the first postcanine of *Adalatherium*. Likewise, the isometric distal postcanine outline of *Adalatherium* is primitive, with the elongation of the crowns seen in *Vintana* and in the remaining Sudamericidae achieved by developing the neomorphic mesial triangle.

The proposals above rest mainly on the tree topology being basically correct regarding *Adalatherium* as a closely related or basal sudamericid, correct identification of the main (plesiomorphic) cusps of *Adalatherium*, and determination of MACN-RN 23 as a lower mesial premolariform of *Gondwanatherium*. Other inferences, such as serial homology and positional extrapolation are in our view less problematic but certain to introduce errors if confidence is misplaced. A set of similar assumptions could be predicated for the upper dentition, but the lack of maxillary teeth in place for the SA gondwanatherians makes such an attempt even more daring. Uncharacteristically, we will choose discipline over temptation and refrain from developing this further. A single skull, a maxilla with a few teeth, or even some unworn molariforms could easily refute or support our ideas; until then, we wait.

An alternative interpretation could be proposed based on the postcanine dentition of *Vintana* (Krause 2014). The upper molariforms have deep furrows in the labial side of the crown. Consequently, it is inferred that in the lower postcanines the furrows are in the lingual side, opposed to the uppers as occurs in caviomorph rodents. The molariforms of *Magallanodon* were interpreted in this way by Goin et al. (2020), and also extrapolated to ferugliotheriids and *Gondwanatherium* (see Goin et al. 2020). If the jaw of the hypsodont *Sudamerica* is used instead as a model, the opposite orientation for CPAP 5916 becomes likely: the deep furrows would be labial and the shallower ones lingual. As a consequence, this would make the holotype of *Magallanodon* a right lower molariform. Nonetheless, in all the brachyodont-toothed taxa (Argentinean ferugliotheriids and *Magallanodon*) the orientation and position (upper/lower and locus location) is likely uncertain, as well as the relationships of these taxa with the remaining gondwanatherians.

The geographical and temporal distribution of gondwanatherians is also challenging. Ghost lineages suggest a minimal Middle Jurassic age for the origin of euharamiyidans, multituberculates, and gondwanatherians; what biogeographical distribution can we expect from that? Are gondwanatherians Gondwanic or Pangeic in origin? Have they certainly reached the mid-Cenozoic? Is the cranial and postcranial anatomy of *Vintana*/*Adalatherium* what we should expect for SA sudamericids? How closely related are the mid to large-sized sudamericids to the small-sized ferugliotheriids?

Despite that the first known gondwanatherians (*Sudamerica ameghinoi*) were collected from Cenozoic levels of Patagonia, Argentina (Salamanca Formation, Paleocene; Scillato-Yané and Pascual 1985; Gurovich 2008) and that few and sparse findings were reported through the Paleogene (e.g., middle Eocene *Greniodon sylvaticus* from Patagonia and unnamed taxon from the Antarctic Peninsula; Goin et al. 2006, 2012), the group certainly diversified during the Mesozoic and throughout Gondwana, including findings in India, Madagascar, and perhaps Africa (Krause et al. 1997, 2003, 2014a, 2020; Prasad et al. 2007; Verma et al. 2012; O'Connor et al. 2019). The SA Mesozoic record of gondwanatherians/multituberculates includes *Gondwanatherium patagonicum* (Bonaparte 1986c, 1988, 1990), *Ferugliotherium windhauseni* (Bonaparte 1986a, b, 1990; Krause et al. 1992; Krause 1993; Kielan-Jaworowska and Bonaparte 1996; Kielan-Jaworowska et al. 2007; Pascual and Ortiz-Jaureguizar 2007), and *Trapalcotherium matuastensis* (Rougier et al. 2009a). Other taxa claimed to be gondwanatherians, such as *Patagonia peregrina* from the early Miocene of Patagonia, provide inconclusive phylogenetic views with the material at hand (*contra* Chimento et al. 2015), as do some records from the Paleogene of the Neotropics (Campbell et al. 2004; Goin et al. 2004; Antoine et al. 2012, 2016).

With the exception of a few lower jaws, all remaining records of gondwanatherians in SA are based on isolated teeth, most of which were collected by picking in the field and/or screen washing techniques (Chap. 1). So far, postcranial material from SA gondwanatherians is unknown, resting all the available evidence on the recently discovered skeleton of *Adalatherium hui* from Madagascar (Krause et al. 2020).

To elucidate the ecological role of the group in the terrestrial scenarios from past SA ecosystems is another pending project. During the Mesozoic, the Late Triassic traversodontid cynodont *Menadon besairiei* constitutes the oldest evidence of hypsodonty (Melo et al. 2019). During the Jurassic, the mammaliaform *Fruitafossor windscheffeli* also has purported hypsodont cheek teeth (Luo and Wible 2005). By the Cretaceous, hypsodont cheek teeth are present only among gondwanatherians: the enigmatic *Galulatherium* (Turonian–Campanian, Tanzania; O'Connor et al. 2019) and sudamericid gondwanatherians (e.g., Bonaparte 1986c, 1988, 1990; Krause et al. 1997, 2014a; Krause 2013; Verma et al. 2012). Sudamericids are, to date, the oldest mammals with hypsodont dentition in SA (Bonaparte 1986c; Gurovich 2008).

Sudamericids played a relevant role in the Late Cretaceous assemblages of SA as primary consumers. The high-crowned cheek teeth were likely related to the origin of grasses and grasslands in southern continents (Prasad et al. 2005; Wilson et al. 2007; Strömberg 2011). Grasses (i.e., Poaceae) have phytoliths (plant silica bodies) that are highly abrasive for dentition, and hypsodonty maintains functional teeth during the life of the animal. During the Cenozoic, hypsodonty appeared several times in different SA mammalian lineages including SA native ungulates, xenarthrans, rodents, and some metatherians (e.g., Madden 2014). Hypsodonty is particularly developed among caviomorph rodents, first recorded in SA by the late middle Eocene (Antoine et al. 2012; Assemat et al. 2019), and diversified extraordinarily in the continent. Considering that caviomorphs and gondwanatherians may have had similar ecological requirements with comparable body sizes, it is tempting to suggest that caviomorphs replaced taxonomically gondwanatherians during the Cenozoic, perhaps similar to the scenario suggested for multituberculates and rodents in Laurasia (Krause 1986; see also Chap. 10).

## References


Antoine PO, Marivaux L, Croft DA, Billet G, Ganerød M, Jaramillo C, Martin T, Orliac MJ, Tejada-Lara J, Altamirano AJ, Duranthon F, Fanjat G, Rousse S, Salas-Gismondi RS (2012) Middle Eocene rodents from Peruvian Amazonia reveal the pattern and timing of caviomorph origins and biogeography. Proc R Soc Lond B 279:1319–1326

Antoine PO, Abello MA, Adnet S, Sierra AJA, Baby P, Billet G, Boivin M, Calderon Y, Candela AM, Chabain J, Corfu F, Croft DA, Ganerod M, Jaramillo C, Klaus S, Marivaux L, Navarrete RE, Orliac MJ, Parra F, Perez ME, Pujos F, Rage JC, Ravel A, Robinet C, Roddaz M, Tejada-Lara JV, Velez-Juarbe J, Wesselingh FP, Salas-Gismondi RS (2016) A 60-million-year Cenozoic history of western Amazonian ecosystems in Contamana, eastern Peru. Gondwana Res 31:30–59

Assemat A, Boivin M, Marivaux L, Pujos F, Benites-Palomino A, Salas-Gismondi R, Tejada-Lara JV, Varas-Malca RM, Negri FR, Ribeiro AM, Antoine P-O (2019) Restes inédits de rongeurs caviomorphes du Paléogène de la région de Juanjui (Amazonie Péruvienne): Systématique, implications macro-évolutives et biostratigraphiques. Geodiversitas 41:699–730

Beck RMD (2017) The skull of *Epidolops ameghinoi* from the Early Eocene Itaboraí Fauna, Southeastern Brazil, and the affinities of the extinct marsupialiform Order Polydolopimorphia. J Mammal Evol 24:373–414

Bi S, Wang Y, Sheng X, Meng J (2014) Three new Jurassic euharamiyidan species reinforce early divergence of mammals. Nature 514:579–584

Bonaparte JF (1986a) Sobre *Mesungulatum houssayi* y nuevos mamíferos cretácicos de Patagonia. In: 4° Congreso Argentino de Paleontología y Bioestratigrafía, Mendoza, Actas, vol 2, pp 48–61

Bonaparte JF (1986b) History of the terrestrial Cretaceous vertebrates of Gondwana. In: 4°Congreso Argentino de Paleontología y Bioestratigrafía, Mendoza, Actas, vol 2, pp 63–95

Bonaparte JF (1986c) A new and unusual Late Cretaceous mammal from Patagonia. J Vertebr Paleontol 6:264–270

Bonaparte JF (1987) The Late Cretaceous fauna of Los Alamitos, Patagonia, Argentina. Part VIII—The mammals. In: Bonaparte JF (ed) The Late Cretaceous fauna of Los Alamitos, Patagonia, Argentina. Rev Mus Argent Cienc Nat "B Rivadavia", Paleontología, vol 3, pp 163–169

Bonaparte JF (1988) Specialized dentitions in two Cretaceous mammals from Patagonia. In: Russell DE, Santoro J-P, Sigogneau-Russell D (eds) Teeth revisited. Proceedings of the VIIth international symposium on dental morphology, Paris, 1986. Mém Mus Hist Nat Paris (série C), vol 53, pp 207–215

Bonaparte JF (1990) New Late Cretaceous mammals from the Los Alamitos Formation, northern Patagonia. Natl Geogr Res 6:63–93

Bonaparte JF (1994) Approach to the significance of the Late Cretaceous mammals of South America. Berlin Geowiss Abt E 13:31–44

Bonaparte JF (2017) Notas sobre el origen de los gliptodontes. PE-APA 18(2):13R

Bonaparte JF, Pascual R (1987) Los mamíferos (Eotheria, Allotheria y Theria) de la Formación Los Alamitos, Campaniano de Patagonia, Argentina. In: 4° Congreso Latinoamericano de Paleontología, Santa Cruz de la Sierra, vol 1, pp 361–378

Bonaparte JF, Van Valen LM, Kramarz A (1993) La fauna local de Punta Peligro, Paleoceno Inferior, de la Provincia de Chubut, Patagonia, Argentina. Evol Monogr 14:1–61

Butler PM (2000) Review of the early allotherian mammals. Acta Palaeontol Pol 45:317–342

Campbell KE, Frailey CD, Romero-Pittman L (2004) The Paleogene Santa Rosa local fauna of Amazonian Peru: Geographic and geologic setting. In: Campbell KE (ed) The Paleogene mammalian fauna of Santa Rosa, Amazonian Peru. Nat Hist Mus Los Angeles County, Science Series, vol 40, pp 3–14

Chimento NR, Agnolín FL, Novas F (2015) The bizarre 'metatherians' *Groeberia* and *Patagonia*, late surviving members of gondwanatherian mammals. Hist Biol 27:603–623

Chimento NR, Agnolín FL, Tsuihiji T, Manabe M, Novas FE (2020) New record of a Mesozoic gondwanatherian mammaliaform from Southern Patagonia. Sci Nat 107:49

Cope ED (1884) The Tertiary Marsupialia. Am Naturalist 18:686–697

Crompton AW (1971) The origin of the tribosphenic molar. Kermack DM, Kermack KA (eds) Early mammals. Zool J Linn Soc 50(suppl 1):65–87

Csiki-Sava Z, Vremir M, Meng J, Brusatte SL, Norell MA (2018) Dome-headed, small-brained island mammal from the Late Cretaceous of Romania. Proc Natl Acad Sci USA 115:4857–4862

Davis B (2011) Evolution of the tribosphenic molar pattern in early mammals, with comments on the "dual-origin" hypothesis. J Mammal Evol 18:227–224

Flynn J, Wyss A (1999) New marsupials from the Eocene-Oligocene transition of the Andean Main Range, Chile. J Vertebr Paleontol 19:533–549

Goin FJ, Abello MA (2013) Los Metatheria sudamericanos de comienzos del Neógeno (Mioceno Temprano, Edad Mamífero Colhuehuapense): Microbiotheria y Polydolopimorphia. Ameghiniana 50:51–78

Goin FJ, Vieytes EC, Vucetich MG, Carlini AA, Bond M (2004) Enigmatic mammal from the Paleogene of Peru. In: Campbell Jr KE (ed) The Paleogene Mammalian Fauna of Santa Rosa, Amazonian Peru. Nat Hist Mus Los Angeles County, Science Series, vol 40, pp 145–153

Goin FJ, Reguero MA, Pascual R, von Koenigswald W, Woodburne MO, Case JA, Marenssi SA, Vieytes EC, Vizcaíno SF (2006) First gondwanatherian mammal from Antarctica. In: Francis JE, Pirrie D, Crame JA (eds) Cretaceous–Tertiary high-latitude paleoenvironments, James Ross Basin, Antarctica, Spec Publ Geol Soc London, London, pp 145–161

Goin FJ, Tejedor MF, Chornogubsky L, López GM, Gelfo JN, Bond M, Woodburne MO, Gurovich Y, Reguero M (2012) Persistence of a Mesozoic, non-therian mammalian lineage (Gondwanatheria) in the mid-Paleogene of Patagonia. Naturwissenschaften 99:449–463

Goin FJ, Woodburne MO, Zimicz AN, Martin GM, Chornogubsky L (2016) A brief history of South American metatherians: evolutionary contexts and intercontinental dispersals. Springer, Dordrecht

Goin FJ, Martinelli AG, Soto-Acuña S, Vieytes EC, Manríquez LME, Fernández RA, Pino JP, Trevisa C, Kaluza J, Reguero MA, Leppe M, Ortiz H, Rubilar-Rogers D, Vargas AO (2020) First Mesozoic mammal from Chile: the southernmost record of a Late Cretaceous gondwanatherian. Bol Mus Nac Hist Nat Chile 69:5–31

Gurovich Y (2005) Bio-evolutionary aspects of Mesozoic mammals: description, phylogenetic relationships and evolution of the Gondwanatheria (Late Cretaceous and Paleocene of Gondwana). Unpublished PhD thesis, Universidad Nacional de Buenos Aires

Gurovich Y (2008) Additional specimens of sudamericid (Gondwanatheria) mammals from the Early Paleocene of Argentina. Palaeontology 51:1069–1089

Gurovich Y, Beck R (2009) The phylogenetic affinities of the enigmatic mammalian clade Gondwanatheria. J Mammal Evol 16:25–49

Hahn G, Sigogneau-Russell D, Wouters G (1989) New data on Theroteinidae—their relations with Paulchoffatiidae and Haramiyidae. Geol Paleontol 23:205–215

Han G, Mao F-Y, Bi S-D, Wang Y-Q, Meng J (2017) A Jurassic gliding euharamiyidan mammal with an ear of five auditory bones. Nature 551:451–456

Harper T, Parras A, Rougier GW (2019) *Reigitherium* (Meridiolestida, Mesungulatoidea) an enigmatic Late Cretaceous mammal from Patagonia, Argentina: morphology, affinities, and dental evolution. J Mammal Evol 26:447–478

Hoffmann S, O'Connor PM, Kirk EC, Wible JR, Krause DW (2014) Endocranial and inner ear morphology of *Vintana sertichi* (Mammalia, Gondwanatheria) from the Late Cretaceous of Madagascar. In: Krause DW (ed) *Vintana sertichi* (Mammalia, Gondwanatheria) from the Late Cretaceous of Madagascar. Soc Vert Paleont, Mem 14, pp 110–136

Hoffmann S, Beck RMD, Wible JR, Rougier GW, Krause DW (2020). Phylogenetic placement of *Adalatherium hui* (Mammalia, Gondwanatheria) from the Late Cretaceous of Madagascar: implications for allotherian relationships. In: Krause DW (ed) *Adalatherium hui* (Mammalia, Gondwanatheria) from the Late Cretaceous of Madagascar. Soc Vert Paleont, Mem 21, pp 213–234

Huttenlocker AK, Grossnickle DM, Kirkland JI, Schultz JA, Luo Z-X (2018) Late-surviving stem mammal links the lowermost Cretaceous of North America and Gondwana. Nature 558:108–112

Jenkins Jr FA, Gatesy SM, Shubin NH, Amaral WW (1997) Haramiyids and Triassic mammalian evolution. Nature 385:715–718

Kielan-Jaworowska Z (1971) Skull structure and affinities of the Multituberculata: results of the Polish-Mongolian Palaeontological Expeditions, Part III. Palaeontol Pol 25:5–41

Kielan-Jaworowska Z, Bonaparte JF (1996) Partial dentary of a multituberculate mammal from the Late Cretaceous of Argentina and its taxonomic implications. Rev Mus Argent Cienc Nat "B Rivadavia", ns 145:1–9

Kielan-Jaworowska Z, Hurum JH (1997) Djadochtatheria—a new Suborder of multituberculate mammals. Acta Palaeontol Pol 42:201–242

Kielan-Jaworowska Z, Cifelli RL, Luo Z-X (2004) Mammals from the age of dinosaurs. Origins, evolution, and structure. Columbia University Press, New York

Kielan-Jaworowska Z, Ortiz-Jaureguizar E, Vieytes C, Pascual R, Goin FJ (2007) First ?cimolodontan multituberculate mammal from South America. Acta Palaeontol Pol 52:257–262

Kirk EC, Hoffmann S, Kemp AD, Krause DW, O'Connor PM (2014) Sensory anatomy and sensory ecology of *Vintana sertichi* (Mammalia, Gondwanatheria) from the Late Cretaceous of Madagascar. In: Krause DW (ed) *Vintana sertichi* (Mammalia, Gondwanatheria) from the Late Cretaceous of Madagascar. Soc Vert Paleont, Mem 14, pp 203–222

von Koenigswald W, Pascual R (1990) The Schmelzmuster of the Paleogene South American rodentlike marsupials *Groeberia* and *Patagonia* compared to rodents and other Marsupialia. Paläontol Zeit 64:345–358

von Koenigswald W, Krause DW (2014) Enamel microstructure of *Vintana sertichi* (Mammalia, Gondwanatheria) from the Late Cretaceous of Madagascar. In: Krause DW (ed) *Vintana sertichi* (Mammalia, Gondwanatheria) from the Late Cretaceous of Madagascar. Soc Vert Paleont, Mem 14, pp 166–181

von Koenigswald W, Goin FJ, Pascual R (1999) Hypsodonty and enamel microstructure in the Paleocene gondwanatherian mammal *Sudamerica ameghinoi*. Acta Paleontol Pol 44:263–300

Krause DW (1986) Competitive exclusion and taxonomic displacement in the fossil record: the case of rodents and multituberculates in North America. In: Flanagan KM, Lillegraven JA (eds) Vertebrates, phylogeny and philosophy. University of Wyoming Contributions to Geology, Spec Pap 3, pp 119–130

Krause DW (1993) *Vucetichia* (Gondwanatheria) a junior synonym of *Ferugliotherium* (Multituberculata). J Paleontol 67:321–324

Krause DW (2013) Gondwanatheria and ?Multituberculata (Mammalia) from the Late Cretaceous of Madagascar. Can J Earth Sci 50:324–340

Krause DW (2014) Dental morphology of *Vintana sertichi* (Mammalia, Gondwanatheria) from the Late Cretaceous of Madagascar. In: Krause DW (ed) *Vintana sertichi* (Mammalia, Gondwanatheria) from the Late Cretaceous of Madagascar. Soc Vert Paleont, Mem 14, pp 137–165

Krause DW, Bonaparte JF (1990) The Gondwanatheria, a new suborder of Multituberculata from South America. J Vertebr Paleont 10(Suppl):31A

Krause DW, Bonaparte JF (1993) Superfamily Gondwanatherioidea: a previously unrecognized radiation of multituberculate mammals in South America. Proc Natl Acad Sci USA 90:9379–9383

Krause DW, Kielan-Jaworowska Z, Bonaparte JF (1992) *Ferugliotherium* Bonaparte, the first known multituberculate from Gondwanaland. J Vertebr Paleont 12:351–376

Krause DW, Prasad GVR, von Koenigswald W, Sahni A, Grine FE (1997) Cosmopolitanism among Gondwanan Late Cretaceous mammals. Nature 390:504–507

Krause DW, Gottfried MD, O'Connor PM, Roberts EM (2003) A Cretaceous mammal from Tanzania. Acta Palaeont Pol 48:321–330

Krause DW, Hoffmann S, Wible JR, Kirk EC, Schultz JA, von Koenigswald W, Groenke JR, Rossie JB, O'Connor PM, Seiffert ER, Dumont ER, Holloway WL, Rogers RR, Rahantarisoa RJ, Kemp AD, Andriamialison H (2014a) First cranial remains of a gondwanatherian mammal reveal remarkable mosaicism. Nature 515:512–517

Krause DW, Wible JR, Hoffmann S, Groenke JR, O'Connor PM, Holloway WL, Rossie JB (2014b). Craniofacial morphology of *Vintana sertichi* (Mammalia, Gondwanatheria) from the Late Cretaceous of Madagascar. In: Krause DW (ed) *Vintana sertichi* (Mammalia, Gondwanatheria) from the Late Cretaceous of Madagascar. Soc Vert Paleont, Mem 14, pp 14–109

Krause DW, Roger RR, Rahantarisoa LJ, Groenke JR, Andriamialison H (2014c). Introduction, systematic paleontology, and geological context of *Vintana sertichi* (Mammalia, Gondwanatheria) from the Late Cretaceous of Madagascar. In: Krause DW (ed) *Vintana sertichi* (Mammalia, Gondwanatheria) from the Late Cretaceous of Madagascar. Soc Vert Paleont, Mem 14, pp 4–13

Krause DW, Sertich JJW, O'Connor PM, Curry Rogers K, Rogers RR (2019) The Mesozoic biogeographic history of Gondwanan terrestrial vertebrates: insights from Madagascar's fossil record. Annu Rev Earth Planet Sci 47:519–553

Krause DW, Hoffmann S, Hu Y, Wible JR, Rougier GW, Kirk EC, Groenke JR, Rogers RR, Rossie JB, Schultz JA, Evans AR, von Koenigswald W, Rahantarisoa LJ (2020) Skeleton of Cretaceous mammal from Madagascar reflects long-term insularity. Nature 581:421–427

Legendre S (1989) Les communautés de mammirères du Palèogène (Eocéne Supérieur et Oligocène) d'Europe occidentale: structures, milieux et évolution. Münchner Geowiss Abh (A) 16:1–110

Linnaeus C (1758) Systema naturae per regna tria naturae, secundum classes, ordines, genera, species, cum characteribus, differentiis, synonymis, locis, vol 1: Regnum animale. Editio decima, reformata. Laurentii Salvii, Stockholm

Luo Z-X, Wible JR (2005) A Late Jurassic digging mammal and early mammalian diversification. Science 308:103–107

Luo Z-X, Gates SM, Jenkins FA Jr, Amaral WW, Shubin NH (2015) Mandibular and dental characteristics of Late Triassic mammaliaform *Haramiyavia* and their ramifications for basal mammal evolution. Proc Natl Acad Sci USA 112:E7101–E7109

Luo X-Z, Meng Q-J, Grossnickle DM, Lui D, Neander AI, Zhang YG, Ji Q (2017) New evidence for mammaliaform ear evolution and feeding adaptation in a Jurassic ecosystem. Nature 548:326–329

Madden RH (2014) Hypsodonty in mammals. Evolution, geomorphology, and the role of earth surface processes. Cambridge University Press, United Kingdom

Mao F-Y, Meng J (2019) A new haramiyidan mammal from the Jurassic Yanliao Biota and comparisons with other haramiyidans. Zool J Linn Soc 186:529–552

Marenssi SA, Reguero MA, Santillana SN, Vizcaíno SF (1994) Eocene land mammals from Seymour Island, Antarctica: palaeobiological implications. Antarctic Sci 6:3–15

Marsh OC (1880) Notice on Jurassic mammals representing two new orders. Am J Sci 20:235–239

Martinelli AG, Chornogubsky L, Abello A, Goin FJ, Reguero M (2014) The first non-therian dryolestoid from Antarctica. In: 2014 SCAR Open Science Conference, Auckland, Abstracts, p 432

McDermott B, Hunter J, Alroy J (2002) Estimating body mass of multituberculate mammals. J Vertebr Paleont 22(Suppl):86A

McKenna MC (1980) Early history and biogeography of South America's extinct land mammals. In: Ciochon RL, Chiarelli AB (eds) Evolutionary biology of the new world monkeys and continental drift. Plenum Press, Nueva York, pp 43–77

Melo TP, Ribeiro AM, Martinelli AG, Soares MB (2019) Earliest evidence of molariform hypsodonty in a Triassic stem-mammal. Nat Commun 10:2841

Meng J, Mao F, Han G, Zheng X-T, Wang X-L, Wang Y (2019) A comparative study on auditory and hyoid bones of Jurassic euharamiyidans and contrasting evidence for mammalian middle ear evolution. J Anat 236:50–71

Mones A (1987) Gondwanatheria, un nuevo Orden de mamíferos sudamericanos (Mammalia: Edentata: ?Xenarthra). Comun Paleontol Mus Hist Nat Montevideo 18:237–240

O'Connor PM, Krause DW, Stevens NJ, Groenke JR, MacPhee RDE, Kalthoff DC, Roberts EM (2019) A new mammal from the Turonian-Campanian (Upper Cretaceous) Galula Formation, southwestern Tanzania. Acta Palaeont Pol 64:65–84

Osborn HF (1907) Evolution of the mammalian molar teeth, to and from the triangular type: including collected and revised researches trituberculy and new sections on the forms and homologies of the molar teeth in the different orders of mammals. In: Gregory WK (ed) Biological Studies and Addresses, vol. I. The MacMillan Company, London, pp 1–250

Pascual R, Carlini AA (1987) A new superfamily in the extensive radiation of South American Paleogene marsupials. Field Zool 39:99–110

Pascual R, Ortiz-Jaureguizar E (2007) The Gondwanan and South American episodes: two major moments in the history of South American mammals. J Mammal Evol 14:75–137

Pascual R, Goin FJ, Carlini AA (1994) New data on the Groeberiidae: unique Late Eocene-Early Oligocene South American marsupials. J Vertebr Paleontol 14:247–259

Pascual R, Goin FJ, Krause DW, Ortiz-Jaureguizar E, Carlini AA (1999) The first gnathic remains of *Sudamerica*: implications for gondwanathere relationships. J Vertebr Paleontol 19:373–382

Patterson B (1956) Early Cretaceous mammals and the evolution of mammalian molar teeth. Fieldiana Geol 13:1–105

Prasad V, Stromberg CAE, Alimohammadian H, Sahni A (2005) Dinosaur coprolites and the early evolution of grasses and grazers. Science 310:1177–1180

Prasad GVR, Verma O, Sahni A, Krause DW, Khosla A, Parmar V (2007) A new Late Cretaceous gondwanatherian mammal from Central India. Proc Indian Natl Sci Acad 73:17–24

Reguero MA, Marenssi SA, Santillana SN (2002) Antarctic Peninsula and South America (Patagonia) Paleogene terrestrial faunas and environments: biogeographic relationships. Palaeogeogr Palaeoclimatol Palaeoecol 179:189–210

Reig OA (1981) Teoría del origen y desarrollo de la fauna de mamíferos de América del Sur. Monogr Nat 1:1–162

Rougier GW, Chornogubsky L, Casadio S, Paéz Arango N, Giallombardo A (2009a) Mammals from the Allen Formation, Late Cretaceous, Argentina. Cretac Res 30:223–238

Rougier GW, Forasiepi AM, Hill RV, Novacek MJ (2009b) New mammalian remains from the Late Cretaceous La Colonia Formation, Patagonia, Argentina. Acta Palaeontol Pol 54:195–212

Rougier GW, Sheth AS, Spurlin BK, Bolortsetseg M, Novacek MJ (2016) Craniodental anatomy of a new Late Cretaceous multituberculate mammal from Udan Sayr, Mongolia. Palaeontol Pol 67:197–248

Sánchez-Villagra MR, Smith KK (1997) Diversity and evolution of the marsupial mandibular angular process. J Mammal Evol 4:119–144

Sánchez-Villagra MR, Kay RF, Anaya-Daza F (2000) Cranial anatomy and palaeobiology of the Miocene marsupial *Hondalagus altiplanensis* and a phylogeny of argyrolagids. Palaeontology 43:287–301

Schultz JA, Krause DW, von Koenigswald W, Dumont ER (2014) Dental function and diet of *Vintana sertichi* (Mammalia, Gondwanatheria) from the Late Cretaceous of Madagascar. In: Krause DW (ed) *Vintana sertichi* (Mammalia, Gondwanatheria) from the Late Cretaceous of Madagascar. Soc Vert Paleont, Mem 14, pp 181–202

Scillato-Yané GR, Pascual R (1984) Un peculiar Paratheria, Edentata (Mammalia) del Paleoceno de Patagonia. 1° Jornadas Argentinas de Paleontología de Vertebrados, La Plata, Abstracts:15

Scillato-Yané GR, Pascual R (1985) Un peculiar Xenarthra del Paleoceno Medio de Patagonia (Argentina). Su importancia en la sistemática de los Paratheria. Ameghiniana 21:173–176

Sereno PC (2006) Shoulder girdle and forelimb in multituberculates: Evolution of parasagittal forelimb posture in mammals. In: Carrano MT, Gaudin TJ, Blob RW, Wible JR (eds) Amniote paleobiology: perspectives on the evolution of mammals, birds, and reptiles. University of Chicago Press, Chicago, Illinois, pp 315–366

Sigogneau-Russell D, Bonaparte JF, Frank RM, Escribano V (1991) Ultrastructure of dental tissues of *Gondwanatherium* and *Sudamerica* (Mammalia, Gondwanatheria). Lethaia 24:27–38

Strömberg CAE (2011) Evolution of grasses and grassland ecosystems. Annu Rev Earth Planet Sci 39:517–544

Verma O, Prasad GVR, Khosla A, Parmar V (2012) Late Cretaceous gondwanatherian mammals of India: distribution, interrelationships and biogeographic implications. J Palaeontol Soc Ind 57:95–104

Wible JR, Rougier GW (2000) The cranial anatomy of *Kryptobaatar dashzevegi* (Mammalia, Multituberculata), and its bearing on the evolution of mammalian characters. Bull Am Mus Nat Hist 247:1–124

Wible JR, Shelley S, Bi S (2019) New genus and species of djadochtatheriid multituberculate (Allotheria, Mammalia) from the Upper Cretaceous Bayan Mandahu Formation of Inner Mongolia. Ann Carnegie Mus 85:285–327

Wilson GP, Das Sarma DC, Anantharaman S (2007) Late Cretaceous sudamericid gondwanatherians from India with paleobiogeographic considerations of Gondwanan mammals. J Vertebr Paleont 27:521–531

Zimicz NA, Goin F (2020) A reassessment of the genus *Groeberia* Patterson, 1952 (Mammalia, Metatheria): functional and phylogenetic implications. J Syst Palaeontol 18:975–992

# Chapter 9
# Ichnology: Mammalian/Mammal-Like Footprints

*If we walk far enough, said Dorothy, I am sure we shall sometime come to some place.*
Lyman Frank Baum
*The Wonderful Wizard of Oz*, 1900

**Abstract** The ichnogenus *Ameghinichnus patagonicus* described by Casamiquela in 1960, from the Middle Jurassic of Patagonia was the first evidence of a Mesozoic mammal in the continent. The trackways showed a sophisticated, ricochetal locomotion pattern, largely unsuspected at the time and with a surprisingly numerical abundant in the fossil site. Later on, other ichnotaxa were reported from the Middle Jurassic of Argentina and the Early Cretaceous of Brazil, complementing the earlier discoveries. We summarize in this chapter, the footprint records from Jurassic and Cretaceous rocks of South America, that have been putatively related to mammaliaforms, although these assignations can be uncertain to some degree, and perhaps related to non-mammaliaform cynodonts.

**Keywords** Footprints · Ichnospecies · Mammalia · Mammaliamorpha · Argentina · Brazil

## 9.1 Introduction

The ichnogenus *Ameghinichnus patagonicus* from the Middle Jurassic La Matilde Formation (Santa Cruz Province, Patagonia) provided the first evidence of a Mesozoic mammal in SA (Casamiquela 1960, 1961, 1964, 1974). Several specimens of *Ameghinichnus* were initially collected by Rodolfo Casamiquela and Galileo Scaglia (Bonaparte and Migale 2010, 2015), together with dinosaur tracks, marks of other vertebrates and invertebrates, as well as root casts (Casamiquela 1964, 2002; Bonaparte 1978; de Valais 2009, 2011; Bonaparte and Migale 2010, 2015). The collection was later enlarged by Bonaparte and collaborators and more recently by de Valais and colleagues, resulting in more than a hundred specimens (de Valais 2009). Recent studies have recognized two different ichnospecies for *Ameghinichnus* footprints: *A. patagonicus* and a set of larger tracks, *A. manantialensis* (de Valais 2009).



G. W. Rougier et al., *Mesozoic Mammals from South America and Their Forerunners*, Springer Earth System Sciences, https://doi.org/10.1007/978-3-030-63862-7_9

After the discovery of *Ameghinichnus* but before the first findings of osteological remains of Mesozoic mammals in SA, Leonardi (1981; see also Leonardi 1980), described mammalian footprints from the Early Cretaceous of southeastern Brazil, naming them *Brasilichnium elusivum.* This ichnospecies was based on several sandstone slabs with tracks collected from the now abandoned commercial São Bento quarry, in the vicinity of Araraquara town (São Paulo State). More recently, other morphotypes differentiated by gross morphology, size, and presumably behavior permitted recognition of three other ichnospecies from the same Early Cretaceous São Bento quarry. Two of those are small and regarded as ichnospecies of *Brasilichnium*: *B. saltatorium* (Buck et al. 2017a) and *B. anaiti* (D'Orazi Porchetti et al. 2018). The third is a large morphotype termed *Aracoaraichnium leonardii* (Buck et al. 2017b).

In addition to the already mentioned Jurassic–Cretaceous likely mammaliaform footprints, the ichnogenus *Dicynodontipus* gathers a dozen ichnospecies from the Permian–Triassic of worldwide distribution (e.g., Melchor and de Valais 2006; Francischini et al. 2018; Díaz-Martínez et al. 2019), including records in Argentina and Brazil, that were related to non-mammaliaform therapsids. Most of the trackways were thought to be produced by dicynodont therapsids (Francischini et al. 2018; Díaz-Martínez et al. 2019), while a few morphotypes were regarded as of cynodont origin (e.g., Marsicano and Barredo 2004; Silva et al. 2008). In particular, small-sized trackways recovered from the Late Triassic *Hyperodapedon* Assemblage Zone, Candelária Sequence, Santa Maria Supersequence (see Chaps. 2 and 3) of Rio Grande do Sul, southern Brazil, were termed *Dicynodontipus protherioides* and considered produced by a non-mammaliaform cynodont (Silva et al. 2008). The *D. protherioides* trackways represent a walking gait with the tail touching the ground, from a very small quadrupedal and cursorial (plantigrade to semi-plantigrade) creature, with alternate pace and half-erect gait, and small lateral movement of the vertebral column. In the *Hyperodapedon* Assemblage Zone, several small-sized non-mammaliaform probainognathians are known (e.g., Martinelli and Soares 2016; see Chap. 3), and perhaps one of them is the producer of the trackways.

In addition to tracks, the SA ichnological record includes different other traces of activity presumably left by early mammals or their closest relative forerunners. These include burrow complexes, coprolites, and tooth marks, mostly recovered from Triassic rocks of western Argentina and southern Brazil. Burrow systems, with complex architecture including branching tunnels and chambers, were described for several units of the Ischigualasto-Villa Unión Basin (Colombi et al. 2012; Krapovickas et al. 2013; Fiorelli et al. 2018). Moreover, there is mention of a burrow in the Botucatu Formation, which is compatible with *Brasilichnium elusivum* as the producer (Manes et al. 2017; Peixoto et al. 2020). The burrows from the Triassic of Argentina were attributed to non-mammaliaform cynodonts: small-sized basal traversodontids or basal probainognathians (Colombi et al. 2012; Krapovickas et al. 2013; Fiorelli et al. 2018). Some non-mammalian cynodonts probably had digging capabilities, based on the gross morphology of their limbs (Martinelli et al. 2005; Bordy et al. 2017; Guignard et al. 2019). Burrows can be a suitable space for habitation or temporal haunt (for reproduction, resting, and feeding), used to avoid adverse weather conditions or predators. These are behavioral features observed in many tetrapod groups but mostly among mammals (Reichman and Smith 1990; Hasiotis et al. 2007). Coprolites are common traces preserved in Triassic rocks and many records have been related to non-mammaliaform cynodont producers, especially coprolites from the Santa Maria Supersequence (southern Brazil; e.g., Francischini et al. 2017) and a few from the Chañares Formation (western Argentina; Mancuso et al. 2018). Tooth marks produced by mammals have not been frequently described. Reptile bones (sphenodontians and small-sized theropod dinosaurs) collected from the lower Upper Cretaceous Candeleros Formation at La Buitrera locality (Río Negro Province, Patagonia, Argentina) showed several small-sized gnawing or bite marks (de Valais et al. 2012; Gianechini and de Valais 2016). Some of these tooth marks indicate opposing, procumbent teeth that are a fraction of a millimeter and suggest they were produced by mammals (de Valais et al. 2012; Gianechini and de Valais 2016). Two mammals are so far present in this locality (see Chaps. 2 and 6; de Valais et al. 2012; Gianechini and de Valais 2016), the meridiolestidan *Cronopio* and a larger undescribed taxon. The tooth marks are viable regarding size and morphology for *Cronopio*, but too small for the yet undescribed taxon. If this is correct, a scavenger behavior for meridiolestidans can be inferred, and although unsurprising, documents the ubiquitous nature and early presence of scavenging in mammals (de Valais et al. 2012).

## 9.2 Systematics

Mammalia *incertae sedis*
*Ameghinichnus* Casamiquela 1961

Type ichnospecies: *Ameghinichnus patagonicus* Casamiquela 1961.

Included ichnospecies: The type ichnospecies and *A. manantialensis* de Valais 2009.

*Ameghinichnus patagonicus* Casamiquela 1961
(Figs. 9.1, 9.2, 9.3 and 9.9)

Holotype: MLP 60-X-31-14A, trackway with six manus/pes sets with opposite arrangement about the midline (associated with paratype MLP 60-X-31-14B).

Locality and horizon: Laguna Manantiales, Santa Cruz, Argentina; La Matilde Formation, Bajocian–Callovian (Middle Jurassic).

Diagnosis (from de Valais 2009): Manus and pes tracks with length/width ratio of 0.7; manus digit impressions subequal in length, with the following relationship; I < II = V < IV < III; angles between III–IV < I–II≈II–III≈IV–V, I–V ~ 151°; manus impressions with an inward rotation of about −16° in relation to the midline, and located aligned-to-posterior to pes impressions; pes digit impressions subequal in length, with relative length: I < IV≈III = V < II; angles between III–IV < II–III < I–II < IV–V, I–V ~ 152°; pes with outward rotation of about 31° in relation to

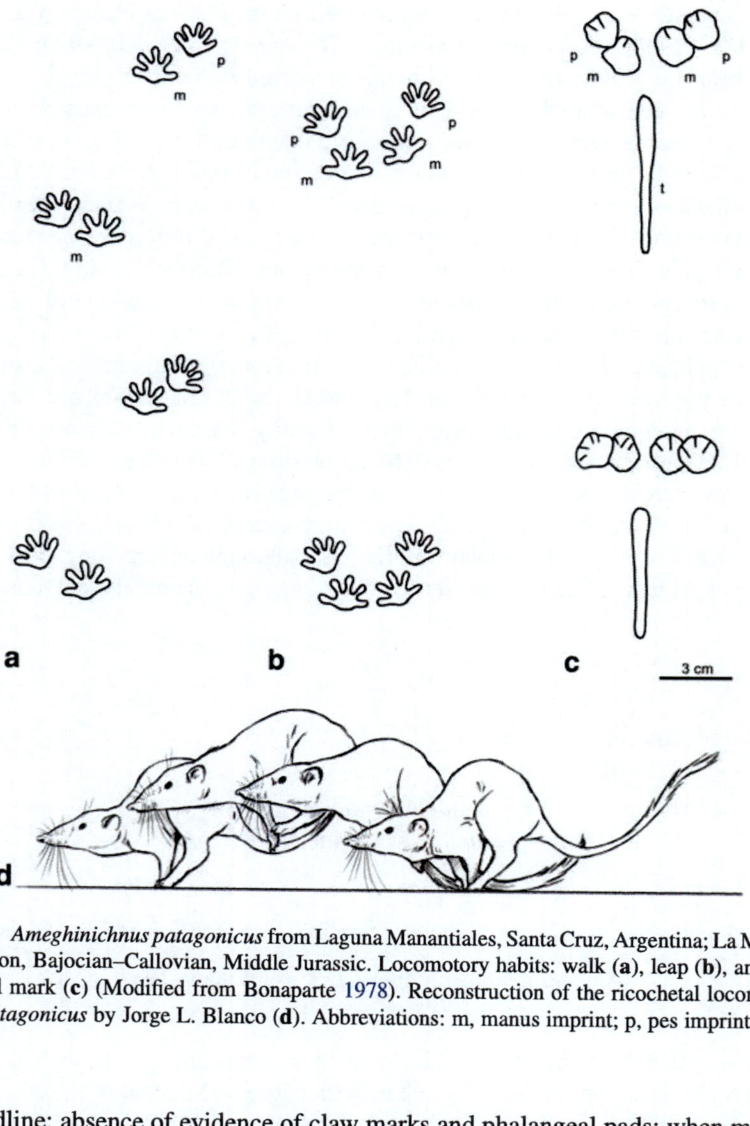


**Fig. 9.1** *Ameghinichnus patagonicus* from Laguna Manantiales, Santa Cruz, Argentina; La Matilde Formation, Bajocian–Callovian, Middle Jurassic. Locomotory habits: walk (**a**), leap (**b**), and leap with tail mark (**c**) (Modified from Bonaparte 1978). Reconstruction of the ricochetal locomotion of *A. patagonicus* by Jorge L. Blanco (**d**). Abbreviations: m, manus imprint; p, pes imprint; t, tail mark

the midline; absence of evidence of claw marks and phalangeal pads; when manus-pes sets with alternate arrangement about the midline, manus and pes angulation is about 143° and 114°, respectively; manus and pes stride length is about 83 mm and 81.2 mm, respectively.

Comments: *Ameghinichnus* trackways were produced by a quadrupedal trackmaker, leaving nearly homopod, symmetric, pentadactyl hand, and foot impressions (Figs. 9.1, 9.2). The handprints are medial to the footprints and with inward rotation from the midline. The footprints are slightly larger than the handprints and with outward rotation from the midline. Hand and foot impressions are wider than long, with digits leaving straight impressions (Fig. 9.1a, b). The length and angles between

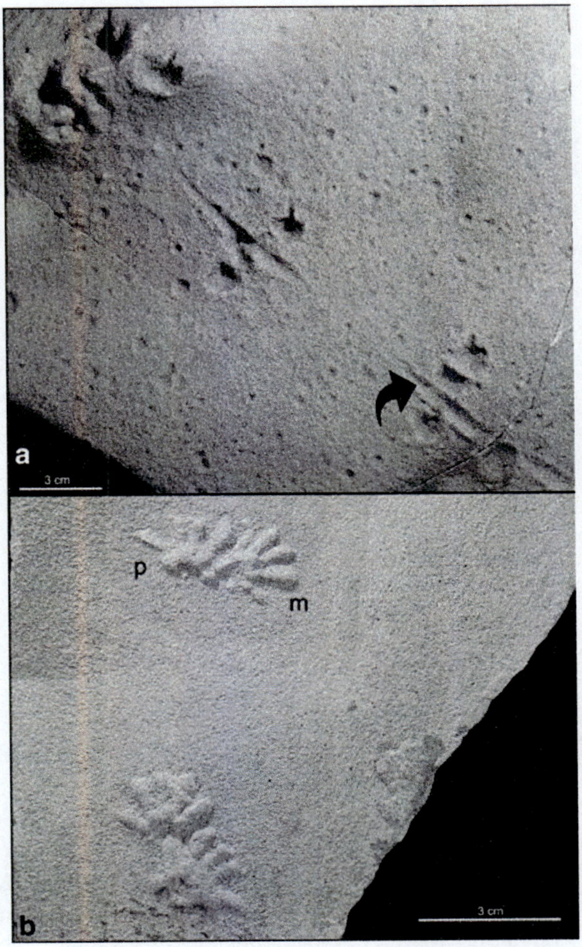


**Fig. 9.2** *Ameghinichnus patagonicus* from Laguna Manantiales, Santa Cruz, Argentina; La Matilde Formation, Bajocian–Callovian, Middle Jurassic. MPEF-IC 1015, footprints with the tail mark (**a**). *Ameghinichnus manantialensis*. Holotype, MPEF-IC 1028, footprints (**b**). Abbreviations: m, manus imprint; p, pes imprint. Arrow indicates the tail mark

the digit imprints are subequal, except between digits I and V, which are the largest (de Valais 2009). Tail marks are often present (Fig. 9.1c). Two different tail impression patterns were described: sinuous and continuous or straight and discrete (de Valais 2009).

*Ameghinichnus* has been interpreted as of mammalian origin, with a sophisticated locomotion system including different capabilities, such as regular alternating quadrupedal cursoriality and ricochetal movements (Casamiquela 1964; Leonardi and Oliveira 1990; Leonardi 1994; de Valais 2009) (Fig. 9.3). Recently, Kuznetsov and Panyutina (2018), reported tracks of *Ameghinichnus* showing tail marks deviated from the body midline. This evidence was used to suggest that the maker was carrying offspring on its back (baby riding). In tracks produced by opossums and other living mammals, the deviation of the tail is consistent with tracks in which the load is asymmetrically distributed by, for example, the weight of carried young (Fig. 9.9).

**Fig. 9.3** *Ameghinichnus patagonicus*, artistic reconstruction by Jorge L. Blanco

If the hypothesis is correct, this represents the oldest evidence of parental care for Mammaliaformes/Mammalia in SA (depending on the affinities of the trackmaker; see below).

Different alternatives have been considered for the trackmaker: a basal therian (Casamiquela 1974), a multituberculate (Kielan-Jaworowska and Gambaryan 1994; Bonaparte and Migale 2010, 2015), or an indeterminate Mammalia (Bonaparte 1978), the latter perhaps better understood as Mammaliaformes, as currently construed (Rougier et al. 2011). *Ameghinichnus* track-type has also been discovered from the Elliot Formation (Lower Jurassic), Karoo Basin of South Africa (Ellenberger 1970, 1972, 1974), the upper Towaco Formation (Lower Jurassic), Newark Basin of North America (Olsen and Rainforth 2001), potentially from the Zagaje Formation (Lower Jurassic), Poland (Gierliński et al. 2004; de Valais 2009), and from the Lower Cretaceous of Angola (Marzola et al. 2014). Some authors, in part because the large size, referred those tracks to non-mammalian cynodonts (Olsen and Galton 1984; Olsen and Rainforth 2001).

Regardless of its systematic affinities, *Ameghinichnus* records the activity of a mammal or mammal-like form in the Middle Jurassic of SA (Fig. 9.3), which is compatible with contemporaneous records in NA, Africa, Europe, and Asia (Lillegraven et al. 1979; Kielan-Jaworowska et al. 2004; de Valais 2009).

*Ameghinichnus manantialensis* de Valais 2009
(Fig. 9.2b)

Holotype: MPEF-IC 1028, trackways with three manus-pes sets with alternate arrangement about the midline (Fig. 9.2b).

Locality and horizon: Laguna Manantiales, Santa Cruz, Argentina; La Matilde Formation, Bajocian–Callovian (Middle Jurassic).

Diagnosis (from de Valais 2009): Manus and pes tracks with length/width ratio of 0.7 and 0.6, respectively. Manus digit impressions subequal in length, with the following relationship: I < V < IV < II≈III; angles between II–III < III–IV < I–II≈IV–V. Manus impressions inward rotation of about −38° in relation to the midline and aligned or anterior to the pes. Pes digit impressions subequal in length, with relative length: I < III < IV≈V < II, angles between III–IV < II–III < I–II < IV–V. Pes impressions outward rotation of about 26° in relation to the midline. Both I–V angles for pes and manus about 107°. Absence of evidence of claw marks and phalangeal pads. When manus and pes sets with alternating arrangement about midline, manus-pes pace angulation averages 156° and 110°, respectively; manus and pes stride length is about 116 mm and 119 mm, respectively.

Comments: *Ameghinichnus manantialensis* is larger than *A. patagonicus*, with the impressions of the hands about 63% longer and 60% wider and the footprints about 50% longer and 72% wider. It also differs from *A. patagonicus* by having different digit length proportions and divarication angles, and the manus located aligned-to-anterior to the pes impression (de Valais 2009).

Mammaliamorpha *incertae sedis*
*Brasilichnium* Leonardi 1981

Type ichnospecies: *Brasilichnium elusivum* Leonardi 1981.

Included ichnospecies: The type ichnospecies, *B. saltatorium* Buck et al. 2017a, and *B. anaiti* D'Orazi Porchetti et al. 2018.

*Brasilichnium elusivum* Leonardi 1981
(Figs. 9.4, 9.6a and 9.7)

Holotype: MNRJ 3902-V, slab with trackway containing 15 manus/pes sets (Fig. 9.4).

Locality and horizon: São Bento quarry, Araraquara municipality, São Paulo State, Brazil; Botucatu Formation, São Bento Group, Berriasian–Valanginian (Lower Cretaceous).

Diagnosis (taken from Leonardi 1981 and Fernandes and Carvalho 2008): Quadrupedal, small-sized trackway; average glenoacetabular distance of 7.5 cm; extreme difference between very small hands and large feet; hand-foot distance increases as speed slows down; relatively high pitch angle for a quadruped; heteropodia is directly proportional to the positive rotation of the longitudinal axis of the feet (values > 75°); longitudinal axes of the feet parallel to the runway axis and with pitch angles of > 125°, without heteropodia; elliptical contoured foot, with almost transverse major axis and anteroposterior axis slightly directed into the track;

**Fig. 9.4** *Brasilichnium elusivum* from São Bento quarry, Araraquara municipality, São Paulo State, Brazil; Botucatu Formation, São Bento Group, Berriasian–Valanginian, Lower Cretaceous. Holotype MNRJ 3902-V, slab with trackway containing 15 manus/pes sets (picture from Heitor Francischini)

short toes, usually rounded, with probable phalangeal formula 2–3–3–3–3; posterior footprints are ectaxonic and tetradactyle (digits II, III, IV, and V) in semi-plantar condition; larger Hypex V-digit in slight abduction; anterior autopods with at least four digits with claws, evident or not, due to preservation conditions; tail strokes always absent.

*Brasilichnium saltatorium* Buck et al. 2017a
(Figs. 9.5, 9.6b)

Holotype: LPP-IC-0001, slab with five sets of manus and pes (LPP-IC-0002, part of the counterpart of LPP-IC-0002) (Fig. 9.5a).

Locality and horizon: São Bento quarry, Araraquara municipality, São Paulo State, Brazil; Botucatu Formation, São Bento Group, Berriasian–Valanginian (Lower Cretaceous).

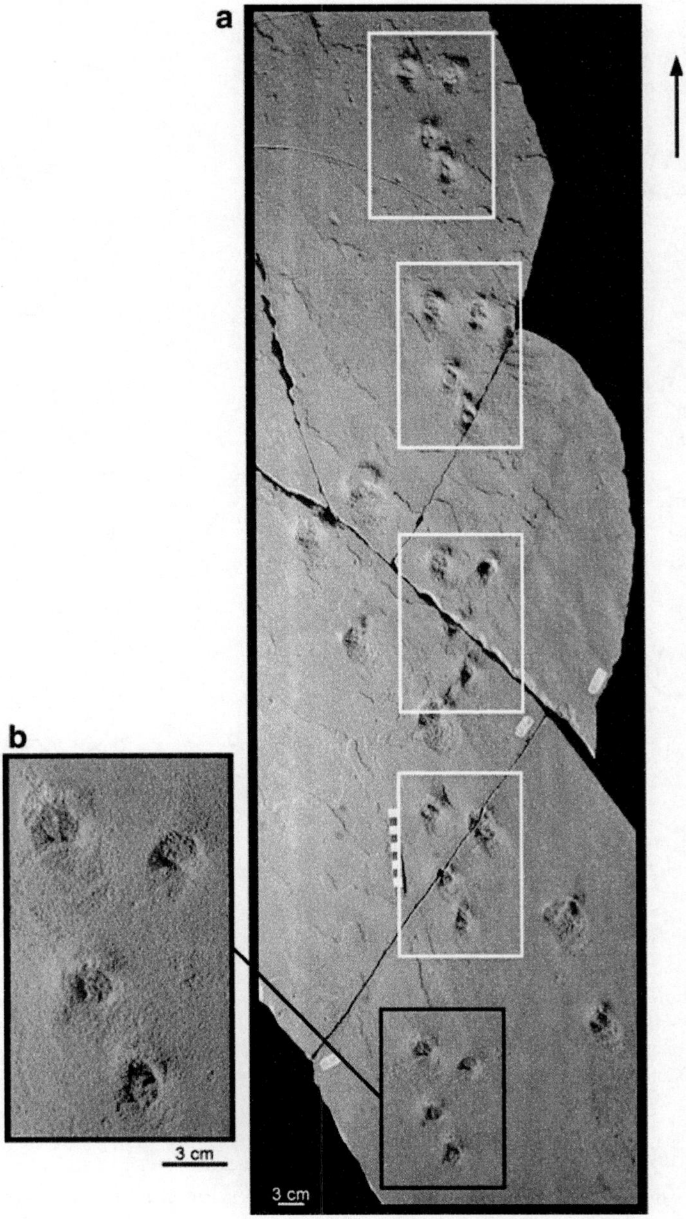


**Fig. 9.5** *Brasilichnium saltatorium* from São Bento quarry, Araraquara municipality, São Paulo State, Brazil; Botucatu Formation, São Bento Group, Berriasian–Valanginian, Lower Cretaceous. Holotype LPP-IC-0001, slab with five sets of manus and pes (**a**). Detail of one set of manus and pes imprint (**b**). The arrow indicates the direction of movement (pictures from Pedro V. Buck)

**Fig. 9.6** *Brasilichnium* ichnospecies from Brazil. *Brasilichnium elusivum*, diagram of the holotype MNRJ 3902-V (**a**). *Brasilichnium saltatorium*, diagram of the holotype LPP-IC-0001 (**b**). *Brasilichnium anaiti*, diagram of the holotype URC-R 70 (**c**). The drawings (**a**) and (**b**) were based on Buck et al. (2017a) and (**c**) on D'Orazi Porchetti et al. (2018)

Diagnosis (taken from Buck et al. 2017a): Quadruped trackway, with small dimensions and hopping locomotion in phases; pes impression rounded to transversally oval; manus, when present, usually elongated in the anteroposterior axis due to the displacement of individual; heteropody not so evident due to preservational aspects related to direction of movement; when the track is descending the inclined plane, manus and pes sets are more evident whereas when the trail goes up the inclined plane, manus sets are unclear; the pes surpasses the manus during the movement cycle; pes horizontally aligned in a subparallel way and laterally spaced; manus vertically aligned and laterally spaced, minimally apart from the trackway midline;

**Fig. 9.7** *Brasilichnium elusivum*, artistic reconstruction by Jorge L. Blanco

taking the trackway midline as a reference, the right pes and manus are posteriorly located to their left counterpart; tetradactyl hindfoot impressions (digits II, III, IV, and V), and mesaxonic condition when digits are visible; digits III and IV longer in relation to the others; tail impressions/drag are always absent.

*Brasilichnium anaiti* D'Orazi Porchetti et al. 2018
(Fig. 9.6c)

Holotype: URC-R 70, slab with imprints of a sequence of 11 pes and 12 manus.

Locality and horizon: São Bento quarry, Araraquara municipality, São Paulo State, Brazil; Botucatu Formation, São Bento Group, Berriasian–Valanginian (Lower Cretaceous).

Diagnosis (taken from D'Orazi Porchetti et al. 2018): Quadrupedal trackway, with neat heteropody, pes print larger than manus print; tetradactyl, paraxonic-to-slightly ectaxonic pes print; foot digit marks are short, usually separated from metapodials by a stenosis or groove (especially on the central two digits), giving the footmark a paw-like appearance; first and fourth pes digit marks distally blunt, whereas the second and third appear distally sharp; digit marks I–III more deeply impressed compared to the lateral one, with the deepest area corresponding to the metatarsal-phalangeal joints; pes print wider than long, with an average FL/FW ratio of 0.72; manus print with at least three digit marks, always preceding the footmarks; no tail marks printed.

Comments: The *Brasilichnium* ichnofacies have a wide geographical and temporal distribution and have also been recognized in the Upper Triassic–Jurassic aeolian deposits of the southwestern USA (e.g., Lockley and Hunt 1995; Lucas et al. 2010; Lockley 2011; Engelmann and Chure 2017), and the Lower Cretaceous Twyfelfontein Formation in northwestern Namibia (D'Orazi Porchetti and Wagensommer 2015). In particular, the latter unit in Namibia is considered to have paleogeographic continuity with the Botucatu desert of Brazil (D'Orazi Porchetti and Wagensommer 2015). *Brasilichnium* is associated with marginal sand paleodunes from arid environments, such as those represented by the desert borders of the Botucatu Formation, outcropping at the Araraquara region, São Paulo. Moreover, *Brasilichnium* tracks were preserved in upward and downward directions relative to paleodunes, with variations (e.g., shape, stance, and gait) in the recovered tracks (Buck et al. 2017a; D'Orazi Porchetti et al. 2017). Among the known sample of *Brasilichnium* from Brazil, *B. elusivum* and *B. saltatorium* represent trackways produced by small-sized animals, whereas the tracks of *B. anaiti* represent a large trackmarker (about four times bigger; D'Orazi Porchetti et al. 2018). Because of its large size, the maker of *B. anaiti* was regarded as a non-mammalian therapsid (e.g., Leonardi 1980; Leonardi and Godoy 1980); however, the overall foot morphology is consistent with both that of non-mammalian therapsids or early mammals (D'Orazi Porchetti et al. 2018). Regarding *B. elusivum* and *B. saltatorium*, both ichnospecies belonged to tetradactyl animals of similar size with different locomotion patterns. The tracks referred to *Brasilichnium elusivum* were characterized by different locomotory capabilities, including walking, half-bounding, bipedal skipping, and running (Leonardi 1981; D'Orazi Porchetti et al. 2017) (Fig. 9.7), whereas those of *B. saltotorium* were characterized by the phases of hopping locomotion (Buck et al. 2017a).

*Aracoaraichnium* Buck et al. 2017b

Type ichnospecies: *Aracoaraichnium leonardii* Buck et al. 2017a.

Included ichnospecies: The type only.

*Aracoaraichnium leonardii* Buck et al. 2017b
(Fig. 9.8)

Holotype: LPP-IC-0015, slab with three sets of manus and pes impressions (Fig. 9.8a).

Locality and horizon: São Bento quarry, Araraquara municipality, São Paulo State, Brazil; Botucatu Formation, São Bento Group, Berriasian–Valanginian (Lower Cretaceous).

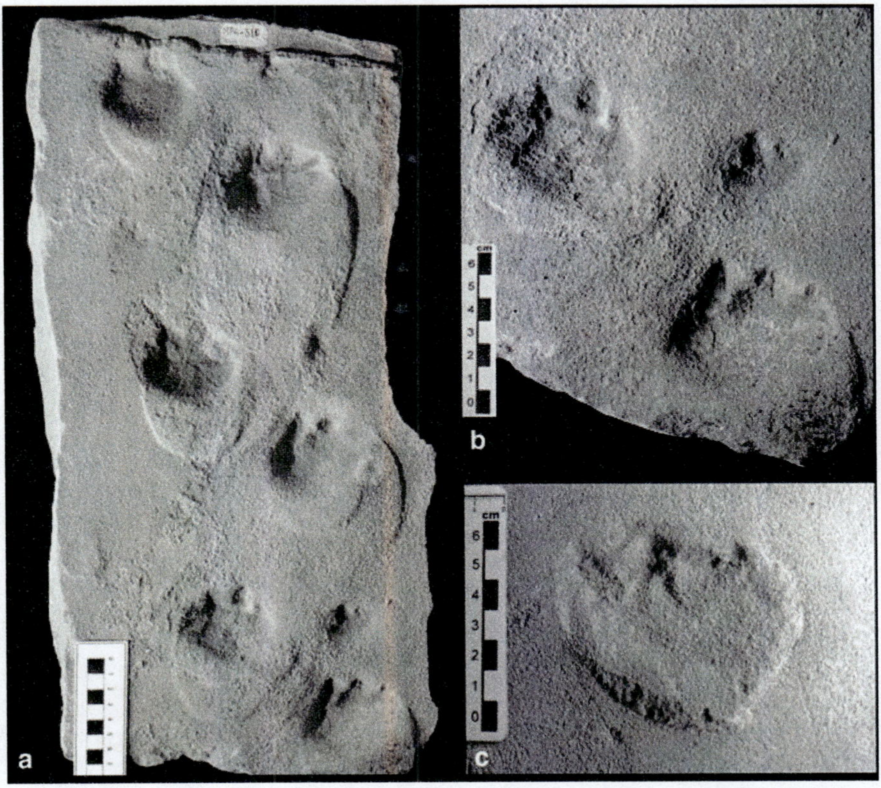


**Fig. 9.8** *Aracoaraichnium leonardii* from São Bento quarry, Araraquara municipality, São Paulo State, Brazil; Botucatu Formation, São Bento Group, Berriasian–Valanginian, Lower Cretaceous. Holotype LPP-IC-0015, slab with three sets of manus and pes (**a**), with detail of first and second pes tracks (**b**). Detail of pes imprint of LPP-IC-0016 (**c**) (pictures from Pedro V. Buck)

Diagnosis (taken from Buck et al. 2017b): Quadruped trackway with walking locomotion; narrow trackway with the interior margin of the pes very close to the midline; footprints semi-plantigrade, mesaxonic, with digits III and IV more anteriorly registered; manus significantly smaller than pes; when manus are imprinted, are always anterior to the pes in the locomotion cycle; both manus and pes are tetradactyl, with short digits; the proximal traces of the digits are wide, but taper distally, ending in claw traces; digit III is the longest, followed by digit IV, which are subequal in length; digits II and V are smaller, but also subequal to each other in length; footprints are wider than they are long, with an oval heel pad, sometimes with the anteroposterior axis directed slightly inward; when the anteroposterior axis is directed inwards, it occurs for both the manus and pes in similar degree; tail impressions/drag are always absent.

Comments: *Aracoaraichnium leonardii* was named based on large-sized trackways collected in the São Bento quarry, Araraquara, the same locality where the holotypes

**Fig. 9.9** *Ameghinichnus patagonicus*, tracks with tail groove shifted to the right of footprints, suggesting an animal walking with load, such as carrying offspring on its back (based on inferences made by Kuznetsov and Panyutina 2018). Artistic reconstruction by Jorge L. Blanco

of *Brasilichnium elusivum*, *B. saltatorium,* and *B. anaiti* were found. They altogether illustrate the diversity of the Botucatu fauna in Brazil, represented by two small (*B. elusivum* and *B. saltatorium*) and two large-sized taxa (*B. anaiti* and *A. leonardii*).

Some of the specimens referred by Buck et al. (2017b), to *A. leonardii* were previously interpreted as having possible trithylodontoid affinities (Leonardi and Oliveira 1990; Leonardi 1994; Leonardi and Carvalho 2002; Leonardi et al. 2007), as in the case for *B. anaiti*, mainly because of the large size, which is unusual but does not exclude Mesozoic mammal species (e.g., *Repenomamus* and *Adalatherium*; Hu et al. 2005; Krause et al. 2014, 2020). In contrast, their overall foot morphology is consistent with both non-mammalian therapsids and early mammals (Buck et al. 2017b; D'Orazi Porchetti et al. 2018). The differences between *B. anaiti* and *A. leonardii* are few, including paraxonic to slightly ectaxonic pes prints in *B. anaiti* to mesaxonic in *A. leonardii.* Other features, such as manus significantly smaller than pes, footprints wider than long, with an oval heel pad and no tail marks on the trails, are shared in both ichnotaxa (Buck et al. 2017b; D'Orazi Porchetti et al. 2018). These differences may justify the recognition of two ichnospecies, but at the very least indicate two animals of similar size and similar manual/pedal morphology living at the same time in the place.

## 9.3 Concluding Remarks

Despite its lack of systematic discrimination, paleoichnology is often the only direct line of evidence about behavior, the interaction of organisms among themselves and with their ancient environments (Seilacher 2007; Buatois and Mángano 2011). The first hints of Mesozoic mammals in SA were the footprint tracks left by small, mouse-sized mammaliaforms discovered in the Middle Jurassic of Patagonia, Argentina (Casamiquela 1960, 1961, 1964, 1974) and the Early Cretaceous of southern Brazil (Leonardi 1981). Osteological remains of SA Mesozoic mammals (see Chaps. 1, 6, 7, and 8) followed almost twenty years after the first ichnological report (Casamiquela 1960, 1961).

Frequently, paleoenvironments preserving footprint tracks are inadequate for the preservation of bones and, in consequence, a direct relationship between the track and its maker is impossible to establish. This is the case for the ichnotaxa described for the Middle Jurassic La Matilde Formation, *Ameghinichnus patagonicus* and *A. manantialensis*, and for the Lower Cretaceous Botucatu Formation, *Brasilichnium elusivum*, *B. saltatorium*, *B. anaiti* (Leonardi 1981; Buck et al. 2017a; D'Orazi Porchetti et al. 2018), and *Aracoaraichnium leonardii* (Buck et al. 2017b). In both formations, there is no osteological record of mammals. Ghost lineages in most current phylogenies (see Fig. 1.7 for a summary), predict a minimal Early Jurassic age for Mammalia. The diversity of lineages attested in the Lower–Middle Jurassic Cañadón Asfalto Formation provides compelling evidence of an earlier diversification that it is hard to conceive as taking place any later than the earliest Jurassic, but more likely in the Triassic (see Fig. 10.4). The affinities of the trackmakers reviewed here do not need, therefore, to be restricted to taxa known in the Jurassic record of SA or for that matter in the Jurassic across the world. Ghost lineage predictions make clear that a number of lineages must, and many others could have been present when the trackways were produced. Having no way to falsify previous systematic attributions, we follow precedence and regard these trackways as mammalian in nature, but they could very well correspond to a variety of non-mammalian relatives, some of which are known to survive in other parts of the world up to the Early Cretaceous (Matsuoka et al. 2016).

As shown in previous chapters, mammals and their stem taxa are unevenly distributed in SA, and the fossil remains from the Jurassic and the Early Cretaceous are less abundant in terms of species diversity and number of specimens when compared to the remaining periods (see also Fig. 2.1). Nonetheless, the ichnospecies mentioned in this chapter, demonstrates that mammals (or mammal-like taxa) in the Mesozoic had a broad distribution, thriving even in extreme ecosystems, as shown by the tracks recorded in the Lower Cretaceous paleodesert of the Botucatu Formation. In the Cenozoic, mammals diversified in a broad range of ecosystems, as exemplified by their disparate morphotypes—e.g., whales, dolphins, bats, cats, sloths, armadillos, jerboas, rats, bears, etc. The Mesozoic, however, was traditionally viewed as a time of more conservative, generalized terrestrially-adapted little mammals (Clemens et al. 1979). The combined weight of the osteological and ichnological data support a previously unfathomable taxonomic diversity and ecological disparity for the Mesozoic (e.g., Kielan-Jaworowska et al. 2004; Luo 2007; Grossnickle and Polly 2013; Krause et al. 2014, 2020; Luo et al. 2015, 2017; Benevento et al. 2019).

The footprint tracks represented by *Ameghinichnus* spp., *Brasilichnium* spp., and *Aracoaraichnium leonardii* clearly suggest a broad range of behaviors and locomotor modalities, including walking, half-bounding, bipedal skipping, running, and hopping (Casamiquela 1964; Leonardi 1981; de Valais 2009; Buck et al. 2017a; D'Orazi Porchetti et al. 2017), which vary according to features of the subtracts (sands, clays) or even during an upward or downward displacement of a knoll. Interestingly, trackways of *Ameghinichnus* have been used to suggest the producer was carrying offspring on its back (baby riding) (Fig. 9.9), as in opossum marsupials and other mammals (see above; Kuznetsov and Panyutina 2018). If this is the case, it represents the earliest evidence of rearing (Middle Jurassic) for mammaliaforms, including an extended period of parental care and a relatively late weaning, behaviors further elaborated in and characterizing living mammals.

## References

Benevento GL, Benson RBJ, Friedman M (2019) Patterns of mammalian jaw ecomorphological disparity during the Mesozoic/Cenozoic transition. Proc R Soc Lond B 286:20190347

Bonaparte JF (1978) El Mesozoico de América del Sur y sus tetrápodos. Op Lilloana 26:1–596

Bonaparte JF, Migale LA (2010) Protomamíferos y mamíferos Mesozoicos de América del Sur, 1st edn. Museo de Ciencias Naturales Carlos Ameghino de Mercedes, Buenos Aires

Bonaparte JF, Migale LA (2015) Protomamíferos y mamíferos Mesozoicos de América del Sur, 2nd edn. Fundación de Historia Natural Félix de Azara, Buenos Aires

Bordy E, Sciscio L, Abdala NF, Mcphee B, Choiniere J (2017) First Lower Jurassic vertebrate burrow from southern Africa (upper Elliot Formation, Karoo Basin, South Africa). Palaeogeogr Palaeoclimat Palaeoecol 468:362–372

Buatois LA, Mángano MG (2011) Ichnology. Organism-substrate interactions in space and time. Cambridge University Press, Cambridge

Buck PV, Ghilardi AM, Peixoto BCPM, dos Reis Fernandes LB, Fernandes MA (2017a) A new tetrapod ichnotaxon from Botucatu Formation, Lower Cretaceous (Neocomian), Brazil, with comments on fossil track preservation on inclined planes and local paleoecology. Palaeogeogr Palaeoclimatol Palaeoecol 466:21–37

Buck PV, Ghilardi AM, dos Reis Fernandes LB, Fernandes MA (2017b) A new ichnotaxon classification of large mammaliaform trackways from the Lower Cretaceous Botucatu Formation, Paraná Basin, Brazil. Palaeogeogr Palaeoclimatol Palaeoecol 485:377–388

Casamiquela RM (1960) El hallazgo del primer elenco (icnológico) Jurásico de vertebrados terrestres de Latinoamérica (noticia). Rev Asoc Geol Argent 15:5–14

Casamiquela RM (1961) Sobre la presencia de un mamífero en el primer elenco (icnológico) de vertebrados del Jurásico de la Patagonia. Physis 22:225–233

Casamiquela RM (1964) Estudios icnológicos. Problemas y métodos de la icnología con aplicación al estudio de pisadas mesozoicas (Reptilia, Mammalia) de la Patagonia. Colegio Industrial Pío IX, Buenos Aires

Casamiquela RM (1974) Sobre la significación de *Ameghinichnus patagonicus*, un mamífero brincador del Jurásico Medio de Santa Cruz (Patagonia). 1° Congreso Argentino de Paleontología y Bioestratigrafía. San Miguel De Tucumán, Actas 2:71–85

Casamiquela RM (2002) Icnitas de vertebrados (Mamíferos, Dinosaurios) del Mesojurásico. In: Haller MJ (ed) Geología y Recursos Naturales de Santa Cruz. 15° Congreso Geológico Argentino, El Calafate, Relatorio 2:433–438

Clemens WA, Lillegraven JA, Lindsay EH, Simpson GG (1979) Where, when, and what—a survey of known Mesozoic mammal distribution. In: Lillegraven JA, Kielan-Jaworowska Z, Clemens WA (eds) Mesozoic mammals: the first two-thirds of mammalian history. University of California Press, Berkeley, pp 7–58

Colombi CE, Fernández E, Currie BS, Alcober OA, Martínez R, Correa G (2012) Large-diameter burrows of the Triassic Ischigualasto Basin, NW Argentina: paleoecological and paleoenvironmental implications. PLoS ONE 7(12):e50662

de Valais S (2009) Ichnotaxonomic review of *Ameghinichnus*, a mammal ichnogenus from the La Matilde Formation (Middle Jurassic), Santa Cruz Province, Argentina. Zootaxa 2203:1–21

de Valais S (2011) Revision of dinosaur ichnotaxa from the La Matilde Formation (Middle Jurassic), Santa Cruz Province, Argentina. Ameghiniana 48:28–42

de Valais S, Apesteguía S, Garrido A (2012) Cretaceous small scavengers: feeding traces in tetrapod bones from Patagonia. Argentina. PLoS One 7(1):e29841

Díaz-Martínez I, Citton P, de Valais S, Cónsole-Gonella C, González SN (2019) Late Permian-Early Jurassic vertebrate tracks from Patagonia: biochronological inferences and relationships with southern African realms. J Afr Earth Sci 160:103619

D'Orazi Porchetti S, Wagensommer A (2015) A vertebrate trackway from the Twyfelfontein Formation (Lower Cretaceous), Damaraland, Namibia. Paläontol Z 89:807–814

D'Orazi Porchetti S, Bertini RJ, Langer MC (2017) Walking, running, hopping. Analysis of gait variability and locomotor skills in *Brasilichnium elusivum* Leonardi, with inferences on trackmaker identification. Palaeogeogr Palaeoclimatol Palaeoecol 465:14–29

D'Orazi Porchetti S, Bertini RJ, Langer MC (2018) Proposal for ichnotaxonomic allocation of therapsid footprints from the Botucatu Formation (Brazil). Ichnos 25:192–207

Ellenberger P (1970) Les niveaux paléontologiques de première apparition des mammifères primordiaux en Afrique du Sud et leur ichnologie. Establissement de zones stratigraphiques détaillées dans le Stormberg du Lesotho (Afrique du Sud) (Trias Supérieur à Jurassique). 2° Gondwana Symposium, South Africa, Abstracts:343–370

Ellenberger P (1972) Contribution à la classification des pistes de vertébrés du Trias: Les types du Stomberg d´Afrique du Sud (I). Palaeovertebrata, Mem Extraord, pp 1–134

Ellenberger P (1974) Contribution à la classification des pistes de vertébrés du Trias: Les types du Stomberg d´Afrique du Sud (II). Palaeovertebrata, Mem Extraord, pp 1–142

Engelmann GF, Chure DJ (2017) Morphology and sediment deformation of downslope *Brasilichnium* trackways on a dune slipface in the Nugget Sandstone of northeastern Utah, USA. Palaeontol Electron 20.2.22A:1–21

Fernandes MA, Carvalho IS (2008) Revisão diagnóstica para a icnoespécie de tetrápode mesozóico *Brasilichnium elusivum* (Leonardi, 1981) (Mammalia) da Formação Botucatu, Bacia do Paraná, Brasil. Ameghiniana 45:167–173

Fiorelli LE, Rocher S, Martinelli AG, Ezcurra MD, Hechenleitner EM, Ezpeleta M (2018) Tetrapod burrows from the Middle–Upper Triassic Chañares Formation (La Rioja, Argentina) and its palaeoecological implications. Palaeogeogr Palaeoclimatol Palaeoecol 496:85–102

Francischini H, Dentzien-Dias P, Schultz CL (2017) A fresh look at ancient dungs: the Brazilian Triassic coprolites revisited. Lethaia 51:389–405

Francischini H, Dentzien-Dias P, Lucas SG, Schultz CL (2018) Tetrapod tracks in Permo-Triassic eolian beds of southern Brazil (Paraná Basin). PeerJ 6:e4764

Gianechini FA, de Valais S (2016) Bioerosion trace fossils on bones of the Cretaceous South American theropod *Buitreraptor gonzalezorum* Makovicky, Apesteguía and Agnolín, 2005 (Deinonychosauria). Hist Biol 28:533–549

Gierliński G, Pienkowski G, Niedźwiedzki G (2004) Tetrapod track assemblage in the Hettangian of Sołtyków, Poland, and its paleoenvironmental background. Ichnos 11:195–213

Grossnickle DM, Polly PD (2013) Mammal disparity decreases during the Cretaceous angiosperm radiation. Proc Biol Sci 280:20132110

Guignard ML, Martinelli AG, Soares MB (2019) Postcranial anatomy of *Riograndia guaibensis* (Cynodontia: Ictidosauria). Geobios 53:9–21

Hasiotis ST, Platt BF, Hembree DI, Everhart MJ (2007) The trace-fossil record of vertebrates. In: Miller W III (ed) Trace Fossils: Concepts, problems, prospects. Elsevier Press, pp 196–218

Hu Y, Meng J, Wang Y, Li C (2005) Large Mesozoic mammals fed on young dinosaurs. Nature 433:149–152

Kielan-Jaworowska Z, Gambaryan PP (1994) Postcranial anatomy and habits of Asian multituberculate mammals. Foss Strat 36:1–92

Kielan-Jaworowska Z, Cifelli RL, Luo Z-X (2004) Mammals from the age of dinosaurs. Origins, evolution, and structure. Columbia University Press, New York

Krapovickas V, Mancuso AC, Marsicano CA, Domnanovich NS, Schultz CL (2013) Large tetrapod burrows from the Middle Triassic of Argentina: a behavioural adaptation to seasonal semi-arid climate? Lethaia 46:154–169

Krause DW, Hoffmann S, Wible JR, Kirk EC, Schultz JA, von Koenigswald W, Groenke JR, Rossie JB, O'Connor PM, Seiffert ER, Dumont ER, Holloway WL, Rogers RR, Rahantarisoa RJ, Kemp AD, Andriamialison H (2014) First cranial remains of a gondwanatherian mammal reveal remarkable mosaicism. Nature 515:512–517

Krause DW, Hoffmann S, Hu Y, Wible JR, Rougier GW, Kirk EC, Groenke JR, Rogers RR, Rossie JB, Schultz JA, Evans AR, von Koenigswald W, Rahantarisoa LJ (2020) Skeleton of Cretaceous mammal from Madagascar reflects long-term insularity. Nature 581:421–427

Kuznetsov AN, Panyutina AA (2018) First paleoichnological evidence for baby-riding in early mammals. Ameghiniana 55:668–676

Leonardi G (1980) On the discovery of an abundant ichno-fauna (vertebrates and invertebrates) in the Botucatu Formation s.s. in Araraquara, São Paulo. Brazil. An Acad Bras Ciênc 52:559–567

Leonardi G (1981) Novo icnogênero de tetrápode Mesozóico da Formação Botucatu, Araraquara, SP. an Acad Bras Ciênc 53:793–805

Leonardi G (1994) Annotated atlas of South America tetrapod footprints (Devonian to Holocene) with an appendix on Mexico and Central America. Brasília, Companhia de Pesquisa de Recursos Minerais, Brasília

Leonardi G, Godoy LC (1980) Novas pistas de tetrápodes da Formação Botucatu no Estado de São Paulo. 31° Congresso Brasileiro de Geologia, 1980. Santa Catarina, Anais 5:3080–3089

Leonardi G, de Oliveira FH (1990) A revision of the Triassic and Jurassic tetrapod footprints of Argentina and a new approach on the age and meaning of the Botucatu Formation footprints (Brazil). Rev Bras Geoc 20:216–229

Leonardi G, Carvalho IS (2002) Jazigo Icnofossilífero do Ouro-Araraquara (SP). In: Schobbenhaus C, Campos DA, Queiroz ET, Winge M, Berbert-Born M (eds) Sítios geológicos e paleontológicos do Brasil. DNPM, Brasília, pp 39–48

Leonardi G, Carvalho IS, Fernandes MA (2007) The desert ichnofauna from Botucatu Formation (Upper Jurassic-Lower Cretaceous), Brazil. In: Carvalho IS, Cassab RCT, Schwanke C, Carvalho MA, Fernandes ACS, Carvalho MSS, Arai M, Oliveira MEQ (eds) Paleontologia: Cenários da vida. Interciência, Rio de Janeiro, pp 379–391

Lillegraven JA, Kielan-Jaworowska Z, Clemens WA (1979) Mesozoic mammals. The first two-thirds of mammalian history. University of California Press, Berkeley

Lockley MG (2011) The ichnotaxonomic status of *Brasilichnium* with special reference to occurrences in the Navajo Sandstone (Lower Jurassic) in the western USA. In: Sullivan RM, Lucas SG, Spielmann JA (eds) Fossil record 3. Bull N M Mus Nat Hist Sci 53:306–315

Lockley MG, Hunt AP (1995) Dinosaur tracks and other fossil footprints of the Western United States. Columbia University Press, New York

Lucas SG, Spielmann JA, Klein H, Lerner AJ (2010) Ichnology of the Upper Triassic (Apachean) Redonda Formation, East-Central New Mexico. Bull N M Mus Nat Hist Sci 47:1–75

Luo Z-X (2007) Transformation and diversification in early mammal evolution. Nature 450:1011–1019

Luo Z-X, Meng Q-J, Ji Q, Liu D, Zhang Y-G, Neander AI (2015) Evolutionary development in basal mammaliaforms as revealed by a docodontan. Science 347:760–764

Luo Z-X, Meng Q-J, Grossnickle DM, Liu D, Neander AI, Zhang Y-G, Ji Q (2017) New evidence for mammaliaform ear evolution and feeding adaptation in a Jurassic ecosystem. Nature 548:326–329

Mancuso AC, Benavente CA, Previtera E, Arcucci AB, Irmis RB (2018) Carnivore coprolites from the lower Carnian (Upper Triassic) Chañares Formation, northwestern Argentina. Palaeogeogr Palaeoclimat Palaeoecol 489:15–28

Manes MIL, Silva RC, Scheffler SM (2017) Registro de uma paleotoca de tetrápode no Cretáceo da Formação Botucatu (Bacia do Paraná) no Mato Grosso do Sul. Paleo RJ/ES 2017, Rio de Janeiro, Livro de Resumos:25

Marsicano CA, Barredo SP (2004) A Triassic tetrapod footprint assemblage from southern South America: palaeobiogeographical and evolutionary implications. Palaeogeogr Palaeoclimat Palaeoecol 203:313–335

Martinelli AG, Soares MB (2016) Evolution of South American cynodonts. Mus Argent Cienc Nat "B Rivadavia." Contribuciones 6:183–197

Martinelli AG, Bonaparte JF, Schultz CB (2005) A new tritheledontid (Therapsida, Eucynodontia) from the Late Triassic of Rio Grande do Sul (Brazil) and its phylogenetic relationships among carnivorous non-mammalian eucynodonts. Ameghiniana 42:191–208

Marzola M, Mateus O, Schulp A, Jacobs L, Polcyn M, Pervov V (2014) Early Cretaceous tracks of a large mammaliamorph, a crocodylomorph, and dinosaurs from an Angolan diamond mine. 74°Annual Meeting, Society of Vertebrate Paleontology, Cincinnati, Abstract:181A

Matsuoka H, Kusuhashi N, Corfe IJ (2016) A new Early Cretaceous tritylodontid (Synapsida, Cynodontia, Mammaliamorpha) from the Kuwajima Formation (Tetori Group) of central Japan. J Vertebr Paleont 36:e1112289

Melchor RM, de Valais S (2006) A review of Triassic tetrapod track assemblages from Argentina. Palaeontology 49:355–379

Olsen PE, Galton PM (1984) A review of the reptile and amphibian assemblage from the Stormberg of Southern Africa, with special emphasis on the footprints and the age of the Stormberg. Palaeontol Afr 25:87–110

Olsen PE, Rainforth EC (2001) The "Age of Dinosaurs" in the Newark Basin, with special reference to the Lower Hudson Valley. New York State Geological Association, New York

Peixoto BCPM, Mángano MG, Minter NJ, dos Reis Fernandes LB, Fernandes MA (2020) A new insect trackway from the Upper Jurassic-Lower Cretaceous eolian sandstones of São Paulo State, Brazil: implications for reconstructing desert paleoecology. PeerJ 8:e8880

Reichman OJ, Smith SC (1990) Burrows and burrowing behaviour by mammals. In: Genoways HH (ed) Current mammalogy. Plenum Press, New York, pp 369–416

Rougier GW, Gaetano L, Drury BR, Colella R, Gómez RO, Páez Arango N (2011) A review of the Mesozoic mammalian record of South America. In: Calvo J, Porfiri J, González Riga B, Dos Santos D (eds) Dinosaurios y Paleontología desde América Latina. Editorial de la Universidad Nacional de Cuyo, Mendoza, pp 195–214

Seilacher A (2007) Trace fossil analysis. Springer, Berlin, Heidelberg, New York

Silva RC, Carvalho IS, Fernandes ACS, Ferigolo J (2008) Pegadas teromorfóides do Triássico Superior (Formação Santa Maria) do Sul do Brasil. Rev Bras Geociênc 38:100–115

# Chapter 10
# The South American Mesozoic Record and Early Evolution of Mammals

*...inside every turning leaf is the pattern of an older tree*
Sting
*I was brought to my senses*
*Mercury Falling*, 1996

**Abstract** The sparse record of archaic Mesozoic South American mammals extends from the latest Early Jurassic to the latest Cretaceous, involving about 115 Ma, which can be further extended to about 160 Ma, including the post-K/Pg evidence. We review here the distribution, predicted time of origin, and likely place of origin for the lineages covered in the preceding chapters during that span of time and against the evolving geological backdrop of continental drift and paleogeography. Size, dental diversity, and likely dietary specializations of the Mesozoic South American mammals are discussed in the context of Mesozoic mammals in general. A few of the many surprising advances in comparative genetic and molecular evolution are discussed as part of a holistic view of early mammalian evolution to which fossils can, and should, be integrated. Social, financial, and geographical issues affecting paleontological research in South America, early mammals, in particular, are highlighted. We recognize that we are still in the early stages of development and that much of what we know about Mesozoic South American mammals is likely to be drastically altered by finds in the continent or underrepresented areas from formely Gondwanan landmasses such as Antarctica or Africa. Their scarce mammalian fossil record has hampered their full incorporation into an integrated view of early mammalian evolution. The relatively robust paleontological community present in several South American countries, relatively inexpensive nature of the discipline, and extensive outcrops are likely to ensure continuity of a synergistic research agenda. The potential for novel data, regional strengths in systematics, and the global resurgent importance of time as integral to model-based phylogenies are auspicious signs for the future of Mesozoic mammal research in South America.

**Keywords** Mammalia · South America · Paleogeography · Mesozoic



G. W. Rougier et al., *Mesozoic Mammals from South America and Their Forerunners*, Springer Earth System Sciences, https://doi.org/10.1007/978-3-030-63862-7_10

## 10.1 Paleogeography, Distribution, and Paleoecology

Plate tectonics is a driving factor of evolutionary change (Large et al. 2015; Zerkle 2018) affecting climate, nutrient cycles, and geographical physical continuity. SA retained close connections even after the progressive opening of the South Atlantic during the Cretaceous (Granot and Dyment 2015; Woelders et al. 2017; Foulger 2018), particularly with Africa via the Brazilian highlands (Ezcurra and Agnolín 2012; Holz 2015; Dunhill et al. 2016; Müller et al. 2019) (Fig. 10.1). The link with Antarctica was not severed until much later in the Eocene–late Oligocene with the establishment of the Antarctic Circumpolar Current and deep water circulation between the Pacific and Atlantic Oceans (Pfuhl and McCave 2005; Sarkar et al. 2019). The idea of this landmass, SA, acting as a recognizable unit is a priori highly unlikely, though some degree of provincialism is present in the fauna and flora, even at times of broad and apparently unimpeded connection between the filial subunits of Gondwana unity (McLoughlin 2001; Ezcurra and Agnolín 2012; Césari and Colombi 2013; Frazão et al. 2015; Dunhill et al. 2016). The tectonic history of SA makes it difficult to advance continental-scale vicariant events prior to the earliest Late Cretaceous, particularly because a high level of intercontinental connectivity seems to have survived geographical fragmentation (Dunhill et al. 2016; Button et al. 2017). After the Permian–Triassic extinction event and during the first half of the Triassic, cynodont synapsids and archosauromorph diapsids exhibited a rapid radiation with main clades having a Pangeic distribution (e.g., Colbert 1973; Kemp 2005; Sues 2019; Allen et al. 2020). However, the latest Late Triassic shows a marked provincialism that after the global Triassic–Jurassic extinction event (McElwain et al. 1999; Ezcurra 2010; Dunhill and Wills 2015), is replaced by an increasing cosmopolitan biota (Button et al. 2017), reflecting a "bottleneck" effect. The foundational influence of a smaller number of surviving lineages increases their proportional representation after severe extinction. The Gondwanan partition of Pangea preserves a distinct fauna to the very Late Triassic, which includes a variety of sauropsid lineages (Sues 2019) and the non-mammaliaform prozostrodontian cynodonts (Bonaparte and Migale 2010, 2015; Martinelli and Soares 2016; Wallace et al. 2019; see Chap. 3). Through these sophisticated, small-sized Late Triassic non-mammaliaform cynodonts from Brazil and Argentina (e.g., Bonaparte and Barberena 1975, 2001; Bonaparte et al. 2003, 2005; Martinelli et al. 2005, 2016, 2017; Martinelli and Rougier 2007; Soares et al. 2014; Wallace et al. 2019; Fig. 10.1), we see the rise of mammaliaform anatomy and the biology we consider typically mammalian (e.g., Bonaparte et al. 2005; Rodrigues et al. 2013, 2014, 2019; Guignard et al. 2019).

The relatively stable Mesozoic climate (Hallam 1985; Sellwood and Valdes 2006), and the continued dominance of araucarias and podocarps (McLoughlin 2001; Césari and Colombi 2013), may have provided a continuity and stability to the Southern landmasses that appear to be absent in the North. Although landmass connectedness during the Mesozoic diminishes from a maximum achieved during the Late Triassic–Early Jurassic to a minimum reached at the latest Cretaceous, biogeographical connectedness among dinosaurs shows a steady decline not fully explained by

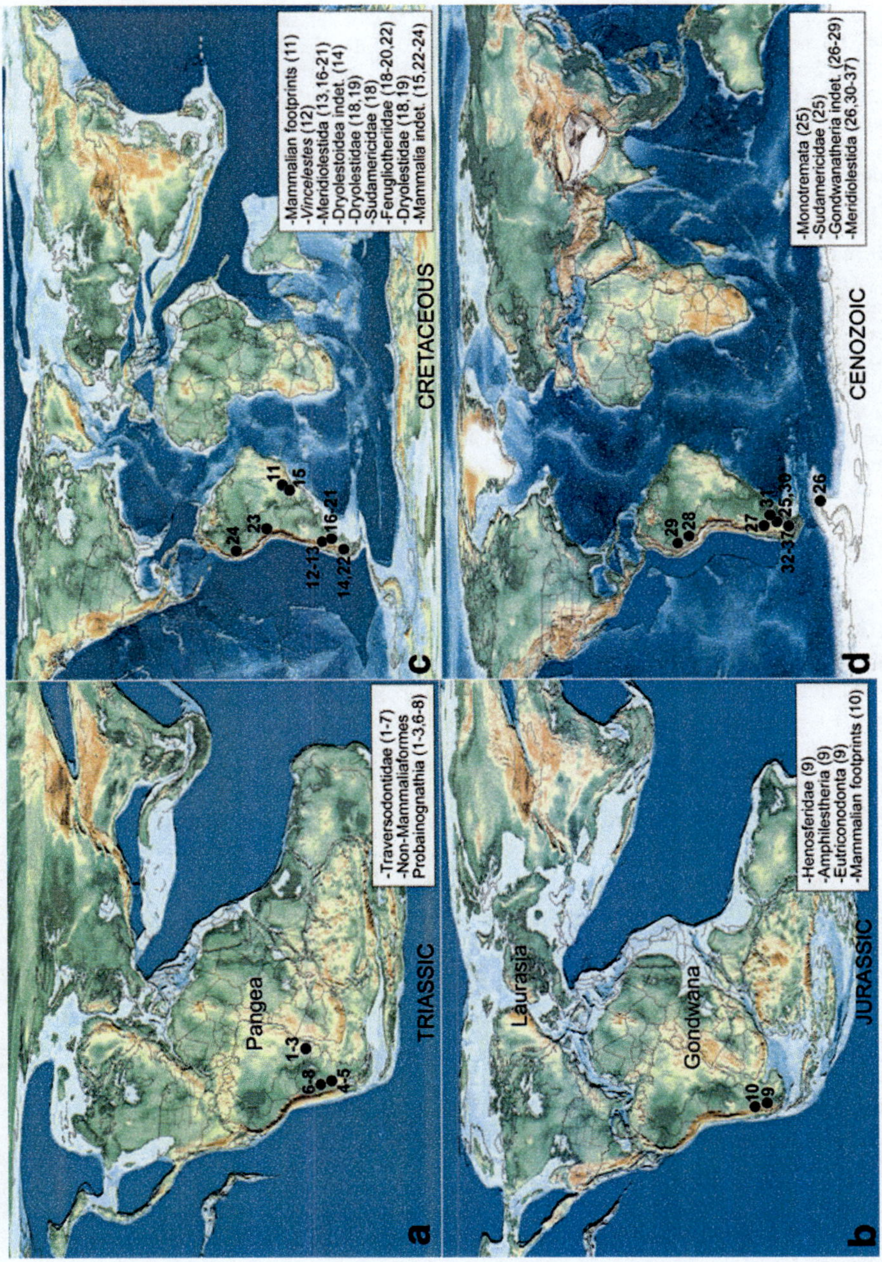
a
TRIASSIC
Pangea
-Traversodontidae (1-7)
-Non-Mammaliaformes Probainognathia (1-3,6-8)
b
JURASSIC
Laurasia
Gondwana
-Henosferidae (9)
-Amphilestheria (9)
-Eutriconodonta (9)
-Mammalian footprints (10)
c
CRETACEOUS
-Mammalian footprints (11)
-Vincelestes (12)
-Meridiolestida (13,16-21)
-Dryolestoidea indet. (14)
-Dryolestidae (18,19)
-Sudamericidae (18)
-Ferugliotheriidae (18-20,22)
-Dryolestidae (18, 19)
-Mammalia indet. (15,22-24)
d
CENOZOIC
-Monotremata (25)
-Sudamericidae (25)
-Gondwanatheria indet. (26-29)
-Meridiolestida (26,30-37)

◀**Fig. 10.1** The high-level SA fossil record in its paleogeographic context. South America as part of the supercontinent Pangea at the beginning of the Mesozoic (Late Triassic map taken from Scotese 2013a) (**a**); SA as part of the supercontinent Gondwana (Middle Jurassic map taken from Scotese 2013b) (**b**) and as part of Occidental Gondwana at the end of the Mesozoic (Late Cretaceous map taken from Scotese 2013c) (**c**); SA isolated from other landmasses at the mid-Cenozoic (middle/late Miocene map taken from Scotese 2013d) (**d**). Abbreviations for localities: TRIASSIC: 1, Candelária region, Rio Grande do Sul, Brazil; 2, Faxinal do Soturno region, Rio Grande do Sul, Brazil; 3, Santa Maria region, Rio Grande do Sul, Brazil; 4, Puesto Viejo, Mendoza, Argentina; 5, Uspallata, Mendoza, Argentina; 6, Ischigualasto-Talampaya Parks, San Juan and La Rioja, Argentina; 7, Los Colorados, La Rioja, Argentina; 8, El Carrizal, San Juan, Argentina. JURASSIC: 9, Queso Rallado, Chubut, Argentina; 10, Laguna Manantiales, Santa Cruz, Argentina. CRETACEOUS: 11, São Bento, Araraquara, São Paulo, Brazil; 12, La Amarga, Neuquén, Argentina; 13, La Buitrera, Río Negro, Argentina; 14, Tres Lagos, Santa Cruz, Argentina; 15, Santo Anastácio, São Paulo, Brazil; 16, Los Barreales Lake, Neuquén, Argentina; 17, Paso Córdoba, Río Negro, Argentina; 18, Ea. Los Alamitos, west slope of Cerro Cuadrado, Río Negro, Argentina; 19, Cerro Tortuga, Río Negro, Argentina; 20, Mirasol Chico Canyon, Chubut, Argentina; 21, Ingeniero Jacobacci, Río Negro, Argentina; 22, Río de Las Chinas Valley, Última Esperanza Province, Chile/Alta Vista and La Anita farms, Santa Cruz, Argentina; 23, Pajcha Pata, Cochabamba, Bolivia; 24, Synclinal de Bagua, Peru. CENOZOIC: 25, Punta Peligro, Chubut Argentina; 26, Seymour Locality IAA 90/1, Seymour Island, Antarctic Peninsula; 27, La Barda, Chubut, Argentina; 28, Santa Rosa, Ucayali, Peru; 29, Contamana, Loreto, Peru; 30, Gran Barranca and 31, Gaiman Chubut, Argentina; 32, Monte Observación; 33, La Cueva; 34, Monte León; 35, Estancia La Costa; 36, Killik Aike Norte, and 37, 8 km south of Coy Inlet, Santa Cruz, Argentina

geographical fragmentation but influenced by it (Dunhill et al. 2016). In the case of mammals, the sampling density is inferior to that of dinosaurs, with large-scale temporal and geographical gaps that are likely to bias results even more severely (Figs. 10.1, 10.2; see also Chap. 2).

A priori, Jurassic SA faunas should be integral members of global ecosystems where highly specialized and local groups are not to be expected. Pascual and Ortiz-Jaureguizar (2007) reviewed the evidence and recognized a "Gondwanan Episode" for the Mesozoic mammalian faunas of SA; however, they did so with little to no reference to phylogenetic structure or sister group distribution. At the time, there was a single reported Jurassic mammal, the australosphenidan *Asfaltomylos patagonicus* (Rauhut et al. 2002; Martin and Rauhut 2005; see Chap. 4), which led Pascual and Ortiz-Jaureguizar (2007), to consider the Mesozoic record *in toto*. Gaetano and Rougier (2012) discussed the biogeographic pattern of eutriconodonts and amphilestherian "triconodonts" as pertinent to the description and study of the taxa from the Lower–Middle Jurassic Cañadón Asfalto Formation (Chap. 5). The results are highly sensitive to sampling, and therefore, susceptible to rapid change; however, based on the topology of the phylogenetic trees, both *Argentoconodon fariasorum* and *Condorodon spanios* suggest "triconodont" high-level cosmopolitanism during the Early Jurassic from a common Late Triassic–earliest Jurassic distribution and a limited degree of local endemism (Gaetano and Rougier 2012). Study of surprisingly complete and well-preserved material from the Early Cretaceous of Spain described later (Martin et al. 2015), arrived at a similar tree topology for "triconodonts". These phylogenetic hypotheses imply very long ghost lineages for all major monophyletic "triconodont" lineages, such as amphilestherians and eutriconodonts, predicting an Early Jurassic minimal age (Fig. 10.4).

Australosphenidans, and in particular, henosferids paint a similar picture. The henosferids are currently known only in SA with sister groups distributed in the southern continents (Chap. 4). They seem to be a local SA product of a group more widely distributed, at the very least, throughout Gondwana. However, the frequent recovery of *Shuotherium*, from the Late Jurassic of China (Chow and Rich 1982) and the Middle Jurassic of England (Sigogneau-Russell 1998), at the base of the clade (Luo et al. 2002, and following analyses; see Chap. 4), suggests australosphenidans took origin from an ancient and widespread group. On that basis alone, the association of Australosphenida with southern continents is likely to be artifactual. Additionally, the phylogenetic position of henosferids is far from settled. The dentitions are remarkably progressive, with fully differentiated, basined talonids, but a protocone might (Rougier et al. 2007a), or might not (Martin and Rauhut 2005; Davis 2011), be present (Fig. 10.3). The lower jaws retain a plesiomorphic dentary morphology and the inference of a full set of postdentary elements (Rougier et al. 2007a). The position of this clade within the mammalian tree is the result of the interplay of these competing sets of characters, with the dentition favoring stem therian affinities and the overall dentary morphology a basal mammalian position; in most interactions they cluster with prototherians, having monotremes as crown members. Krause et al. (2014, not included in Krause et al. 2020) recovered henosferids as stem therians instead of as stem monotremes. If that were the case, the expectation of australosphenids being restricted to southern continents is made further unlikely. As in the case of the "triconodonts", *Henosferus molus* and the closely related *Asfaltomylos patagonicus* may also support a local radiation of a lineage of cosmopolitan distribution.

The Early Cretaceous *Vincelestes neuquenianus* (Bonaparte 1986a; Bonaparte and Rougier 1987; Rougier et al. 1992), most consistently recovered as a stem therian (Fig. 10.2; Chap. 7), argues a similar case as the henosferids if they are proven to be related to the origin of therians. *Vincelestes* has no close relatives either in or out of SA; it is highly specialized and autapomorphic, reflecting an artificially long branch that arguably would be shortened once our Late Jurassic–Early Cretaceous sampling improves. As such, *Vincelestes* is ill-suited to paint a detailed picture of the dentition of the last common ancestor of *Vincelestes* and therians beyond the most basic and fundamental elements of its molar pattern (Fig. 10.3). *Vincelestes* is, however, deeply nested amidst Laurasian taxa (Fig. 7.1; Chap. 7), arguing once again for a local product of a group of broader distribution. As long as *Juramaia* (Luo et al. 2011), reported to come from the lower Upper Jurassic Tiaojishan Formation (Zhang et al. 2008; Wang et al. 2017), is recovered as a therian (Bi et al. 2018), the likely minimal age for Prototribosphenida (the taxon formed by *Vincelestes* plus Theria and their descendants) is Middle Jurassic (Fig. 10.4). However, doubts were raised regarding *Juramaia*'s age by King and Beck (2020), because of poor fit in dated Bayesian phylogenies when the age is not constrained, indicating a faulty provenance for the specimen or a very precocial basal therian taxon. It appears likely that more than endemism and disjointed biogeographic patterns, the sparse Mesozoic record of pre-Late Cretaceous mammals from SA is more the reflection of the amount of missing

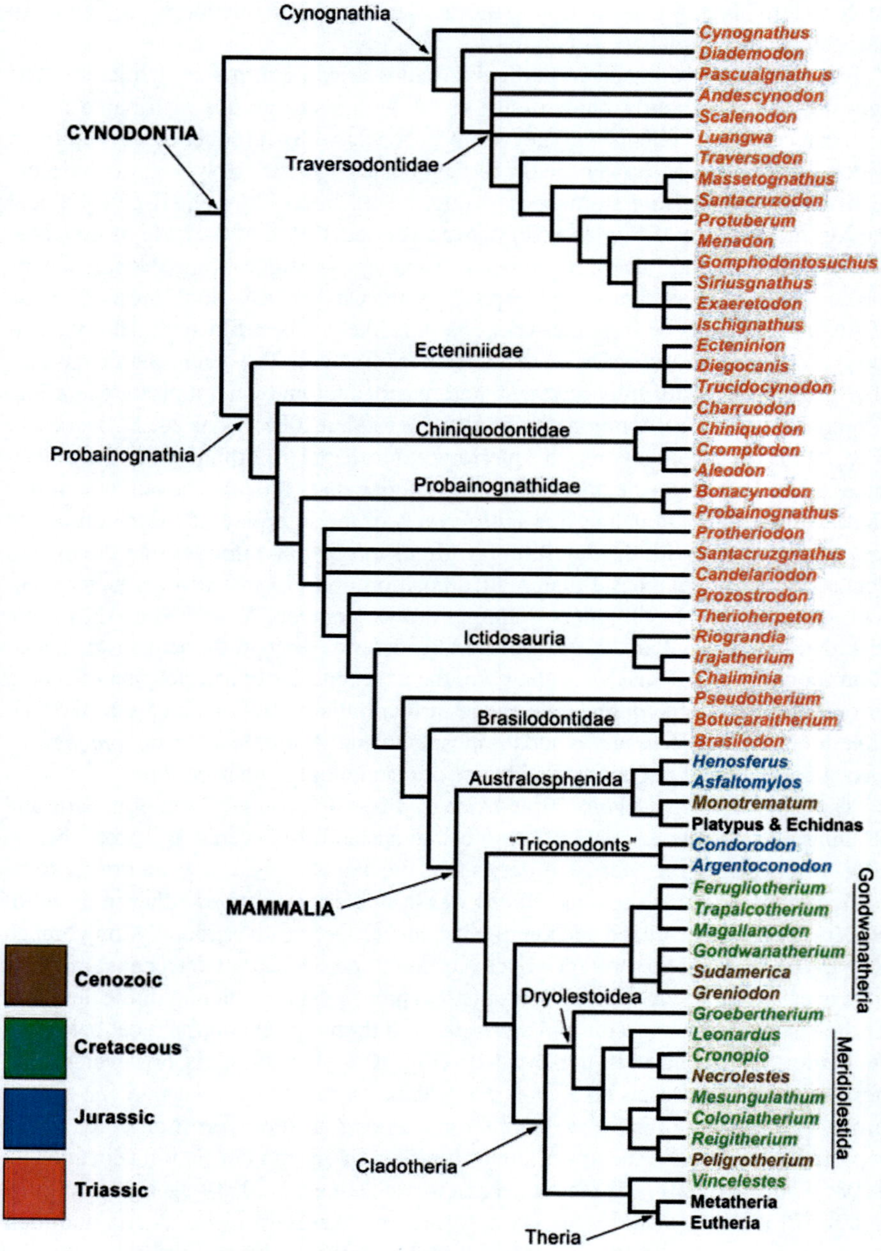


**Fig. 10.2** Phylogenetic relationships of most taxa discussed throughout the book (see previous Chapters), including the non-mammaliaform cynodonts and Mesozoic mammal lineages from SA. Phylogeny follows publications and discussion in each specific chapter

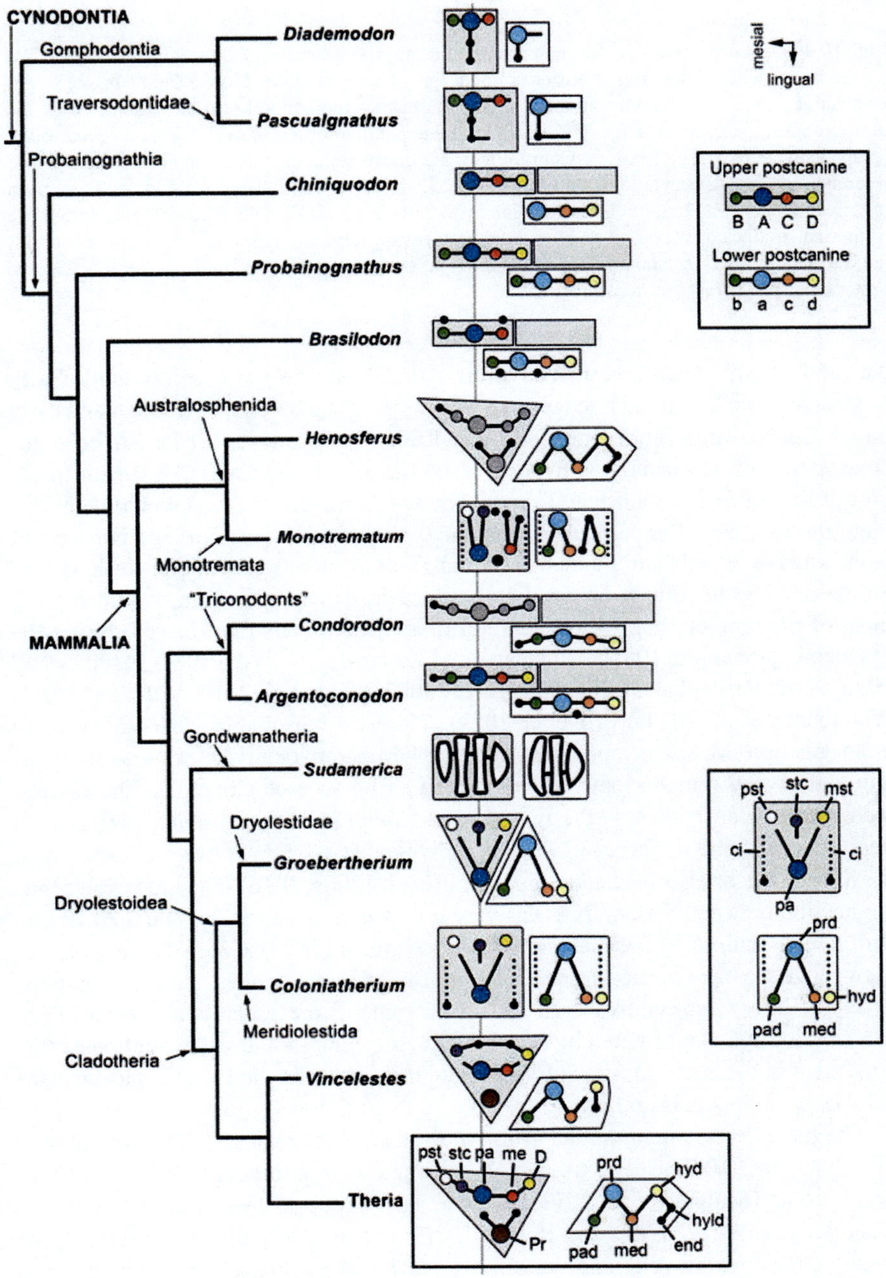

CYNODONTIA
Gomphodontia
Traversodontidae
Probainognathia
Diademodon
Pascualgnathus
Chiniquodon
Probainognathus
Brasilodon
Australosphenida
Henosferus
Monotrematum
Monotremata
MAMMALIA
"Triconodonts"
Condorodon
Argentoconodon
Gondwanatheria
Sudamerica
Dryolestidae
Groebertherium
Dryolestoidea
Coloniatherium
Meridiolestida
Cladotheria
Vincelestes
Theria
mesial
lingual
Upper postcanine
B A C D
Lower postcanine
b a c d
pst stc mst
ci ci
pa
prd
hyd
pad med
pst stc pa me D
Pr
prd hyd hyld
pad med end

◂**Fig. 10.3** Dental diversity of SA non-mammaliaform cynodonts and Mesozoic mammals. We reconstructed a linear diagram of posterior upper (grey background) and lower (white background) postcanine teeth in occlusal view in selected taxa to illustrate tooth morphologies. In some cases, the morphologies are hypothetical as is the case for the upper dentition of *Henosferus* and *Condorodon* (cusps indicated in grey). Main cusp homologies are traced along the teeth using colored cusps and reference inserts. Abbreviations: A–a; B–b; C–c; D–d, upper and lower cusp nomenclature used in "triconodonts" and basal non-mammalian cynodonts. Hypothetical cusp homologies with those of therians indicated by colors; ci, cingulum; end, entoconid; hyd, hypoconid; hyld, hypoconulid; me, metacone; med, metaconid; mst, metastyle; pa, paracone; pad, paraconid; pr, protocone; prd, protoconid; pst, parastyle; stc, stylocone

data and random collection than anything else. Ghost lineages extend into the Early to Mid-Jurassic for all the lineages in SA (Fig. 10.4; see below). As recognized early on by Bonaparte (1986b; Bonaparte and Kielan-Jaworowska 1987), SA becomes connected with Laurasia, putatively NA, by the end of the Cretaceous, with major groups spreading both north and south. The Late Cretaceous fauna would be in this view the product of endemism and isolated evolution in SA, perhaps portions of Gondwana divorced from Laurasia. Therians would enter SA from the North, spread South, and by the advent of the K/Pg events, or shortly afterward, would replace most of the archaic, non-tribosphenic groups of mammals that thrived during the Mesozoic (Bonaparte 1986b, Bonaparte and Kielan-Jaworowska 1987; Pascual et al. 1996, 2001; Pascual and Ortiz-Jaureguizar 2007; Goin et al. 2012; Wilf et al. 2013; Woodburne et al. 2014a, b; Defler 2019). This is a reasonable view; so far there is not a single Mesozoic specimen that can be unequivocally referred to therians, although a few remains have been claimed to be so (see Chap. 2). The earliest record of therians in SA is the metatherian *Cocatherium lefipanum*, from Danian deposits (earliest Paleocene) of Patagonia, collected at ~5 m above the level where the K/Pg boundary has been identified (Goin et al. 2006a, 2016). The Late Cretaceous connection between SA and NA was very unfortunately called the "First American Biotic Interchange" by Pascual and Ortiz-Jaureguizar (2007). Given the very limited knowledge we have of the Mesozoic faunas of earlier exchanges, particularly in the Jurassic–Early Cretaceous, earlier faunal movements are possible and even likely. The possibility is highlighted by taxa long documented in Laurasian continents that are absent in the Early Jurassic of Patagonia, but present in the Late Cretaceous (see below), such as dryolestoids.

The most diverse mammalian group in the Late Cretaceous of SA is the Meridiolestida (Fig. 10.2; Chap. 6), which we regard as a monophyletic branch of dryolestoids (Rougier et al. 2011), or as a sister group to them closer to therians (Rougier et al. 2012; Wible and Rougier 2017; but see alternative interpretation for some of the taxa by Averianov et al. 2013). The oldest known meridiolestidan is *Cronopio dentiacutus* (Rougier et al. 2011), from the Cenomanian (early Late Cretaceous) of Northern Patagonia. *Groebertherium* (Bonaparte 1986a; Rougier et al. 2009a), from the Campanian–Maastrichtian (Late Cretaceous), also from northern Patagonia, appears to cluster with Laurasian Dryolestidae (Rougier et al. 2011; Averianov et al. 2013; O'Meara and Thompson 2014). The minimal age for the last

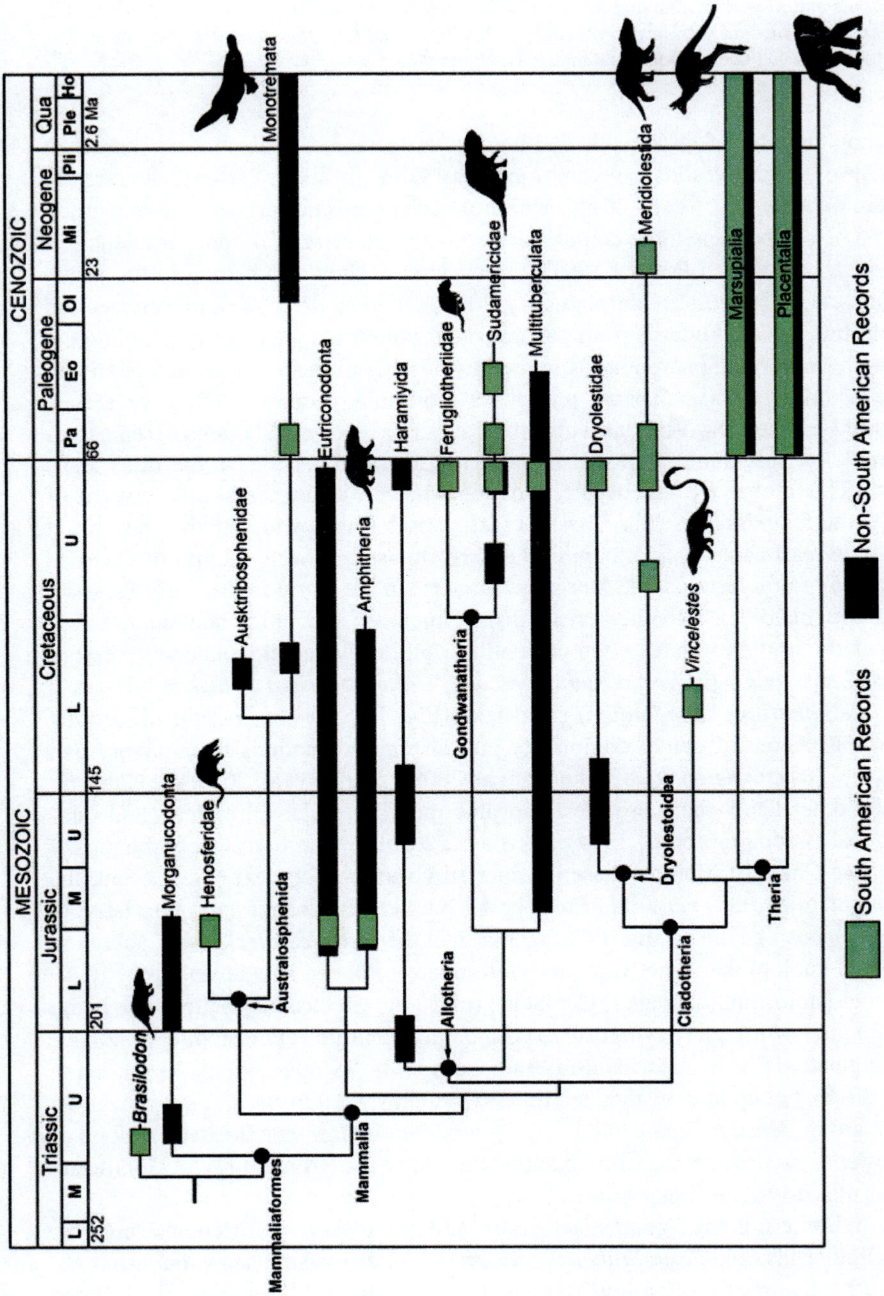
MESOZOIC
CENOZOIC
Triassic
Jurassic
Cretaceous
Paleogene
Neogene
Qua
L
M
U
L
M
U
L
U
Pa
Eo
Ol
Mi
Pli
Ple
Ho
252
201
145
66
23
2.6 Ma
Brasilodon
Morganucodonta
Henosferidae
Australosphenida
Ausktribosphenidae
Monotremata
Eutriconodonta
Amphitheria
Haramiyida
Allotheria
Gondwanatheria
Ferugliotheriidae
Sudamericidae
Multituberculata
Dryolestidae
Dryolestoidea
Meridiolestida
Vincelestes
Cladotheria
Theria
Marsupialia
Placentalia
Mammaliaformes
Mammalia
South American Records
Non-South American Records

◂**Fig. 10.4** Time-calibrated phylogenetic relationships of main groups of non-mammaliaform cynodonts and Mesozoic mammal lineages, with emphasis on the SA record. Abbreviations: Eo, Eocene; Ho, Holocene; L, Lower; M, Middle; Mi, Miocene; Ol, Oligocene; Pa, Paleocene; Ple, Pleistocene; Pli, Pliocene; Qua, Quaternary; U, Upper

common ancestor of dryolestids and meridiolestidans is Middle Jurassic based on both the oldest dryolestids (Averianov et al. 2014) and the likely earliest Late Jurassic therian *Juramaia* (Luo et al. 2011). Meridiolestidans are currently an endemic group from SA, but perhaps, they could also be present in Africa, if some taxa such as *Donodon* (Sigogneau-Russell 1991), are eventually shown to belong to the group (Bonaparte 1994, 2002; Chornogubsky 2011; see Chap. 6). Meridiolestidans and dryolestids would originate from generalized dryolestoids, possibly resembling the more plesiomorphic paurodontids in the low number of molariforms and relatively large talonids. *Groebertherium* presents a morphology unremarkable for a Late Jurassic Holarctic Dryolestidae, but in the Late Cretaceous of Patagonia, it represents a remarkable case of morphological stasis and survival in SA of a Dryolestidae sensu stricto. The inference is that dryolestoids should be present in the Middle Jurassic of the southern landmasses (Fig. 10.4); in fact, considering how distinctive they are, it would be reasonable to find them in the Early Jurassic as well. At present, there is no record of dryolestoids in SA or anywhere else in the world in the Early Jurassic.

Mesungulatoidea (Rougier et al. 2011), includes the more bunodont, thick-enameled meridiolestidans, often presenting well-developed cingula and accessory cusps. Large (e.g., *Peligrotherium tropicalis*, *Coloniatherium cilinskii*) and small (e.g., *Reigitherium bunodontum*) sized taxa (Fig. 10.5), were present, illustrating a varying degree of dental complexity and distinct adaptations to omnivory and herbivory (Páez Arango 2008; Rougier et al. 2009b; Harper et al. 2019; see Chap. 6). On the other hand, the more plesiomorphic meridiolestidans are small and thin-enameled (Crompton et al. 1994), such as *Leonardus cuspidatus* (Chornogubsky 2011; see Chap. 6). Both of these lineages survived the K/Pg extinction event, the mesungulatoids (i.e., *Peligrotherium*) being relatively abundant and quite large in the Paleocene of Punta Peligro (Patagonia) and the rare *Necrolestes* spp. sounding the death knell of the insectivore meridiolestidans in the early Miocene (Fig. 10.4). Mesungulatoids minimal age is Coniacian (mid-Late Cretaceous) represented by an edentulous jaw that, however, can be confidently identified (Forasiepi et al. 2012). Mesungulatoids have superficial resemblances with "symmetrodonts" (Averianov et al. 2013), but up to now, they are limited in their record to SA and possibly are a native group. *Mesungulatum* and relatives would arise after the effective break up of Gondwana and the opening of the South Atlantic, on the SA part of west Gondwana during the early Late Cretaceous.

Ferugliotheriids and gondwanatherians present two very different scenarios. Ferugliotheriids (i.e., *Ferugliotherium windhauseni*, *Trapalcotherium matuastensis*, and perhaps *Magallanodon baikashkenke*) as a group are known exclusively from the Late Cretaceous of Patagonia (Bonaparte 1986a; Krause et al. 1992; Rougier et al. 2009a; Goin et al. 2020); they are distinct from any other Mesozoic mammals

CENOZOIC
MESOZOIC
Neogene
Paleogene
Cretaceous
Jurassic
Triassic
23
66
145
201
252 Ma

◂**Fig. 10.5** Selected SA non-mammaliaform cynodonts and Mesozoic mammal lineages illustrating shape and size disparity through time. **Triassic** (left to right): *Exaeretodon argentinus* (Cynodontia, Traversodontidae) from the upper Carnian Ischigualasto Formation, Argentina; *Pascualgnathus polanskii* (Cynodontia Traversodontidae), from the upper Ladinian-lower Carnian Río Seco de la Quebrada Formation, Argentina; *Probainognathus jenseni* (Cynodontia, Probainognathia), from the lower Carnian Chañares Formation, Argentina (below); *Brasilodon quadrangularis* (Cynodontia, Probainognathia) from the lower Norian Candelária Sequence, Brazil (above); *Chaliminia musteloides* (Cynodontia, Probainognathia), from the Norian Los Colorados Formation, Argentina. **Jurassic** (left to right): *Henosferus molus* (Australosphenida, Henosferidae), *Condorodon spanios* (Amphilestheria) and *Argentoconodon fariasorum* (Eutriconodonta, Triconodontidae), the three taxa from the Toarcian–Bajocian Cañadón Asfalto Formation, Argentina. **Early Cretaceous**: *Vincelestes* (Cladotheria) from the Barremian–lower Aptian La Amarga Formation, Argentina. **Late Cretaceous**: (left to right): *Cronopio dentiacutus* (Dryolestoidea, Meridiolestida) from the Cenomanian Candeleros Formation, Argentina; *Leonardus cuspidatus* (Dryolestoidea, Meridiolestida) from the Campanian–Maastrichtian Los Alamitos Formation, Argentina; *Coloniatherium cilinskii* (Dryolestoidea, Meridiolestida, Mesungulatoidea) from the Campanian–Maastrichtian La Colonia Formation, Argentina; *Reigitherium bunodontum* (Dryolestoidea, Meridiolestida, Mesungulatoidea) from the Campanian–Maastrichtian Los Alamitos and La Colonia formations, Argentina; *Gondwanatherium patagonicum* (Gondwanatheria, Sudamericidae) from the Campanian–Maastrichtian Los Alamitos Formation, Argentina; *Ferugliotherium windhauseni* (Gondwanatheria or Multituberculata, Ferugliotheriidae) from the Campanian–Maastrichtian Los Alamitos and La Colonia formations, Argentina. **Paleogene** (left to right): *Peligrotherium tropicalis* (Dryolestoidea, Meridiolestida, Mesungulatoidea), *Sudamerica ameghinoi* (Gondwanatheria, Sudamericidae), *Monotrematum sudamericanum* (Monotremata, Ornithorhynchidae), the three taxa from the Danian Salamanca Formation. **Neogene**: *Necrolestes patagonensis* (Dryolestoidea, Meridiolestida) from the lower Miocene Sarmiento Formation, Argentina. Silhouettes by Jorge L. Blanco

and at present, the easiest interpretation is that they are an endemic SA clade. The question becomes more complicated once their relationships to other groups are considered. Bonaparte et al. (1989) and Krause et al. (1992) built the case for multituberculate affinities. If the Late Cretaceous plagiaulacoid premolars described from Patagonia (Kielan-Jaworowska and Bonaparte 1996; Kielan-Jaworowska et al. 2007), are indeed *Ferugliotherium,* as argued previously (Gurovich and Beck 2009; Rougier et al. 2009a) and here (Chap. 8), the case becomes quite strong. If this were the case, missing diversity of multituberculates in the southern continents becomes evident. A cosmopolitan distribution of multituberculates in the southern continents has been argued on the basis of another plagiaulacoid premolar, in this case, from the Aptian of Australia. Rich et al. (2009) described *Corriebaatar marywaltersae* as a cimolodontan multituberculate, a group that is otherwise known from the Early Cretaceous–Eocene of Laurasia and possibly the Late Cretaceous of Argentina. Other putative remains of multituberculates from Gondwana are much less convincing, a fragmentary Late Cretaceous tooth from Madagascar (Krause et al. 2006; Krause 2013), and some possible multituberculates from the Late Cretaceous of Morocco (Sigogneau-Russell 1991; Hahn and Hahn 2003, 2007), or perhaps haramiyidans (Butler and Hooker 2005). The Late Cretaceous ferugliotheriids cannot be regarded as early immigrants from NA where the much better multituberculate record shows no similar morphology. A pre-Late Cretaceous presence of cimolodontan multituberculates in the Gondwana filial masses is to be expected if they are to reflect a vicariant event with regard to the northern members of the group. A dispersal during the Early Cretaceous appears biogeographically more difficult and earlier dates are not yet supported by specimens.

Alternatively, ferugliotheriids can be part of the gondwanatherian radiation representing a basal and somewhat plesiomorphic group recovered as the sister group to the remaining gondwanatherians (Krause and Bonaparte 1993; Krause et al. 2014, 2020; Goin et al. 2020). Gondwanatherians (e.g., *Gondwanatherium patagonicum*, *Sudamerica ameghinoi*, and perhaps *Greniodon sylvaticus* from SA, *Bharattherium bonapartei* from India, *Lavanify miolaka*, *Vintana sertichi*, and *Adalatherium hui* from Madagascar, perhaps *Galulatherium jenkinsi* from Tanzania, and other indeterminate material) have, so far, a purely Gondwanan record (reviewed recently by Krause et al. 2019, 2020; Goin et al. 2020), all of them being Late Cretaceous or younger (see Chap. 8). The most complete specimens are those from the Maastrichtian of Madagascar (Krause et al. 2014, 2020), which have recently provided a remarkably complete picture of the bizarre *Vintana sertichi* and *Adalatherium hui*. Phylogenetic analyses of these relatively complete specimens recovered in most instances gondwanatherians as a nested group of allotherians. The sister group to gondwanatherians is the Jurassic Euharamiyida (Krause et al. 2020) or the Triassic Haramiyida (Krause et al. 2014 and some analyses of Krause et al. 2020), rendering the minimal age for the origin of the group at least Middle Jurassic (Fig. 10.4). As in the case of other SA groups, gondwanatherians may be the result of a local endemism in Gondwana, but their origin likely took place against a paleogeographical backdrop that in theory would be amenable to a more widespread Jurassic distribution.

The presence of gondwanatherians in SA, Antarctica, Madagascar, and India is well-supported. The restudy of a fragmentary jaw (O'Connor et al. 2019), that had remained unnamed for a few years (Krause et al. 2003), resulted in the erection of a new taxon of a likely gondwanatherian, *Galulatherium jenkinsi*. This form is very peculiar and unlike other gondwanatherians, which are characterized by a rodentiform jaw and molariforms with heavy enamel furrows and fossae; *Galaulatherium* lacks enamel, the teeth are columnar with no furrows or fossae, and lacks a rodentifom incisor. The tall columnar teeth do resemble some gondwanatherians like *Gondwanatherium*, *Sudamerica*, and *Bharattherium* (Bonaparte 1986c; Pascual et al. 1999; Prasad et al. 2007). It is possible that *Galulatherium* is a gondwanatherian, but at present, we remain skeptical; the characters and taxa sampled tend to cluster together rodentiform morphologies. *Galulatherium* would be the only gondwanatherian from continental Africa, and if its affinities are confirmed, it implies an early, likely Jurassic origin for Gondwanatheria as outlined above. If the Tanzanian mammal is not a Gondwanatheria, or if it is an immigrant from Madagascar or another former Gondwanan landmass, the origin of Gondwanatheria could be an event occurring post-Gondwana break up. India/Madagascar had been separated from Africa and Antarctica, by 130 Ma in the Hauterivian–Barremian (Matthews et al. 2016), so gondwanatherians must have been on board at the start of the long journey of the Indian subcontinent. At this time, the Southern Atlantic was still incompletely opened, and Africa and

SA shared extensive contact. Antarctica and Africa certainly hold the keys to further refine this scenario, which at present can only be tentatively advanced.

Monotremes in SA have been interpreted as representing an early Cenozoic migration from Australia, a "reverse migration" (Pascual et al. 1992b; Rich and Vickers-Rich 2012; Rich et al. 2009; Goin et al. 2016; Krause et al. 2019). At its core, the argument rested on the absence of monotremes in the relatively abundant Upper Cretaceous collection of the Los Alamitos Formation, a case that could also be made with the more recent discoveries from the similarly aged Allen and La Colonia formations. The monotremes, or the prototherian lineage, is present in Australia, at least since the Early Cretaceous (Rowe et al. 2008; Rich et al. 2016). However, it is hard to provide much weight to negative evidence when the record is so incomplete. If the Australian fauna were randomly sampled, it would be unlikely monotremes would be recorded, even more unlikely, if the sampling corresponded to a relatively circumscribed area (as is the case regarding the mostly Patagonian record for the Late Cretaceous of SA) that fails to record favorable ecological conditions. *Monotrematum* is dentally (Pascual et al. 1992a, b; Fig. 10.3; see also Chap. 4), similar to ornithorhynchids, a fact also receiving limited support from fragments of maxilla (Carlini et al. 2002) and femora (Forasiepi and Martinelli 2003). It is, nevertheless, uncertain if morphological similarities to fossil platypuses are not in fact plesiomorphic monotreme features (Rowe et al. 2008; Phillips et al. 2009), given the relatively modern split of echidnas and platypuses predicted by molecular phylogenies (Phillips et al. 2009), and a common amphibious ancestry. The near-littoral conditions for the Los Alamitos, Allen, and La Colonia formations may well be unfavorable for the preservation of monotremes. The added fact that at least under some assumptions, the henosferids are a basal member of the prototherian lineage, makes the possibility of Mesozoic SA monotremes, or close relatives, likely (Fig. 10.4). The role of Antarctica during the Cretaceous as a faunal reservoir, or center of origin, for mammalian clades, is at present wholly unknown, but serving as an inert bridge communicating the southern fragments of Gondwana is currently a necessary but unrealistic notion.

Therians, though unknown in the Mesozoic of SA (see Chap. 2), are nonetheless integral parts of the discussion on the origins of the mammalian fauna of SA. The establishment of a nested Marsupialia within metatherian groups of Asian-NA distribution (Rougier et al. 1998, and later versions), implied southward dispersal of metatherians and the likely origin of crown group marsupials locally in SA, with the Australian marsupials reaching their current homeland via Antarctica (e.g., Woodburne and Case 1996; Muizon et al. 1997; Goin et al. 2016). Most analyses recognize the carnivorous Sparassodonta as non-marsupial metatherians (e.g., Rougier et al. 1998, 2004; Forasiepi 2009; Engelman and Croft 2014; Forasiepi et al. 2015; Wilson et al. 2016; Muizon et al. 2018; Prevosti and Forasiepi 2018), whose stem lineages entered SA in the earliest Paleocene or more likely in the yet undocumented Late Cretaceous. The Paleocene Bolivian locality of Tiupampa (~ 64 Ma; Woodburne et al. 2014a), yielded about a dozen species each of metatherians and eutherians (e.g., Marshall et al. 1985; Muizon 1991a; Muizon et al. 2015, 2018). These therians from Tiupampa and the polydolopimorphian *Cocatherium*, stratigraphically very close to the K/Pg boundary (Goin et al. 2006a, 2016), argue strongly in favor of a pre-Paleocene ingression of metatherians, which unlike most of the eutherian clades, show the beginning of endemic diversification (e.g., Croft et al. 2020). It has been argued based on extremely fragmentary material that the origin and diversification of several non-marsupial metatherian groups can be traced to the Late Cretaceous of NA (e.g., Case et al. 2005), a result that can also be deduced from ghost lineage analyses of some of the recent phylogenetic results (e.g., Williamson et al. 2014; Wilson et al. 2016). Phylogenies also suggest that not one, but several different lineages entered SA (e.g., Horovitz et al. 2008; Forasiepi 2009; Williamson et al. 2012, 2014; Wilson et al. 2016; Ladevèze et al. 2020), more easily explaining the documented Paleocene diversity. Metatherians radiated in SA into different taxonomic groups and morphotypes that include mainly small to large-sized carnivores (more than 200 kg), and small to medium-sized herbivores including: frugivorous, granivorous, folivorous, and omnivorous (e.g., Abello et al. 2012; Zimicz 2012, 2014a, b; Goin et al. 2016; Prevosti and Forasiepi 2018). SA metatherians include several stem groups (e.g., Jaskhadelphyidae, Pucadelphyidae, Protodidelphidae, Derorhynchidae, Sternbergiidae; Marshall 1987; Muizon 1991a; Muizon et al. 1997; Goin et al. 2016; Muizon et al. 2018; collectively referred to as Didelphimorphia or "Ameridelphia" *incertae sedis*; Williamson et al. 2014; Goin et al. 2016). Sparassodonta, a stem-marsupial group of predators (e.g., Prevosti and Forasiepi 2018 and references there) and Polydolopimorphia (e.g., Chornogubsky 2010) likely another stem-marsupial group (e.g., Beck 2017; but see alternative placement for Polydolopimorphia in Goin et al. 2009, 2016; Chornogubsky and Goin 2015), which includes some taxa with multicusp bunodont molars proceeded by a plagiaulacoid-like premolar. Similar dental architecture has also been reported in abderitids and caenolestid paucituberculatans, and phalangerid, burramyid, and macropodid diprotodont marsupials (e.g., Simpson 1933; Archer 1984; Abello 2013; Black et al. 2012). The crown group Marsupialia in SA is represented by three groups: (1) Paucituberculata, which includes several fossil species and the living shrew opossums (e.g., Abello 2007, 2013; Goin et al. 2009); (2) Microbiotheria, with the Monito del Monte and relatives, crucial to understanding the early Australian radiation (Szalay 1982, 1994; Kirsch et al. 1991; Springer and Kirsch 1991; Springer et al. 1998; Horovitz and Sánchez-Villagra 2003; Meredith et al. 2009); and (3) Didelphimorphia, including extinct and living opossums (e.g., Voss and Jansa 2009; Beck and Taglioretti 2019).

In contrast to metatherians, SA eutherians show very close phylogenetic ties with northern ungulate-like taxa (Muizon and Cifelli 2000; O'Leary et al. 2013; Billet et al. 2015; Muizon et al. 2015), which has been used to argue for a more recent arrival of those eutherians, perhaps post-K/Pg (Flynn et al. 2012; Goin et al. 2012, 2016; Muizon et al. 2015; Croft 2016; Krause et al. 2019). Eutherians diversified quickly in SA, mostly occupying the ecological role of omnivores and herbivores in all its types: browsers and grazers (e.g., Vizcaíno et al. 2012a; Madden 2015; Cassini 2013), small to large-sized and exceptionally reaching more than one ton body mass, representing components of the SA megafauna (e.g., Vizcaíno et al. 2012b; Fariña et al. 2013). Early in the Paleogene, placentals in SA were represented by native ungulates—collectively, the Sout American Native Ungulates, or SANUs—(Notoungulata, Litopterna, Astrapotheria, Pyrotheria, Xenungulata), xenarthrans (Pilosa and Cingulata) (e.g., Patterson and Pascual 1968, 1972; Simpson 1980; Marshall and Cifelli 1990; Croft et al. 2020), and several lineages of basal groups (e.g., Pantodonta, Palaeoryctidae, and archaic ungulate-like forms, such as Didolodontidae and Mioclaenidae "condylarths") (e.g., Simpson 1967; Muizon and Marshall 1987a, b; Muizon 1991b; Muizon and Cifelli 2000; Muizon et al. 2015; Gelfo et al. 2007, 2009; Gelfo and Sigé 2011). The ancestry of these lineages rests among NA taxa, migrating into SA sometime in the very latest Cretaceous or beginning of the Cenozoic. SANUs (exemplified by notoungulates and litopterns) may have had close phylogenetic affinities with Perissodactyla (Buckley 2015; Welker et al. 2015; Westbury et al. 2017), and their origin has been estimated at the very beginning of the Cenozoic (Westbury et al. 2017). Phylogenetic position of xenarthrans is still unsettled, with alternative views in dispute (dos Reis et al. 2012; Gaudin and Croft 2015; Gibb et al. 2016; Bargo and Nyakatura 2018, and references therein). If the topology that supports xenarthrans as an early branch of the placental tree, with all other placental groups jointed in a monophyletic Epitheria, is correct (McKenna 1975; O'Leary et al. 2013), the cladogenetic event involved in the origin for the group is certainly an early event, in agreement with molecular estimates, going back to the latest Cretaceous (e.g., Bininda-Emonds et al. 2007; Delsuc et al. 2016) or the earliest Paleocene (e.g., Delsuc et al. 2004; Presslee et al. 2019). A SA origin for xenarthrans cannot be excluded but is currently unsupported.

To this group of archaic SA placentals, immigrants from Africa (rodents and monkeys), were added later in the Paleogene. Fossil evidence establishes that caviomorph rodents and platyrrhine primates have been in SA since the middle Eocene (Antoine et al. 2012, 2016, 2017; Bond et al. 2015; Assemat et al. 2019). This endemic fauna of SA, which would be the consequence of the continent's isolation for millions of years (e.g., Simpson 1980; but see Wilf et al. 2013), ended during the late Neogene, following the connection of SA to NA, which allowed the arrival of new placental groups (Carnivora, Artiodactyla, Perissodactyla, Gomphotheriidae, Cricetidae, and Hominidae) from NA and in relation to the Great American Biotic Interchange (GABI) (e.g., Webb 1976, 1985, 1991; Woodburne 2010; Cione et al. 2015).

Compared with the Cenozoic, the Mesozoic record of mammals in SA is still too sparse to provide a coherent and detailed biogeographical picture. The very general proposal of Bonaparte (1986b, Bonaparte and Kielan-Jaworowska 1987), of a distinct SA Cretaceous fauna, still holds sway, but it is unclear if the mammalian taxa are in fact of Patagonian, SA, or Gondwanan distribution. The pre-Late Cretaceous record suggests the presence of a cosmopolitan fauna with low-level local diversification events, in agreement with a higher degree of connectivity among the precursors of the current continental masses. The survival across the K/Pg extinction horizon of gondwanatherians, two lineages of meridiolestidans, and likely native Mesozoic monotremes (but see contrary views in Pascual et al. 1992b; Rich et al. 2009; Goin et al. 2016), suggests the transition in SA, and likely Gondwana as a whole, was less severe (Pascual et al. 2001; Pascual and Ortiz-Jaureguizar 2007; Rougier et al. 2011), than the well-documented cataclysmic disruption in NA (Lyson et al. 2019).

The survival of these archaic lineages into the Cenozoic, however, had little to no impact on the later fauna, surviving as minor components until at least the early Miocene in SA, Eocene of Antarctica, and reaching the present day in Australia. Both metatherians and eutherians arrived from NA (either latest Late Cretaceous or earliest Paleocene) and rapidly diversified, giving rise to a variety of non-marsupial metatherians, marsupials, and the earliest representatives of the SA native placentals. The crown group Marsupialia likely originated in SA; the lack of tropical and equatorial record of Mesozoic/earliest Paleocene SA mammals allows, however, for a modicum of equivocation.

Some stem marsupials in the early Paleogene and microbiotherians seem to have served the role of small insectivores and faunivorous mammals, an ecological role occupied by dryolestids and the more generalized meridiolestidans during the Late Cretaceous and in much-diminishing numbers in the Cenozoic of SA and Antarctica. Other metatherians (e.g., polydolopimorphians and paucituberculatans) may have served the role of small-sized herbivores and frugivores, as ferugliotheriids did in the Late Cretaceous; however, ferugliotheriids do not appear to reach the Paleocene.

Among placentals, archaic-ungulates such as "condylarths" and SANUs may have replaced in the ecosystems the specialized mesungulatoid meridiolestidans, and perhaps also the rodentiform gondwanatherians, as the main herbivores of the post-Cretaceous SA faunas. Mesungulatoids did not survive the Paleocene, while gondwanatherians appear to have survived until the middle Eocene (Goin et al. 2006b, 2012), or perhaps the Oligocene, relying on the putative findings from some Neotropical sites (Goin et al. 2004; Antoine et al. 2012, 2016). The arrival into SA of stem Caviomorpha from Africa, in the middle Eocene, may have played a role in the extinction of gondwanatherians, where perhaps not only ecological replacement but competitive displacement might also happen, as was proposed for the decline and extinction of multituberculates in NA (Krause 1986). If the claim made by Antoine et al. (2012, 2016), of gonwanatherians in the site of Contamana is correct, the SA Neotropics may record both rodentiform groups simultaneously, setting a paleoecological test case between these two largely convergent lineages of mammals.

In sum, therians, both metatherians and eutherians, formed the bulk of the Cenozoic SA mammals, with the fauna completed by a few relicts of Mesozoic lineages. Metatherians and less likely eutherians are to be expected in the late Mesozoic of SA, particularly in the largely unsampled northern SA. Therians occupied in the Cenozoic a variety of adaptive zones, which we believe were exploited by archaic mammalian lineages during the Mesozoic. In SA, the last gondwanatherians date to the middle Eocene or perhaps Oligocene, while the youngest meridiolestidan dryolestoid is from the early Miocene as a minoritarian component. The consolidation of the mammalian fauna in SA, as we know it at present, is a much more recent event. The late Neogene climatic changes, extinction event, and faunal interchange (GABI) had a pivotal impact and today placentals are the major component of the total SA mammalian fauna.

## 10.2 Paleobiological Corollaries from the SA Mammals

The fossil record of Mesozoic mammals in SA is geographically very disparate and temporally skewed toward the Late Cretaceous (Fig. 10.1), conceivably giving us only a partial view of the fauna. The answer to the question: is there anything peculiar about the Mesozoic mammals of SA?, is by and large equivalent to: is there anything peculiar about the Cretaceous mammals of Patagonia?

In Recent mammals, it has been long known that phylogenetic proximity is a reliable predictor of body size correlation for larger mammals (Smith et al. 2004), which in their study is regarded as heavier than 18 gr. Smaller mammals do not appear to have a strong phylogenetic signal but are shown to be the most speciose taxa. On the other hand, Cope's Rule (or Depéret's Rule) posits that mammalian lineages tend to increase in size over time (Depéret 1907, 1909). Support for this idea has been found in paleontological studies and in combined studies including both Recent and extinct taxa (Slater 2013; Saarinen et al. 2014; Bokma et al. 2016), but often not in data sets using only Recent species, or when they were analyzed separately (Alroy 1998; Smith et al. 2010; Monroe and Bokma 2010; Raia et al. 2012); however, Baker et al. (2015), did support Cope's rule based on an updated methodological approach of older databases. The most recent review of the problem provides additional support for a general validity of Cope's rule in combined datasets of fossil and living mammals (Bokma et al. 2016), where the size trend in a lineage would not be caused by the anagenetic change in a species, but by speciation events and cladogenesis. However, patterns of change within each lineage seem to be different from each other and most of the rate variations (either increase or decrease) can be correlated with a few ecological and environmental variables (Cooper and Purvis 2010). Higher estimated rates of body size evolution were found to be present in cold, low-lying mainland ecosystems. Baker et al. (2015) concluded that an increase in body size is more than twice as likely as a decrease in a randomly given lineage.

The increase of body size is near exponential following the K/Pg transition (Smith et al. 2010), following a similar trajectory and eventually leveling to a similar plateau across all major continents. The best-fit model for body size evolution has been defended as the Ornstein-Uhlenbeck process until the K/Pg event and a Brownian motion process from the Cenozoic onwards. Cooper and Purvis (2010) proposed instead an "Early Burst" model. Herbivores do not become substantially larger than omnivores and faunivorous mammals until later in the Cenozoic (Alroy 1998; Smith et al. 2010), and most Mesozoic taxa (with the possible exception of gondwnatherians) were not strict herbivores, but within a range of omnivory that included a varying degree on non-animal food sources, similar to rodent models (Martin et al. 2016).

Mesozoic mammals are small in general (see Kielan-Jaworowska et al. 2004), as were mammaliaforms and their closest relatives, the non-mammaliform prozostrodontians (with the exception of tritylodontids) (Chap. 3). Disruptions on the size structure of communities following extinction events, resulting in smaller-sized descendent faunas, has been daubed the "Lilliput" effect (Urbanek 1993; Twitchett

2007; Huttenlocker 2014), and it has played a role in the achievement of the "small size" as a distinctive plesiomorphic condition for the origin of Mammalia (Rowe 1988; Lautenschlager et al. 2019). Body size is a powerful predictor of a multitude of biological properties in mammals, such as population density (Darmuth 1981; Stephens et al. 2019), home range and habitat (Ofstad et al. 2016), lifestyle (Sikes and Ylönen 1998; Sibly and Brown 2007), metabolic rate, and body temperature (Avaria-Llautureo et al. 2019).

The known Mesozoic mammals from SA are not unusual regarding body size; most are relatively small (Fig. 10.5), with molariforms in the vicinity of a millimeter. This is true for all the mammals from the Lower–Middle Jurassic Cañadón Asfalto Formation, which includes henosferids and "triconodonts", with *Argentoconodon* being significantly larger and robust, but still on par with other mid-sized "triconodonts" (Cifelli and Madsen 1998; Butler and Sigogneau-Russell 2016). However, the ichnogenus *Ameghinichnus* from the Middle Jurassic Laguna Manatiales Formation includes the species *Ameghinichnus manantialensis* that reaches a stride length of 10 cm and hand/foot impression of approximately 17 mm (de Valais 2009; see Chap. 9). This is a larger mammal, if it is indeed a mammal, than any other known by osteological evidence in SA. The Early Cretaceous *Vincelestes* is large for Mesozoic mammals (Fig. 10.5), roughly similar to an opossum, with the bodyweight averaging 1 kg (650–1200 gr estimate; Rougier 1993; Macrini et al. 2007; see Chap. 7). Only the faunivorous gobiconodontids and relatives from the Early Cretaceous of Asia and NA are in some instances larger, in a few cases much larger (over 12 kg), such as *Repenomamus giganticus* (Hu et al. 2005).

The Jurassic and Early Cretaceous lineages are not represented in later SA sediments, and therefore, it is not known how these lineages changed over time or what trends may have been present. *Argentoconodon* and *Vincelestes* appear to be highly specialized with a large cumulative number of derived traits telegraphing a long phyletic history. Regardless of the potential morphology of the last common ancestor of these taxa and the rest of Mammalia, they fit nicely within the general framework of Mesozoic mammals regarding their size and morphology. What is strange about *Vincelestes* is its age and geographical position, not its morphology. Had *Condorodon*, *Argentoconodon*, or *Vincelestes* been found in Laurasia, they would be incorporated into the ongoing phylogenetic research and would complete much needed missing information, but not alter the general outline of our understanding of early mammalian evolution. Henosferids, on the other hand, are a problem altogether different. The contrasting mandibular and dental morphologies represent a novel morphotype with a generalized "tribosphenic" dentition (Fig. 10.3), acquired very early in the Jurassic and a primitive mandibular morphology present in successive mammalian sister groups. The putatively related *Ausktribosphenos* and other relatives possess complex teeth that make cusp homologies uncertain; the mandibular structure on the other hand is less radically plesiomorphic. *Asfaltomylos* and *Henosferus* are two taxa that could not have been predicted a priori; if they are a poorly known stage of the evolution of basal mammals in general or just an oddity of SA is not yet fully known. Resolution of this problem depends on the recovery of specimens preserving other character systems able to disentangle the contradicting phylogenetic signals of the dentition and lower jaw. Henosferids and *Vincelestes* attest to the early presence of small to medium size tribosphenic, or nearly so, mammals that appear not to have closely related forms in the Late Cretaceous or Cenozoic. Why tribosphenic forms were dominant in Laurasia but disappeared in the Mesozoic of Gondwana is a difficult question to answer.

The basal meridiolestidan *Cronopio* (Cronopioidea nov.; see Chap. 6) is a small mammal (skull about 2.5 cm length), as are the Late Cretaceous non-mesungulatoid meridiolestidans (Fig. 10.5). These sharp-toothed meridiolestidans, such as *Leonardus,* embody the image of the archetypical Mesozoic mammal: tall, sharp, unbasined teeth (Fig. 10.3), long skinny snout, and likely with a nearby insect, a grub, or a small vertebrate for a snack. The bunodont and more omnivorous mesungulatoids are of a variety of sizes ranging from the minute Late Cretaceous *Reigitherium* to the very large and robust Paleocene *Peligrotherium* (Fig. 10.5). *Coloniatherium* from the Late Cretaceous is morphologically similar to the patronymic *Mesungulatum* but substantially larger (the size of a large opossum). A dated morphocline can be recognized between *Mesungulatum, Coloniatherium,* and *Peligrotherium,* with a progressive increase in size and crown complexity (Harper et al. 2019), in the younger taxa. *Reigitherium* and *Peligrotherium* show a degree of dental complexity functionally associated with omnivory and herbivory comparable to that of small therians. *Peligrotherium* is the largest known non-therian mammal in SA culminating the mesungulatoid expansion away from the primitive sharp insectivorous dentitions of basal meridiolestidans, which continued with relatively minor nuances until the Miocene as represented by the Antarctic record (Martinelli et al. 2014) and the Patagonian *Necrolestes* (Rougier et al. 2012; Wible and Rougier 2017). The mesungulatoid radiation, profound dental complexity, and the relatively colossal size achieved by *Peligrotherium* are uniquely SA events for which we lack, at present, any other parallel.

The enigmatic gondwanatherians are the most herbivorous group (Bonaparte 1986c; Koeniswald et al. 1999; Gurovich 2008) of the Mesozoic of SA and possibly of the Mesozoic in general. The complex crown surface, transverse lophs, thick enamel, and member taxa of large to very large body size support that inference. The Malagasy *Vintana* and *Adalatherium* have an estimated weight of about 3 kg and 9 kg, respectively (Krause et al. 2014, 2020). The SA gondwanatherians are not small, but do not reach the size of *Vintana.* The recently described *Magallanodon* (Goin et al. 2020) and *Sudamerica* (Pascual et al. 1999), have first lower molariforms approximately 10 mm long, while *Gondwanatherium* would be approximately 2/3 the size (Bonaparte 1986c). If ferugliotheriids are related to gondwanatherians, they represent the plesiomorphic morphology for the group, with low crowns, distinct cusps, and modestly developed crests. The convoluted and sturdy hypsodont molariforms of *Gondwanatherium* and relatives represent the earliest known occurrence of this dental type in mammals. Simple high-crowned molariforms were reported in the unrelated Jurassic *Fruitafossor* (Luo and Wible 2005). The achievement of large, thickly enameled and hypsodont cheek teeth in the later members of the group mirrors to some degree the event occurring among mesungulatoids. Remarkably, hypsodont molariforms are also achieved by the Miocene meridiolestidan *Necrolestes* (Scott 1905; Goin et al. 2007; Rougier et al. 2012; Wible and Rougier 2017).

The Mesozoic mammals of SA appear to be structured along similar lines as those from other continents, with most taxa fulfilling the iconic "small insectivore" niche even if done by mammals showing an unusual combination of features, such as henosferids. The lack of definite omnivore–herbivores of gliriform shape during the Jurassic–Early Cretaceous is likely artefactual. Multituberculates, haramiyidans, and euharamiyidans are common on other continental masses, present in the Jurassic, and given the paleogeography and minimal clade ages, are to be expected as members of the early communities in SA. The Late Cretaceous continues to have a variety of small meridiolestidan "insectivores" as dominant, records the presence of the omnivore-insectivore feruglıotheriids, and the rise of the more strictly herbivore hypsodont gondwanatherians. Mesungulatoid meridiolestidans expand into the omnivore niche by the development of prominent cingula, thick enamel, and complex, more bunodont molariforms. A relative increase in size and adaptations reflecting an emphasis on omnivory and herbivory appears to be a distinctive trend in the Late Cretaceous–Paleocene archaic mammals of SA.

Fossils are a prime source of hard morphology: teeth, bones, and the functional inferences deriving from ossified structures, a classical view epitomized by G.G. Simpson's "Fossils and the History of Life" (Simpson 1983), which to the present day embodies the bulk of the paleontological research agenda. Other aspects of the phenomic spectrum are much harder to extract from the fossilized past: soft tissue, life history, behavior, physiology, etc., are intractable and inferentially problematic (Ravosa et al. 2016).

Not all is a lost cause, as osteological or biochemical correlates of morphological characters can allow inferences predicting the condition in ancient mammals. The study of the ear region, for example, has long been promising given the tight osteological enclosure of the main auditory organs and the peculiar mammalian auditory performance so distinct from that of other vertebrates. A narrative of the rise of mammalian hearing drawing on both modern physiology and paleontological osteology has been advanced and refined (Manley 2000, 2001, 2017, 2018; Grothe and Pecka 2014; Köppl and Manley 2018; Harper and Rougier 2019). Similarly, other biological attributes of early mammals can be extrapolated from a combination of fossilized specimens and comparative anatomy, physiology, etc. Growth rates, life span, and basal metabolic rates have been estimated based on enamel/dentin deposition rates (O'Meara et al. 2018; Newham et al. 2020). Bone histology, fur impressions, and dental complexity can be used as proxies to infer other aspects of archaic mammalian physiology and development (Chinsamy and Hurum 2006; Jasinoski and Chinsamy 2012; Kolb et al. 2015; Botha-Brink et al. 2018; LeBlanc et al. 2018). In the case of Mesozoic taxa, promising comparisons and inferences are often undermined by the lack of appropriate models; that is, if a feature is "mammalian" in nature, it follows that it is absent in non-mammals. This apparent platitude is, however, relevant: crown group Mammalia is the sister group of the remaining amniotes, each one with an independent evolutionary history of about 320 Ma (see Ford and

Benson 2020 for a recent review). While understanding the osteological transformations (such as those affecting the middle ear) constitutes one of the triumphs of comparative anatomy (Maier and Ruf 2016; Anthwal et al. 2020), the physiological and functional implications of those events are much more poorly established. "Transitional" forms and peculiar morphologies abound for which no modern models exist, raising the spectrum of futility; if something is known in a modern model the fossil may not be informative, and if it is unknown in modern taxa, the interpretation can only be an inferential and tentative, detracting from the informational content. This may be an overly conservative view of the value of specimens falling outside the inference of phylogenetic brackets (Witmer 1995, 1998), particularly those that represent very long branches, as is the case for mammals among amniotes. Within mammals, monotremes represent an ancient group preserving a large number of features we expect to have been present in the last common ancestor of mammals, like their reproduction, body temperature, overall osteology, brain morphology, etc. (Griffiths 1978; Ashwell 2013), thus providing a vital outgroup for comparing fossil Mesozoic mammals. We would be at a loss if therians were the sole source and model of non-osteological data for archaic mammals.

Application of evolutionary logic to the origin of iconic mammalian molecules and gene evolution have, in some instances, helped propose a series of molecular events/physiological consequences that can be used to explain potentially correlated features seen in fossils. A primary example is the origin of the most iconic mammalian feature, mammary glands and their product: the milk. A paleontological approach to the origin of mammae and milk appears, at least at present, unrewarding; however, comparative anatomy has long established the underlying homology between mammary and cutaneous glands (Gegenbaur 1875; Klaatsch 1884). The precise nature of the glandular precursor of mammary glands was still considered unsettled at the time of the influential review by Blackburn (1991); nevertheless, more recent research suggests apocrine glands associated with hair as the primitive type gland from which mammary glands arose (Oftedal 2002a, b, 2012). The immune antimicrobial and some enzymatic components of milk are also found in sweat glands or share with those products great similarity (Oftedal 2013; Skibiel et al. 2013; Oftedal et al. 2014; Urashima et al. 2014; Power and Shulkin 2016). The origin of the nutritional components of the milk, particularly the protein fraction, was more obscure. Caseins are a uniquely mammalian family of phosphoproteins, universally present in all extant mammals (Lefèvre et al. 2009) and forming up to 80% of all protein in milk. Caseins form micellae stabilized by calcium ions and hydrophobic interactions, behaving as a colloid in the milk solution (de Korth et al. 2011; Bhat et al. 2016). Loss of functionality of the three vitellogenin genes ancestral to amniotes in therians and reduction to a single functional gene in monotremes (Brawand et al. 2008; Robinson 2008; Warren et al. 2008), can be correlated to the increased role of caseins as calcium suppliers to developing therians. Monotreme membranous eggs are small (between 2–4 mm; Griffiths 1978), permeable, and known to absorb uterine substances (Oftedal 2002a, b). The hatchlings are extremely altricial, requiring prolonged milk nurturing (Griffiths 1978), that takes place by licking milk from the base of the hairs of the female's nippleless areola (mammary patch). Milk replaces the yolk as the main source of calcium in the evolution of mammals, and caseins, present in monotremes and therians, become ubiquitous (e.g., Rijnkels 2002; Lefèvre et al. 2009, 2010; Enjapoori et al. 2014). A strong case can be made for the origin of the caseins by duplication and modification of the enamel matrix proteins (Kawasaki et al. 2011). Caseins, enamel matrix proteins, and salivary proteins are all part of a family of secretory calcium binding phosphoproteins-SCPP (Hall 2015), which serve a central role in the degree of mineralization of vertebrate tissues (Kawasaki 2009). Enamel matrix proteins function as a patterning network for the deposition of the inorganic hydroxyapatite crystals and are ancient proteins preceding the origin of amniota (Sire et al. 2007; Gasse et al. 2015; Gasse and Sire 2015).

We suggest that the origin of caseins and mammalian salivary proteins by diversification of the SCPP gene family may be related to another important change in mammalian enamel microstructure: the acquisition of prismatic enamel. Unlike most non-mammalian amniotes, the mammalian enamel is spatially highly organized (Koenigswald and Sanders 1997; Wood and Rougier 2005; Spoutil et al. 2010), with a variety of characteristics that have often been somewhat imprecisely characterized as "prismatic enamel" or "rod enamel". Non-mammaliaform probainognathians (Chap. 3) lack prismatic enamel structure; however, early mammaliaforms show various degrees of enamel organization, which led to contradictory descriptions of the enamel microstructure (Wood et al. 1999). As a result, enamel structure appeared to evolve multiple times among lineages in what has been dubbed "underlying homology" (Koenigswald and Clemens 1992). Wood and Rougier (2005) argued that the idea of "prismatic enamel" could be better approached by deconstructing this character complex in simpler units, each one with its independent phylogenetic history. It is nevertheless clear that the enamel of mammaliaforms is more organized in its deposition than the plesiomorphic synapsid columnar enamel, prevalent even among specialized more basal synapsids (Sander 1997; Wood et al. 1999; O'Meara et al. 2018; Whitney and Sidor 2019). A higher degree of enamel ultrastructure must be mediated by corresponding changes in the enamel matrix proteins, given the central role they play in Recent taxa (Koenigswald and Sanders 1997; Mathur and Polly 2000; Fukumoto et al. 2014). The expansion and modification of the SCPP gene family was taking shape among basal mammaliaforms; it is likely that the origin of caseins and salivary proteins are also part of the dramatic reorganization and expansion of the gene domain substantially predating the origin of Mammalia. If this proves to be the case, several of the most iconic mammalian features such as milk, enamel structure, and the digestive role of oral saliva may all be part of a complex transformation of a gene family eventually affecting seemingly unrelated character systems that predate the origin of Mammalia.

Hairs, evolution of brown fat, origin of the diaphragm, diversification of opsins and mammalian vision, etc., are susceptible to similar treatment via a comparative molecular and genetic approach (Wu et al. 2008; Oelkrug et al. 2013; Borges et al. 2018; Sefton et al. 2018; Dhouailly et al. 2019; Jastroch and Seebacker 2020). No typical mammalian feature evolved abruptly with no homology to an ancestral form; gene duplication and modification of genes and expression pathways already present in ancestral taxa serve as the natural base for the origin of new molecules and their function, as shown above for casein proteins (Kawasaki et al. 2011) or for α-Lactalbumin from lysozyme genes (Blackburn et al. 1989). The affordability of high output gene sequencing coupled with the relative ease of alignment and recognition of homologous sequences promises to afford us a much deeper understanding of the series of molecular events leading to the origin of the major distinguishing features of mammals. Functional predictions of hypothetical intermediate proteins resulting from known or expected genes is a burgeoning field making use of the rapidly accumulating gene sequence databases (Ijaq et al. 2019; Sureyya Rifaioglu et al. 2019). Computational exploration of the biological properties of these expected products could provide further clues on the sequence of events and function leading to the origin of mammalian features.

Fossils will likely never be able to provide a granular understanding of complex molecular transformations, even if they have reliable morphological correlates. They do, however, serve to tangibly illustrate unpredicted but viable combinations of characters associated with biogeographical data of a time in the past (e.g., Slater and Harmon 2013). The advent of dated Bayesian analysis capable of producing time-calibrated phylogenies incorporating combined molecular and phenomic data using a Fossilized Birth-Death model (Heath et al. 2014; Pyron 2016; Drummond and Stadler 2016; Zhang et al. 2016; Stadler et al. 2018), has radically changed the role of fossil evidence in a wide range of questions in evolutionary biology. The intuitive idea that the age of fossils is a valuable information is fully incorporated into Bayesian FBD models that can tackle massive total-evidence datasets (Zhang et al. 2016; Kealy and Beck 2017; Casali et al. 2020). Fossils no longer serve as an antagonistic foil for a molecular-based narrative but are essential data with an important influence on the topological outcome (Barido-Sottani et al. 2020). Discrepancies between morphological and molecular phylogenetic estimates persist, but they appear to be due to relative data sampling and not to an inherent larger amount of homoplasy in the morphological datasets (Zou and Zhang 2016). In fact, an improved fossil record can arguably improve resolution and diminish conflict between morphological and molecular resolutions of the mammalian tree (Beck and Baillie 2018). Parsimony-based phylogenies efficiently explore tree space by minimizing the amount of ad hoc change; Bayesian methods instead model the probability of these changes to occur, which may be philosophically attractive and in some instances desirable (Ramírez-Chavez et al. 2016). Fossils are today an integral part of any attempt to use time, morphology, and geography to try to untangle the early history of mammals. We can incorporate information into readily available models that will produce falsifiable predictions to be addressed by new studies or specimens. Discovery of a new group or a taxon inaccessible to molecular sampling that breaks a long ghost lineage is likely to be more consequential than one more species of an already heavily sampled group (equivalent to ameliorating long-branch attraction in molecular studies). If lucky, such a fossil is a time capsule carrying a surprising combination of characters, the product of a biogeographic scenario that may no longer exist. Therein lies the interdisciplinary promise and attraction of paleontology.

## 10.3 Perspectives and Future Directions

It has been 35 years since the rather serendipitous finding of an isolated upper molar of *Mesungulatum* in the Upper Cretaceous sediments of the Los Alamitos Formation (Bonaparte and Soria 1985). This iconic locality yielded a very rich vertebrate association from the late Mesozoic (Fig. 10.6) and the first bony remains from Mesozoic mammals for SA (Fig. 10.7). Since then, sustained effort steadily increased and improved the number, temporal, and geographical distribution of the fossil record. New discoveries of more specimens and localities followed in rapid succession during the late eighties–early nineties, setting the basic framework of SA Mesozoic mammals and resulting in the first detailed studies (Bonaparte 1986b, c, 1990, 1992, 1994; Bonaparte et al. 1989; Rougier et al. 1992; Hopson and Rougier 1993; Krause et al. 1992). The 2000's witnessed the description of the first osteological remains of Jurassic mammals (Rauhut et al. 2002; Rougier et al. 2007a, b; Gaetano and Rougier 2011, 2012), and further exceptional Cretaceous specimens (Rougier et al. 2011). An auspicious future for the research on SA early mammals and close relatives seems at hand. Several of the long-known localities are still being actively explored and yielding additional novel material; even more encouraging are the very recent discoveries from the Cretaceous of Chile (Goin et al. 2020), expanding the geographical range of Cretaceous mammals. The tantalizing, but still not very informative records of Mesozoic mammals from Brazil, Bolivia, and Peru, are bound to eventually be improved with better specimens clarifying their affinities (Castro et al. 2018; Gayet et al 2001). New techniques are unveiling microvertebrate remains previously inaccessible; advances in technology, particularly micro-computerized tomographic imaging, permit the gathering of new information from specimens previously studied via traditional methods. The paleontological community interested in non-mammaliaform cynodonts and early mammals has steadily grown in the recent past, particularly in Brazil and Argentina, which together command the vast majority of the relevant record. Given the relatively small number of senior scientists and students taking up the discipline, the personalities involved have had a large impact. The influential and foundational nature of José F Bonaparte's work is still evident in every aspect of Mesozoic SA mammals and close relatives. Bonaparte, a great collector but a divisive mentor and colleague, did not oversee a flourishing of the discipline akin to what his contemporary Rosendo Pascual achieved under similar circumstances for the study of Cenozoic mammals.

A resilient core of specialists, however, has prospered through the vagrancies of support for the sciences in Latin America, often associated with public Museums/Universities and perhaps more importantly, centralized National Research Agencies (e.g., CONICET and ANPCyT in Argentina, CNPq and CAPES in Brazil). These research programs are laced by international collaborations, often providing much needed funding and grants (e.g., National Science Foundation-USA, National Geographic Society, Paleontological Society, and Fulbright Scholar Program); explorations and research programs are, however, frequently led by local scientists. The relatively inexpensive and simple technology necessary in vertebrate paleontology

◂**Fig. 10.6** Reconstruction of the paleoenvironment and selected vertebrates found in the Campanian-Maastrichtian of the Los Alamitos Formation, Río Negro Province, Argentina. Illustrated vertebrates include the titanosaur dinosaur *Aeolosaurus*, birds, abelisaurid theropod, the hadrosaur *Secernosaurus*, chelid turtles, and pipid frogs; fishes are represented by dipnoi, siluriforms, and lepisosteiforms. Three mammals are shown in the foreground, including *Gondwanatherium*, *Leonardus* and *Mesungulatum*; no postcrania confidently referred to those taxa is known from Los Alamitos, therefore, body shapes are inferred from related taxa. Art work by Jorge L. Blanco

**Fig. 10.7** Close-up of the mammals in Fig. 10.6, Los Alamitos Formation, Campanian-Maastrichtian, Río Negro Province, Argentina. The gondwanatherian *Gondwanatherium* (**a**), and the meridiolestidans *Leonardus* (**b**) and *Mesungulatum* (**c**). Not to scale. Art work by Jorge L. Blanco

to produce valuable scientific information has made the discipline viable even in settings where academic development can be challenging. Resourcefulness, enthusiasm, and not a small degree of doggedness are necessary characteristics of anyone attempting to leave a mark in the field amidst the seemingly perennial turmoil of Latin American institutions.

The Mesozoic outcrops are extensive and relatively easy to access throughout the southern cone of SA, such as Patagonia. It is unsurprising the first teeth and jaws came from there; the unavoidable focus on the Patagonian record is likely producing a heavily distorted view of the Mesozoic in SA in general. Northern SA was tropical during both the Jurassic and the Cretaceous (Fig. 10.1; reviewed in Torsvik and Cocks 2017), so it is to be expected that the bulk of the mammalian diversity is missed by the latitudinally-skewed Patagonian record. Using the Patagonian record to represent the mammalian fauna of SA is probably analogous to the past use of the Laurasian evidence to interpret Mesozoic mammals as a whole: mandated by necessity but patently artifactual.

The eventual Mesozoic finds from tropical SA are liable to upend the admittedly rudimentary general ideas regarding Mesozoic mammals in SA. Completing the Mesozoic fossil record of other related Gondwanan landmasses, particularly Antarctica, Africa, and India, is also expected to alter our view regarding the distribution and origin of major groups. What is Pangeic, Gondwanan, SA, etc., is at present established mostly by the presence of remains in a single or few localities in a continent that cannot be very easily compared to similarly aged localities in other continents. The Late Cretaceous is the best documented Mesozoic Period in SA; however, there are no equivalent localities (regarding time and paleoenvironment) in Antarctica and Australia, and only a handful in Africa, Madagascar, and India. Antarctica, a drawbridge of central position among the filial Gondwanan landmasses, is the big question mark; a few discoveries there can radically alter much of what we know.

Much certainly is left to discover in SA; new groups, we do not even suspect are sure to appear. Currently, we recognize a minimum of 20 Mesozoic mammalian genera (Fig. 10.2), spanning from the latest Early Jurassic to the Late Cretaceous, an average of 0.15 taxa per million years, a paltry record compared with the roughly 6500 living mammalian genera (Burgin et al. 2018). Certainly, past diversity may have been substantially different than that of the Cenozoic, or present-day; the case is, however, undeniable that we "for now see through a glass darkly" trying to collage what we know into a coherent picture. The intrinsic incompleteness of fossil data makes the phylogenetic position of many taxa uncertain; phylogeny, or more precisely the ancestor-descendant relationship, is the only unifying property of life (Darwin 1859). The topology of a tree determines the minimal age for a group, its biogeographic history, its evolutionary trends, its heyday, and time of demise. Incompletely known taxa and incompletely known groups exacerbate these problems. Portions of the Mesozoic mammal tree are relatively stable; we are confident that dryolestoids are relatively close to stem therians, "triconodonts" are stem therian mammals basal to the clade formed by dryolestoids and therians, etc. (Figure 10.2). Other groups, like henosferids, monotremes, and gondwanatherians are controversial or tentative regarding their high-level relationships. Given their great antiquity, henosferids have an oversized impact on the minimal age of Mammalia, as early members of either the monotreme or therian stem lineage (Fig. 10.3). Achieving stability in the phylogenetic relationships of multituberculates, gondwanatherians, "haramiyidans", monotremes, and australosphenidans would go a long way to help narrow viable options regarding biogeography and time of origin of major SA Mesozoic clades.

We argue that Mesozoic mammals from SA remain mostly unknown; we have a few precious and tantalizing specimens, but they cannot be more than a teaser of things to come. The scaffold of the early mammalian tree rests heavily on the Laurasian record; SA is currently the best positioned of the southern continents to significantly contribute to achieve a more balanced view until other places of the former Gondwanan landmasses unveil more fully their riches of Mesozoic mammals.

We invite you to rewrite some, much, or everything we say here by studying museum collections if you must, but hopefully by heading out to the field and finding that specimen that solves a stubborn question. Science will be grateful for the new data, and you can keep the excitement and happiness that comes from discovery and playing with ideas: the best deal in the world!

## References

Abello MA (2007) Sistemática y bioestratigrafía de los Paucituberculata (Mammalia, Marsupialia) del Cenozoico de América del Sur. Unpublished PhD thesis, Universidad Nacional de La Plata

Abello MA (2013) Analysis of dental homologies and phylogeny of Paucituberculata (Mammalia: Marsupialia). Biol J Linn Soc 109:441–465

Abello MA, Ortiz-Jaureguizar EO, Candela AM (2012) Paleoecology of the Paucituberculata and Microbiotheria (Mammalia, Marsupialia) from the late Early Miocene of Patagonia. In: Vizcaíno SF, Kay RF, Bargo MS (eds) Early Miocene Paleobiology in Patagonia. Cambridge University Press, Cambridge, High-latitude paleocommunities of the Santa Cruz Formation, pp 156–172

Allen BJ, Wignall PB, Hill DJ, Saupe EE, Dunhill AM (2020) The latitudinal diversity gradient of tetrapods across the Permo-Triassic mass extinction and recovery interval. Proc R Soc Lond B 287:20201125

Alroy J (1998) Cope's Rule and the dynamics of body mass evolution in North American fossil mammals. Science 280:731–734

Anthwal N, Fenelon JC, Johnston SD, Renfree MB, Tucker AS (2020). Transient role of the middle ear as a lower jaw support across mammals. eLife 2020:e57860

Antoine PO, Marivaux L, Croft DA, Billet G, Ganerød M, Jaramillo C, Martin T, Orliac MJ, Tejada-Lara J, Altamirano AJ, Duranthon F, Fanjat G, Rousse S, Salas-Gismondi RS (2012) Middle Eocene rodents from Peruvian Amazonia reveal the pattern and timing of caviomorph origins and biogeography. Proc R Soc Lond B 279:1319–1326

Antoine P-O, Abello MA, Adnet S, Altamirano Sierra AJ, Baby P, Billet G, Boivin M, Calderon Y, Candela A, Chabin J, Corfu F, Croft DA, Ganerød M, Jaramillo C, Klaus S, Marivaux L, Navarrete RE, Orliac MJ, Parra F, Pérez ME, Pujos F, Rage J-C, Ravel A, Robinet C, Roddaz M, Tejada-Lara JV, Vélez-Juarbe J, Wesselingh FP, Salas-Gismondi R (2016) A 60 million-year Cenozoic history of western Amazonian ecosystems in Contamana, eastern Peru. Gondwana Res 31:30–59

Antoine P-O, Salas-Gismondi R, Pujos F, Ganerød M, Marivaux L (2017) Western Amazonia as a hotspot of mammalian biodiversity throughout the Cenozoic. J Mammal Evol 24:5–17

Archer M (1984) The Australian marsupial radiation. In: Archer M, Clayton G (eds) Vertebrate zoogeography and evolution in Australia. Hesperian Press, Carlisle, pp 633–708

Ashwell KWS (ed) (2013) Neurobiology of monotremes: Brain evolution in our distant mammalian cousins. CSIRO Publishing, Collingwood, Victoria

Assemat A, Boivin M, Marivaux L, Pujos F, Benites-Palomino A, Salas-Gismondi R, Tejada-Lara JV, Varas-Malca RM, Negri FR, Ribeiro AM, Antoine P-O (2019) Restes inédits de rongeurs caviomorphes du Paléogène de la région de Juanjui (Amazonie Péruvienne): Systématique, implications macro-évolutives et biostratigraphiques. Geodiversitas 41:699–730

Avaria-Llautureo J, Hernández CE, Rodríguez-Serrano E, Venditti C (2019) The decoupled nature of basal metabolic rate and body temperature in endotherm evolution. Nature 572:651–654

Averianov AO, Martin T, Lopatin AV (2013) A new phylogeny for basal Trechnotheria and Cladotheria and affinities of South American endemic Late Cretaceous mammals. Naturwissenschaften 100:311–326

Averianov AO, Martin T, Lopatin AV (2014) The oldest dryolestid mammal from the Middle Jurassic of Siberia. J Vertebr Paleontol 34:924–931

Baker J, Meade A, Pagel M, Venditti C (2015) Adaptive evolution toward larger size in mammals. Proc Natl Acad Sci USA 112:5093–5098

Bargo MS, Nyakatura JA (2018) Morphology and evolution of the Xenarthra: An introduction. J Mammal Evol 25:445–447

Barido-Sottani J, van Tiel NMA, Hopkins MJ, Wright DF, Stadler T, Warnock RCM (2020) Ignoring fossil age uncertainty leads to inaccurate topology and divergence time estimates in time calibrated tree inference. Front Ecol Evol 8:183

Beck RMD (2017) The skull of *Epidolops ameghinoi* from the Early Eocene Itaboraí Fauna, Southeastern Brazil, and the affinities of the extinct marsupialiform Order Polydolopimorphia. J Mammal Evol 24:373–414

Beck RMD, Baillie C (2018) Improvements in the fossil record may largely resolve current conflicts between morphological and molecular estimates of mammal phylogeny. Proc R Soc Lond B 285:20181632

Beck RMD, Taglioretti ML (2019) A nearly complete juvenile skull of the marsupial *Sparassocynus derivatus* from the Pliocene of Argentina, the affinities of "sparassocynids" and the diversification of opossums (Marsupialia; Didelphimorphia; Didelphidae). J Mammal Evol: https://doi.org/10.1007/s10914-019-09471-y

Bhat MY, Dar TA, Singh LR (2016) Casein proteins: Structural and functional aspects. In: Gigli I (ed) Milk proteins—From structure to biological properties and health aspects. IntechOpen Limited, London, pp 3–17

Bi S, Zheng X, Wang X, Cignetti NE, Yang S, Wible JR (2018) An Early Cretaceous eutherian and the placental-marsupial dichotomy. Nature 558:390–395

Billet G, de Muizon C, Schellhorn R, Ruf I, Ladevèze S, Bergqvist L (2015) Petrosal and inner ear anatomy and allometry amongst specimens referred to Litopterna (Placentalia). Zool J Linn Soc 173:956–987

Black KH, Archer M, Hand SJ, Godthelp H (2012) The rise of Australian marsupials: A synopsis of biostratigraphic, phylogenetic, palaeoecologic and palaeobiogeographic understanding. In: Talent JA (ed) Earth and life. International year of planet Earth. Springer Science + Business Media, New York, pp 983–1078

Blackburn DG (1991) Evolutionary origins of the mammary gland. Mammal Rev 21:81–96

Blackburn DG, Hayssen V, Murphy CJ (1989) The origins of lactation and the evolution of milk: A review with new hypotheses. Mammal Rev 19:1–26

Bokma F, Godinot M, Maridet O, Ladevèze S, Costeur L, Solé F, Gheerbrant E, Peigné S, Jacques F, Laurin M (2016) Testing for Depéret's Rule (body size increase) in mammals using combined extinct and extant data. Syst Biol 65:98–108

Bonaparte JF (1986a) Sobre *Mesungulatum houssayi* y nuevos mamíferos cretácicos de Patagonia. 4° Congreso Argentino de Paleontología y Bioestratigrafía. Mendoza, Actas 2:48–61

Bonaparte JF (1986b) History of the terrestrial Cretaceous vertebrates of Gondwana. 4° Congreso Argentino de Paleontología y Bioestratigrafía. Mendoza, Actas 2:63–95

Bonaparte JF (1986c) A new and unusual Late Cretaceous mammal from Patagonia. J Vertebr Paleontol 6:264–270

Bonaparte JF (1990) New Late Cretaceous mammals from the Los Alamitos Formation, northern Patagonia. Natl Geogr Res 6:63–93

Bonaparte JF (1992) Una nueva especie de Triconodonta (Mammalia), de la Formación Los Alamitos, Provincia de Río Negro y comentarios sobre su fauna de mamíferos. Ameghiniana 29:99–110

Bonaparte JF (1994) Approach to the significance of the Late Cretaceous mammals of South America. Berlin Geowiss Abt E 13:31–44

Bonaparte JF (2002) New Dryolestida (Theria) from the Late Cretaceous of Los Alamitos, Argentina, and paleogeographical comments. Neues Jahrb Geol Paläontol 224:339–371

Bonaparte JF, Barbarena MC (1975) A possible mammalian ancestor from the Middle Triassic of Brazil (Therapsida-Cynodontia). J Paleontol 49:931–936

Bonaparte JF, Barberena MC (2001) On two advanced carnivorous cynodonts from the Late Triassic of Southern Brazil. Bull Mus Comp Zool 156:59–80

Bonaparte JF, Kielan-Jaworowska Z (1987) Late Cretaceous dinosaur and mammal faunas of Laurasia and Gondwana. In: Currie PM, Koster EH (eds) 4° Symposium on Mesozoic Terrestrial Ecosystems, Short Papers, Occasional Papers of the Tyrrell Museum of Palaeontology 3:24–29

Bonaparte JF, Migale LA (2010) Protomamíferos y mamíferos Mesozoicos de América del Sur. Museo de Ciencias Naturales Carlos Ameghino de Mercedes, Buenos Aires. 1° Edition

Bonaparte JF, Migale LA (2015) Protomamíferos y mamíferos Mesozoicos de América del Sur. Fundación de Historia Natural Felix de Azara, Buenos Aires. 2° Edition

Bonaparte JF, Rougier GW (1987) Mamíferos del Cretácico Inferior de Patagonia. 4° Congreso Latinoamericano de Paleontologia. Santa Cruz de la Sierra 1:343–359

Bonaparte JF, Soria MF (1985) Nota sobre el primer mamífero del Cretácico Argentino, Campaniano– Maastrichtiano (Condylarthra). Ameghiniana 21:177–183

Bonaparte JF, Krause DW, Kielan-Jaworowska Z (1989) *Ferugliotherium windhauseni* Bonaparte, the first known multituberculta from Gondwanaland. J Vertebr Paleontol l 9:14A

Bonaparte JF, Martinelli AG, Schultz CL, Rubert R (2003) The sister group of mammals: Small cynodonts from the Late Triassic of southern Brazil. Rev Bras Paleontol 5:5–27

Bonaparte JF, Martinelli AG, Schultz CL (2005) New information on *Brasilodon* and *Brasilitherium* (Cynodontia, Probainognathia) from the Late Triassic of southern Brazil. Rev Bras Paleontol 8:25–46

Bond M, Tejedor MF, Campbell KE, Chornogubsky L, Novo N, Goin FJ (2015) Eocene primates of South America and the African origins of New World monkeys. Nature 520:538–541

Borges R, Johnson WE, O'Brien SJ, Gomes C, Heesy CP, Antunes A (2018) Adaptive genomic evolution of opsins reveals that early mammals flourished in nocturnal environments. BMC Genom 19:121

Botha-Brink J, Soares MB, Martinelli AG (2018) Osteohistology of Late Triassic prozostrodontian cynodonts from Brazil. PeerJ 6:e5029

Brawand D, Wahli W, Kaessmann H (2008) Loss of egg yolk genes in mammals and the origin of lactation and placentation. PLoS Biol 6:e63

Buckley M (2015) Ancient collagen reveals evolutionary history of the endemic South American "ungulates". Proc R Soc Lond B 282:20142671

Burgin CJ, Colella JP, Kahn PL, Upham NS (2018) How many species of mammals are there? J Mammal 99:1–14

Butler PM, Hooker JJ (2005) New teeth of allotherian mammals from the English Bathonian, including the earliest multituberculates. Acta Paleontol Pol 50:185–207

Butler PM, Sigogneau-Russell D (2016) Diversity of triconodonts in the Middle Jurassic of Great Britain. Palaeontol Pol 67:35–65

Button DJ, Lloyd GT, Ezcurra MD, Butler RJ (2017) Mass extinctions drove increased global faunal cosmopolitanism on the supercontinent Pangaea. Nat Commun 8:733

Carlini AA, Ortiz Jaureguizar E, Rougier GW (2002) El primer resto craneano de Monotremata del Paleoceno temprano de Patagonia (Argentina). 1° Congreso Latinoamericano de Paleontología de Vertebrados, Santiago de Chile, Abstracts

Casali DM, Dos Santos JE, Jr MF, Santos FR, Perini FA (2020) Total-evidence phylogeny and divergence times of Vermilingua (Mammalia: Pilosa). Syst Biodivers 18:216–227

Case JA, Goin FJ, Woodburne MO (2005) "South American" marsupials from the Late Cretaceous of North America and the origin of marsupial cohorts. J Mammal Evol 12:461–494

Cassini GH (2013) Skull geometric morphometrics and paleoecology of Santacrucian (Late Early Miocene; Patagonia) native ungulates (Astrapotheria, Litopterna, and Notoungulata). Ameghiniana 50:193–216

Castro MC, Goin FJ, Ortiz-Jaureguizar E, Vieytes EC, Tsukui K, Ramezani J, Batezelli A, Marsola JCA, Langer MC (2018) A Late Cretaceous mammal from Brazil and the first radioisotopic age for the Bauru Group. R Soc Open Sci 5:180482

Césari SN, Colombi CE (2013) A new Late Triasssic phytogeographical scenario in westernmost Gondwana. Nat Commun 4:1889

Chinsamy A, Hurum JH (2006) Bone microstructure and growth patterns of early mammals. Acta Palaeontol Pol 51:325–338

Chornogubsky L (2010) Sistemática de la familia Polydolopidae (Mammalia, Marsupialia, Polydolopimorphia) de América del Sur y la Antártica. Unpublished PhD thesis, Universidad Nacional de Buenos Aires

Chornogubsky L (2011) New remains of the dryolestoid mammal *Leonardus cuspidatus* from the Los Alamitos Formation (Late Cretaceous, Argentina). Paläontol Z 85:343–350

Chornogubsky L, Goin FJ (2015) A review of the molar morphology and phylogenetic affinities of *Sillustania quechuense* (Metatheria, Polydolopimorphia, Sillustaniidae), from the early Paleogene of Laguna Umayo, southeastern Peru. J Vertebr Paleontol 35:e983238

Chow M, Rich TH (1982) *Shuotherium dongi* n. gen. and sp., a therian with pseudo-tribosphenic molars from the Jurassic of Sichuan. China. Aust Mammal 5:127–142

Cifelli RL, Madsen SK (1998) Triconodont mammals from the medial Cretaceous of Utah. J Vertebr Paleontol 18:403–411

Cione AL, Gasparini GM, Soibelzon E, Soibelzon LH, Tonni EP (2015) The Great American Biotic Interchange. A South American perspective, Springer, Dordrecht

Colbert EH (1973) Continental drift and the distributions of fossil reptiles. In: Tailing DH, Runcorn SK (eds) Implications of continental drift for the Earth sciences. Academic Press, London, pp 395–412

Cooper N, Purvis A (2010) Body size evolution in mammals: Complexity in tempo and mode. Am Nat 175:727–738

Croft DA (2016) Horned armadillos and rafting monkeys: The fascinating fossil mammals of South America. Indiana University Press, Bloomington

Croft DA, Gelfo JN, López GM (2020) Splendid innovation: The extinct South American native ungulates. Annu Rev Earth Planet Sci 48:259–290

Crompton AW, Wood CB, Stern DN (1994) Differential wear of enamel: A mechanism for maintaining sharp cutting edges. Adv Comp Environ Physiol 18:321–346

Darmuth J (1981) Population density and body size in mammals. Nature 290:699–700

Darwin C (1859) On the origin of species by means of natural selection, or the preservation of favoured races in the struggle for life. John Murray, London

Davis B (2011) Evolution of the tribosphenic molar pattern in early mammals, with comments on the "dual-origin" hypothesis. J Mammal Evol 18:224–227

de Kort E, Minor M, Snoeren T, van Hooijdonk T, van der Linden E (2011) Effect of calcium chelators on physical changes in casein micelles in concentrated micellar casein solutions. Int Dairy J 21:907–913

de Valais S (2009) Ichnotaxonomic review of *Ameghinichnus*, a mammal ichnogenus from the La Matilde Formation (Middle Jurassic), Santa Cruz Province, Argentina. Zootaxa 2203:1–21
Defler T (2019) History of terrestrial mammals in South America. How South American mammalian fauna changed from the Mesozoic to Recent times. Springer, Dordrecht
Delsuc F, Vizcaíno SF, Douzery EJP (2004) Influence of Tertiary paleoenvironmental changes on the diversification of South American mammals: a relaxed molecular clock study within xenarthrans. BMC Evol Biol 4:11
Delsuc F, Gibb GC, Kuch M, Billet G, Hautier L, Southon J, Rouillard J-M, Fernícola JC, Vizcaíno SF, MacPhee RDE, Poinar HN (2016) The phylogenetic affinities of the extinct glyptodonts. Curr Biol 26:R155–R156
Depéret CJJ (1907) Les transformations du monde animal. Flammarion, Paris
Depéret CJJ (1909) The transformations of the animal world. Kegan Paul, Trench, Trübner & Company Limited, London
Dhouailly D, Godefroit P, Martin T, Nonchev S, Caraguel F, Oftedal O (2019) Getting to the root of scales, feather and hair: As deep as odontodes? Exp Dermatol 28:503–508
dos Reis M, Inoue J, Haseqawa M, Asher RJ, Donoghue PCJ, Yang Z (2012) Phylogenetic datasets provide both precision and accuracy in estimating the timescale of placental mammal phylogeny. Proc R Soc Lond B 279:3491–3500
Drummond AJ, Stadler T (2016) Bayesian phylogenetic estimation of fossil ages. Phil Trans R Soc B 371:20150129
Dunhill AM, Wills MA (2015) Geographic range did not confer resilience to extinction in terrestrial vertebrates at the end-Triassic crisis. Nat Commun 6:7980
Dunhill AM, Bestwick J, Narey H, Sciberras J (2016) Dinosaur biogeographical structure and Mesozoic continental fragmentation: A network-based approach. J Biogeogr 43:1691–1704
Engelman RK, Croft D (2014) A new species of small-bodied sparassodont (Mammalia, Metatheria) from the Middle Miocene locality of Quebrada Honda, Bolivia. J Vertebr Paleontol 34:672–688
Enjapoori AK, Grant TR, Nicol SC, Lefèvre CM, Nicholas KR, Sharp JA (2014) Monotreme lactation protein is highly expressed in monotreme milk and provides antimicrobial protection. Genome Biol Evol 6:2754–2773
Ezcurra MD (2010) Biogeography of Triassic tetrapods: Evidence for provincialism and driven sympatric cladogenesis in the early evolution of modern tetrapod lineages. Proc R Soc Lond B 277:2547–2552
Ezcurra MD, Agnolín FL (2012) A new global palaeobiogeographical model for the late Mesozoic and early Tertiary. Syst Biol 61:553–566
Fariña RA, Vizcaíno SF, De Iuliis G (2013) Megafauna: Giant beasts of Pleistocene South America. Indiana University Press, Bloomington
Flynn JJ, Charrier R, Croft DA, Wyss AR (2012) Cenozoic Andean faunas: Shedding new light on South American mammal evolution, biogeography, environments, and tectonics. In: Patterson BD, Costa LP (eds) Bones, clones, and biomes. University of Chicago Press, Chicago, pp 51–75
Forasiepi AM (2009) Osteology of *Arctodictis sinclairi* (Mammalia, Metatheria, Sparassodonta) and phylogeny of Cenozoic metatherian carnivores from South America. Monografías Mus Arg Sci Nat "Bernardino Rivadavia" (n.s.) 6:1–174
Forasiepi AM, Martinelli AG (2003) Femur of a monotreme (Mammalia, Monotremata) from the Early Paleocene Salamanca Formation of Patagonia, Argentina. Ameghiniana 40:625–630
Forasiepi AM, Coria RA, Hurum J, Currie PJ (2012) First dryolestoid (Mammalia, Dryolestida) dentary from the Coniacian of Patagonia, Argentina. Ameghiniana 49:497–504
Forasiepi AM, Babot MJ, Zimicz N (2015) *Australohyaena antiqua* (Mammalia, Metatheria, Sparassodonta), a large predator from the late Oligocene of Patagonia, Argentina. J Syst Palaeontol 13:503–525
Ford DP, Benson RBJ (2020) The phylogeny of early amniotes and the affinities of Parareptilia and Varanopidae. Nat Ecol Evol 4:57–65
Foulger GR (2018) Origin of the South Atlantic igneous province. J Volcanol Geotherm Res 355:2–20

Frazão A, da Silva HR, de Moraes Russo CA (2015) The Gondwana breakup and the history of the Atlantic and Indian Oceans unveils two new clades for early neobatrachian diversification. PLoS ONE 10(11):e0143926

Fukumoto S, Nakamura T, Yamada A, Arakaki M, Saito X, Fukumoto E, Yamada Y (2014) New insights into the functions of enamel matrices in calcified tissues. Jpn Dent Sci Rev 50:47–54

Gaetano LC, Rougier GW (2011) New materials of *Argentoconodon fariasorum* (Mammaliaformes, Triconodontidae) from the Jurassic of Argentina and its bearing on triconodont phylogeny. J Vertebr Paleontol 31:829–843

Gaetano LC, Rougier GW (2012) First amphilestid from South America: A molariform from the Jurassic Cañadón Asfalto Formation, Patagonia, Argentina. J Mamm Evol 19:235–248

Gasse B, Sire J-Y (2015) Comparative expression of the four enamel matrix protein genes, amelogenin, ameloblastin, enamelin and amelotin during amelogenesis in the lizard *Anolis carolinensis*. EvoDevo 6:29

Gasse B, Chiari Y, Silvent J, Davit-Béal T, Sire J-Y (2015) Amelotin: An enamel matrix protein that experienced distinct evolutionary histories in amphibians, sauropsids and mammals. BMC Evol Biol 15:47

Gaudin TJ, Croft DA (2015) Paleogene Xenarthra and the evolution of South American mammals. J Mammal 96:622–634

Gayet M, Marshall LG, Sempere T, Meunier FJ, Cappetta H, Rage J-C (2001) Middle Maastrichtian vertebrates (fishes, amphibians, dinosaurs and other reptiles, mammals) from Pajcha Pata (Bolivia). Biostratigraphic, palaeoecologic and palaeobiogeographic implications. Palaeogeogr Palaeoclimatol Palaeoecol 169:39–68

Gegenbaur C (1875) Zur genaueren Kenntnis der Zitzen der Säugethiere. Morphol Jahrb 1:266–281

Gelfo JN, Sigé B (2011) A new didolodontid mammal from the late Paleocene–earliest Eocene of Laguna Umayo, Peru. Acta Palaeontol Pol 56:665–678

Gelfo JN, Ortiz-Jaureguizar E, Rougier GW (2007) New remains and species of the "condylarth" genus *Escribania* Mammalia: Didolodontidae) from the Early Paleocene of Patagonia, Argentina. Earth Env Sci Trans Roy Soc Edinb 98:127–138

Gelfo JN, Goin FJ, Woodburne MO, de Muizon C (2009) Biochronological relationships of the earliest South American Paleogene mammalian faunas. Palaeontology 52:251–269

Gibb GC, Condamine FL, Kuch M, Enk J, Moraes-Barros N, Superina M, Poinar HN, Delsuc (2016). Shotgun mitogenomics provides a reference phylogenetic framework and timescale for living xenarthrans. Mol Biol Evol 33:621–642

Goin FJ, Vieytes EC, Vucetich MG, Carlini AA, Bond M (2004) Enigmatic mammal from the Paleogene of Peru. In: Campbell KE (ed) The Paleogene mammalian fauna of Santa Rosa, Amazonian Peru. Nat Hist Mus Los Angeles County, Sci Ser 40:145–153

Goin FJ, Pascual R, Tejedor MF, Gelfo JN, Woodburne MO, Case JA, Reguero MA, Bond M, López GM, Cione AL, Udrizar Sauthier D, Balarino L, Scasso RA, Medina FA, Ubaldón MC (2006a) The earliest Tertiary therian mammal from South America. J Vertebr Paleontol 26:505–510

Goin FJ, Reguero MA, Pascual R, Koenigswald W von, Woodburne MO, Case JA, Vieytes C, Marenssi SA, Vizcaíno SF (2006b) First gondwanatherian mammal from Antarctica. In: Francis JE, Pirrie D, Crame JA (eds) Cretaceous–Tertiary high-latitude palaeoenvironments, James Ross Basin, Antarctica. Geological Society of London, Spec Publ 258:135–144

Goin FJ, Abello A, Bellosi E, Kay R, Madden R, Carlini AA (2007) Los Metatheria sudamericanos de comienzos del Neógeno (Mioceno Temprano, Edad-Mamífero Colhuehuapense). Parte 1: Introducción. Didelphimorphia y Sparassodonta. Ameghiniana 44:29–71

Goin FJ, Candela AM, Abello A, Oliveira EO (2009) Earliest South American paucituberculatans and their significance in understanding of "pseudodiprotodont" marsupial radiations. Zool J Linn Soc 155:867–884

Goin FJ, Tejedor MF, Chornogubsky L, López GM, Gelfo JN, Bond M, Woodburne MO, Gurovich Y, Reguero M (2012) Persistence of a Mesozoic, non-therian mammalian lineage (Gondwanatheria) in the mid-Paleogene of Patagonia. Naturwissenschaften 99:449–463

Goin FJ, Woodburne MO, Zimicz AN, Martin GM, Chornogubsky L (2016) A brief history of South American metatherians: Evolutionary contexts and intercontinental dispersals. Springer, Dordrecht

Goin FJ, Martinelli AG, Soto-Acuña S, Vieytes EC, Manríquez LME, Fernández RA, Pino JP, Trevisa C, Kaluza J, Reguero MA, Leppe M, Ortiz H, Rubilar-Rogers D, Vargas AO (2020) First Mesozoic mammal from Chile: The southernmost record of a Late Cretaceous gondwanatherian. Bol Mus Nac Hist Nat Chile 69:5–31

Granot R, Dyment J (2015) The Cretaceous opening of the South Atlantic Ocean. Earth Planet Sci Lett 414:156–163

Griffiths M (1978) The biology of the monotremes. Academic Press, New York, San Francisco, London

Grothe B, Pecka M (2014) The natural history of sound localization in mammals—A story of neuronal inhibition. Front Neural Circuits 8:116

Guignard ML, Martinelli AG, Soares MB (2019) The postcranial anatomy of *Brasilodon quadrangularis* and the acquisition of mammaliaform traits among non-mammaliaform cynodonts. PLoS ONE 14(5):e0216672

Gurovich Y (2008) Additional specimens of sudamericid (Gondwanatheria) mammals from the Early Paleocene of Argentina. Palaeontology 51:1069–1089

Gurovich Y, Beck RMD (2009) The phylogenetic affinities of the enigmatic mammalian clade Gondwanatheria. J Mammal Evol 16:25–49

Hahn G, Hahn R (2003) New multituberculate teeth from the Early Cretaceous of Morocco. Acta Palaeontol Pol 48:349–356

Hahn G, Hahn R (2007) Catalogus Haramiyorum cum figuris (Allotheria Mesozoica). Fossilium Catalogus I: Animalia. Pars 143. Backhuys Publishers, Leiden

Hall BK (2015) Hall book bone and cartilage. Developmental and evolutionary skeletal biology. Academic Press, New York, San Francisco, London. 2° Edition

Hallam A (1985) A review of Mesozoic climates. J Geol Soc 142:433–445

Harper T, Rougier GW (2019) Petrosal morphology and cochlear function in Mesozoic stem therians. PLoS ONE 14(8):e0209457

Harper T, Parras A, Rougier GW (2019) *Reigitherium* (Meridiolestida, Mesungulatoidea) an enigmatic Late Cretaceous mammal from Patagonia, Argentina: Morphology, affinities, and dental evolution. J Mammal Evol 26:447–478

Heath TA, Huelsenbeck JP, Stadler T (2014) The fossilized birth-death process for coherent calibration of divergence-time estimates. Proc Natl Acad Sci USA 111:E2957–E2966

Holz M (2015) Mesozoic paleogeography and paleoclimates—A discussion of the diverse greenhouse and hothouse conditions of an alien world. J S Am Earth Sci 61:91–107

Hopson JA, Rougier GW (1993) Braincase structures in the oldest known skull of a therian mammal: Implications for mammalian systematics and cranial evolution. Am J Sci 293:268–299

Horovitz I, Sánchez-Villagra MR (2003) A morphological analysis of marsupial mammal higher-level phylogenetic relationships. Cladistics 19:181–212

Horovitz I, Ladevèze S, Argot C, Macrini T, Martin T, Hooker JJ, Kurz C, de Muizon C, Sánchez-Villagra MR (2008) The anatomy of *Herpetotherium fugax* Cope 1873, a metatherian from the Oligocene of North America. Palaeontographica 284:109–141

Hu Y, Meng J, Wang Y, Li C (2005) Large Mesozoic mammals fed on young dinosaurs. Nature 433:149–152

Huttenlocker AK (2014) Body size reductions in nonmammalian eutheriodont therapsids (Synapsida) during the End-Permian Mass Extinction. PLoS ONE 9(2):e87553

Ijaq J, Malik G, Kumar A, Das PS, Meena N, Bethi N, Sundararajan VS, Suravajhala P (2019) A model to predict the function of hypothetical proteins through a nine-point classification scoring schema. BMC Bioinformatics 20:14

Jasinoski SC, Chinsamy A (2012) Mandibular histology and growth of the nonmammaliaform cynodont Tritylodon. J Anat 220:564–579

Jastroch M, Seebacher F (2020) Importance of adipocyte browning in the evolution of endothermy. Phil Trans R Soc Lond B 375:20190134

Kawasaki K (2009) The SCPP gene repertoire in bony vertebrates and graded differences in mineralized tissues. Dev Genes Evol 219:147–157

Kawasaki K, Lafont A-G, Sire J-Y (2011) The evolution of milk casein genes from tooth genes before the origin of mammals. Mol Biol Evol 28:2053–2061

Kealy S, Beck RMD (2017) Total evidence phylogeny and evolutionary timescale for Australian faunivorous marsupials (Dasyuromorphia). BMC Evol Biol 17:240

Kemp TS (2005) The origin and evolution of mammals. Oxford University Press, Oxford

Kielan-Jaworowska Z, Bonaparte JF (1996) Partial dentary of a multituberculate mammal from the Late Cretaceous of Argentina and its taxonomic implications. Rev Mus Argent Cienc Nat "B Rivadavia", ns 145:1–9

Kielan-Jaworowska Z, Cifelli RL, Luo Z-X (2004) Mammals from the age of dinosaurs. Origins, evolution, and structure. Columbia University Press, New York

Kielan-Jaworowska Z, Ortiz-Jaureguizar E, Vieytes C, Pascual R, Goin FJ (2007) First ?cimolodontan multituberculate mammal from South America. Acta Palaeontol Pol 52:257–262

King B, Beck RMD (2020) Tip dating supports novel resolutions of controversial relationships among early mammals. Proc R Soc Lond B 287:20200943

Kirsch JAW, Dickerman AW, Reig OA, Springer MS (1991) DNA hybridization evidence for the Australian affinity of the American marsupial *Dromiciops australis*. Proc Natl Acad Sci 88:10465–10469

Klaatsch H (1884) Zur Morphologie der Säugethierzitzen. I. Theil. Die Taschen-und Beutelbildungen am Driisenfeld der Monotremen. Morphol Jahrb 9:253–324

von Koenigswald W, Clemens WA (1992) Levels of complexity in the microstructure of mammalian enamel and their application in studies of systematics. Scanning Microsc 6:195–218

von Koenigswald W, Sanders PM (1997) Tooth enamel microstructure. AA Balkema, Rotterdam

von Koenigswald W, Goin FJ, Pascual R (1999) Hypsodonty and enamel microstructure in the Paleocene gondwanatherian mammal *Sudamerica ameghinoi*. Acta Palaeontol Pol 44:263–300

Kolb C, Scheyer TM, Veitschegger K, Forasiepi AM, Amson E, Van der Geer AAE, Van den Hoek Ostende LW, Hayashi S, Sanchez-Villagra MR (2015) Mammalian bone palaeohistology: A survey and new data with emphasis on island forms. PeerJ 3:e1358

Köppl C, Manley GA (2018) A functional perspective on the evolution of the cochlea. Cold Spring Harb Perspect Med 9:a033241

Krause DW (1986) Competitive exclusion and taxonomic displacement in the fossil record: The case of rodents and multituberculates in North America. In: Flanagan KM, Lillegraven JA (eds) Vertebrates, phylogeny and philosophy. University of Wyoming Contributions to Geology, Spec Pap 3:119–130

Krause DW (2013) Gondwanatheria and ?Multituberculata (Mammalia) from the Late Cretaceous of Madagascar. Can J Earth Sci 50:324–340

Krause DW, Bonaparte JF (1993) Superfamily Gondwanatherioidea: A previously unrecognized radiation of multituberculate mammals in South America. Proc Natl Acad Sci USA 90:9379–9383

Krause DW, Kielan-Jaworowska Z, Bonaparte JF (1992) *Ferugliotherium* Bonaparte, the first known multituberculate from Gondwanaland. J Vertebr Paleont 12:351–376

Krause DW, Gottfried MD, O'Connor PM, Roberts EM (2003) A Cretaceous mammal from Tanzania. Acta Palaeontol Pol 48:321–330

Krause DW, O'Connor PM, Curry Rogers K, Sampson SD, Buckley GA, Rogers RR (2006) Late Cretaceous terrestrial vertebrates from Madagascar: Implications for Latin American biogeography. Ann Missouri Bot Gard 93:178–208

Krause DW, Hoffmann S, Wible JR, Kirk EC, Schultz JA, von Koenigswald W, Groenke JR, Rossie JB, O'Connor PM, Seiffert ER, Dumont ER, Holloway WL, Rogers RR, Rahantarisoa RJ, Kemp AD, Andriamialison H (2014) First cranial remains of a gondwanatherian mammal reveal remarkable mosaicism. Nature 515:512–517

Krause DW, Sertich JJW, O'Connor PM, Curry Rogers K, Rogers RR (2019) The Mesozoic biogeographic history of Gondwanan terrestrial vertebrates: Insights from Madagascar's fossil record. Annu Rev Earth Planet Sci 47:519–553

Krause DW, Hoffmann S, Hu Y, Wible JR, Rougier GW, Kirk EC, Groenke JR, Rogers RR, Rossie JB, Schultz JA, Evans AR, von Koenigswald W, Rahantarisoa LJ (2020) Skeleton of Cretaceous mammal from Madagascar reflects long-term insularity. Nature 581:421–427

Ladevèze S, Selva C, de Muizon C (2020) What are "opossum-like" fossils? The phylogeny of herpetotheriid and peradectid metatherians, based on new features from the petrosal anatomy. J Syst Palaeontol: https://doi.org/10.1080/14772019.2020.1772387

Lautenschlager S, Gill P, Luo Z-X, Fagan MJ, Rayfield EJ (2019) The role of miniaturization in the evolution of the mammalian jaw and middle ear. Nature 561:533–537

Large RR, Halpin JA, Lounejeva E, Danyushevsky LV, Maslennikov VV, Gregory D, Sack PJ, Haines PW, Long JA, Makoundi C, Stepanov AS (2015) Cycles of nutrient trace elements in the Phanerozoic ocean. Gondwana Res 28:1282–1293

LeBlanc ARH, Brink KS, Whitney MR, Abdala F, Reisz RR (2018) Dental ontogeny in extinct synapsids reveals a complex evolutionary history of the mammalian tooth attachment system. Proc R Soc Lond B 285:20181792

Lefèvre CM, Sharp JA, Nicholas KR (2009) Characterisation of monotreme caseins reveals lineage-specific expansion of an ancestral casein locus in mammals. Reprod Fertil Dev 21:1015–1027

Lefèvre CM, Sharp JA, Nicholas KR (2010) Evolution of lactation: Ancient origin and extreme adaptations of the lactation system. Annu Rev Genomics Hum Genet 11:219–238

Luo Z-X, Wible JR (2005) A Late Jurassic digging mammal and early mammalian diversification. Science 308:103–107

Luo Z-X, Kielan-Jaworowska Z, Cifelli RL (2002) In quest for a phylogeny of Mesozoic mammals. Acta Palaeontol Pol 47:1–78

Luo Z-X, Yuan C-X, Meng Q-J, Ji Q (2011) A Jurassic eutherian mammal and divergence of marsupials and placentals. Nature 476:442–445

Lyson TR, Miller IM, Bercovici AD, Weissenburger K, Fuentes AJ, Clyde WC, Hagadorn JW, Butrim MJ, Johnson KR, Fleming RF, Barclay RS, Maccracken SA, Lloyd B, Wilson GP, Krause DW, Chester SGB (2019) Exceptional continental record of biotic recovery after the Cretaceous-Paleogene mass extinction. Science 366:977–983

Macrini TE, Rougier GW, Rowe T (2007) Description of a cranial endocast from the fossil mammal *Vincelestes neuquenianus* (Theriiformes) and its relevance to the evolution of endocranial characters in therians. Anat Rec 290:875–892

Madden RH (2015) Hypsodonty in mammals: Evolution, geomorphology and the role of earth surface processes. Cambridge University Press, Cambridge

Maier W, Ruf I (2016) Evolution of the mammalian middle ear: A historical review. J Anat 228:270–283

Manley GA (2000) Cochlear mechanisms from a phylogenetic viewpoint. Proc Natl Acad Sci USA 97:11736–11743

Manley GA (2001) Evidence for an active process and cochelar amplifier in nonmammals. J Neurophysiol 86:541–549

Manley GA (2017) The mammalian Cretaceous cochlear revolution. Hear Res 352:23–29

Manley GA (2018) Travelling waves and tonotopicity in the inner ear: A historical and comparative perspective. J Comp Physiol A Neuroethol Sens Neural Behav Physiol 204:773–781

Marshall LG (1987) Systematics of Itaboraian (middle Paleocene) age "opossum-like" marsupials from the limestone quarry at São José de Itaboraí, Brazil. In: Archer M (ed) Possums and opossums—Studies in evolution, vol 1. Surrey Beatty and Sons, Sydney, pp 91–160

Marshall LG, Cifelli RL (1990) Analysis of changing diversity patterns in Cenozoic Land Mammal Age faunas, South America. Palaeovertebrata 19:169–210

Marshall LG, de Muizon C, Sigé B (1985) The "Rosetta stone" for mammalian evolution in South America. Nat Geogr Res 1:274–288

Martin T, Rauhut OWM (2005) Mandible and dentition of *Asfaltomylos patagonicus* (Australosphenida, Mammalia) and the evolution of tribosphenic teeth. J Vertebr Paleontol 25:414–425

Martin SA, Alhajeri BH, Steppan SJ (2016) Dietary adaptations in the teeth of murine rodents (Muridae): A test of biomechanical predictions. Biol J Linn Soc 119:766–784

Martin T, Marugan-Lobon J, Vullo R, Martin-Abad H, Luo Z-X, Buscalioni AD (2015) A Cretaceous eutriconodont and integument evolution in early mammals. Nature 526:380–384

Martinelli AG, Rougier GW (2007) On *Chaliminia musteloides* (Eucynodontia: Tritheledontidae) from the Late Triassic of Argentina and a phylogeny of Ictidosauria. J Vertebr Paleontol 27:442–460

Martinelli AG, Soares MB (2016) Evolution of South American cynodonts. Mus Argent Cienc Nat "B Rivadavia". Contribuciones 6:183–197

Martinelli AG, Bonaparte JF, Schultz CB (2005) A new tritheledontid (Therapsida, Eucynodontia) from the Late Triassic of Rio Grande do Sul (Brazil) and its phylogenetic relationships among carnivorous non-mammalian eucynodonts. Ameghiniana 42:191–208

Martinelli AG, Chornogubsky L, Abello A, Goin FJ, Reguero M (2014) The first non-therian dryolestoid from Antarctica. 2014 SCAR Open Science Conference, Auckland, Abstracts:432

Martinelli AG, Soares MB, Schwanke C (2016) Two new cynodonts (Therapsida) from the Middle-early Late Triassic of Brazil and comments on South American probainognathians. PLoS ONE 11(10):e0162945

Martinelli AG, Eltink E, Da-Rosa ÁAS, Langer MC (2017) A new cynodont (Therapsida) from the *Hyperodapedon* Assemblage Zone (upper Carnian–Norian) of southern Brazil improves the Late Triassic probainognathian diversity. Pap Palaeontol 3:401–423

Mathur AK, Polly PD (2000) The evolution of enamel microstructure: How important is amelogenin? J Mammal Evol 7:23–42

Matthews KJ, Maloney KT, Zahirovic S, Williams SE, Seton M, Muller RD (2016) Global plate boundary evolution and kinematics since the late Paleozoic. Glob Planet Change 146:226–250

McElwain JC, Beerling DJ, Woodward FI (1999) Fossil plants and global warming at the Triassic-Jurassic boundary. Science 285:1386–1390

McKenna MC (1975) Toward a phylogenetic classification of the Mammalia. In: Luckett WP, Szalay FS (eds) Phylogeny of the Primates. Plenum Publishing, New York, pp 21–46

McLoughlin S (2001) The breakup history of Gondwana and its impact on pre-Cenozoic floristic provincialism. Aust J Bot 49:271–300

Meredith RW, Westerman M, Springer MS (2009) A phylogeny of Diprotodontia (Marsupialia) based on sequences for five nuclear genes. Mol Phylogenet Evol 1:554–571

Monroe MJ, Bokma F (2010) Little evidence for Cope's rule from Bayesian phylogenetic analysis of extant mammals. J Evol Biol 23:2017–2021

Muizon C de (1991a) La fauna de mamíferos de Tiupampa (Paleoceno inferior, Formación Santa Lucía), Bolivia. In: Suárez-Soruco R (ed) Fósiles y Fácies de Bolivia, vol I, Vertebrados. Santa Cruz: Revista Técnica, Yacimientos Petrolíferos Fiscales Bolivianos, pp 575–624

Muizon C de (1991) Nouveaux Condylarthres du Paléocène inférieur de Tiupampa (Bolivie). Bull Mus Natl Hist Nat 13:201–227

Muizon C de, Cifelli RL (2000) The "condylarths" (archaic Ungulata, Mammalia) from the early Palaeocene of Tiupampa (Bolivia): Implications on the origin of the South American ungulates. Geodiversitas 22:47–150

Muizon C de, Marshall LG (1987a) Le plus ancien Condylarthre (Mammalia) sud-américain (Crétacé Supérieur, Bolivie). C R Acad Sci 304:771–774

Muizon C de, Marshall LG (1987) Deux nouveaux Condylarthres (Mammalia) du Maastrichtien de Tiupampa (Bolivie). C R Acad Sci 304:947–950

Muizon C de, Cifelli RL, Cespedes Paz R (1997) The origin of borhyaenoids, South American dog-like marsupials. Nature 389:486–489

Muizon C de, Billet G, Argot C, Ladevèze GF (2015) *Alcidedorbignya inopinata*, a basal pantodont (Placentalia, Mammalia) from the early Palaeocene of Bolivia: Anatomy, phylogeny and palaeobiology. Geodiversitas 37:397–634

Muizon C de, Ladevèze S, Selva C, Vignaud R, Goussard F (2018) *Allqokirus australis* (Sparassodonta, Metatheria) from the early Palaeocene of Tiupampa (Bolivia) and the rise of the metatherian carnivorous radiation in South America. Geodiversitas 40:363–459

Müller RD, Zahirovic S, Williams SE, Cannon J, Seton M, Bower DJ, Tetley MG, Heine C, Le Breton E, Liu S, Russel SHJ, Yang T, Leonard J, Gurnis M (2019) A global plate model including lithospheric deformation along major rifts and orogens since the Triassic. Tectonics 38:1884–1907

Newham E, Gill PG, Brewer P, Benton MJ, Fernandez V, Gostling NJ, Haberthür D, Jernvall J, Kankanpää T, Kallonen A, Navarro C, Pacureanu A, Richards K, Brown KR, Schneider P, Suhonen H, Tafforeau P, Williams KA, Zeller-Plumhoff B, Corfe IJ (2020) Reptile-like physiology in Early Jurassic stem-mammals. Nat Commun 11:5121

O'Connor PM, Krause DW, Stevens NJ, Groenke JR, MacPhee RDE, Kalthoff DC, Roberts EM (2019) A new mammal from the Turonian-Campanian (Upper Cretaceous) Galula Formation, southwestern Tanzania. Acta Palaeont Pol 64:65–84

O'Leary MA, Bloch JI, Flynn JJ, Gaudin TJ, Giallombardo A, Giannini NP, Goldberg SL, Kraatz BP, Luo ZX, Meng J, Ni X, Novacek MJ, Perini FA, Randall ZS, Rougier GW, Sargis EJ, Silcox MT, Simmons NB, Spaulding M, Velazco PM, Weksler M, Wible JR, Cirranello AL (2013) The placental mammal ancestor and the post-K-Pg radiation of placentals. Science 339:662–667

O'Meara RN, Thompson RS (2014) Were there Miocene meri-dolestidans? Assessing the phylogenetic placement of *Necrolestes patagonensis* and the presence of a 40 million year merdiolestidan ghost lineage. J Mammal Evol 21:271–284

O'Meara RN, Dirks W, Martinelli AG (2018) Enamel formation and growth in non-mammalian cynodonts. R Soc Open Sci 5:172293

Oelkrug R, Goetze N, Exner C, Lee Y, Ganjam GK, Kutschke M, Müller S, Stöhr S, Tschöp MH, Crichton PG, Heldmaier G, Jastroch M, Meyer CW (2013) Brown fat in a protoendothermic mammal fuels eutherian evolution. Nat Commun 4:2140

Ofstad EG, Herfindal I, Solberg EJ, Sæther B-E (2016) Home ranges, habitat and body mass: Simple correlates of home range size in ungulates. Proc R Soc Lond B 283:20161234

Oftedal OT (2002a) The mammary gland and its origin during synapsid evolution. J Mammary Gland Biol Neoplasia 7:225–252

Oftedal OT (2002b) The origin of lactation as a water source for parchment-shelled eggs. J Mammary Gland Biol Neoplasia 7:253–266

Oftedal OT (2012) The evolution of milk secretion and its ancient origins. Animal 6:355–368

Oftedal OT (2013) Origin and evolution of the major constituents of milk. In: McSweeney P, Fox P (eds) Advanced dairy chemistry. Springer, Boston, pp 1–42

Oftedal OT, Nicol SC, Davies NW, Taufik E, Fukuda K, Sekii N, Taufik E, Fukuda K, Saito T, Urashima T (2014) Can an ancestral condition for milk oligosaccharides be determined? Evidence from the Tasmanian echidna (*Tachyglossus aculeatus setosus*). Glycobiology 24:826–839

Páez Arango N (2008) Dental and craniomandibular anatomy of *Peligrotherium tropicalis*: The evolutionary radiation of South American dryolestoid mammals. Unpublished Ms thesis, University of Louisville

Pascual R, Ortiz-Jaureguizar E (2007) The Gondwanan and South American episodes: Two major moments in the history of South American mammals. J Mammal Evol 14:75–137

Pascual R, Archer M, Ortiz-Jaureguizar E, Prado JL, Godthelp H, Hand SJ (1992a) First discovery of monotremes in South America. Nature 356:704–705

Pascual R, Archer M, Ortiz-Jaureguizar E, Prado JL, Godthelp H, Hand SJ (1992b) The first non-Australian monotreme: An Early Paleocene South American platypus (Monotremata, Ornithorhynchidae). In: Augee ML (ed) Platypus and echidnas. The Royal Society of New South Wales, Sydney, pp 1–14

Pascual R, Ortiz-Jaureguizar E, Prado JL (1996) Land Mammals: Paradigm for Cenozoic South American Geobiotic Evolution. Münchner Geowiss Abh 30:265–319

Pascual R, Goin FJ, Krause DW, Ortiz-Jaureguizar E, Carlini AA (1999) The first gnathic remains of *Sudamerica*: Implications for gondwanathere relationships. J Vertebr Paleontol 19:373–382
Pascual R, Balarino ML, Udrizar Sauthier DE (2001) The K/T and Tertiary-Pleistocene South American mammalian turnovers. Similar phenomena? 7° International Symposium on Mesozoic Terrestrial Ecosystems. Asociación Paleontológica Argentina, Buenos Aires, Publicación Especial 7:151–156
Patterson B, Pascual R (1968) Evolution of mammals in southern continents. V. Fossil mammal fauna of South America. Quart Rev Biol 43:409–451
Patterson B, Pascual R (1972) The fossil mammal fauna of South America. In: Keast A, Erk FC, Glass B (eds) Evolution of mammals and Southern continents. State Univesity of New York Press, Albany
Pfuhl HA, McCave IN (2005) Evidence for Late Oligocene establishment of the Antarctic Circumpolar Current. Earth Planet Sci Lett 235:715–728
Phillips MJ, Bennett TH, Lee MS (2009) Molecules, morphology, and ecology indicate a recent, amphibious ancestry for echidnas. Proc Natl Acad Sci USA 106:17089–17094
Power ML, Schulkin J (2016) Milk: The biology of lactation. Johns Hopkins University, Baltimore
Prasad GVR, Verma O, Sahni A, Krause DW, Khosla A, Parmar V (2007) A new Late Cretaceous gondwanatherian mammal from Central India. Proc Indian Natl Sci Acad 73:17–24
Presslee S, Slater GJ, Pujos F, Forasiepi AM, Fischer R, Molloy K, Mackie M, Olsen JV, Kramarz A, Taglioretti M, Scaglia F, Lezcano M, Lanata JL, Southon J, Feranec R, Bloch J, Hajduk A, Martin FM, Gismondi RS, Reguero M, de Muizon C, Greenwood A, Chait BT, Penkman K, Collins M, MacPhee RDE (2019) Palaeoproteomics resolves sloth relationships. Nat Ecol Evol 3:1121–1130
Prevosti FJ, Forasiepi AM (2018) Evolution of South American mammalian predators during the Cenozoic: Paleobiogeographic and paleoenvironmental contingencies. Springer, Dordrecht
Pyron RA (2016) Novel approaches for phylogenetic inference from morphological data and total-evidence dating in squamate reptiles (lizards, snakes, and amphisbaenians). Syst Biol 66:38–56
Raia P, Carotenuto F, Passaro F, Fulgione D, Fortelius M (2012) Ecological specialization in fossil mammals explains Cope's rule. Am Nat 179:328–337
Ramírez-Chaves HE, Weisbecker V, Wroe S, Phillips MJ (2016) Resolving the evolution of the mammalian middle ear using Bayesian inference. Front Zool 13:1–10
Rauhut OW, Martin T, Ortiz-Jaureguizar E, Puerta PF (2002) A Jurassic mammal from South America. Nature 416:165–168
Ravosa MJ, Menegaz RA, Scott JE, Daegling DJ, McAbee KR (2016) Limitations of a morphological criterion of adaptive inference in the fossil record. Biol Rev Camb Philos Soc 91:883–898
Rich TA, Vickers-Rich P (2012) Paleobiogeography of Mesozoic mammals revisited. In: Talent JA (ed) Earth and life. International year of planet earth. Springer Science + Business Media, New York, pp 913–934
Rich TH, Vickers-Rich P, Flannery TF, Kear BP, Cantrill DJ, Komarower P, Kool L, Pickering D, Trusler P, Morton S, Nvan K, Fitzgerald EMG (2009) An Australian multituberculate and its palaeobiogeographic implications. Acta Palaeontol Pol 54:1–6
Rich TH, Hopson JA, Gill PG, Trusler P, Rogers-Davidson S, Morton S, Cifelli RL, Pickering D, Kool L, Siu K, Burgmann FA, Senden T, Evans AR, Wagstaff BE, Seegets-Villiers D, Corfe IJ, Flannery TF, Walker K, Musser AM, Archer M, Pian R, Vickers-Rich P (2016) The mandible and dentition of the Early Cretaceous monotreme *Teinolophos trusleri*. Alcheringa 40:475–501
Rijnkels M (2002) Multispecies comparison of the casein gene loci and evolution of casein gene family. J Mammary Gland Biol Neoplasia 7:327–345
Robinson R (2008) For mammals, loss of yolk and gain of milk went hand in hand. PLoS Biol 6:e77–e77
Rodrigues PG, Ruf I, Schultz CL (2013) Digital Reconstruction of the otic region and inner ear of the non-mammalian cynodont *Brasilitherium riograndensis* (Late Triassic, Brazil) and its relevance to the evolution of the mammalian ear. J Mammal Evol 20:291–307

Rodrigues PG, Ruf I, Schultz CL (2014) Study of a digital cranial endocast of the non-mammaliaform cynodont *Brasilitherium riograndensis* (Later Triassic, Brazil) and its relevance to the evolution of the mammalian brain. Paläontol Z 88:329–352

Rodrigues PG, Martinelli AG, Schultz CL, Corfe IJ, Gill PG, Soares MB, Rayfield EJ (2019) Digital endocast of *Riograndia guaibensis* (Late Triassic, Brazil) and the evolution of the brain in non-mammalian cynodonts. Hist Biol 31:1195–1212

Rougier GW (1993) *Vincelestes neuquenianus* Bonaparte (Mammalia, Theria), un primitivo mamífero del Cretácico Inferior de la Cuenca Neuquina. Unpublished PhD thesis, University of Buenos Aires

Rougier GW, Wible JR, Hopson JA (1992) Reconstruction of the cranial vessels in the Early Cretaceous mammal *Vincelestes neuquenianus*: Implications for the evolution of the mammalian cranial vascular system. J Vertebr Paleontol 12:188–216

Rougier GW, Wible JR, Novacek MJ (1998) Implications of *Deltatheridium* specimens for early marsupial history. Nature 396:459–463

Rougier GW, Wible JR, Novacek MJ (2004) New specimen of *Deltatheroides cretacicus* (Metatheria, Deltatheroida) from the Late Cretaceous of Mongolia. Bull Carnegie Mus Nat Hist 36:245–266

Rougier GW, Martinelli AG, Forasiepi AM, Novacek MJ (2007a) New Jurassic mammals from Patagonia, Argentina: A reappraisal of australosphenidan morphology and interrelationships. Am Mus Novit 3566:1–54

Rougier GW, Garrido AC, Gaetano L, Puerta P, Corbitt C, Novacek MJ (2007b) First Jurassic triconodont from South America. Am Mus Novit 3580:1–17

Rougier GW, Chornogubsky L, Casadio S, Paéz Arango N, Giallombardo A (2009a) Mammals from the Allen Formation, Late Cretaceous, Argentina. Cretac Res 30:223–238

Rougier GW, Forasiepi AM, Hill RV, Novacek MJ (2009b) New mammalian remains from the Late Cretaceous La Colonia Formation, Patagonia, Argentina. Acta Palaeontol Pol 54:195–212

Rougier GW, Apesteguía S, Gaetano LC (2011) Highly specialized mammalian skulls from the Late Cretaceous of South America. Nature 479:98–102

Rougier GW, Wible JR, Beck RMD, Apesteguía S (2012) The Miocene mammal *Necrolestes* demonstrates the survival of a Mesozoic nontherian lineage into the late Cenozoic of South America. Proc Natl Acad Sci USA 109:20053–20058

Rowe TB (1988) Definition, diagnosis, and origin of Mammalia. J Vertebr Paleontol 8:241–264

Rowe TB, Rich TH, Vickers-Rich P, Springer M, Woodburne MO (2008) The oldest platypus and its bearing on divergence timing of the platypus and echidna clades. Proc Natl Acad Sci USA 105:1238–1242

Saarinen JJ, Boyer AG, Brown JH, Costa DP, Ernest SM, Evans AR, Fortelius M, Gittleman JL, Hamilton MJ, Harding LE, Lintulaakso K, Lyons SK, Okie JG, Sibly RM, Stephens PR, Theodor J, Uhen MD, Smith FA (2014) Patterns of maximum body size evolution in Cenozoic land mammals: Eco-evolutionary processes and abiotic forcing. Proc Roy Soc Lond B 281:20132049

Sander PM (1997) Non-mammalian synapsid enamel and the origin of mammalian enamel prisms: The bottom-up perspective. In: von Koenigswald W, Sander PM (eds) Tooth enamel microstructure. AA Balkema, Rotterdam, pp 41–62

Sarkar S, Basak C, Frank M, Berndt C, Huuse M, Badhani S, Bialas J (2019) Late Eocene onset of the Proto-Antarctic Circumpolar Current. Scientific Reports 9:10125

Scotese CR (2013a) Map Folio 44, Late Triassic (210 Ma, Norian), PALEOMAP PaleoAtlas for ArcGIS, vol 3. Jurassic and Triassic, PALEOMAP Project, Evanston, IL

Scotese CR (2013b) Map Folio 37 Middle Jurassic (Bajocian and Bathonian, 169.7 Ma), PALEOMAP PaleoAtlas for ArcGIS, vol 3, Triassic and Jurassic Paleogeographic, Paleoclimatic and Plate Tectonic Reconstructions. PALEOMAP Project, Evanston, IL

Scotese CR (2013c) Map Folio 17, Late Cretaceous, (Maastrichtian, 68 Ma), PALEOMAP PaleoAtlas for ArcGIS, vol 2, Cretaceous Paleogeographic, Paleoclimatic and Plate Tectonic Reconstructions. PALEOMAP Project, Evanston, IL

Scotese CR (2013d) Map Folio 5, Middle/Late Miocene, (Serravallian & Tortonian, 10.5 Ma), PALEOMAP PaleoAtlas for ArcGIS, vol 1, Cenozoic. PALEOMAP Project, Evanston, IL

Scott WB (1905). Mammalia of the Santa Cruz beds. Part II. Insectivora. In: Scott WB (ed) Reports of the Princeton University Expedition to Patagonia, 1896–1899. Vol. 5, Paleontology II. Princeton University Press, Princeton, pp 365–383

Sefton EM, Gallardo M, Kardon G (2018) Developmental origin and morphogenesis of the diaphragm, an essential mammalian muscle. Dev Biol 440:64–73

Sellwood BW, Valdes PJ (2006) Mesozoic climates: General circulation models and the rock record. Sed Geol 190:269–287

Sibly RM, Brown JH (2007) Effects of body size and lifestyle on evolution of mammal life histories. Proc Natl Acad Sci USA 104:17707–17712

Sigogneau-Russell D (1991) Nouveaux mammifères theriens du Crétacé Inférieur du Maroc. C R Acad Sci 313:279–285

Sigogneau-Russell D (1998) Discovery of a Late Jurassic Chinese mammal in the upper Bathonian of England. C R Acad Sci 327:571–576

Sikes RS, Ylönen H (1998) Considerations of optimal litter size in mammals. Oikos 83:452–465

Simpson GG (1933) The "plagiaulacoid" type of mammalian dentition: A study of convergence. J Mammal 14:97–107

Simpson GG (1967) The beginning of the age of mammals in South America, Part 2. Bull Amer Mus Nat Hist 137:1–259

Simpson GG (1980) Splendid Isolation: The curious history of South American mammals. Yale University Press, New Haven

Simpson GG (1983) Fossils and the history of life. Scientific American Books—WH Freeman & Co., New York

Sire JY, Davit-Béal T, Delgado S, Gu X (2007) The origin and evolution of enamel mineralization genes. Cells Tissues Organs 186:25–48

Skibiel AL, Downing LM, Orrand TJ, Hood WR (2013) The evolution of the nutrient composition of mammalian milks. J Anim Ecol 82:1254–1264

Slater GJ (2013) Phylogenetic evidence for a shift in the mode of mammalian body size evolution at the Cretaceous-Palaeogene boundary. Methods Ecol Evol 4:734–744

Slater GJ, Harmon LJ (2013) Unifying fossils and phylogenies for comparative analyses of diversification and trait evolution. Methods Ecol Evol 4:699–702

Smith FA, Brown JH, Haskell JP, Lyons SK, Alroy J, Charnov EL, Dayan T, Enquist BJ, Ernest SKM, Hadly EA, Jones KE, Kaufman DM, Marquet PA, Maurer BA, Niklas KJ, Porter WP, Tiffney B, Willig MR (2004) Similarity of mammalian body size across the taxonomic hierarchy and across space and time. Am Nat 163:672–691

Smith FA, Boyer AG, Brown JH, Costa DP, Dayan T, Ernest SKM, Evans AR, Fortelius M, Gittleman JL, Hamilton MJ, Harding LE, Lintulaakso K, Lyons SK, McCain C, Okie JG, Saarinen JJ, Sibly RM, Stephens PR, Theodor J, Uhen MD (2010) The evolution of maximum body size of terrestrial mammals. Science 330:1216–1219

Soares MB, Martinelli AG, Oliveira TV (2014) A new prozostrodontian cynodont (Therapsida) from the Late Triassic *Riograndia* Assemblage Zone (Santa Maria Supersequence) of Southern Brazil. An Acad Brasil Ciênc 86:1673–1691

Spoutil F, Vlcek V, Horacek I (2010) Enamel microarchitecture of a tribosphenic molar. J Morphol 271:1204–2018

Springer MS, Kirsch JAW (1991) DNA hybridization, the compression effect, and the radiation of diprotodontian marsupials. Syst Zool 40:131–151

Springer MS, Westerman M, Kavanagh JR, Burk A, Woodburne MO, Kao DJ (1998) The origin of the Australasian marsupial fauna and the phylogenetic affinities of the enigmatic monito del monte and marsupial mole. Proc R Soc Lond B 265:2381–2386

Stadler T, Gavryushkina A, Warnock RCM, Drummond AJ, Heath TA (2018) The fossilized birth-death model for the analysis of stratigraphic range data under different speciation modes. J Theor Biol 447:41–55

Stephens PA, Vieira MV, Willis SG, Carbone C (2019) The limits to population density in birds and mammals. Ecol Lett 22:654–663

Sues H-D (2019) The rise of reptiles: 320 million years of evolution. Johns Hopkins University Press, Baltimore

Sureyya Rifaioglu A, Doğan T, Martin MJ, Cetin-Atalay R, Atalay V (2019) DEEPred: Automated protein function prediction with multi-task feed-forward deep neural networks. Sci Rep 9:7344

Szalay FS (1982) A new appraisal of marsupial phylogeny and classification. In: Archer M (ed) Carnivorous marsupials. Royal Zoological Society of New South Wales, Sydney, pp 621–640

Szalay FS (1994) Evolutionary history of the marsupials and an analysis of osteological characters. Cambridge University Press, New York

Torsvik TH, Cocks LRM (2017) Earth history and palaeogeography. Cambridge University Press, Cambridge

Twitchett RJ (2007) The Lilliput effect in the aftermath of the end-Permian extinction event. Palaeogeogr Palaeoclimatol Palaeoecol 252:132–144

Urashima T, Messer M, Oftedal OT (2014) Comparative biochemistry and evolution of milk oligosaccharides of monotremes, marsupials, and eutherians. In: Pontarotti P (ed) Evolutionary biology: Genome evolution, speciation, coevolution and origin of life. Springer, Cham, pp 3–33

Urbanek A (1993) Biotic crises in the history of Upper Silurian graptoloids: A paleobiological model. Hist Biol 7:29–50

Vizcaíno SF, Kay RF, Bargo MS (eds) (2012a) Early Miocene paleobiology in Patagonia. Cambridge University Press, Cambridge

Vizcaíno SF, Cassini GH, Toledo N, Bargo MS (2012b) On the evolution of large size mammalian herbivores of the Cenozoic faunas of southern South America. In: Patterson BD, Costa E (eds) Bones, clones, and biomes: The history and geography of recent Neotropical mammals. University of Chicago Press, Chicago, pp 76–101

Voss RS, Jansa SA (2009) Phylogenetic relationships and classification of didelphid marsupials, an extant radiation of New World metatherian mammals. Bull Am Mus Nat Hist 322:1–177

Wallace RVS, Martínez R, Rowe T (2019) First record of a basal mammaliamorph from the early Late Triassic Ischigualasto Formation of Argentina. PLoS ONE 14(8):e0218791

Wang X, Jiang S, Zhang J, Cheng X, Yu X, Li Y, Wei G, Wang X (2017) New evidence from China for the nature of the pterosaur evolutionary transition. Sci Rep 7:42763

Warren W, Hillier L, Marshall Graves J et al (2008) Genome analysis of the platypus reveals unique signatures of evolution. Nature 453:175–183

Webb SD (1976) Mammalian faunal dynamics of the Great American Interchange. Palaeobiology 2:216–234

Webb SD (1985) Late Cenozoic mammal dispersals between the Americas. In: Stheli FG, Webb SD (eds) The Great American Biotic Interchange. Plenum Press, New York, pp 357–386

Webb SD (1991) Ecogeography and the Great American Interchange. Paleobiology 17:266–280

Welker F, Collins MJ, Thomas JA, Wadsley M, Brace S, Cappellini E, Turvay ST, Reguero M, Gelfo JN, Kramarz A, Burger J, Thomas-Oates J, Ashford DA, Ashton PD, Rowsell K, Porter DM, Kessler B, Fischer R, Baessmann C, Kaspar S, Olsen JV, Kiley P, Elliot JA, Kelstrup CD, Mullin V, Hofreiter M, Willerslev E, Hublin J-J, Orlando L, Barnes I, MacPhee RDE (2015) Ancient proteins resolve the evolutionary history of Darwin's South American ungulates. Nature 522:81–84

Westbury, M., Baleka, S., Barlow, A. Hartmann S, Paijmans JLA, Forasiepi AM, Bond M, Gelfo JN, Reguero MA, López-Mendoza P, Taglioretti M, Scaglia F, Rinderknecht A, Jones W, Mena F, Billet G, Muizon C de, Aguilar JL, MacPhee (2017) A mitogenomic timetree for Darwin's enigmatic South American mammal *Macrauchenia patachonica*. Nat Commun 8:15951

Whitney MR, Sidor CA (2019) Histological and developmental insights into the herbivorous dentition of tapinocephalid therapsids. PLoS ONE 14(10):e0223860

Wible JR, Rougier GW (2017) Craniomandibular anatomy of the subterranean Meridolestidan *Necrolestes patagonensis* Ameghino, 1891 (Mammalia, Cladotheria) from the Early Miocene of Patagonia. Ann Carnegie Mus 84:183–251

Wilf P, Cúneo R, Escapa IH, Pol D, Woodburne MO (2013) Splendid and seldom isolated: The paleobiogeography of Patagonia. Annu Rev Earth Planet Sci 41:561–603

Williamson TE, Brusatte SL, Carr TD, Weil A, Standhardt BR (2012) The phylogeny and evolution of Cretaceous-Palaeogene metatherians: Cladistic analysis and description of new early Palaeocene specimens from the Nacimiento Formation, New Mexico. J Syst Palaeontol 10:625–651

Williamson TE, Brusatte SL, Wilson GP (2014) The origin and early evolution of metatherian mammals: The Cretaceous record. ZooKeys 465:1–76

Wilson GP, Ekdale E, Hoganson J, Calede JJ, Linden AV (2016) A large carnivorous mammal from the Late Cretaceous and the North American origin of marsupials. Nat Commun 7:13734

Witmer LM (1995) The extant phylogenetic bracket and the importance of reconstructing soft tissues in fossils. In: Thomason JJ (ed) Functional morphology in vertebrate paleontology. Cambridge University Press, New York, pp 19–33

Witmer LM (1998) Application of the extant phylogenetic bracket (EPB) approach to the problem of anatomical novelty in the fossil record. J Vertebr Paleont 18(Suppl):87A

Woelders L, Vellekoop J, Kroon D, Smit J, Casadío S, Prámparo M, Dinarès-Turell J, Peterse F, Sluijs A, Lenaerts JTM, Speijer R (2017) Latest Cretaceous climatic and environmental change in the South Atlantic region. Paleoceanography 32:466–483

Wood CB, Rougier GW (2005) Updating and recoding enamel microstructure in Mesozoic mammals: In search of discrete characters for phylogenetic reconstruction. J Mammal Evol 12:433–460

Wood CB, Dumont ER, Crompton AW (1999) New studies of enamel microstructure in Mesozoic mammals: A review of enamel prisms as a mammalian synapomorphy. J Mammal Evol 6:177–213

Woodburne MO (2010) The Great American Biotic Interchange: Dispersals, tectonics, climate, sea level and holding pens. J Mammal Evol 17:245–264

Woodburne MO, Case JA (1996) Dispersal, vicariance and the late Cretaceous to early Tertiary land mammal biogeography from South America to Australia. J Mammal Evol 3:121–161

Woodburne MO, Goin FJ, Bond M, Carlini AA, Gelfo JN, López GM, Iglesias A, Zimicz AN (2014a) Paleogene land mammal faunas of South America: A response to global climatic changes and indigenous floral diversity. J Mammal Evol 21:1–73

Woodburne MO, Goin FJ, Raigemborn MS, Heizler M, Gelfo JN, Oliveira EV (2014b) Revised timing of the South American Early Paleogene land mammal ages. J S Amer Earth Sci 54:109–119

Wu D-D, Irwin DM, Zhang Y-P (2008) Molecular evolution of the keratin associated protein gene family in mammals, role in the evolution of mammalian hair. BMC Evol Biol 8:241

Zerkle AL (2018) Biogeodynamics: Bridging the gap between surface and deep Earth processes. Philos T R Soc A 376:20170401

Zhang C, Stadler T, Klopfstein S, Heath TA, Ronquist F (2016) Total-evidence dating under the fossilized birth-death process. Syst Biol 65:228–249

Zhang H, Wang M, Liu X (2008) Constraints on the upper boundary age of the Tiaojishan Formation volcanic rocks in West Liaoning-North Hebei by LA-ICP-MS dating. Chinese Sci Bull 53:3574–3584

Zimicz AN (2012) Ecomorfología de los marsupiales paleógenos de América del Sur. Unpublished PhD thesis, Universidad Nacional La Plata

Zimicz AN (2014a) Paleoecología de los Bonapartherioidea (Marsupialia, Polydolopimorphia, Bonapartheriiformes). Ameghiniana 51:106–128

Zimicz AN (2014b) Avoiding competition: the ecological history of metatherian carnivores. J Mammal Evol 21:383–393

Zou Z, Zhang J (2016) Morphological and molecular convergences in mammalian phylogenetics. Nat Commun 7:12758